FACILITY DESIGN

MANUFACTURING ENGINEERING

SECOND EDITION

STEPHAN KONZ

KANSAS STATE UNIVERSITY

Publishing Horizons, Inc.
Scottsdale, Arizona

Editor: Nils Anderson
Developmental Editor: Gay L. Pauley
Production Manager: A. Colette Kelly
Sales & Marketing: Don DeLong
Cover Design: Kevin Kall
Typesetting: Ash Street Typecrafters, Inc.

Copyright © 1994, Publishing Horizons, Inc.
An Affiliate of
Gorsuch Scarisbrick, Publishers
8233 Via Paseo del Norte, Suite F-400
Scottsdale, AZ 85258

10   9   8   7   6   5   4   3   2   1

Printed in the United States of America.

Copyrights and Acknowledgments appear on page 347 and constitute a
continuation of the copyright page.

Library of Congress Cataloging-in-Publication Data

Konz, Stephan A.
      Facility design : manufacturing engineering / Stephan Konz.—2nd
ed.
         p.   cm.
      Includes bibliographical references and index.
      ISBN 0-942280-64-4
      1. Factories—Design and construction.   2. Facility management.
I. Title.
TS177.K66   1993
725'.4—dc20   93-35405
                                                CIP

CONTENTS

PREFACE

As instructors of design know, design is a tremendously complex undertaking. Designers must take into account any number of concepts, criteria, and alternatives, many of which a novice designer may not even be aware. The problem of information overload may be especially acute for manufacturing engineers, who must work to resolve whole strings of design challenges. These designers need to reckon with cost analysis, manufacturing processes, material handling, operations research, ergonomics, production scheduling, quality assurance, cost reductions of energy, materials and labor, safety, plant layout, architecture, lean production, waste minimization, buffers, robotics, group technology, and so on.

This new edition of *Facility Design: Manufacturing Engineering* offers this diverse design information. The text is broken into four main parts:

1. Evaluation of Alternative Layouts
2. Building Shell and Specialized Areas
3. Material Handling Among Workstations
4. Services and Environment

Each part includes several chapters (see the table of contents). Because each chapter covers a great deal of information, the book makes every effort to present material concisely, in a logical, systematic format. In addition, several features help to make the book more accessible to student readers. For example, the two column format permitted inclusion of a great number of figures and tables; this illustrative material should be a great help to students as they read and apply the text discussion. Each chapter begins with a list of key concepts, and all chapter headings are numbered to help students follow the flow and interrelationship of material. Finally, each chapter concludes with *Review Questions* and *Problems and Projects*.

This textbook is designed for students of industrial engineering who are doing a "capstone" design course. Instructors often employ one of two strategies for teaching the course:

- Instructors may choose to assign a series of related design projects, leading to a final integrated design of a particular facility. (The teacher's manual for this book includes sample design projects.)
- Instructors may choose to have students work with a specific industrial facility to solve a specific problem. For example, a student may strive to "improve the assembly process in Dept. XX" or to "reduce hazardous waste in XX plant." This focus on an actual facility exposes students to the complex real world; as in real life, they encounter vague information, vague criteria, and "people problems."

Facility Design has been updated and revised to incorporate new concerns, methodologies, and terminology. We asked many of you who used the first edition of the text to offer suggestions for ways to update and improve the book, and many of your ideas have been incorporated. My own experience in teaching this material for 25 years has helped me recognize potential problem areas for students, and I strive to address those concerns in this text. Student feedback regarding the text has been extremely positive.

The Instructor's Manual that accompanies this text includes exam questions and answers and a variety of laboratory exercises. This manual is available through the publisher.

Acknowledgments

I would like to thank all those individuals who offered ideas on how the book might be improved in its second edition. A special thanks to the following individuals for their constructive suggestions: Robert J. Graves, University of Massachusetts, Amherst; G. Fred Sheets, Jr., California State Polytechnic University at Pomona; and Floyd Olson, Mankato State University.

Stephan Konz

I EVALUATION OF ALTERNATIVE LAYOUTS

CHAPTER

1 | INTRODUCTION

IN PART I EVALUATION OF ALTERNATIVE LAYOUTS

1 INTRODUCTION

2 CRITERIA

3 BASIC PLANNING INFORMATION

4 ALTERNATIVE MACHINE ARRANGEMENTS

5 FLOW LINES

6 LOCATION OF ITEMS: MATH MODELS

7 PRESENTATION OF LAYOUTS

CHAPTER PREVIEW

The text is designed to integrate material from a variety of areas into a capstone design course. A brief summary is given of engineering design and project scheduling.

CHAPTER CONTENTS

1 Final Integrating Design Course
2 Engineering Design
3 Project Scheduling

KEY CONCEPTS

design stages
milestones
PERT
solution in principle
system integration

A designer needs to consider many aspects of facility design. Ergonomics and safety are woven throughout the text. Topics which are covered in entire textbooks or series of books are summarized in a chapter; examples are illumination, noise, ventilation, waste management, and energy management. There is a only a brief mention of topics such as economic evaluation, production scheduling, Just-In-Time, project scheduling (PERT), work measurement, metal cutting, etc. Articles are summarized in a paragraph or even just a sentence.

Students already familiar with a specific topic should skim that material; students unfamiliar with a specific topic should realize that the coverage in this book is just the "tip of the iceberg" and prepare to dig deeper.

1 FINAL INTEGRATING DESIGN COURSE

1.1 Introduction
Traditionally, engineering educators have used facility design as an integrating course for seniors. This text is designed for that course.

Although there is nothing so practical as a good theory, the emphasis in the text is not on theory but on design and application. It is anticipated that the instructor will assign a series of connected design projects that will lead to a final integrated design of some facility, either a hypothetical one or an actual one from which the student gathers data. Generally the students will work in teams.

1.2 Planning Divisions
Figure 1.1 shows an analogy of a factory versus a person. Note the five planning areas: layout, materials handling, communications, utilities, and building structure. Since "everything touches," the design should consider all five areas and their interactions, not just a single area such as materials handling.

Planning also goes through phases: orientation, overall plan, detail plan, and implementation.

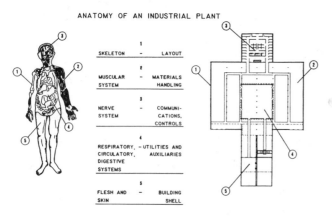

ANATOMY OF AN INDUSTRIAL PLANT

SKELETON	—	LAYOUT
MUSCULAR SYSTEM	—	MATERIALS HANDLING
NERVE SYSTEM	—	COMMUNICATIONS, CONTROLS
RESPIRATORY, CIRCULATORY, DIGESTIVE SYSTEMS	—	UTILITIES AND AUXILIARIES
FLESH AND SKIN	—	BUILDING SHELL

FIGURE 1.1 Facilities and people are analogous in that both are complex systems composed of complex subsystems (Muther and Hales, 1979). From *Systematic Planning of Industrial Facilities*, Vol. 1, 1979, R. Muther and L. Hales, Management & Ind. Research Publications, Kansas City, MO.

Table 1.1 shows the meshing of the five areas and four phases. In preplanning and orientation, compile the basic company needs, develop future needs, and develop a plan for the design project. In the overall plan, each of the five areas has a **solution in principle**. For an office building, the solution in principle might be to use a new two-story building at an industrial park location and to place accounting next to the executive offices on the second floor. An example of detail planning would be that the receptionist's desk will face the door at a 30° angle. Specific details are planned for lighting, electrical outlets, etc. The final step is implementation. Here the proposals are detailed on drawings and in reports, approvals received (probably including modifications of the proposals), and construction and installation done.

2 ENGINEERING DESIGN

2.1 Design Stages
Engineering **design stages** can be summarized in the acronym DAMES (Konz, 1990), where D = Define the problem, A = Analyze, M = Make search, E = Evaluate alternatives, and S = Specify and Sell solution. See Table 1.2.

2.11 Define the Problem.
Project definition should include the number of units of each type to be made, the criteria upon which the design will be evaluated, and when the design is due. Note that criteria are multiple; that is, the criteria would not only include building a factory at lowest cost but also reducing pollution, improving quality, maximizing flexibility, etc. What resources are available—both for the product and for the design team? Who will do what by when?

2.12 Analyze in Detail.
Next, gather specific details. One key is to determine what the limits are (also called constraints and restrictions). Gather specifications of the components and assemblies. What process capabilities are needed? What are the skills and availabilities of the workers? The preferred concept is to fit the process to the people, rather than to force the people to adjust to the process. A key concept is of a world with variability (stochastic) rather than a world without variability (deterministic). That is, process times have means and variances, not just a mean; part dimensions have a mean and a variance, not just a mean; and so forth.

2.13 Make Search of Solution Space.
The goal is to get an optimum solution, not just a feasible solution. To select the best solution from feasible solutions requires multiple solutions from which to select. However, designers tend to be "satisfiers" and stop designing as soon as they have a solution which meets the criteria. One of the virtues of math models is that they force attention on

TABLE 1.1 Meshing of the five areas and four planning phases (adapted from Hales, 1984). At each phase, the five areas are coordinated (system integration) and approval is given (phase is "locked") before going on to the next phase.

	Orientation	Overall Plan	Detail Plan	Implementation
Layout	Locate area to be planned	Do overall (block) layout	Detail layout of machinery and equipment	Detailed drawings of layout
Handling	Integrate external handling	Do overall handling plan	Detail material handling	Procure and install equipment; train operators
Communications	Tie-in external communications/controls	Do basic communications/controls plan	Detail communications and controls	Procure and install equipment; train operators
Utilities	Plan access and egress	Do distribution plan	Details of utilities	Procure and install equipment; debug
Building	Plan site	Do preliminary building plan	Details of building	Construct or rehabilitate building

determining which variables are relevant and what specific values of the coefficients should be used.

2.14 Evaluate Alternatives. Minimizing cost is a good objective, but there are different types of cost with different units: capital cost ($ in year 1), operating cost ($ in year 1, 2, 3 . . .), pollution cost (ppm of compound X in the waste water), and quality costs (field failure rates of the product in year 2, 3, 4 . . .), among others. A difficult challenge is to tradeoff these multiple criteria; see Table 6.7 for a popular technique.

2.15 Specify and Sell Solution. The designer must give specific detail in addition to broad concepts. Then the design must be evaluated and accepted by others. Because the goal is implementation of a design rather than an elegant but rejected concept, design is a back-and-forth (iterative) process between designers and the client. Experience has shown that 2D or 3D models of the proposed design are important communication and selling tools. After the design is approved, additional detail will be needed from the designer to implement the design.

2.2 Project Management Part of good design is management of the design process. Five points will be emphasized (Konz, 1990).

2.21 Scope Defined. The scope of the project should be agreed upon ahead of time. Who will do what by when? What resources (budget) are available? What is the "work product" (i.e., report, layout, computer analysis, etc.)? Give **milestones** (completion dates), tentative at first,

TABLE 1.2 The five steps of the engineering design procedure can be remembered by the acronym DAMES—define, analyze, make search, evaluate, specify, and sell.

Step	Comments	Example
Define the problem broadly	Make statement broad and detail-free Give criteria, number of replications, schedule	Design, within five days, a workstation for assembly of 10,000/yr of unit Y with reasonable quality and low manufacturing cost
Analyze in detail	Identify limits (constraints, restrictions) Include variability in components and users Make machine adjust to man, not converse	Obtain specification of components and assembly Obtain skills and availability of personnel Get restrictions in fabrication and assembly techniques and sequence Obtain more details on cost accounting, scheduling, and tradeoffs of criteria
Make search of solution space	Don't be limited by constraints Try for optimum solution, not feasible solution Have more than one solution	Seek a variety of assembly sequences, layouts, fixtures, units/h, hand tools, etc.
Evaluate alternatives	Trade off multiple criteria Calculate benefit/cost	Alternative A: installed cost $1000; cost/unit = $1.10 Alternative B: installed cost $1200; cost/unit = $1.03
Specify and sell solution	Specify solution in detail Sell solution Accept a partial solution rather than nothing Follow up to see that design is implemented and that design reduces the problem	Recommend Alternative B Install Alternative B1, a modification of B suggested by the supervisor

for the various events. It is important that the *project scope be put in writing*. If your supervisor will not put the scope in writing, you should do it and submit it for approval. It is important to minimize potential misunderstandings as soon as possible.

2.22 Clients' Servant.

This point is attitudinal. You are trying to help the clients, the customers. You are not a master, giving commands. You are trying to satisfy their needs, not they trying to satisfy yours.

2.23 Frequent Consultation.

Frequent consultation not only provides you with more facts but also improves the attitudes and emotions of the clients toward you. Consult with "users" not just with "staff." Information can be obtained from face-to-face consultation and tours (take notes) and from videotapes of existing operations. For videotapes of existing operations, be sure to take several minutes of tape with multiple views. Views of multiple operators are good because operators often use considerably different micromotions. It may be possible to visit similar operations in other facilities. Inspection of records tends to give relatively little information, because recorded data generally are not complete and in convenient form. Written requests for information are fairly useless since no one wants to fill out forms, and if they do, you get "bare bones" answers.

The decision makers should be consulted at multiple approval points, not just at the end of the project. This allows them time to consider your ideas, change their minds on various points early (rather than after you have done irrelevant work), and change priorities. Early discussion of alternatives also tends to bring out potential problems that you may not have considered.

Frequent consultation also implements a "no-surprises" policy. Decision makers do not like to be surprised. Therefore, discuss various aspects of the project with the relevant people on an informal basis; later they can be set into the "concrete" of a formal written and oral presentation at a meeting. The goal of frequent consultation with affected groups is not only to improve the technical quality of the proposal but also to improve acceptance.

2.24 Use Experts.

Both line and staff people consider themselves experts in their jobs. Whether they are expert is not as important as their own concept of themselves as expert. You want them on your side, not against you.

Consult these experts *during* the project. For example, cost-accounting departments consider themselves experts in cost justifications. Even if you are able to do a cost justification by yourself, it is better to have the cost accountants go over your project *during* the project. You want to go into the final decision stages with them on your side, not with them unfamiliar with the project or, even worse, opposing you.

2.25 Give the Buyer a Choice.

Decision makers want to make decisions. They do not like to be confronted with no alternatives and just rubber stamp something. The choices can be "yes" or "no" for the entire project, for pieces of the project, for the implementation schedule, etc., but if you make the alternative all or nothing, it may be nothing.

3 PROJECT SCHEDULING

Project planning is simple in concept but difficult in execution. There are two stages: (1) plan the work, and (2) work the plan. In the first step (a) list the project *tasks*, (b) list the *sequence* (precedence) requirements of the tasks, and (c) allocate resources to the tasks so that task *time* is determined. Then the tasks, sequences, and times are combined

TABLE 1.3   Tasks for replacing machine X with machine Y and robot Z.

Number	Activities	Follows	Precedes	Days
1	Get cost of machine X		4	3
2	Get cost of machine Y		4	2
3	Get cost of robot Z		4	8
4	Compare X versus $Y + Z$	1, 2, 3	5	2
5	Decide on $Y + Z$	4	6	1
6	Get approval on $Y + Z$	5	7, 12	6
7	Order machine Y	6	8	1
8	Receive machine Y	7	9	41
9	Remove machine X	8	10	2
10	Install machine Y	9	11	1
11	Train operator on Y	10	16	1
12	Order robot Z	6	13	2
13	Receive robot Z	12	14	35
14	Install robot Z	13	15	6
15	Train operator on Z	14	16	4
16	System operational	11, 15		

into a plan. The second step is to execute the plan. As the project progresses, the time, sequence, and even the tasks themselves change; thus, the plan needs to be continually revised and updated.

Table 1.3 and Figure 1.2 give the tasks, sequence, and times for a small project—replacing machine X with machine Y and robot Z. The example will be used to demonstrate the use of the Gantt chart, a simplified version of **PERT** (Program Evaluation Review Technique), and an enhanced version of PERT.

3.1 Gantt Charts

Figure 1.3 shows a Gantt chart of the project. The chart is relatively simple to construct (many vendors sell wall-mounted units with pegs, tapes, and magnetic bars in a variety of colors). The Gantt chart is useful not only for planning the work but also for working the plan. It forces a plan to be made, shows the situation (planned versus done) at a glance, and is easy to modify; modifications do not require computers or complicated mathematics. It is used far more often than either the simplified or enhanced PERT.

3.2 Simplified PERT

In the 1950s a variety of techniques were introduced which extended the Gantt chart. They generally are known now as PERT or critical path techniques. (PERT stands for Program Evaluation Review Technique, but the words are rarely used anymore; people simply say PERT.) The critical path label came from the realization that some parts of the network were more important than others—they are the "bottleneck" tasks. These bottleneck tasks are said to be "on the critical path." The network techniques also are more useful for complex problems—say over 100 tasks.

What was shown in Figure 1.2 is known as the *forward pass*. What was discovered in the 1950s (and seems so simple now) was the *backward pass*. Figure 1.4 shows the forward and backward passes combined.

The difference between the forward and backward pass is the *slack*. When slack is zero, the task is on the critical path. (There can be multiple critical paths.)

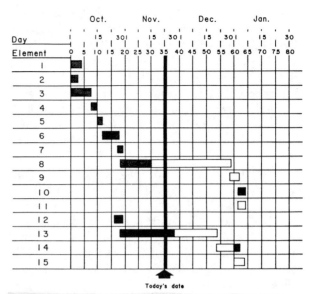

FIGURE 1.3 Gantt chart representation of Figure 1.2 and Table 1.3. There is one row for each task. The horizontal axis is *scaled* time. The tasks are located to show the beginning and finish time for each task—they are *scaled*. Hanging from the top of the chart is a movable cord or bar showing today's date. In the figure, "today" is November 22. As tasks are completed, they are shaded. In this figure, tasks 1, 2, 3, 4, 5, 6, 7, and 12 have been completed. Task 8 is about 5 days behind schedule; task 13 is about 3 days ahead of schedule.

The knowledge of which tasks are on the critical path is very useful in stage 2—working the plan. In terms of shortening the total time, expediting is useful only for tasks on the critical path. If a task is not on the critical path, then it will not affect the final schedule, as long as there is still slack time. Thus knowledge of the critical path allows the supervisor to know which tasks to worry about. This information is not available in the Gantt chart.

Advantages of PERT (Moore, 1973) are as follows: (1) putting the tasks, sequences, and times into the network shows potential problems while the project is still in the planning stage, (2) it shows which tasks should and should not be expedited, (3) it sets up progress checkpoints, and (4) it tends to be self-fulfilling since people work toward

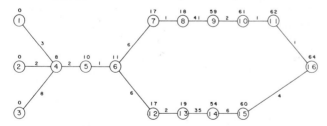

FIGURE 1.2 Tasks of Table 1.1 are shown as "nodes" (circles; events; milestones). Task times are on the "arrows" (lines) following the node. (There is an alternative graphical technique in which tasks are placed on arrows.) The network shows task precedence (left to right); it is assumed that a task cannot start until its predecessors have been completed. The number above each node is the cumulative time to reach that node. Node 4 shows that when several paths join in a node, the node time is the maximum of the path times. Node 6 shows how one task time can affect more than one node. The lines between nodes (times) are not drawn to scale.

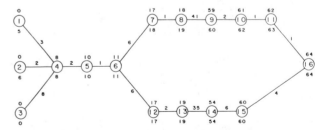

FIGURE 1.4 Forward and backward paths are shown in one figure (Figure 1.2 shows just the forward path). The total from the forward pass at the final task (64 days) is put on the bottom of the final node. Then, working backward, subtract task time from the total, enter it at the bottom of the node to the left. When two paths converge (as at node 6), the smaller number is entered (11 instead of 12). The difference between the forward pass time (above the node) and the backward pass time (below the node) equals *slack* time. When slack time is zero, the task is on the *critical path*.

known schedules. Disadvantages are (1) that PERT takes time and effort to set up and (especially) maintain and (2) that the times may be too long. Since people work toward schedules, the project may take longer where there was a PERT analysis. If a person is asked how long it will take to do a project, the person tends to give an inflated estimate. That is, if you can do the task in three weeks, you tell the boss four weeks to be on the safe side. Then, when four weeks goes into the formal schedule, you take four weeks.

3.3 Enhanced PERT   Two enhancements will be discussed: (1) time variability and (2) time/cost tradeoffs.

In simplified PERT, one estimate of the task time is used. In enhanced PERT, three times are used: (1) an optimistic task completion time, (2) a most likely time, and (3) a pessimistic time. Table 1.4 shows three estimates for the example problem. The average of the three times is called *expected time*.

$$t_e = (t_o + 4t_m + t_p)/6 \qquad (1)$$

where

t_e = expected time

t_o = optimistic time

t_m = most likely time

t_p = pessimistic time

The first assumption is that 4 is the appropriate weighting for t_m.

Figure 1.5 shows the network with the expected time values (rounded to the nearest day). Note that the project time now is estimated as 67 days instead of 64. This is because the pessimistic time is almost always farther from the mostly likely time than the optimistic time. Technically this result is known as a *positive skew to the distribution*. It may be viewed as a more realistic schedule but also as a

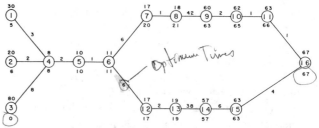

FIGURE 1.5   Expected times (from Table 1.4) arrived at when three times are used tend to be greater than when a single estimate is used. Compare against Figure 1.4.

longer schedule. Since it is a longer schedule (and people work toward goals), working your plan may take longer just because of the techniques used to plan your work.

Three time estimates instead of one give another possibility. They permit an estimate of time variability. Four additional assumptions are made: (1) The range of the times (i.e., pessimistic minus optimistic) is equal to 6 standard deviations; (2) the expected time for a node is the sum of the times for the path to that node; (3) the variability of each task is independent of the variability of all other tasks (i.e., task times are independent), and thus the total variance is the sum of the variances of the task times on the path to that node; and (4) the distribution of the cumulative variances can be approximated by a normal distribution. That is,

$$\sigma_i = (t_i - t_i)/6 \qquad (2)$$
$$t_i = \Sigma (t_{i-1} + t_{i-2} + \ldots) \qquad (3)$$
$$\sigma_i^2 = \sigma_{i-1}^2 + \sigma_{i-2}^2 + \ldots \qquad (4)$$

For example, the variance of task 3 is $(3/6)^2 = 0.25$ day²; the variance of task 4 is $(1/6)^2 = 0.03$ day². Then the variance of the time to node 5 = 0.25 + 0.03 = 0.28 day²; the standard

TABLE 1.4   Tasks for replacing machine X with machine Y and robot Z.

				Time (days)			
Number	Task	Follows	Precedes	Opt.	Most Likely	Pes.	Expected
1	Get cost of machine X		4	2	3	5	3
2	Get cost of machine Y		4	1	2	4	2
3	Get cost of robot Z		4	6	8	9	8
4	Compare X versus $Y + Z$	1, 2, 3	5	2	2	3	2
5	Decide on $Y + Z$	4	6	1	1	2	1
6	Get approval on $Y + Z$	5	7, 12	4	6	8	6
7	Order machine Y	6	8	1	1	2	1
8	Receive machine Y	7	9	30	41	60	42
9	Remove machine X	8	10	2	2	3	2
10	Install machine Y	9	11	1	1	2	1
11	Train operator on Y	10	16	1	1	3	1
12	Order robot Z	11	13	2	2	4	2
13	Receive robot Z	12	14	31	35	60	38
14	Install robot Z	13	15	5	6	10	6
15	Train operator on Z	14	16	3	4	6	4
16	System operational	11, 15					

TABLE 1.5   PERT times and costs for the example problem.

| Task | Normal | | Crash | | Expedited |
	Time (days)	Cost ($)	Time (days)	Cost ($)	Cost/Day ($/day)
1	3	50	2	100	50
2	2	50	1	75	25
3	8	125	6	150	12
4	2	60	1	90	30
5	1	40	1	40	—
6	6	100	3	50	17
7	1	40	1	40	—
8	41	4500	35	4000	83
9	2	400	1	600	200
10	1	200	1	200	—
11	1	40	1	40	—
12	2	50	1	60	10
13	35	600	32	900	100
14	6	1200	5	1400	200
15	4	150	3	190	40
16					

Check standard deviation &

deviation is 0.53 days. Then, using the normal distribution, 95% of the times would fall between 1.96 (0.53) = 1.04 days of the mean. Rounding 1.04 off to 1.0, it might be said that the 95% confidence limits for node 5 is 10 − 1 = 9 days to 10 + 1 = 11 days. (You can turn the problem around and ask what the probability of node 5 by 9 or 11 days is.) This looks quite impressive when done by a computer—especially when it predicts to several decimal points. But don't bet any money that that is what will happen in reality.

The second version of enhanced PERT, PERT/COST, is to trade off various task times against cost. It requires four estimates for each task: (1) a normal time, (2) the task cost for the normal time, (3) a "crash" (expedited) time, and (4) the task cost for the crash time. Table 1.5 shows values for the example project.

The last column in the table shows the cost of saving a day for each task. Assumptions are (1) the times are accurate, (2) the costs are accurate, and (3) the tradeoff between time and cost is linear. Naturally it is worthwhile to expedite tasks only on the critical path. If it is decided to cut the project time from 64 days to 63 days, tasks 3, 4, 5, 6, 12, 13, 14, and 15 are on the critical path. Task 12 looks like the best bet. Note that changing the time for task 12 to 1 day makes the total time 63 days, but also makes the upper branch a critical path. That is, as you consider expediting alternatives, the critical path may change.

DESIGN CHECKLIST: INTRODUCTION

Project definition
    Number of units of each type
    Design criteria
    Design due when?
    Resources available
    Scope in writing
Analysis
    Constraints
    Variability on numbers
    Machine adjusts to man
Alternative designs
Improve implementation by
    Defined scope

    Being client's servant, not master
    Frequent consultation
    Aid from experts
    Choices for decision makers
    Many approval points

Plan the work
    List of project tasks
    Sequence of tasks
    Time to tasks
    Gantt, simplified PERT, or enhanced PERT

Work the plan
    Who does updating?

REVIEW QUESTIONS

1. Table 1.1 shows the four phases of planning and, under each, the five areas. Give them.
2. Give the five stages of engineering design summarized by DAMES.
3. What is the difference between a stochastic and a deterministic world? Give some examples of each.
4. When evaluating a design, how can multiple criteria be traded off?
5. Design projects need to be managed. Briefly discuss the five points emphasized in the text.
6. Give the two stages of project planning.
7. Are activities drawn to a time scale in a Gantt chart or in a PERT diagram?
8. Make a simple network of six tasks. Show how the critical path is calculated.
9. What are the advantages and disadvantages of simplified PERT?
10. When three times are used in PERT, what are the five assumptions?
11. Why will the scheduled time be longer for a project using three PERT times/task than one?
12. In PERT/COST what four estimates are needed for each task?

PROBLEMS AND PROJECTS

1.1 Revise the data of Table 1.2. Make a Gantt chart.
1.2 Revise the data of Table 1.2. Calculate time at each node. Identify the critical path.
1.3 Make a table similar to Table 1.2 for the task of a lab report/project for this course. Calculate time at each node. Identify the critical path.

1.4 Read 3 to 5 articles on a topic related to this chapter. Summarize them in a 500-word executive summary. List the references. Do not use articles cited in the text.

REFERENCES

Gould, L. Is material handling the Achilles' heel of the FMS? *Modern Material Handling*, 56–58, July 1990.

Hales, H. *Computer-Aided Facilities Planning*. New York: Marcel Dekker, 1984.

Konz, S. *Work Design: Industrial Ergonomics*, 3rd ed. Scottsdale, Ariz.: Publishing Horizons, 1990.

Moore, F. *Production Management*, 6th ed. Homewood, Ill.: Irwin, 1973.

Muther, R., and Hales, H. *Systematic Planning of Industrial Facilities*: Vol. 1. Kansas City, Mo.: Management and Ind. Research Publications, 1979.

CHAPTER PREVIEW

After applying the six ergonomic criteria for job design, the designer will need to tell management the bottom line. Various ratios can be used to evaluate the merit of a design or of an operating system.

CHAPTER CONTENTS

1 Ergonomic Criteria
2 Economic Analysis
3 Resource Utilization Ratios
4 Management Control Ratios
5 Operating Efficiency Ratios

KEY CONCEPTS

bottom line
ergonomic criteria
internal rate of return
net present value
planning horizon
request for proposal

1 ERGONOMIC CRITERIA

Box 2.1 gives six **ergonomic criteria** (human factors considerations) for job design.

2 ECONOMIC ANALYSIS

Management wants to know the **bottom line**—the last line of an economic analysis. However, there are different numbers that can be on the bottom line. A full discussion is beyond the scope of this text; for a more complete discussion, see an engineering economics text such as White et al. (1989).

There are two dimensions to the problem: (1) obtaining good data and (2) calculating the bottom line. The secret of good decisions is to spend your time getting good data. Even complex bottom line calculations should take less than an hour. But if you have bad data, even complex computations of garbage input give garbage output. Even worse, the complex calculations of the garbage may give a face validity to it and cause decision makers to assume the garbage is good stuff.

2.1 Obtaining Good Data
There are three basic numbers to obtain: (1) life of the project, (2) annual savings, and (3) initial capital cost.

2.11 Project Life. The application life can be limited by the physical life of the equipment (rarely) or by life of the application (usually). The world (especially the business world) is very dynamic, and therefore management has good reason to doubt anyone's ability to predict the far future—and perhaps even the near future. Thus, **planning horizons** usually are less than 20 years, and often a 10- or even 5-year planning horizon is used. The planning horizon may vary with the product, type of investment (building versus machinery), rate of change of technology in the industry, political stability, etc. An alternative to changing the planning horizon is to require a higher figure of merit on the bottom line as uncertainty increases.

2.12 Annual Savings. Determining annual savings is a difficult step. Many different factors need to be considered:

- direct labor
- downtime
- maintenance
- utilities
- scrap and rework
- medical and safety
- product quality
- absenteeism
- any other factors that may impact savings

The problems are complex. For example, how do you put a specific dollar value on absenteeism for alternatives *A* and *B*? for utility cost for alternatives *A* and *B*? for maintenance for alternatives *A* and *B*? Note that most products are made in batches, with a factory making 10–50 different products; it is therefore difficult to allocate costs for one specific product. Nevertheless, the analyst needs to determine the annual savings of the proposal. Resist the temptation to identify some costs as intangible (meaning they are difficult to quantify). An example of an intangible would be flexibility. It may help to make the intangible more specific. For example, for a materials handling project, it could be flexibility of path, flexibility of carrier used, flexibility of throughput rates, or flexibility to changed product. The

---

BOX 2.1   *Job design criteria (Konz, 1990)*

1. *Design first for safety.* No job design that endangers the worker's safety or health is acceptable. However, life does not have infinite value. Management must take reasonable precautions. Naturally, the definition of *reasonable* is debatable; however, our society is putting increased emphasis on safety. After designing for safety, design for performance, and then for worker comfort. Finally, consider "higher wants," such as job satisfaction and feeling of accomplishment.
2. *Make the machine user-friendly.* The machine is to adjust to the worker; the process is to adjust to the person, not the converse. If the machine, process, or system does not function well, redesign the machine, process, or system rather than blaming the operator.
3. *Reduce the percentage excluded from the design.* Permit everyone to use the machine or procedure. Gender, age, strength, and so forth, should not prevent people from participating.
4. *Design jobs to be cognitive and social.* Physical and procedural work can now be done by machines. In addition, due to wage differentials, if a manufacturing process has a high labor content, it probably will not be done in the United States or other "first-world" country.
5. *Emphasize communication.* We communicate to machines with controls and receive information from machines by displays. As output is increasingly determined by machines, it becomes more critical to determine the proper use of the machines and to minimize errors. Many of these errors are people errors. For example, computers don't make mistakes, programmers do.
6. *Use machines to extend human performance.* The choice is not man or machine—it is which machine to use. Small machines (word processor, electric drill) tend to have costs (total of capital, maintenance, and power) of less than $.25/h. Even large machines (automobiles, lathes) tend to cost only $1 to $2/h. Labor costs, however, run $5 to $25/h. Thus, the real question is how many "slaves" (machines) the human will supervise and how to design the system to use the output of the slaves.

goal is to (1) compare equal alternatives and (2) give costs for all factors that do not make them equal.

2.13 Capital Costs.
It takes money to make money. That is, in order to reduce annual operating expenses, the firm will have to invest money. This initial one-time cost can be put into three categories: (1) equipment costs, including installation and jigs and fixtures; (2) retraining costs; and (3) engineering costs of making the study. In many firms the retraining and engineering costs are not charged to the project but are absorbed in overhead.

2.2 Calculation of the Bottom Line
Four different bottom lines will be discussed ("How to justify," 1991). Assume the ABC Company is considering a materials handling project. The assumed project life is 5 years. Operating costs are $60,000/year for the present method and $10,000 under the proposed for an annual savings of $50,000. The project requires $100,000 for equipment; the equipment will have 0 scrap value at the end of the 5 years. The company's required rate of return is 20%. (The time value of money for a firm depends on the firm's alternative uses for the capital, the firm's credit history, the inflation rate of the economy, and so on. In the United States, typical required rates are 20 to 30%.) See Figure 2.1.

2.21 Payback.
The simplest approach is payback. That is, the firm gets its money back in two years.

$$\text{Payback period} = \frac{\text{net cash outlay}}{\text{cash inflow/year}} \qquad (1)$$

$$= \frac{100,000}{50,000} = 2 \text{ years}$$

The disadvantage of this simple calculation is that it does not consider the time value of money. Thus, it may give different rankings of alternatives than more precise methods.

2.22 Savings/Investment Ratio (SIR).
If the time value of money is calculated, the ratio of savings to investment gives the *discounted* payback period.

$$SIR = \frac{\text{discounted cash inflows}}{\text{discounted cash outflows}} \qquad (2)$$

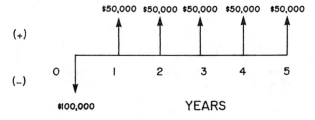

FIGURE 2.1 Cash flow diagrams show negative flows downward and positive flows upward.

$$= \frac{\dfrac{C_1}{(1+r)} + \dfrac{C_2}{(1+r)^2} + \cdots \dfrac{C_n}{(1+r)^n}}{C_o}$$

where

$$C_1, C_2 \ldots = \text{net periodic cash flow for an investment}$$
$$r = \text{rate of return (required)}$$
$$C_o = \text{initial cost of investment}$$

$$= \frac{\dfrac{50,000}{1.2} + \dfrac{50,000}{1.44} + \dfrac{50,000}{1.728} + \dfrac{50,000}{2.074} + \dfrac{50,000}{2.489}}{100,000}$$

$$= \frac{149,525}{100,000} = 1.49$$

If the ratio is less than 1, then the predicted return is less than the required return.

2.23 Net Present Value.
The **net present value** (NPV) approach gives the *difference* between benefits and costs instead of the *ratio*. All benefits and costs are brought to one time (the present); using the firm's desired rate of return, the net present value should be greater than zero.

$$NPV = -C_o + \frac{C_1}{(1+r)} + \frac{C_2}{(1+r)^2} \cdots \frac{C_n}{(1+r)^n} \qquad (3)$$

$$= -100,000 + \frac{50,000}{1.2} + \frac{50,000}{1.44} + \frac{50,000}{1.728} + \frac{50,000}{2.074} + \frac{50,000}{2.489}$$

$$= -100,000 + 149,525$$

$$= 49,525$$

2.24 Internal Rate of Return (IRR).
In the **internal rate of return** approach, the internal rate of return of the project (the rate at which lifetime savings equal lifetime costs) is calculated and then the value is compared with the desired value (20% in this example). The net present value is assumed to be zero.

$$NPV = 0 = -C_o + \frac{C_1}{(1+IRR)} + \frac{C_2}{(1+IRR)^2} \cdots \frac{C_n}{(1+IRR)^n} \qquad (4)$$

$$100,000 = \frac{50,000}{(1+IRR)} + \frac{50,000}{(1+IRR)^2} \cdots \frac{50,000}{(1+IRR)^5}$$

The value of *IRR* is determined by trial and error. For example, if *IRR* = 20%, the sum of the annual saving terms is $149,525. This is larger than $100,000, so a larger *IRR* would be used—say 30%. Eventually the identity balances at an *IRR* of 41%. (Technical note: For a single root to the polynomial, the value must be positive on both sides of the equal sign. See White et al., 1989, for a more extensive discussion.)

2.25 Comments.
The decision of whether to implement a specific project depends not only on the project itself but also on what other uses the firm has for

the money. The payback method is popular with management due to its simplicity. It is satisfactory for projects in which the time value of money is not decisive, for example, short projects and projects with low cash flows. The net present value and profitability index are computationally correct but not very well understood by managers. Internal rate of return is thus the method of choice for projects in which the time value of money is important.

Note that the above analysis does not include any consideration of inflation or income taxes. Income taxes vary with depreciation schedules, special tax breaks, etc. Note also that the savings were assumed equal in each year. In reality, projects usually have a startup or debugging period before the savings actually occur. The planning horizon of 5 years probably was quite arbitrary. If the equipment has a salvage value at the end of the application life, treat it as a positive cash flow. It is good practice to do a sensitivity analysis of the project. That is, rework the problem but assume project life was 6 years, or annual savings was $8,000, or initial capital cost was $125,000, etc. Finally, some projects need to be done regardless of bottom line calculations. An example would be a safety requirement, such as punch press guards.

See Box 2.2 for comments on owning versus renting. See Box 2.3 for checkpoints in writing a **Request for Proposal** (RFP)—a statement of project requirements to be met by a vendor.

3 RESOURCE UTILIZATION RATIOS

Resource utilization will be divided into people, equipment, space, and energy (White, 1979b).

3.1 Resource Utilization: People

When designing, there is a natural tendency to reduce labor cost per unit. However, a key question is how much the worker's activity is contributing to the bottom line of service to the customer.

For example, a phone system could be designed with fewer phones—reducing serving cost/person served. But if 20 employees (e.g., costing $18/h or 30 cents/minute) now wait for 10 minutes/day instead of 5 minutes, then waiting cost is up by $1.50/day/person (20) = $30/day. The reduced phone cost needs to be balanced with the increased idle time.

Highly paid knowledge workers often use computers. However, this has resulted in many of them performing clerical tasks (report writing, duplication, etc.) formerly done by lower paid staff. Thus, it is false economy to reduce the support staff of highly paid workers, because someone still has to do the clerical work, and it ends up being done by the highly paid person.

A fruitful approach is to reduce errors, because the cost of correcting errors is very large—especially if the error is not detected until it reaches the customer.

BOX 2.2   Own or rent?

After the decision is made that a piece of equipment (machine tool, conveyor, etc.) is needed, then the decision of whether to own it must be made.

Own

In the majority of cases, the decision is made to own the equipment, and the equipment is purchased from operating funds. The bank account is reduced, and the firm owns the equipment.

What if the firm does not have enough money in the bank? One alternative is to buy used equipment. Used equipment has several advantages besides price. It is often available immediately, and it is often available because of application obsolescence rather than physical obsolescence. That is, the previous owner disposed of it because product requirements changed rather than because the machine itself was a problem.

Another possibility is to borrow the money. The interest cost can be included in the cash flow analysis, decreasing the annual profit of the proposal. Thus, using the example of Section 2.1, if the firm furnished its own capital of $100,000 and paid 12% interest to a bank, then the first year interest would be $12,000 and the first year income would be $50,000 – 12,000 = $38,000. If $20,000 of principal was paid off, the interest for the second year would be .12 ($80,000) = $9,600.

Rent

There are a number of renting alternatives. Leasing is renting for a specified time period—usually several years. With leasing, the vendor retains title to the equipment but the lessee can use the equipment for the specified length of time. For example, consider a lease of a $20,000 automobile for 4 years. The lessor might assume that the vehicle would retain 20% of its value after four years. In addition to the $16,000 expense, the lessor would add maintenance expense, insurance, cost of interest, and profit. Then the lessee would agree to pay a monthly payment of (say) $500. Although the lessee pays more in the long run than if it were bought, the immediate capital cost is low. Since the lessor may have specialized ability to sell used equipment or obtain capital, the premium for leasing over buying may not be large. To qualify for tax benefits, the lease must be a true lease, without option to buy.

Another alternative is just to rent the equipment. In this case the lessor does not have the protection of a long-term contract, so the monthly payment naturally will be higher. However, renting may be the best alternative for short-term or intermittent use. For example, if a large lift truck is needed only two weeks a year (e.g., during annual maintenance of the plant), then renting probably is the best option.

Another choice is to rent with an option to buy. It really is a contract to buy, with the seller repossessing the item if payments are not made.

BOX 2.3   *RFP requirements guidelines (Gould, 1990)*

When giving a vendor a request for proposal, clear communication minimizes the problems. Here are 10 checkpoints:

1. Give a *quantified* statement of the system objective. For example, move *XXX* units/h of product *YY* with downtime of less than .5% and cost of less than *ZZ*.
2. Give the facility characteristics, for example, floor type, aisle widths, and accessibility to loads.
3. Give the load characteristics, for example, load type, weight, "footprint," stability, overhang, storage method, and queues.
4. Give the load flow, for example, a flow diagram and a from-to movement table.
5. Give the control system, for example, computer tracking requirements and interfaces, operator interfaces, and paperwork.
6. Give the methods for evaluating the design concept. Examples would be a computer simulation or experience at another site.
7. Give the cost criteria. For example, list staffing (direct and indirect), labor costs (including fringes), overhead costs, energy costs, weighting of cost-justification factors, and required internal rate of return. Give who will pay for shipping, installation, and plant modifications such as sprinklers, feeders, guards, and Heating, Ventilation and Air Conditioning (HVAC).
8. Give any applicable standards. Examples might be local building codes, American National Standards Institute (ANSI) standards, and Occupational Health and Safety Administration (OSHA) standards.
9. Give the project schedule. Key dates are bidding dates; contract signature date; and dates of installation of hardware, installation of software, initial demonstration of system operation, and demonstration of debugged system.
10. Give the acceptance test criteria. For example, the system will operate for 10 working days with less than .5% downtime before final payment is made to the vendor.

Utilization of people in material handling can be evaluated with two ratios: Material Handling Labor (*MHL*) and Direct Labor Material Handling (*DLMH*).

$$MHL = IMHD/TD \qquad (5)$$

where

MHL = material handling labor ratio

$IMHD$ = indirect material handling, dollars

TD = total annual payroll, dollars

$$DLMH = DMHD/DLD \qquad (6)$$

where

$DLMH$ = direct labor material handling ratio

$DMHD$ = direct labor material handling, dollars

DLD = direct labor total, dollars

Typically *IMHD* would include full-time material handling-equipment operators (trucks, cranes, manual labor), related activities (dock workers, storage operators, waste-removal-operators), and supporting activities (tool room, material handling, maintenance, inventory control records, traffic). Much of the supporting activities charged would be less than full time; the proper percentage would be determined by occurrence (work) sampling. A poor *MHL* would be 40%, fair 25%, and good 10% (Singh, 1982).

Not all material handling is done by indirect labor; thus the use of the *DLMH*. Generally it is desirable to have a low *DLMH*, as "production labor is to concentrate on production." The *DMHD* would be established through occurrence sampling stratified by department and perhaps even by operator. A poor *DLMH* would be 25%, fair 15%, and good 5% (Singh, 1982).

Some improvement possibilities include: (1) moving supplies to the worker rather than the person to the item, (2) having automated data entry of part numbers, (3) arranging machines so one person can tend multiple machines, and (4) reducing the number of moves required.

3.2 Resource Utilization: Equipment

Utilization of material handling equipment can be evaluated with two ratios: Handling Equipment Utilization (*HEU*) and Busy Percentage (*BP*).

$$HEU = (PS)(PW) \qquad (7)$$

where

HEU = handling equipment utilization ratio

PS = speed moved, % of "standard"

PW = weight moved, % of "standard"

$$BP = W/OBS \qquad (8)$$

where

BP = busy percent, %

W = number of observations working

OBS = number of total observations

Both ratios are determined through occurrence sampling and judgment. Equipment is observed; the observer rates the equipment speed as a percentage of "standard" speed for that situation and the weight carried as a percentage of standard that could be carried. Depending on the organization's desires, working can be defined as moving with full load, moving with either partial load or full load, or moving loaded or empty, or working can even include loading/unloading. Different organizations use different definitions of working, so comparisons between organizations must be made with caution. A poor BP would be 30%, fair 60%, and good 80% (Singh, 1982). If the busy percentage is above 75% for trucks and cranes, then it probably is too high for this "random service" equipment; consider fixed path equipment such as conveyors or driverless tractors.

Some improvement possibilities include: (1) putting couplers on lift trucks to permit use of various attachments, (2) real-time dispatching for trucks to reduce amount of deadhead trips, and (3) batching work orders.

3.3 Resource Utilization: Space

Utilization of space can be evaluated with two ratios: Storage Space Utilization (SSU) and Aisle Space (ASP).

$$SSU = OSS/TSS \qquad (9)$$

where

SSU = storage space utilization ratio

OSS = storage space occupied by material, ft³

TSS = total storage space, ft³

$$ASP = AS/TS \qquad (10)$$

where

ASP = aisle space, proportion

AS = aisle space, ft³

TS = total space, ft³

Both ratios are based on *cubic* feet rather than square feet to encourage use of the vertical dimension. A poor *SSU* would be 25%, fair 50%, and good 75% (Singh, 1982). Storage can be five or more levels high; aisles can be bridged with storage above the cross aisles. If the storage ratio is too high or aisle ratio too low, there will be problems of access to material. Be careful to measure the actual situation rather than working from drawings.

Some improvement possibilities include: (1) narrow (less than 8 ft) aisles, (2) mezzanines over low-bay activities, and (3) random rather than dedicated storage. See Chapter 11.

3.4 Resource Utilization: Energy

Utilization of energy can be evaluated with the Energy Utilization Index (EUI).

$$EUI = BTU/ft^2 \qquad (11)$$

Table 2.1 gives the conversions for various types of fuels to BTU and KWH. These units are used rather than dollars (1) to eliminate the effect of inflation and (2) to permit

TABLE 2.1 Conversion to BTU and KWH for various fuels.

If Energy Source Is	And Units Are	Then to Get	
		Units in KWH, Multiply by	Units in BTU, Multiply by
Bituminous coal	pounds	3.66	12,500
Diesel oil	gallons	39.8	136,000
Electric power	KWH	1.	3,412
Fuel oil (#2)	gallons	41.	140,000
Fuel oil (#4)	gallons	42.8	146,000
Gasoline	gallons	35.2	120,000
Natural gas	cubic ft	.293	1,000
Propane	pounds	6.3	21,500
Steam (purchased)	pounds	.293	1,000

comparison with facilities with different utility rates. (Chapter 24 covers energy minimization in more detail.)

4 MANAGEMENT CONTROL RATIOS

Management control will be divided into materials, movement, and loss.

4.1 Management Control: Materials

Inventory control of materials is needed, especially in work in progress as well as for raw materials and finished goods. Two indices are Inventory Turnover Ratio (IT) and Inventory Fill Ratio (IF).

$$IT = ASALES/INV \qquad (12)$$

where

IT = inventory turnover ratio

$ASALES$ = annual sales, dollars

INV = inventory at end of period, dollars

$$IF = OF/TO \qquad (13)$$

where

IF = inventory fill ratio

OF = order lines filled in a period, order lines/period

TO = total order lines received in a period, order lines/period.

For inventory turnover, a convenient breakdown is the *ABC*, where *A* items (the giants) account for 10% of the numbers but about 50% of the value, *C* items (the midgets) account for 50% of the numbers but about 1% of the value, and *B* items (the rest) are 40% of the items and 49% of the value. For inventory fill, it is better to count lines on an order rather than number of orders. If only orders filled were calculated, an order would show the same ratio for both 5 of 10 lines or 19 of 20 lines filled.

Some improvement possibilities include: (1) automated identification of items (bar codes) to aid location of items (see Chapter 16) and (2) standardization of design to reduce the number of part numbers required (see group technology in Chapter 4).

4.2 Management Control: Movement

Movement by itself is not productive. More expensive are equipment and space, which consume time and energy and require people. Two indices are the Movement/Operation Ratio (MO) and the Average Distance/Move (ADM).

$$MO = M/O \qquad (14)$$

where

MO = movement/operation ratio

M = total number of moves for a product, moves

O = total number of operations for a product, operations

$$ADM = TD/M \qquad (15)$$

where

TD = total distance moved, ft

M = total number of moves for a product, moves

The number of moves, operations, and distances are commonly developed with a flow chart. The MO ratio focuses on the production system while the ADM focuses on the layout. A poor MO would be 7, fair 5, and good 3 (Singh, 1982). Data for the distances especially must be measured as done, not as imagined from an office desk.

Some improvement possibilities include: (1) plot flow for material, equipment, and people, and development of alternative layouts to reduce distances; (2) batching process orders to reduce equipment deadheading; and (3) handling, storing, and controlling different things differently (i.e., washers versus paint versus engines).

4.3 Management Control: Loss

Loss control requirements can be illustrated from "as much as 3 to 5% of all products handled are damaged" and "the cost of property damage ranges from 5 to 50 times the cost of personnel injuries." Two indices are the Damaged Loads Ratio (DLR) and the Inventory Shrinkage Ratio (ISR).

$$DLR = NDL/N \qquad (16)$$

where

DLR = damaged load ratio

NDL = number of damaged loads

N = number of loads

$$ISR = VI/EI \qquad (17)$$

where

ISR = inventory shrinkage, ratio

VI = verified inventory investment, dollars

EI = expected inventory investment, dollars

Damaged loads should be determined by occurrence sampling, not by operator self-report. Any part of any product being damaged is considered a damaged load. The ratio should be determined for different stages, such as receiving, manufacturing, and shipping. The inventory shrinkage loss is due to pilferage, recording errors, shipping errors, and so on.

Some improvement possibilities include (1) installing sprinklers and smoke detectors, (2) having people enter and leave the building past a "choke point" with human surveillance (see Chapter 10), and (3) making material handlers receive formal training, with periodic review.

5 OPERATING EFFICIENCY RATIOS

Operating efficiency will be divided into manufacturing, storage and retrieval, and receiving and shipping.

5.1 Operating Efficiency: Manufacturing

In 1980 approximately 75% of U.S. metalworking was performed in job shops. Approximately 40% of total manufacturing labor worked in job shops. In a typical job shop, a typical workpiece spends 95% of its time waiting and 5% of its time on a machine. There is much room for improvement in the "95% problem," especially since managers and engineers for many years have focused on their cost reduction efforts on the 5%. Two ratios to use are Manufacturing Cycle Efficiency (MCE) and Job Lateness (JL).

$$MCE = MT/ET \qquad (18)$$

where

MCE = manufacturing cycle efficiency ratio

MT = machine time, h

ET = elapsed time, h

$$JL = NL/NC \qquad (19)$$

where

JL = job lateness ratio

NL = number of jobs late in a week, jobs/week

NC = number of jobs completed in a week, jobs/week

Machine time would include all time on machines, heat treating, and inspection, including load/unload time. Elapsed time would be from when the unit entered the department (e.g., 8 A.M. June 4) until it left (e.g., 9 A.M. June 20). Only working days count and only 8 h/day (assuming one shift is used), and so in this example, elapsed time would be 12 days $\times$ 8 h/day = 96 + 1 = 97 h. A poor MCE would be 1%, fair 20%, and good 50% (Singh, 1982).

NL and NC are relatively self-explanatory. Don't over-emphasize the importance of job lateness. While lateness is to be avoided, it should not be avoided regardless of cost.

Some improvement possibilities include (1) having automatic identification systems to improve materials control data entry (see Chapter 16), (2) standardizing material handling methods and containers, and (3) considering group technology and cell layout. (However, note the conflict between group technology, which recommends running parts in the same family together, and material resource planning (MRP), which recommends running parts just before they are needed.)

5.2 Operating Efficiency: Storage and Retrieval

Although much has been written about warehousing theory, there is a large gap between theory and practice. A ratio to use is the Order-Picking Ratio (OPR).

$$OPR = LP/H \qquad (20)$$

where

OPR = order-picking ratio

LP = lines (of orders) picked per day, lines/day

H = hours worked/day, hours/day

Depending upon the application, order-picking time can be just the time for order picking or also could include time for stocking and packing.

Some improvement possibilities include (1) evaluating single-order picking versus batch picking versus zone picking, (2) evaluating moving picker to load versus moving load to picker, and (3) evaluating dedicated storage (each part number having a specific location always assigned to it) versus random storage (part number being stored anywhere). See Chapter 11.

5.3 Operating Efficiency: Receiving and Shipping

There are many opportunities for improvement in receiving and shipping productivity. Part of the challenge is that the work is greatly influenced by the vendors and suppliers and by the internal user departments. That is, much is beyond the control of the local supervisor and needs a "systems" approach. A ratio to use is Receiving Ratio (RR).

$$RR = S/H \tag{21}$$

where

RR = receiving ratio

S = pounds (or pallets of ft^3) shipped/day, lb/day

H = hours worked, h/day

Naturally a Shipping Ratio (SR) also could be used.

Some improvement possibilities include (1) scheduling arrival and departure of trucks; (2) automatic identification systems, label printers, and scanners to facilitate data entry and reduce errors; and (3) extendable conveyors that go into trailers. See Chapter 12.

White (1979a) felt that special attention should be paid to in-process handling (Manufacturing Cycle Efficiency and Job Lateness ratios, Eq. 18, 19), receiving (Receiving Ratio, Eq. 21), energy (Energy Utilization Index, Eq. 11), and damage and pilferage (Damaged Loads Ratio and Inventory Shrinkage Ratio, Eq. 16, 17).

Once these ratios are calculated, a desired value (standard) can be set, improvement opportunities can be determined, and projects can be begun. Remember that improving one ratio may hurt others.

DESIGN CHECKLIST: CRITERIA

Criteria to be used
Weight for each criterion

REVIEW QUESTIONS

1. Give the six ergonomic criteria for job design.
2. When making an economic analysis, what are the three basic numbers needed?
3. Briefly discuss the four different techniques for calculating the bottom line.
4. Give the three major classes of ratios for plant design and operation.
5. Discuss a specific ratio for a specific application. Do you have any suggestions to improve productivity?

PROBLEMS AND PROJECTS

2.1 Meet with a supervisor of a factory. Which criteria given in this chapter does this supervisor consider important?
2.2 Visit a specific existing industrial facility. What was done to maximize flexibility? What more could be done? How does the supervisor of the facility like your ideas?
2.3 Read 3 to 5 articles on a topic related to this chapter. Summarize them in a 500-word executive summary. List the references. Don't use articles cited in the text.

REFERENCES

Gould, L. "Ten tips for a good RFP." *Modern Materials Handling*, Vol. 45, 55, Dec. 1990.

How to justify a materials handling project. *Modern Materials Handling*, Vol. 46, 26–28, Mid March 1991.

Konz, S. *Work Design: Industrial Ergonomics*, 3rd ed. Scottsdale, Ariz.: Publishing Horizons, 1990.

Singh, P. Designing underhung crane and monorail systems. *Plant Engineering*, Vol. 36, 19, 105–108, Sept. 16, 1982.

White, J. Productivity—Four areas of special opportunity. *Modern Materials Handling*, Vol. 34, 10, 74–77, Oct. 1979a.

White, J. *Yale Management Guide to Productivity*. Philadelphia: Ind. Truck Division Eaton Corp, 1979b.

White, J., Agee, M., and Case, K. *Principles of Engineering Economic Analysis*, 3rd ed. New York: Wiley, 1989.

3 BASIC PLANNING INFORMATION

CHAPTER PREVIEW

For Systematic Layout Planning, information about PQRST (Product, Quantity, Routing, Supporting Services, and Time) must be obtained and the relationships of the various departments (activities) must be established. Finally, information is needed concerning space (available and desired). From this information, alternatives can be considered and a design selected and implemented.

CHAPTER CONTENTS

1 Input Data (PQRST)
2 Relationships
3 Space Requirements

KEY CONCEPTS

assembly drawings	Just-In-Time	relationship chart
bill of material	lean production	rework
debugging	learning curves	routing sheet
economic lot size	organization chart	standardization
external setup	permanent part time	standard time
flow diagram	PQ diagram	telecommuting
footprint	PQRST	yield
from-to chart	product life	

1 INPUT DATA (PQRST)

First we will consider "product flow" relationships and then "service flow" relationships. See Figure 3.1 and Chapter 6 for more on Systematic Layout Planning.

1.1 Product

Although layouts are made of nonproduct areas, such as offices and laboratories, in most layout situations there is a product. However, even when a product is considered, nonproduct activities, such as offices, toilets, and stairs, are important.

Product information begins with the component drawing; see Figure 3.2.

Assembly drawings (scaled drawings identifying components and illustrating assembly) are the next level; see Figure 3.3.

The **bill of material** is made from this collection of drawings; see Figure 3.4. It lists the total number of each item needed to make one final assembly. Generally several parts are made, either sequentially or simultaneously. Total requirements come from the bill of material of each product times production rate. Thus, to make 1 two-door automobile requires 1 left door and 1 right door. Each door requires a window crank; each crank requires a knob.

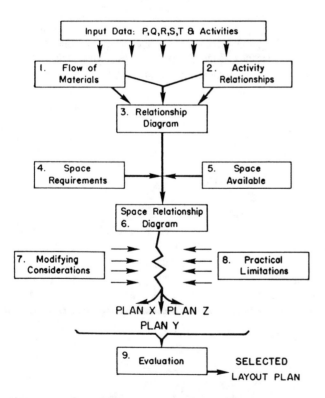

FIGURE 3.1   Systematic Layout Planning (Muther, 1973) requires input concerning **PQRST** (Product, Quantity, Routing, Supporting Services, and Time).

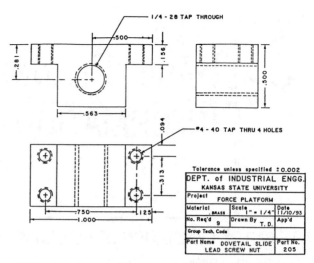

FIGURE 3.2   Component drawings include a scaled, dimensioned, toleranced drawing of the component, the raw material needed, and a part number.

1.2 Quantity

Quantity (the amount of goods and services produced) has six subdivisions: (1) make/buy, (2) standardization, (3) product life, (4) lot size, (5) yield, and (6) product-quantity relationship.

1.21 Make/Buy.

A decision must be made whether to make the item in your facility or elsewhere. Certain items are "always buy" (standard specialty items, e.g., light bulbs, bolts, screws, and motors, and processes for which you do not have expertise, e.g., plating, foundry, or forging). Certain items (generally assemblies) will be "always make." Items will be make or buy depending on the price,

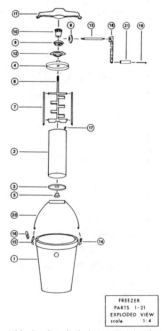

FIGURE 3.3   Assembly drawings include a scaled drawing of the assembly or subassembly with the components identified. The subassembly or assembly will have a part number as well as a list of all the component part numbers and quantities needed. Usually there are a number of subassembly drawings.

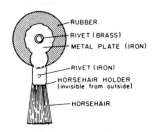

Bill of Materials (Parts List)				
Number in Unit	Name of Item	Part Number	Make	Name of Vendor, If Vendor Item
1	Rubber (eraser)	454		Goodyear
2	Metal plates (iron)	221	Yes	
1	Rivet (brass)	134		National Rivet
1	Rivet (iron)	122		National Rivet
30	Horsehair brush	---		Nebit Brothers
1	Holder (brass)		Yes	

FIGURE 3.4 Bills of materials list the quantities of each part number needed to make 1 unit (Rago, 1963). This simplified example does not include packaging materials and customer instructions.

quality, and delivery schedule of your factory and the various vendors.

1.22 Standardization.

Standardization uses the same component or assembly in multiple products to give economies of scale in design, manufacturing, and sales. Standardization is used especially on inside parts, which the final customer never sees, rather than on "style" items. For example, an auto manufacturer may use the same front seat frame, battery connector, wheel rim, and other inside parts in many different models. Parts from the same vendor often are used in products made by different firms. Sometimes firms even sell components to competitors. For example, Saginaw Steering Division of GM sells steering components to non-GM firms.

Using existing items instead of new items reduces design costs. The greater production volume permitted by standardization yields lower manufacturing costs/unit. Another benefit is simplification of spare parts requirements, because the number of items that have to be stocked is reduced and there is less chance that the spare part is out of stock.

1.23 Product Life.

Product life is like any life. Products are born, grow, mature, and die. Some products have shorter lives than others; for example, style or fashion items tend to have short lives. Good design will keep the basic item (e.g., a dress) almost constant and change only the style features (trim, pockets, button size and location, etc.). Certain staple items, such as furniture, automobiles, and food, change slowly—especially the inside (nonstyle) parts.

However, even if product design does not change much from year to year, that does not mean that industry sales are constant from year to year: The economic cycle (boom, recession) changes sales; the reaction to domestic and foreign competitors affects the sales of a specific firm

in an industry; and management of a firm also can shift allocation of production between various plants. Even if annual demand is constant, sales will fluctuate with the season of the year. One cannot assume that production of any product (or even a combination of products) will be constant for a specific factory. Designers should therefore build flexibility into the production facility and its operation. Two techniques of building manufacturing flexibility are (1) to build flexibility into software (computer integrated manufacturing) instead of into hardware and (2) to minimize setup cost.

See Box 3.1 on releasing items to manufacturing.

1.24 Lot Size.

One possibility is to have a factory which makes only one model of one product and makes it continuously. Although such factories exist, they are rare. The vast majority of factories make multiple products and multiple models of each product.

Manufacturing cost depends on setup cost, run cost/unit, and their ratio. See Table 3.1 and Figure 3.5. Three points can be made:

1. Regardless of setup cost, larger lot sizes reduce manufacturing cost.
2. Regardless of lot size, manufacturing cost is lower if setup cost is lower.
3. After a certain level, increases in lot size have relatively little influence on manufacturing cost. This critical level decreases as setup cost decreases and, for low-cost setups, lot size has relatively little effect on manufacturing cost.

Point 1 has been conventional wisdom for many years. Points 2 and 3 have been emphasized only in the last few years (since the mid-1980s in the United States). Emphasizing points 2 and 3 is known as **Just-In-Time** (JIT)

TABLE 3.1 Manufacturing cost (setup cost (T_S) + unit run cost (T_r)) for various setup/run cost ratios and lot sizes (N).

T_r	T_s	T_s/T_r	N	T_s/N	$T_r + T_s/N$
1	100	100	1	100	101
			10	10	11
			25	4	5
			50	2	3
			100	1	2
1	10	10	1	10	11
			10	1	2
			25	.4	1.4
			50	.2	1.2
			100	.1	1.1
1	1	1	1	1	2
			10	.1	1.1
			25	.04	1.04
			50	.02	1.02
			100	.01	1.01
1	.1	.1	1	.1	1.1
			10	.01	1.01
			25	.004	1.004

BOX 3.1   *Releasing a part for manufacturing*

Good practice follows five steps for new parts or assemblies.

1. *Prerelease review.* Before design engineering releases an item to manufacturing, it should have a formal conference with representatives from manufacturing. Typical representation would be from industrial engineering, tool engineering, quality assurance, production control, and (especially) purchasing. The group challenges the details on the drawings. Can a specially designed item be replaced by a standard, purchased item or an item already used on a different model? Can standard threads be used instead of precision threads? Should component tolerances be tightened (to help assembly) or reduced (to help fabrication)? The purpose of the conference is to change the design before the designers set it in concrete by officially releasing the drawings.

Rather than discussing manufacturability only when design engineering is ready to release an item, many firms have a manufacturing consultant permanently assigned to help design engineers tap into manufacturing expertise early in the design cycle. Although this concept has been used since at least the 1950s, it has been rediscovered by consultants and marketed as part of the Taguchi method.

2. *Production system outline.* The manufacturing engineering team needs to decide on the general production concept. What will be made and what purchased? What machines will be used? Will a job shop or production line be used? What will be the typical batch size?

3. *Detail.* In the detail step, specific tools and fixtures are designed and feeds and speeds are selected. What size conveyor or truck will be used? Will the operator sit or stand? The overall layout as well as the detailed workstation layout must be planned.

4. *Installation.* The concepts need to become reality in this stage. Often they will need to be modified to conform to reality or to changed requirements. Time standards need to be established for cost accounting, scheduling, acceptable day's work, and so on.

5. **Debugging** (finding and eliminating problems). Although learning how to operate the equipment will go on for considerable time, the initial period will be the most intense. Part-assembly drawings may have to be revised, vendor's promises may be found to be optimistic, training may be found inadequate, and so forth. Depending on the amount of change required and the skill of the designers, this shakedown period can last from 1 day to several years. On complex projects such as flexible manufacturing systems (FMS), installation may be prolonged. For FMS, utilization might be 40% after 8 months and 75% after 14 months (Gould, 1990). The longer debugging lasts, the more "heads will roll."

---

manufacturing or **lean production**. See Chapter 5 for more on JIT.

The real question then becomes what batch size should be produced for each model of each product. This problem is addressed by the **economic lot size** formula, in which the challenge is to balance one-time costs (setup and paperwork) versus continuing costs (inventory). See Figure 3.6. It can be formulated either within an organization (how many to make) or for purchasing (how many to buy).

$$STPU = \frac{\text{annual setup cost, \$/year}}{\text{total units/year}}$$

$$= \frac{ASC}{A} = \frac{S(N)}{A}$$

$$= \frac{S(N)}{Q(N)} \quad S/Q$$

where

$STPU$ = setup cost, \$/unit
ASC = annual setup cost, \$/year
= $S(N)$
S = setup cost, \$/lot
= $PC + ST(SW)$
PC = paperwork cost, \$/lot
ST = setup time, hour
SW = setup wage, \$/hour
N = number of lots/year
A = annual requirements, units/year
Q = quantity (number) of units/lot

Fig. 3.6 shows setup costs/unit getting smaller and smaller as the number/lot gets larger. Note how steeply the curve rises for small lot sizes. If setup was the only cost, we would make large lots. Reducing cost/setup (shifting the curve vertically) reduces setup cost/unit but does not change benefits of large lots.

However, there is another cost: inventory. Inventory cost is (amount in inventory) (inventory cost/unit) or, using symbols:

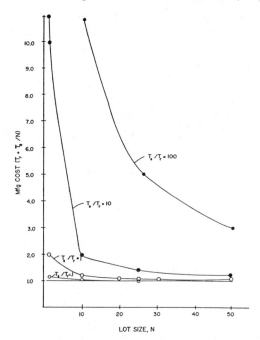

FIGURE 3.5 Manufacturing cost per unit declines with larger lot sizes and with smaller setup costs. In addition, as the ratio of setup cost to run cost decreases, the effect of additional units in the lot becomes less important.

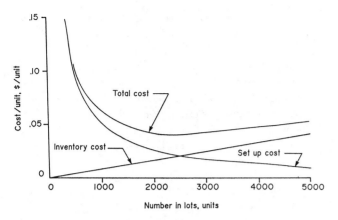

FIGURE 3.6 Total costs are composed of setup costs + inventory cost. The minimum of the total cost curve is the quantity at which inventory cost equals setup cost.

$INCPU$ = inventory cost/unit, $/unit
      = $INCPL/A$
where
$INCPL$ = inventory cost per lot, $
      = $Q/2\,(C)\,(p)$
   Q = order quantity (Average inventory is $Q/2$ if zero safety stock is assumed and the withdrawal rate from inventory is constant (linear).)
   C = cost/unit, $
   p = annual inventory cost, %/100 (Typical rates are 20–30%, so p = .2 to .3.)

As can be seen in Figure 3.6, inventory rises from zero (assuming no safety stock) in a straight line. A safety stock adds a constant inventory cost (horizontal line), regardless of lot size. Then $TCPU$ = Total cost/unit, $/unit.

Generally we are interested in the minimum of the total curve, $QMIN$, the point where the slope of the total curve is zero. From calculus, this means setting the first derivative equal to zero and solving for Q:

$$QMIN = \sqrt{\frac{2AS}{Cp}} \qquad (1)$$

For a two-curve situation *where one curve is a straight line*, the minimum is at the point where the two component curves are equal.

There are some critical assumptions in the economic lot size formula:

- The formula is applicable only to *equal-size* lots. If the lots are not equal, $QMIN$ might be too high or too low.

- Obsolescence, part of inventory carrying cost, often is overestimated. If a product has a five-year life, obsolescence is not 20% but zero, until at least the end of the model production and probably longer if spare parts are considered. Overestimating obsolescence tends to make predicted $QMIN$ smaller than actual $QMIN$.

- Consider your shipping pattern. If you ship most of your production shortly after you produced it, your average inventory is not $Q/2$ but $Q/3$, $Q/4$, or even $Q/10$. Overestimating the number in storage or the time in storage tends to make predicted $QMIN$ smaller than actual $QMIN$.

- The equation does not consider overage, the extra units for setup units and scrap. For a lot of 20, you might start with 25 units. But for a lot of 100, you also might add 5 units for a total of 105 units. Since overage is not in the equation, predicted $QMIN$ is smaller than actual $QMIN$.

- The equation assumes quality is independent of lot size. In practice, once a process is set up correctly, errors are relatively few. Many setups result in many opportunities for errors. Thus, predicted $QMIN$ is smaller than actual $QMIN$.

- The equation does not consider learning. That is, time/unit is the same for lot sizes of 10, 100, or 1,000. In practice, time/unit is lower for larger lots. Thus, predicted $QMIN$ is smaller than actual $QMIN$.

- The equation does not consider inflation. It assumes that future costs are equal to present costs, which is unlikely. Thus, predicted $QMIN$ is smaller than actual $QMIN$.

- Note the shape of the total cost curve in Figure 3.6. It is saucer shaped, not cup shaped. That is, departures from $QMIN$ become expensive only if large and especially if departures are to small lot sizes.

The concept of Just-In-Time production scheduling (see Chapter 5) involves frequent setups, which has led to intensive work to reduce the cost of each setup. If setup cost is reduced by R%, then $QMIN$ is reduced by the following (Esrock, 1985):

$$PCQMIN = 100\,(1 - \sqrt{1 - .01\,R})$$
where
$PCQMIN$ = change in $QMIN$, percent   (2)
   R = reduction in setup cost, percent

For example, a 75% reduction of setup cost/lot yields

$$100\,(1 - \sqrt{1 - .75}) = 100(1/2) = 50\% \text{ reduction in } QMIN.$$

Since the amount put into inventory is $QMIN$, equation 2 also applies to inventory costs; that is, a 75% decrease in setup cost/lot gives a 50% reduction in inventory cost.

Because there are frequent setups, lead time is reduced. Because lead time is reduced, lead time variability is reduced (it is easier to predict 1 week ahead than 10 weeks ahead). This in turn reduces the size of the required safety stock, because safety stock is required by the variance of lead time rather than the mean lead time.

The change in productivity is as follows:

$$PINC = 100\,(1 - \sqrt{1 - .01R})\,/\,(B - 1 + \sqrt{1 - .01R}) \quad (3)$$

where

$PINC$ = productivity increase, pieces/h

B = total time to setup and run batch/setup time

R = percent reduction in setup time

For example, assuming .5 h for setup and 2.0 h to run a batch of 1,000, then total time = 2.5 h and B = 5; output = 1,000/2.5 h = 400 pieces/h. Assume setup time is reduced by 75% to .125 h. Then a 75% change in setup time gives a new lot size of 50% of the original, or 500. The new run time is 1 h and total time = 1.125 h; output = 500/1.125 = 444 pieces/h. Productivity now is 444 pieces/h/400 pieces/h = 111% of before.

Setup reduction does not come automatically. A good approach is to consider setup changes as similar to equipment changes for racers at the Indianapolis 500 race. The keys are preplanning, getting the proper tools and items ready before the change (**external setup**; see Table 3.2 and Box 3.2), and training of the quick change team. The Society of Manufacturing Engineers (SME) has a number of books, clinics, and videotapes on setup reduction. Contact the SME Customer Service Center at 1-800-733-4SME.

An example of preplanning is determining what fixture strategy will be used (Young and Bell, 1991). Possibilities include side clamping (vice), through clamping (bolts), and down clamping (surface clamp). What are the reference location surfaces? A bolt which takes 1 turn to fasten or unfasten is better than one which takes 15 turns. Can clamp heights be standardized (e.g., with shims or spacers)? Do all screws use the same hex socket size so only one hand tool is needed? Are all hand tools, screws, punch holders, spacers, etc., stored close to the using location?

Table 3.2 gives steps to single-minute setup (Cochran and Swinehart, 1991). The keys are (1) to use external setup rather than internal setup and (2) to eliminate adjustment. Setup adjustment is needed to determine offset (the difference between programmed dimension and actual dimension). The conventional way to determine offset is to make a trial cut on a piece, enter offset into controller, make trial

TABLE 3.2   Steps to single-minute setup (adapted from Cochran and Swinehart, 1991).

1. **Separate internal setup actions from external setup actions.**
   - Internal setup = setup while machine is stopped.
   - External setup = setup while machine is operational. After video-taping setups, classify and analyze data.

2. **Convert internal setup actions to external setup actions.** Don't interrupt internal setup actions for external setup actions. Make fixture setup, tool setup, and gauge acquisition external. Have common tool packets resident at the machine.

3. **Eliminate adjustment.** Adjustment consumes as much as 50% of setup time (time from completion of last good part A to acceptance of first good part B). Use quick change tooling, such as sine key location of fixtures, angle plates, and other work holding devices to establish predefined locations on the worktable. Use part probes to minimize tool cutting position errors.

4. **Achieve single-minute setup.**

cut on a second piece, enter offset into controller, and so on, until the machined and programmed dimensions are equal. Cochran and Swinehart recommend using a probe (sensor) to determine locations; that is, fixtures are used to clamp parts, and probes are used to locate parts. The fixture positions the part within a tolerance compatible for probing; then the probe is used to locate the part. This is closed loop machining (see Box 4.1). Note that parts should be designed so that locating datums correspond to parts features suitable for probing.

Consider the quick change team as "industrial athletes," complete with a coach, practices, videotapes, competitions, T-shirts, team banquets, team photos, and so on.

1.25 Yield.   The **yield** (number of good items per number of items started) section is divided into (1) yield equation and (2) scrap and rework rates.

Yield equation.   Modify the theoretical order quantity by the yield of the process. At each station, an item is acceptable, is scrapped, or is **reworked** (repaired and inserted back into work flow). See Figures 3.8 and 3.9. At each station, the rework can be fed back into the flow (upstream or downstream of the station) or can be scrapped.

The station output is

$$OUPUT = INPUT(YIELD) \qquad (4)$$

$OUTPUT$ = number of good units from station

$INPUT$ = number of good units into station

$YIELD$ = immediate output + output from rework downstream + output from rework upstream

   = $1 - P_{r1} - P_{s1}$

P_{r1} = Proportion of units reworked at station 1

P_{s1} = proportion of units scrapped at station 1

$NP_{r1}P_{d1}$ = number of rework units inserted downstream

$NP_{r1}P_{u1}$ = number of rework units inserted upstream

P_{d1} = proportion of reworked units inserted downstream

P_{u1} = proportion of rework units inserted upstream

To illustrate with a simple numerical example, assume N = 1,000 units.

P_{r1} = .5%

P_{s1} = .1%

P_{d1} = 90%

P_{u1} = 9.5% (i.e., since $100 - 90 - .5 = 9.5\%$, 9.5% were scrapped at the rework station)

Output of the station is:

$$1,000 \, (1 - .005 - .001) = \quad 994$$

$$+ 5\,(.9) = \quad 4.5$$
$$+ 5\,(.095) = \frac{.475}{998.975}$$

In theory, the reworked units entering upstream of the station might not be processed properly at station 1 (i.e., have a rework and scrap rate, which probably would not be the same as for units that did not go through rework). In practice, units are discrete and .475 units cannot be inserted upstream—it must be 0 or 1. Thus, problems using the equation 4 approach should be solved by computer simulation. If you wish to use a calculator, use equation 5. As a simple approximation

$$OUTPUT = INPUT\,(YIELD) \qquad (5)$$
$$YIELD = 1 - S + R$$

BOX 3.2   *Case study of setup time reduction (Lesnet, 1983)*

The Burlington, Iowa, switchgear plant cut its work-in-process inventories to less than 1/3 of its previous level—disposing of 1,000 linear feet of 20-ft-high skid towers and 1,000 pallet-size part storage bins. It did this because it was able to reduce setup times sharply, in one example from 47 min to 2 min. Some highlights of its procedure in the press area:

- They videotaped actual die changes. Because the tape was in real time, it was easy to identify and discuss the actual setup.
- The videotape was discussed by the operator and supervisor first and then by the staff. Suggested changes were listed.

Some problems and changes were as follows:

Problems

Partially stripped threads in hold-down holes.
Die operating height was so high that the operator had to stand on a skid.
Each die had a different shut height, requiring much adjustment.
No two dies were bolted down in the same manner.

The die coming from the press was moved to a storage rack before the new die was inserted.

There were alignment problems with the rollers in the press bed T slots and the die parallels.
The die sequence was not easily available.

Changes

Replace all marginal threads.
Height was reduced 5 in.

All dies within a press size were standardized for shut height.

A standardized mounting technique was devised and applied to each die.
A second level was added to the die cart (see Figure 3.7) so the new die could be slid into place, after the old die was removed, with just a simple vertical adjustment.
Rollers were put on the die itself.

A daily die sequence is now available at each press so the next die, material, and support can be prepositioned and ready for change.

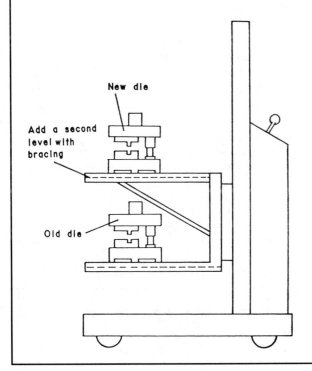

FIGURE 3.7   Adding a second shelf to a standard electric die cart reduced setup time, because after the old die has been removed, the new die can be added by adjusting the platform level (Lesnet, 1983). Copyright © Institute of Industrial Engineers, 25 Technology Park, Norcross, GA 30092. (404) 449-0460.

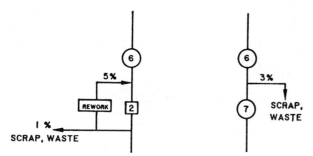

FIGURE 3.8   Rework (left figure) and scrap (right figure) should be included in flow process charts. Rework and scrap often are more difficult to handle than good product. For a detailed analysis, scrap can be divided into scrap (material lost due to human error), waste (loss due to the design, e.g., sawdust, metal chips), and shrinkage (loss due to theft or physical deterioration, e.g., food).

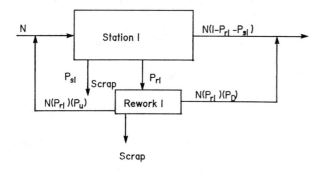

FIGURE 3.9   Stations have good output, scrap, and rework. The rework (often done "under the table") can re-enter the flow either upstream or downstream of the station.

where

S = scrap proportion

R = rework proportion

All rework is inserted downstream. Note that the yield accumulates along a series of stations. Using equation 5, assume $N = 100$ good items; the scrap rate at station 1, 2, and 3 was 4%, 3%, and 3%, respectively; and the rework rate was 3%, 2%, and 1%, respectively. Then Table 3.3 shows the output of station 3 is only 96 units (assuming outputs were rounded to discrete units at each stage).

In determining quantity, you know how many you want to finish with; the unknown is how many you want to start with. Table 3.4 shows that you need to start with 104 units (assuming discretization at each stage).

In practice, people tend to add a few units for a safety margin, that is, release 106, instead of 104, to get 100 good units. The reason is that actual scrap on an operation may be worse than the estimate. If, for example, there were 4 scrap units on operation 2 instead of the 3 you had planned on, the total order will be 1 unit short. Then you would need to release a second lot to the shop (which also has potential for scrap). How big should the second lot be—2? 3? It is hard to say. Thus, it is easier to release a few extra on the first lot.

The order size also depends on what happens to excess good units. Assume a customer orders 500 units; the order is released with 510 units and 505 good units are completed. If you can ship the 5 excess units and the customer will pay for them or the 5 can be put into inventory for the next order, you can allow larger releases than if the 5 excess must be scrapped.

Scrap and rework rates. Determining the specific numerical values for scrap and rework rates for each specific station is difficult. Rejects are defined by the traditional philosophy that all parts within the "goalposts" of the tolerance limits are equally satisfactory. Calculations become more complex if the Taguchi concept (that quality is measured by deviation from the target center) is used.

Setup will automatically cause scrap on some operations but not on others. However, each setup is an opportunity for error, so total setup scrap probably will be higher for small lots.

Scrap occurring once an operation is up and running is not as common. Once a process is running, it tends to make zero or little scrap until it goes out of control and makes more and more scrap. That is, scrap usually is not a random process (e.g., with 1 unit in 100 randomly being a reject). A more common situation would be the first unit of a run is scrap, then the process runs without scrap for 2,000 units, then drifts out of control with 10% of the next 100 units being scrap, then the process is corrected and no scrap is produced for the next 3,000 units, then it drifts out of control with 6% of the next 50 units being out of control and scrapped, and so forth.

The rate prediction is made more difficult by social problems. Scrap is socially unacceptable. If someone reports scrap to a supervisor, the supervisor will react negatively, yelling, getting mad, or complaining. People don't want to leave a "paper trail" showing their mistakes. Therefore, they don't report scrap; they rework it without any incriminating paperwork. The supervisor may know about the rework and assign people to do it, but the supervisor doesn't leave a paper trail either. Thus, the real flow of product through a plant is not a smooth-flowing river but a

TABLE 3.3   Predicted output with known input quantity. The quantity is made discrete at each workstation.

Operation	Quantity
1	$100 (1 - .04 + .03) = 99$
2	$99 (1 - .03 + .02) = 98$
3	$98 (1 - .03 + .01) = 96$

TABLE 3.4   Predicted input with known output quantity. The quantity is made discrete at each workstation.

Operation	Quantity
3	$x (1 - .03 + .01) = 100; \ x = 100/.98 = 102$
2	$y (1 - .03 + .02) = 101; \ y = 101/.99 = 103$
1	$z (1 - .04 + .03) = 103; \ z = 103/.99 = 104$

river with many backwaters and eddies; see Figure 3.10. Improvements in quality may therefore be much greater than they appear in the formal paperwork system, because a change in the reported scrap rate from .4% to .3% may reflect a change in the real rework rate from 2% to 1.5%.

In planning the facility, the engineer should allow space for rework operations and for scrap and should realize that the actual amount of work being done at a workstation may be considerably more than appears in the paperwork system.

1.26 Product-Quantity Relationship. When planning the overall facility, use a product-quantity (PQ) data sheet (see Figure 3.11) to accumulate all the information for one product in one place. Then, from the collection of PQ sheets, construct a **PQ diagram** (see Figure 3.12). PQ diagrams also are called *volume-variety* diagrams. The *Y* axis should be in hours/year. If the curve is steep, a few products are quite important and a line layout should be considered. If the curve is flat, no product stands out and a job shop should be considered. If several products are similar, some operations might be combined and group technology should be considered. Another possibility is a line layout for some products and a job shop for others. (See Chapter 4 for more on lines, job shops, and group technology.)

1.3 Routing

How will routing be made? Once it has been decided to make the items rather than purchase them,

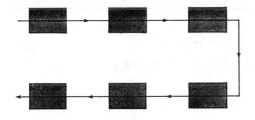

Idealized production flow

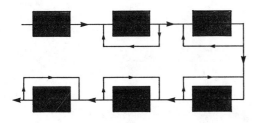

Real production flow

FIGURE 3.10 Ideal flow of product through a factory is shown above; a more realistic flow is shown below. Rework will cause "eddies" in the flow, substantially increasing the number of items processed at each station.

the various manufacturing steps are determined. The sequence of steps is shown in a **flow diagram**. Figure 3.13 shows flow diagram symbols. Figure 3.14 shows a flow diagram following a single product; the accompanying process chart is Figure 3.15. Figure 3.16 shows an assembly flow diagram.

In a line (product) layout, the steps are put in sequence on the factory floor just as they are on the piece of paper. In a job shop (process) layout, there isn't sufficient production volume to justify arranging the workstations just for one product. Thus, the workstations that are similar are grouped together to get the advantage of specialization of labor and equipment; each move of a product, however, now follows a complex path. (See Chapters 4 and 5 for more on product versus process layout). Although a **routing sheet** is prepared for both types of layout, for the line layout the sequence is obvious from the physical arrangement and so the routing sheet need not accompany the parts. For a job shop layout, the sequence varies with each product and therefore a routing sheet accompanies each order so the material handler can tell where to send the item next; see Table 3.5 for an example of a job shop routing sheet.

A routing sheet should have

- component name and number
- raw material requirements
- operation description and number
- equipment and tooling requirements
- speeds and feeds
- unit times (run and setup)

Challenge every step on the routing sheet *before* using it for the layout. This is the time to make changes in methods and sequences, not after the equipment is "set in concrete."

Flow of product and people (product flow) is not sufficient for making a good layout; you also must consider the service areas—the "service flow" relationships.

1.4 Supporting Services

Facilities contain service areas as well as production areas. Although services are necessary, they don't directly change the product. Examples might be the nurses' station, offices, toilets, cafeterias, battery recharging areas for lift trucks, tool cribs, and so forth. Storage areas are also part of supporting services. The layout needs to include all these areas since service areas often use more space than production areas. In many facilities (such as offices and labs) there may be no product flow, only service flow.

1.5 Time

First, determine how long it will take to make the product. Then, considering the hours available/year, decide how to staff the facility.

When operating a facility, three general guidelines for staffing are to

```
PRODUCT-QUANTITY DATA SHEET                        Plant  Manhattan          Project
                                                   Data gathered by
                                                   Date  1-26-85             Sheet        of

Fill-in as applicable.                    PRODUCT INFORMATION    PRODUCTION REQUIREMENTS
For ONE PRODUCT - Form and/or Treat only
Product Name & Description  Bales of Waste Paper            Quantity produced this year  2,000,000   Source  J. Elliott
Finished condition (fluid, delicate, hazardous, etc.)  Baled   Quantity anticipated next year  2,000,000  Approved  J. Elliott
Size-shape  2.5 ft wide by 5.5 ft long by 3 ft high         Quantity anticipated in 5 yrs.             Est. by
Normal unit of measure  Bale        Weight/unit  1500#       Length of time present product or model will be produced
Starting material condition  Shredded or scrapped           Seasonal variation  Negligible
Size-shape  4 ft square boxes    Weight/unit  50#           Expansion Plans  None
Normal container: as received  ---    as shipped  ---

                                                            Trends in product:
For ONE PRODUCT - Assemble and/or Disassemble involved          Size         None      Diversification    None
Product Name & Description                                      Weight       None      Simplification     None
Finished condition                                             Materials     None
Size-shape                        Weight/unit                  Rec'g. & Shipping amounts and frequencies   None
Major Components:   Material Condition   Size-shape   Weight/unit  Refinements   None
a.                                                             Other
b.
c.
d.                                                            Operating hours  8/10       per shift        18     per day
e.                                                                                                        80     per week
          See Parts List(s) or Component Breakdown [ ]       Plan Layout for (no. of units)
                                                                          (hour, day, week)
For SEVERAL PRODUCTS                                         Quantity:                    Per     % of    Plan
                                                                                          Order   Produc- Layout
Name of Product or Group  Condition  Size-shape  Weight/unit  This Yr.  Last Yr.  Next Yr.  5 Yrs.  or Lot  tion  for
A.
B.
C.
D.
E.
F.
G.

Trends in Product:
Seasonal Variation:
Expansion Plans:

      NOTES:
```

FIGURE 3.11   Product-quantity data sheets summarize the information for each product (Muther, 1973).

- minimize idle capacity
- consider shiftwork
- use filler jobs or filler people

See Chapter 15 for a discussion of these three points.

1.51 Time/Unit. Determine **standard time** and actual time (Konz, 1990):

$$STDTIM = (OBTIM)(PACE)/(1 - ALLOW) \qquad (6)$$

where

TABLE 3.5   Routing sheets (also called operations charts) vary depending on the organization, but the table below is typical. Operation numbers are given in multiples of ten so that operations originally omitted can be added without changing the numbers of the other operations. Some routing sheets also include bill of material information.

Part name Punch   Part number 541-675   Raw material   1040 10 mm round

Analyst SK   Date 20 Jan 84   Used on Model 80

Operation Number	Operation Name	Machine	Tooling	Feed (mm/rev)	Speed (rev/min)	Std. h/unit	Remarks
10	Turn 4 mm dia Turn 3 mm dia	J & L T. Lathe	#642 Box	.225	318		
	Cut off to length	J & L	#6 cutoff	Hand	318	.008	
20	Mill 5 mm radius	#1 Milwaukee	Tool 84	Hand		.004	
30	Heat-treat	#4 furnace				.006	
35	Degrease	Vapor degreaser					
40	Measure hardness	Rockwell tester				.002	
50	Store						

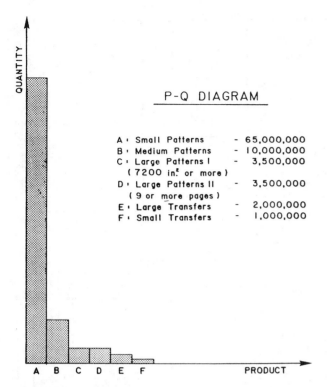

P-Q DIAGRAM

A · Small Patterns	-	65,000,000
B · Medium Patterns	-	10,000,000
C · Large Patterns I	-	3,500,000
(7200 in.² or more)		
D · Large Patterns II	-	3,500,000
(9 or more pages)		
E · Large Transfers	-	2,000,000
F · Small Transfers	-	1,000,000

FIGURE 3.12 PQ diagrams list products made on the X-axis, arranged in order of quantity. The quantity axis (Y-axis) can be scaled in units (300 units of model A, 4,000 of model B, 50 of model C); then B would plot at 4,000, A at 300, and C at 50. Do not put a break in the scale to make widely different quantities fit. That eliminates the visual impact that the figure is trying to emphasize. The quantity axis also can be put in labor-hour units. Assume A required 5 h/unit, B was 1, and C was 100. Then A hours are 5 × 300 = 1,500, B hours are 4,000 × 1 = 4,000, and C hours are 50 × 100 = 5,000. The C would plot first at 5,000, the B at 4,000, and the A at 1,500. If the curve is steep, consider product (line) layout; if the curve is flat, consider process (job shop) layout.

STDTIM = standard time, h/unit

OBTIM = observed time, h/unit

PACE = pace rating of operator, proportion

ALLOW = allowances (personal, fatigue, and delay), proportion

Assume a time study technician observes an operator doing task *A*. If the observed time is .05 h, the operator was working at a 110% pace, and allowances were 15%, then over the total shift the operator should take (.05)(1.1)/(.85) = .065 h/unit and, in an 8 h shift, make 8/.065 = 123 units.

If there is no operator presently doing the job, then standard time can be estimated from standard data.

The standard time is the time a process *should take*. However, for planning purposes, use actual time—the time it *does take*.

$$ACTTIM = STDTIM/EFF \qquad (7)$$

where

ACTTIM = actual time, h/unit

STDTIM = standard time, h/unit

EFF = efficiency, proportion

Thus, if the operator produces at 90% efficiency, then $ACTTIM = .065/.9 = .072$ h/unit and, in an 8 h shift, 8/.072 = 111 units will be made.

Efficiency varies from 100% because of inaccurate time standards, operator learning (see Box 3.3), changes in method not reflected in the time standard, scrap and rework, absenteeism, and so forth. If efficiency is reported as 100%, something is probably wrong.

See Box 3.4 for comments on machining times and Box 3.5 for comments on number of machines/operator.

1.52 Hours Available per Year. There are 8,760 hours in a year, but most facilities do not use their equipment 24 h/day for 365 days/year. The number of h/year that the equipment will be used must be determined. See Table 3.8. The determination compares capital cost (equipment and building) to labor cost. Equipment generally does not wear out; it becomes obsolete. Thus, if equipment cost can be spread over more units of production or the same units of production can be produced with less equipment, then capital cost/unit will be reduced. But using more than one shift may increase labor cost, due to shift pay differential (usually 2 to 5%). For example, assume wage cost was $15/h and shift differential was 10%. Then extra shifts would cost a premium of $1.50/h. If equipment cost/person (including the building) was $60,000 and equipment life was 15 years, then capital cost/year would be $4,000. If the equipment was operated 1 shift of 1,900 h/year, capital cost would be 4,000/1,900 = $2.10/h. A double shift of 3,800 h would give a capital cost of 4,000/3,800 = 1.05/h—saving $1.05/h. But the labor cost of $1.50 is greater than the benefit of $1.05, so a single shift would be better. However, the double shift decision could be made to reduce total costs per unit (considering marketing, engineering, administrative costs, etc.) even though increasing manufacturing costs.

More refined calculations include greater maintenance costs on equipment run twice as many hours/year, changes in utility costs due to a second shift, time value of money, use of 1,800 or 1,850 h/shift instead of 1,900, greater employee turnover on evening and night shifts (and thus greater training, quality, and administrative costs), and so forth.

In addition, multiple shifts may be used for customer service reasons (being open longer hours for customer

FIGURE 3.13 Flow diagram symbols have been standardized as a circle for operation, an arrow for move (transportation), a square for inspection, a triangle (upside down pile) for storage, and a capital *D* for delay (unplanned storage). Inspection with change in the item (an operation) is a circle inside a square. Some people put a number inside each symbol (identifying operation 1, 2, 3, or inspection 1, 2, 3), and others do not. Some people also darken the circle for "do" operations but not for "get ready" and "put away" operations.

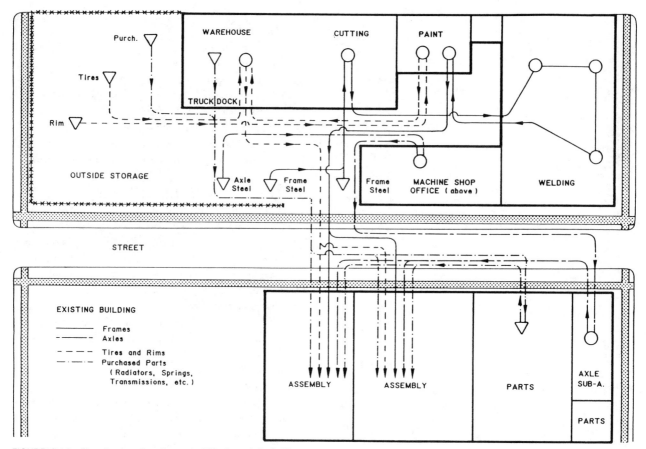

FIGURE 3.14   Flow diagrams show the product flow through the facility.

convenience, as with stores, hospitals, or transportation or for meeting market demand) or difficulty in stopping and starting the process (steel melting, chemical processing).

Related to the multiple shift decision is the break time decision. There are breaks for vacations, for holidays, for meals, and for coffee. What happens to output during these breaks? Is the equipment shut down or does a relief worker operate it? In manufacturing (but not retailing),

working during holidays may require a 50 to 100% premium in pay. Working during vacations, meals, and coffee breaks usually does not require a pay differential, although the replacement worker may be less proficient than the regular worker, resulting in an output loss. There also are scheduling problems in finding work for the relief worker during nonbreak times.

Time available for working also is reduced by personal, fatigue, and delay allowances. See equation 6.

BOX 3.3   *Organizational and individual learning*

As organizations and individuals produce more and more of a product, they get better and better at doing it, so time/unit decreases. Quantification of this decrease began in the 1930s and the technique was well known by World War II. The following is a brief summary.

Learning is given as a rate:

$LCRATE = 2XTIME/XTIME$

where

$LCRATE$ = **learning curve** rate, %

$2XTIME$ = time/unit at quantity $2X$

$XTIME$ = time/unit at quantity X

Thus, if time/unit at unit 50 was 1.0 and time/unit at unit 100 was .9, then the learning curve rate is 90%. The key concept is that time/unit decreases in a straight line when log of

time/unit is plotted versus log of cumulative quantity. In practice, time and quantity are plotted on log-log paper.

Typical learning curve rates are 90 to 95%—that is, when the quantity doubles, the time at the doubled unit is 90% of the time at the original unit. (See Konz, 1990, Chapter 25 for more detail, including learning rates for various processes.)

Learning can drastically change the time/unit and thus capacity requirements and amount scheduled per release. For example, assume a 90% curve and a standard time/unit of 1.0 h established after 50 units are produced. Then time/unit at unit 100 = 1 (.9) = .9 h, at unit 200 = .9 (.9) = .81 h, and unit 400 = .81 (.9) = .729 h, etc. See Table 3.6.

A key implication is that production and the labor force cannot both remain constant. That is, if the facility is producing a constant 500 units/period month after month, then the work hours, and thus employment, should decrease. If the labor force is constant, then output should increase.

1.53 Staffing.   Related to the decision of how many hours a facility will be operated per week is the staffing decision. Staffing options include not only full time but part time, telecommuting, temporary, and subcontracting.

Part-time workers.   When staff work less than 30 h/week, they are **permanent part time**. Advantages to the organization include lower cost/hour, better fit to fluctuating demand, less effect of stress, and, possibly higher quality workers.

In general, part-time workers cost less than full-time workers. The wage itself may be lower, because the workers lack experience. This especially applies if part-time workers

FLOW CHART										Exception No. ————		
SUBJECT HOUSING 882 FABRICATION						FORM NO.				DATE		
FILE NO.		PAGE NO. 1		OF 1		PAGES	CHARTED BY					
SUMMARY OF STEPS IN PROCESS												
	OPERATIONS	TRANSPORTS	INSPECTIONS	DELAYS		STORAGE		TOTAL STEPS		TOTAL DIST.	TOTAL MIN.	
PRESENT	11	13	2	12						225 ft		
PROPOSED												
SAVINGS												

LINE	DETAILS OF PRESENT/PROPOSED METHOD (CIRCLE ONE)	Operation	Transport	Inspection	Delay	Storage	TIME	DIST	POSSIBILITIES					NOTES
1	INSPECT	O	⇒	☑	D	▽	1.00	10						
2	TRANSPORT TO R.D.P.	O	⇒	☐	D	▽		30						
3	DRILL	Ø	⇒	☐	D	▽	4.95							
4	TRANSPORT H.M.	O	⇒	☐	D	▽		35						
5	MILL	Ø	⇒	☐	D	▽	5.38							OPERATION 673 3-8-20% - 5.38
6	TRANSPORT TO MILL	O	⇒	☐	D	▽		6						
7	MILL	Ø	⇒	☐	D	▽	5.08							6.35 5.08 MIN. OPERATION 12-20%
8	TRANSPORT TO MILL	O	⇒	☐	D	▽		6						
9	MILL	Ø	⇒	☐	D	▽	2.87							OPERATION 13
10	TRANSPORT TO R.D.P.	O	⇒	☐	D	▽		45						
11	DRILL	Ø	⇒	☐	D	▽	4.8							OPERATION 14
12	TRANSPORT TO DRILL & TAP	O	⇒	☐	D	▽		8						
13	DRILL & TAP	Ø	⇒	☐	D	▽	3.35							15 - 16
14	TRANSPORT TO DRILL	O	⇒	☐	D	▽		8						
15	DRILL 20 HOLES	Ø	⇒	☐	D	▽	6.51							17
16	TRANSPORT TO 18 & 19	O	⇒	☐	D	▽		3						
17	COUNTERBORE & TAP	Ø	⇒	☐	D	▽	2.68							18 - 19
18	TRANSPORT TO 20,21	O	⇒	☐	D	▽		3						
19	DRILL & TAP 20,21	Ø	⇒	☐	D	▽	6.88							20 - 21
20	TRANSPORT TO 22-23 24-25	O	⇒	☐	D	▽		3						
21	DRILL & TAP	Ø	⇒	☐	D	▽	1.93							22-24 SAME AS 18&19 23-25
22	TRANSPORT TO MILL	O	⇒	☐	D	▽		45						
23	MILL OPERATION 26	Ø	⇒	☐	D	▽	2.68							OPERATION 26 SAME AS 13
24	TRANSPORT TO INSPECTION	O	⇒	☑	D	▽		75						
	TOTALS													

FIGURE 3.15   Single-object process charts (as the name indicates) follow a single object through every step. For cost reduction analysis purposes, it may be useful to use this type of chart to follow a *person* rather than an object.

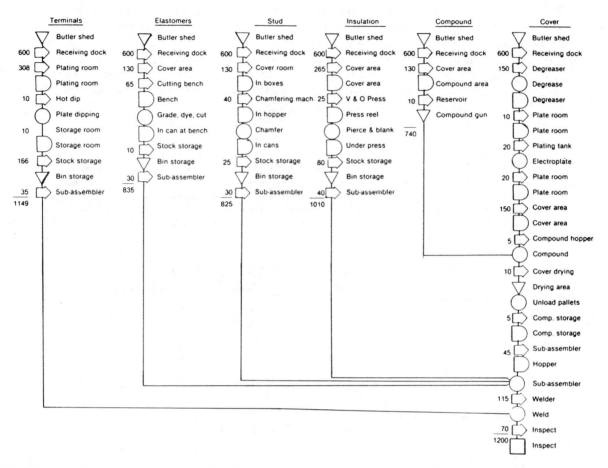

FIGURE 3.16 Assembly flow diagrams are useful for methods analysis; they point out the relationship among components and help emphasize storage problems.

have high turnover. Part-time workers may have lower pension costs, because they may quit before retirement. However, some fringe costs may be allocated on a per hour basis. For example, vacation pay may be earned on the basis of hours worked; a worker may earn 40 h of vacation per 500 hours of work. Some fringe costs may be higher. All employees, full or part time, might be paid for Christmas.

Part-time workers make it easier to schedule for the peaks and valleys of demand—daily, weekly, or monthly. For example, assume that 8 trucks arrive at the shipping dock in the first 4 hours of an 8 h shift and 2 trucks arrive in the last 4 hours. Staffing with 3 full-time employees would give considerable idle time in the second part of the shift. An alternative would be 1 full-time employee, supplemented with 3 employees working 4 h each; this would provide staff for 16 h in the morning and 4 h in the afternoon versus the 12 h and 12 h from 3 full-time employees. The same type of analysis can be done for weekly or monthly peaks. Again, it is less expensive to staff for the peaks with part-time workers rather than have full-time workers idle during the valleys.

An alternative is to have enough full-time workers for the valleys and to schedule overtime for them during peaks. This strategy is viable only if the peaks are not too high and

the duration of the overtime is not too long. What is too high and too long will be interpreted differently by various groups. However, in this society, few manufacturing workers want to work over 50 h/week for very long. (Workers in other fields, such as retailing, construction, and agriculture, often work over 50 h/week, but the work pace is not as high and the effort is often for a short time.) Overtime for 6 months or more also gets "old" for full-time industrial workers.

However, it also is possible to have part-time workers work overtime. If a person works 8 h/day and is put on 2 h overtime, there is a 25% increase in output—assuming output rate/hour does not decline. If a part-time worker works 25% longer, the daily shift goes from 4 h to 5 h. Changing a part-time worker to 6 h/day (50% increase in output) or even 8 h/day (100% increase in output) is realistic, whereas a 50 or 100% increase in the output of an 8 h/day worker is, of course, not realistic. In addition, if a full-time worker works over 8 h/day, there often is a premium on the wage rate, but a premium on the increased h/day of a part-time worker would be very unusual.

Part-time workers should also be able to handle stressful (or boring) jobs better than full-time workers since they have less daily exposure. Stress can be due to cumulative

TABLE 3.6 Demonstration of the effect of learning on performance versus standard. X = amount of experience when the standard time/unit is determined. For example, if time/unit at 125 units is .3 min and the learning rate is 98%, then at 250 units, time/unit = .3(.98) = .294 min and at 500 units, time/unit = .294 (.98) = .288 min.

Learning curve, percent	Time/Unit at					Percent of Standard at				
	.5X	X	2X	4X	32X	.5X	X	2X	4X	32X
98	1.02	1.00	.98	.96	.90	88	100	102	104	111
95	1.05	1.00	.95	.90	.77	95	100	105	111	129
90	1.11	1.00	.90	.81	.59	90	100	111	123	160
85	1.18	1.00	.85	.72	.44	85	100	118	138	225

BOX 3.4 Machining times

Machining (cutting) time is determined as follows:

$$CT = \frac{LENGTH}{FRATE}$$

where

$$
\begin{aligned}
CT &= \text{cutting time, min} \\
LENGTH &= \text{length of cut, inches} \\
FRATE \text{ (lathe, drill, shaper)} &= \text{feedrate, inches/min} \\
&= f(RPM) \\
f &= \text{feed rate, inches/rev} \\
RPM &= \text{revolutions/min of work} \\
&\quad \text{or cutter} \\
&= \frac{12\,V}{\pi D} \cong \frac{4\,V}{D} \\
V &= \text{velocity of surface,} \\
&\quad \text{ft/min} \\
D &= \text{diameter of work or cutter, inches} \\
FRATE\text{(mill)} &= fn(RPM), \text{ inches/min} \\
n &= \text{number of teeth/cutter}
\end{aligned}
$$

For a lathe example, assume a 15-inch long, 3.50-inch dia shaft of B1112 steel was to be turned. Table 3.7 gives recommended cutting speeds and feeds. Tooling companies have detailed recommendations (machining tables, calculators). Carbide is the "workhorse" and has 8 grades (C1–4 for nonferrous; C5–8 for ferrous). Coated carbide inserts run at 50 to 100% higher cutting speeds than uncoated and provide 3 to 5 times as many pieces per cutting edge. Ceramic is more brittle and should be avoided for rough cuts and interrupted cuts. Using a carbide cutter,

$$V = 550 \text{ ft/min and .015 in/rev feed}$$

$$RPM = 4\,(550)/3.50 = 628 \text{ rpm}$$

Assume the closest rpm on the actual lathe was 600 rpm. Then feed rate is $F = .015\,(600) = 9.0$ inches/min. Then cutting time is $CT = 15/9 = 1.67$ min.

For a face milling example, assume a rectangular block of aluminum 8 inches long and 4 inches wide cut by a 6-inch diameter carbide mill with 8 teeth. For a rough cut, distance traveled is the length of the work + the cutter radius (used at start of cut); for a finish cut, distance traveled is the length of the work + cutter radius (at start) + cutter radius (at finish).

Then $V = 80\%$ of 725 = 580 surface ft/min, and assume $f = .010$ in/tooth:

$$RPM = 4(580)/6 = 387 \text{ rev/min}$$

Assume the closest machine rpm is 350. Then table feed is

$$F = .010\,(8)\,(350) = 28 \text{ inches/min}$$

For a single rough cut, cutting time is

$$CT = (8 + 3)/28 = .39 \text{ min}$$

For a single finish cut, cutting time is

$$CT = (8 + 3 + 3)/28 = .500 \text{ min}$$

The recommended cutting speed in the table is based on an optimum tool life of 30 or 60 minutes of cutting time. The Taylor tool life equation describes the relationship between cutting speed and tool life:

$$VT^n = C$$

where

$$
\begin{aligned}
T &= \text{tool life, minutes} \\
n &= .13 \text{ (HSS)} \\
&= .25 \text{ (carbide)} \\
&= .5 \text{ (ceramic)} \\
C &= \text{constant}
\end{aligned}
$$

If you wish to operate at an SFM other than recommended, then you can estimate the new tool life. For example, assume that a HSS tool had $VT^{13} = 47$ (Niebel et al., 1989). When this is plotted on a log–log graph with V on the Y axis and T on the X axis, the plot is a straight line which declines with increasing tool life.

This tool life formula is similar to the standard learning curve formula where $n = .13$ is a 92% curve and $n = .25$ is an 84% curve (Konz, 1990). That is, for HSS you can double tool life from its present value by reducing V to 92% of its present value (a decrease of 8%) or cut tool life in half from its present value by increasing V to 108% of its present value. For carbides, you can double tool life by reducing V to 84% of its present value or cut it in half by increasing V to 117% of its present value. This indicates how sensitive tool life is to cutting speed.

Tools can be considered disposable and thrown away before they are used up. For example, a timer on a CNC machine could record the time the tool is making chips and reject use of the tool at 90% of predicted tool life. (Usually there would be a duplicate tool loaded in the machine's magazine and the alternate tool would then be used.) Another strategy to reduce downtime is to replace all existing tools when the machine is set up.

Total time for a part, in addition to machining time, includes setup time, loading and unloading time, and so forth. Machining time may be less than 20% of total time.

TABLE 3.7 Cutting speeds (ft/min) and feeds for machining. Table speeds are for single-point lathe turning roughing cut with .015 inch/rev feed and .125 inch depth of cut. For finishing cuts use .004–.008 inch/rev and .015 to .040 depth of cut; cutting speeds can be doubled. For milling, use 80% of table values; for drilling use 60%. For milling feeds, use .005 to .010 inch/tooth.

| | | Tool Material | |
| | | Carbide | |
Work Material	HSS	(uncoated)	Ceramic
Brass and aluminum	250	725	Incompatible
B1112 steel (free-cutting, low-carbon steel)	225	550	1,400
4140 steel	130	300	900
Gray cast iron	100	225	800
18–8 stainless	90	275	500

TABLE 3.8 Some of the many utilization alternatives. The utilization percentage assumes zero holidays and vacations.

Utilization (%)	Weekly Hours	Alternative
12	20	5 days of 4 h
12	20	5 days of 8.0 h × 2 weeks/month
20	33.5	5 days of 6.5 h
21	35	5 days of 7.0 h
22	37.5	5 days of 7.5 h
24	40	5 days of 8 h
24	40	4 days of 10 h
24	40	4 days of 9 h; 1 day of 4 h
27	45	5 days of 9 h
29	48	6 days of 8 h
30	50	5 days of 10 h
31	52	6 days of 8 h + 4 on Sunday (retail)
36	60	5 days of 12 h; double shift of 12 h days (alternating 3 and 2 days/week)
43	72	6 days of 12 h; 2 shifts of workers work 3 days of 12 h
48	80	5 days of 16 h; double shift of 8 h
60	100	5 days of 20 h; double shift of 10 h
67	112	7 days of 16 h; double shift of 8 h
71	120	5 days of 24 h; triple shift of 8 h
100	168	7 days of 24 h; triple shift of 8 h or double shift of 12 h days (alternating 3 and 4 days/week)

trauma, chemical exposure, noise, contact with the public, or other factors. For example, if data entry office workers work 8 h/day, they have 16 h/day to recover and have a recovery/work ratio of 16/8 = 2. If they work 4 h/day, they have 20 h/day to recover: 5 h to recover for every hour of work. If the job is boring, a short exposure is better than a long exposure.

Part-time workers may be of higher quality than full-time workers, especially for low-paying jobs. For example, many college students will work part time while they are going to school, accepting rates of pay they would reject if the job were full time and permanent. However, quality produced is a function of experience as well as potential. The greater experience of full-time workers may compensate for their lower potential.

Using part-time workers also has disadvantages—learning, moonlighting, and fixed cost/employee.

Perhaps the largest potential cost of part-time workers is that they have less experience on the job. A full-time worker may get 1,800 h of experience within 1 year, but a person working 4 h/day takes 2 years. The amount of learning needed for a job needs to be considered. The more

---

BOX 3.5   Number of machines/operator

Often it is technically possible to assign more than one machine to an operator because the machines are (1) semi-automatic (require an operator to load/unload them but do the processing automatically) or (2) automatic. (Operators are needed for automatic machines to replenish loading/ unloading mechanisms and to repair breakdowns.) This problem of the proper number of operators/machine is known technically as the *machine interference* problem. Actually it is a broader problem; it need not be just operator versus machine. It's a problem of utilization among multiple activities; these could be left hand and right hand, store counter and customers, waitress and number of tables, technicians and engineers, secretaries and executives, nurses and doctors.

Stecke (1992) has an excellent discussion of the interference problem and solution techniques.

There are a number of service rules:

Serve potential customers in a fixed, predetermined sequence (whether they need service or not).

Serve customers (needing service) randomly.

Serve customers (needing service) with some priority rule (such as those closest, those requiring least time to serve,

those with greatest importance).

Both the time between services (reliability) and the time to service (maintainability) can be assumed as constant or variable. Constant is more appropriate for loading/unloading while variable is better for repairs. Actually assuming times as constant is using a mean time with an accompanying standard deviation of zero, and assuming variable times is using a mean with a nonzero standard deviation. The distribution of times can be assumed as normal, exponential, log-normal, etc. Some design guidelines to follow:

With most problems, some activities are more expensive than others. In the United States machines tend to be cheap/hour and labor expensive. Thus, keep operators busy even if machines are idle.

Pooling is good. That is, assign 20 machines to 4 operators rather than 5 machines per operator. See Principle 1.4 in Chapter 15.

Simulation tends to give better answers than probability and queuing techniques since the assumptions can be more realistic.

complex the job, the more there is to learn and the more benefit to gain from a full-time person. Note that many part-time workers in the fast-food industry learn the job in less than 50 h of work—perhaps in as little as 20 h of work. The benefits of experience result not only in lower time/unit but, even more, in fewer errors; therefore, full-time workers are advantageous when quality is critical. Note, however, that for low-paying jobs, part-time workers may be of higher quality than full-time workers, so quality is not automatically better or worse for part-time workers.

There are certain costs of employment that are constant despite hours of employment. Examples include hiring and training costs. Some fringe benefits (e.g., medical benefits, life insurance) might depend on whether the person was on the payroll for the month, not on how many hours were worked. If hiring a person costs $500 and that person works 1,000 h in a year, then cost is $.50/h; if the person works 2,000 h, then the cost is $.25/h. Part-time workers may have greater turnover rates, so the hiring cost/hour may be magnified for them. If medical benefits cost the firm $400/month and the person works 80 h/month, then cost is $5/h, whereas it would be $2.50/h for a person working 160 h/month.

Telecommuting workers. In **telecommuting**, work may be done by your firm's employees but not at your local facility, through work at a remote company facility or work at home. The worker communicates with the facility through telephone and fax lines, electronic mail, and express mail services.

The facility can be remote. For example, Cigna employs 120 people in Ireland to process medical claims flown in daily from the United States. McGraw-Hill employs 50 people in Ireland to update worldwide circulation files of 16 magazines; they are connected by a direct computer link to the mainframes in New Jersey. In addition to lower labor costs, different time zones can be an advantage. Quarterdeck Office Systems, a California software company, has technical support staff in California and 20 specialists in Ireland. After business hours in California, they throw a switch and all inquiries are routed automatically to Ireland (Wysocki, 1991).

The U.S. Post Office began Remote Video Entry (RVE) in 1991. Formerly, a letter-sorting machine had an operator present. In the new system, the operator sits at a terminal remote from the machine and the information concerning the address of a letter and where it should go are communicated by cable. The length of the cable can vary from 100 feet to 3,000 miles. The primary advantage to the Post Office is staffing flexibility (especially the ability to shift a worker instantaneously from one city to another).

In 1991, an estimated 5,500,000 Americans worked at home but "commuted" to work via telephones connected to computers, fax machines, and electronic mail (Segal, 1991). Typically, these people are in information-intensive jobs such as word processing, data entry, bookkeeping, sales, or marketing. Most (85%) work part time; most (80%) work for firms with fewer than 100 employees.

Telecommuting at home is usually voluntary and requires the agreement of worker and supervisor. Typical rules for workers include the following:

- maintain regularly scheduled work hours
- be as accessible as are on-site counterparts during agreed-upon hours
- maintain designated work spaces
- manage dependent care and personal obligations to meet job responsibilities successfully (e.g., make child-care arrangements as if working in the office).
- have performance measured by objectives and results
- go to the office periodically (e.g., for staff meetings)

Temporary workers. Farmers have hired daily and seasonal labor for thousands of years. Construction firms traditionally have core staff and hire temporary people for specific projects. Starting in about 1940, special firms began furnishing temporary secretaries and office help. Gradually, this service spread to other jobs, such as warehouse labor, then to technical (engineers, accountants) and executive areas.

Temporary staff tend to cost more than core staff on an hourly basis, but the big advantage is that a firm can staff for the bottom of the business cycle and still have a no-layoff policy for core staff. (The famous Japanese lifetime employment policy applies only to a core staff of males, excluding temporary workers and females.) Temporary workers can be removed from the payroll quickly and with no trauma to the organization. Temporary workers (free-lancers) are like entrepreneurs: They have freedom, and possibly high income, but little security.

Subcontracted workers. Certain jobs may be permanently subcontracted to other firms. Examples include food service workers, guards and security workers, custodial workers, yard and building maintenance workers, office machine and computer repair workers, and warehouse workers. Public warehousing rose from 8% of total warehousing to 16% in 1991 (*Management Newsletter*, 1991). The subcontractor uses workers with specialized technical and management skills and may pay lower wages. In addition, a firm may not have enough work to keep a worker with specialized skills occupied (e.g., computer repair, roof repair), but the subcontractor can spread specialized work among a number of customers.

2 RELATIONSHIPS

In the previous steps, you established the types of equipment needed and decided how much of each type of equipment was necessary. The next step is to group equipment into departments. Departments can be combined from similar products (line production), similar processes (job shop), or, most likely, some line departments and some process departments. A good clue for how to organize activities into departments is to look at the **organization chart**, which shows reporting relationships and responsibility levels; see Figure 3.17. Note that some people may be located off-site.

The goal is to group the departments rationally. Grouping is based on relationships. For product relationships among departments, use a from-to chart; for service relationships use a relationship diagram.

Table 3.9, a **from-to chart,** summarizes the flow from one department (or machine) to another; the information comes from the various routing sheets and the production schedule (see also Table 6.9). Note the two sides of the diagonal are not identical; that is, the loads from A to B are not the same as the loads from B to A. The from-to chart usually combines the movements from several products. Thus, the number in the chart should be a common unit such as pallets/week or tote boxes/day rather than mixed units such as pieces of product A, pieces of product B, etc. Aneke and Carrie (1983) have developed a flow complexity index which can be calculated from the from-to chart. When flow complexity = 0, flow sequence is the same for every product; consider line layout. When flow complexity = 1, the flow sequence is completely random; consider job shop layout.

Muther (1973) has developed a "magnitude counting" procedure for converting different size, bulk, and shape loads to equal material handling difficulty. The basic units are the *mag* (a "handful" or 10 in.³) and the *macromag* (a "pallet load" or 1 m³). There are conversion factors for size, bulk, shape, risk, and condition.

Table 3.10, a **relationship chart**, was developed by Muther to handle the relationships between service areas. (It has been found very useful for product flows also and now is often used in place of the from-to chart for products.) That is, instead of making a from-to chart for product flow and a relationship chart for service areas and then

combining them into one overall relationship chart, people just begin by using one relationship chart for products and services.

The relationship chart gives letter grades for "closeness desired between areas." Use A = absolutely necessary, B = important, C = average, D = unimportant, and E = not desirable to be close. (Muther uses six levels instead of five and uses A = absolutely necessary to be close, E = especially necessary to be close, I = important to be close, O = ordinary closeness, U = unimportant closeness, and X = not desirable to be close; I think five levels is easier to use than six). Letters are used rather than numbers, because numbers imply more accuracy than is available. Avoid too many A relations, since if everything is important, nothing is important. A good goal is 10% As, 15% Bs, 25% Cs, and 50% Ds. (For six levels Muther recommends 2–5% As, 3–10% Es, 5–15% Is, and 10–25% Os.) However, if several departments or machines have As, this implies a flow line arrangement instead of a job shop arrangement. For the critical A, B, and E relationships, give a reason for closeness, indicating the reasons with a number. Reasons will vary with the problem but common reasons are 1 = product movement, 2 = supervisory closeness, 3 = personnel movement, 4 = tool or equipment movement, 5 = noise and vibration.

TABLE 3.9 From-to tables are developed from a product-by-product flow analysis. The table summarizes flow in standard units such as pallets/week. If flow occurs "below the diagonal," it is "backtracking."

Product	Dept. Flow				Pallets/Week
1	ABC E				10
2	AB E				5
3	BCDE				8
4	B DE				3

			To		
From	A	B	C	D	E
A	–	15			
B		–	18	3	5
C			–	8	10
D				–	11
E					–

TABLE 3.10 Relationship charts identify the degree of desired relationship between machines or departments with a letter. The reasons for A, B, and E relationships are indicated by a number. Thus, $E/5$ means an E relationship for reason 5. (Departments are identified by letters in most of this example table; however, for reports and presentations to management the entire word on the row and understandable abbreviations for the column headings are used for all departments. Thus, use *shipping* or *SH*, not E.)

	Department				
Dept.	A	B	C	D	SH
A	–	B/2	D	E/5	D
B		–	B/1	D	D
C			–	D	C
D				–	D
Shipping					–

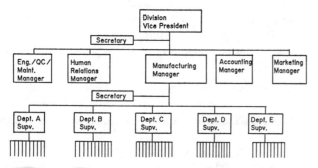

FIGURE 3.17 Organization charts show similar responsibilities at the same level. Positions sometimes are considered line (production) and staff (service to the line). A supervisor may supervise as few as 5 complex jobs (engineering) or as many as 25 simple jobs (routine production work). Typical staff functions are engineering, accounting, human relations (personnel), quality assurance, marketing, and maintenance.

It is essential that, after the relationships are put down, they be reviewed and signed off by the various department heads affected. After all, they are going to have to live with your design. The planner is the clients' servant, not master. The planner's job is to get a layout implemented, not to make drawings that are not used.

3 SPACE REQUIREMENTS

The next step allocates space to departments. Three methods are calculation, conversion, and space standards.

3.1 Calculation

There are three substeps: (1) number of workstations, (2) space/workstation, and (3) space/department.

3.11 Number of Workstations.

Calculation begins with a routing sheet (see Table 3.5) for each item made in the facility. (If a part presently is purchased but may be made in the future, you may wish to analyze it now but keep the data separate.) For space purposes, the key information includes the machines to be used for each operation, the setup time on that operation, and the run time/unit on that operation. If a routing sheet is not available, routing of a similar part can be used, with some loss of accuracy.

Second, for each machine, compile all the items to be made on that machine.

Third, for each part number, determine the lot size in each planning period. This number is especially difficult to quantify accurately. The other numbers in space calculations can be estimated within 10%, or even 1%, but production quantity may have an error of 100%.

Fourth, for each part number, multiply its production quantity by its production time. Include both setup and run times. This time estimate should be actual time rather than standard time. See equation 7.

Fifth, calculate machine capacity for the planning period. See Sections 1.51 and 1.52.

Sixth, calculate the number of machines required. Divide required hours by available hours. Allow a safety factor for service parts orders, future changes in demand (down as well as up), and so forth. (Of course, if your calculations show you need 2.4 machines, you will need 3 machines.) However, capacity can be adjusted by methods other than buying additional machines, for example:

- increase working hours (more shifts, overtime; see Table 3.8)
- improve output/h (better methods, process, equipment)
- revise production scheduling
- buy instead of make

3.12 Space/Workstation.

Each workstation is a miniature factory with its own receiving and shipping, production, and storage areas.

$$WSPACE = MSPACE + OMSPACE + STSPACE \quad (8)$$

where

$WSPACE$ = workstation space, ft²

$MSPACE$ = machine space (length × width), ft² (include travel) (This is the "**footprint.**")

$OMSPACE$ = operator and maintenance space, ft²

Space needs to be allocated on one or more sides of the machine for a seated or standing operator. The operator needs to obtain and dispose of items as well as process items. Allow space for the operator to reach the machine from an aisle. The operator symbol on a layout should be a 24" × 36" rectangle. Allow sufficient operator space rather than the minimum of 24–36".

Maintenance and service space allows access to machines that usually are not movable. Simulate lubrication, changing tools, and repair of broken components to determine space required on each side of the machine.

Subdivisions of storage space are (1) input buffer storage of product; (2) supplies and maintenance materials; (3) tools, dies, and fixtures; (4) rework, scrap, and waste; and (5) output buffer storage of product. Input and output buffers decouple the workstation from shocks and disturbances; see Chapter 5. Supplies include empty containers, packaging materials, tool bits, and so forth. Cleaners and lubricants may be flammable, requiring special storage. Space for maintenance materials depends on the centralization/decentralization decision. The Just-In-Time concept of minimization of setup time emphasizes local (rather than central) staging of tools, dies, and fixtures. Rework and scrap (see Figure 3.10) tend to be greatly underestimated, because people don't want publicity about problems. Problems are handled quietly and without focusing attention on them. Reworked items typically need to be inspected (to see that they are defective) and then stored and reworked in a batch.

Note that storage can be multiple level (i.e., use the cube), so square feet of storage does not equal square feet of floor space. See Chapter 11 and equation 9 in Chapter 2.

3.13 Space/Department.

For each department

$$DSPACE = NWORKS\,(WSPACE) + TSPACE \quad (9)$$

where

$NWORKS$ = number of workstations

$DSPACE$ = department space, ft²

$TSPACE$ = transportation space, ft²

Transportation space primarily is for aisles but could include floor-level conveyors. In addition to within-department space, add an allowance (e.g., 10%) for between-department aisles. See Chapter 10 and equation 10 in Chapter 2.

3.2 Conversion   Conversion, the second method of allocating space, estimates the new space from the present space. Conversion should be done on a department-by-department basis rather than for the facility as a whole. First, decide for each department what the space should be to do the current job (not what the space presently is). Second, by what percentage will the department be expanded or contracted? This will give you desired area for each department. Then compare the desired area with the area available and make adjustments.

3.3 Space Standards   The third method of allocating space is to use space standards. These probably are most appropriate for service and storage areas. For example, if you knew the number of employees, from Chapter 14 you could estimate the amount of toilet, shower, and locker room space. Office space might be allocated as 50 ft² for a secretary, 70 ft² for a professional, and 100 ft² for a manager. Medical facilities might be 1 ft² per person on the largest shift with a minimum of 100 ft².

Finally, after using the calculation, conversion, or space standards method, put the space required for each department in a table indicating any special restrictions, such as outside wall needed, crane, special foundation, waste disposal. For best planning, list all the utilities for each department (110, 220, and 440 power, gas and compressed air lines, sewers, water supply, waste disposal, telephone and computer lines) as well as special problems, such as noise, heat, and special foundations needed.

DESIGN CHECKLIST: BASIC PLANNING INFORMATION

Product information
    Drawings of components
    Assembly drawings
    Bill of material
    Make versus buy
    Product life
Order quantity
    Scrap rate
    Rework rate
    PQ diagram
Routing for each unit
    Operations in sequence

Time/operation
    Run
    Setup
Supporting services
Relationships
    From-to chart
    Relationship chart
Space requirements
    Number of each type of machine
    Space/machine
    Space restrictions

REVIEW QUESTIONS

1. What do PQRST stand for in the systematic layout procedure?
2. Give the five steps when releasing a part from design to manufacturing.
3. Ignoring inventory costs, discuss the effect of setup costs on manufacturing costs.
4. Briefly discuss the three keys to setup time reduction.
5. Overage, quality, learning, and inflation are not considered in the standard economic lot size formula. How does a change in each of the four affect economic lot size?
6. Give at least three ways setup time could be reduced for a punch press.
7. Briefly discuss the yield of a process with multiple stations.
8. Briefly discuss the problem of getting good estimates of scrap and rework rates.
9. Sketch a PQ diagram (a) which indicates a job shop layout is desirable and (b) which indicates a flow line layout is desirable.

10. What should be on a routing sheet?
11. Sketch a process chart showing (a) operations, (b) storage, (c) inspection, (d) delays, (e) transportation. Indicate scrap at one operation, rework at another.
12. Assume three products are made in four departments. Product 1 has 20 pallets/week in sequence ABCD. Product 2 has 10 pallets/week in sequence ACD. Product 3 has 4 pallets/week in sequence BD. Construct a from-to chart.
13. Make a relationship chart for the four departments of question 12. Assume a shipping dock has a B relationship with Department A, and a C relationship with Departments B, C, and D.
14. Briefly describe a job with which you are familiar. How many hours/week are scheduled? Discuss why that schedule is appropriate.
15. The labor force and the production rate should not both remain constant. Discuss.

16. If you know the length × width of each machine, what additional space is needed to determine total space in the department?
17. Give the formula relating standard time and observed time. Give the formula relating standard time and actual time.
18. Discuss the exponent n in the tool life equation.
19. Briefly discuss the advantages and disadvantages of using part-time workers.
20. Briefly discuss telecommuting.

PROBLEMS AND PROJECTS

3.1 Make a bill of material of a sandwich from a fast-food restaurant.
3.2 For the university department in which you are enrolled, calculate the economic order quantity for computer paper. How does actual purchasing policy agree with your calculations? If it differs, see if you can get the supervisor to change. If the supervisor won't change, what are the reasons?
3.3 Using the example of Table 3.5, make a table of the flow of eight products among six departments. Then make a from-to chart.
3.4 Make a PQ diagram for a restaurant or cafeteria.
3.5 Assume a product requires 4 operations. Scrap rate is 2, 2, 6, and 4% respectively. Rework rate is 1, 2, 5, and 4% respectively. If you need 1,000 good items, what lot size will you release? If you need 50, what lot size will you release?

REFERENCES

Aneke, N., and Carrie, A. Flow complexity: A data suitability index for flowline systems. *Int. J. Production Research*, Vol. 21, 3, 425–436, 1983.

Cochran, D., and Swinehart, K. The total source error of adjustment model: A methodology for the elimination of setup and process adjustment. *Int. J. Production Research*, Vol. 29, 7, 1423–1435, 1991.

Esrock, Y. The impact of reduced setup time. *Production and Inventory Management*, 94–101, 4th Quarter 1985.

Konz, S. *Work Design: Industrial Ergonomics*, 3rd ed. Scottsdale, Ariz.: Publishing Horizons, 1990.

Lesnet, D. Facility found that means of implementing quick die changes were readily at hand. *Industrial Engineering*. Vol. 15, 50–54, Nov. 1983.

"Management Newsletter," *Modern Material Handling*, p. 7, 4 July 1991.

Muther, R. *Systematic Layout Planning*, Kansas City, MO: Management and Industrial Research Publications, 1973.

Niebel, B., Draper, A., and Wysk, R. *Modern Manufacturing Process Engineering*. New York: McGraw-Hill, 1989.

Rago, L. *Production Analysis and Control*. Scranton, Penn.: Int. Textbook Co., 1963.

Segal, E. How to commute in your bathrobe. *Wall Street Journal*, 1 July 1991.

Stecke, K. Machine interference. Chapter 57 in *Handbook of Industrial Engineering*, 2nd ed. G. Salvendy, Ed. New York: Wiley, 1992.

Wysocki, B. Overseas calling. *Wall Street Journal*, 14 August 1991.

Young, R., and Bell, R. Fixturing strategies and geometric queries in set-up planning. *Int. J. Production Research*, Vol. 29, 3, 537–550, 1991.

ALTERNATIVE MACHINE ARRANGEMENTS

CHAPTER PREVIEW

Work and workstations can be arranged many ways. Assembly arrangements and component manufacturing arrangements are discussed. Then two popular versions of component manufacturing (group technology and flexible manufacturing systems) are discussed in detail.

CHAPTER CONTENTS

1 Overview of Challenge
2 Assembly
3 Components
4 Group Technology
5 Flexible Manufacturing Systems

KEY CONCEPTS

assembly cells
balance delay
coding
families
flexible manufacturing system
   (FMS)
flow lines
group technology

hierarchical system
job shop
kitted
maintainability
nonprogressive assembly
progressive assembly
reliability

1 OVERVIEW OF CHALLENGE

In *construction*, the materials, machines, and people are brought to the work. In *manufacturing*, the machines and people are generally in a fixed location and only materials are moved. Construction is most appropriate for large items (ships, houses, refineries) made to special order at field locations. It tends to be considerably more expensive than manufacturing. In the last 50 years some manufacturing concepts have been brought to construction—especially the fabrication of components at central sites and only assembly (installation) at the field site. For houses, examples are central fabrication of doors, windows, cupboards, walls, and entire kitchens and bathrooms with field installation. The Japanese have built entire factories in Japan and shipped them overseas in special ships. The remainder of this chapter will deal with manufacturing.

For manufacturing, the first decision is the amount of specialization of labor and specialization of machine. For example, for assembly an entire unit could be assembled at one workstation or each step could be done at a separate workstation. Figure 4.1 illuminates the spectrum of choices for components, assembly, and offices. The proper level of mechanization depends primarily upon the availability and the relative costs of capital and labor. The ratio differs widely by country, year, and industry. Engineers, with an innate love of machines, often tend to be biased toward using machines.

For assembly, we will discuss one specialization extreme (the nonprogressive workstation) versus the other

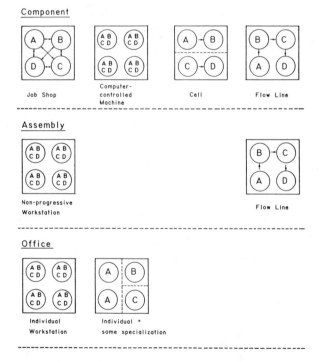

Component

Assembly

Office

FIGURE 4.1 Specialization and mechanization differ among components, assembly, and office work. The letters (*ABCD*) indicate different processes.

approach—the flow line, the 3S (standardization, specialization, simplification) combination. For more on flow lines, see Chapter 5. For components, we will discuss the computer-controlled machine, the job shop, the cell, and the flow line. Within each approach, there are certain advantages and disadvantages. As was pointed out in Chapter 2, there are many possible design criteria.

2 ASSEMBLY

2.1 Problem   Consider an assembly of N elements with sufficient product demand to require m people to work. Should each worker do all N elements (**nonprogressive assembly**, a complete workstation) or should the job be split so each person does m/N elements (**progressive assembly**, an assembly line)? Workstations need not be entirely progressive or nonprogressive; in addition to 15 stations doing all N elements or 15 stations doing N/m elements, solutions can include possibilities such as 3 people at each of 5 stations.

2.2 Advantages and Disadvantages   The discussion will be grouped into advantages, neutral characteristics, and disadvantages.

2.21 Advantages of Nonprogressive.   Five advantages are balance delay, scheduling flexibility, shocks, cumulative trauma, and satisfaction.

Balance delay time is eliminated. Since each worker does a complete job, the job does not need to be divided into equal time segments. On progressive lines, if these segments are not equal (balanced), there is idle time—**balance delay** time—which tends to be 5 to 15% of the work times. Balance delay is increased by more stations, smaller buffers, shorter cycle times, longer element times, and increasing precedence requirements.

Scheduling flexibility is increased. Nonprogressive workstations can make multiple products simultaneously (different stations making different products). Flow lines work best with steady, uninterrupted production of a single product. However, it is possible, with some penalties, to design flow lines to make multiple projects simultaneously; see Chapter 5. Making multiple products at the same time results in numerous scheduling, inventory, and marketing advantages. Flow lines may require more setups to produce the desired flexibility.

Shocks do not have multiple effects. When using nonprogressive stations, shocks (absenteeism, machine breakdowns, interruptions) affect only a single operator or station instead of the entire line. See Chapter 5 for some ways of attenuating shocks on lines.

Cumulative trauma is less of a problem because individuals do fewer repetitions of any specific motion. The reduced repetition of motion patterns allows individual muscles to rest (active rest). Although the person is still working, a specific muscle is not constantly active. However, the greater complexity of the activity makes limiting the use of one muscle, through mechanization or automation, more difficult.

Satisfaction may increase. When jobs are divided on a line, some groupings are more desirable than others; nonprogressive assembly reduces disputes over all workers doing their fair share. Each worker has the satisfaction of completing a unit.

Job enlargement may not make much difference to real assembly workers, due to population selection of factory work. However, first-world societies spend considerable resources educating children to achieve their maximum potential, and many people feel that a 30-second-cycle job does not bring out maximum potential. For the organization, the importance of satisfied workers depends on what alternatives people have to short-cycle work. Dissatisfaction will be reflected in difficulty in hiring and increased turnover and absenteeism.

2.22 Neutral Characteristics.
Quality, training, material handling, and walking can favor nonprogressive or progressive designs.

Quality may be higher or lower. Doing all steps at a single station allows an operator to adjust later stages to the results of previous stages; feedback of information within a single person is simple and direct. In a line, adjustment capability is greatly restricted, because feedback is given to other people and with a time delay. A line tends to be machine paced, which leads to incomplete work at some stations. A line station is often simple and automated, so quality resides in the equipment; when it is good, it is very, very good and when it is bad, it is horrid. Lines also tend to conceal responsibility, because many people handle each item, making it difficult to identify exactly which person or station is responsible for poor quality.

Training is more complex for nonprogressive than progressive because more needs to be learned. However, it is difficult (but not impossible) to train line workers without interfering with the remaining line workers' output. As a compromise, many line workers cross-train to handle multiple jobs on the line.

Material handling of partly processed parts is minimal since all the work is done at one location. Space for bulky components and assemblies may be a problem, however. Components need to be located at many stations instead of one; finished assemblies are taken away from many stations instead of one. Thus, specialization and mechanization of material handling is more difficult.

Walking may increase if the nonprogressive station is larger than can be used conveniently by a single sitting or standing operator. For large units, several stations may be staffed by a single person. The MTM (Methods Time Measurement) predetermined time system gives .2 s/ft for walking. If the hands work while the walking occurs,

walking time does not add to cost. From a fatigue viewpoint, walking occasionally is superior to constant standing.

2.23 Disadvantages of Nonprogressive.
Disadvantages are standard direct labor cost/unit, skill requirements, equipment capital cost, in-process inventory, and supervision.

Standard direct labor cost/unit will be higher for nonprogressive than progressive since the operator will not be as specialized and will not have as much practice (experience). If cycle time is 10 minutes, then an operator will experience about 45 units/day. If the cycle time is 1 minute on a progressive line, then an operator will experience 450 units/day. Actual direct labor time (i.e., including balance delay time, downtime, etc.) may not be as favorable to the line as standard direct labor time.

Skill requirements are higher. Each person needs to know the entire job, not just a small part. With lines, unskilled workers with brief training (and specialized tools and procedures) are able to produce excellent quality; with nonprogressive, the same quality requires skilled workers with long-term experience. However, unskilled workers may have difficulty adjusting to changing products. Wages may be higher for skilled workers, depending on labor market conditions. Skilled workers may not be available. If they are available, they may have high turnover.

Equipment capital costs will be higher since equipment will be duplicated at each station. (An exception is if the progressive line has fixtures mounted on a conveyor instead of at the station; in that case, nonprogressive may have lower fixture costs.) The smaller volume at each station also does not encourage special-purpose equipment.

In-process inventory will be higher, because component stocks will be at many stations instead of a few. Buffer stocks tend to be smaller for the line. (A typical annual inventory cost is 25 to 30% of inventory value; $1,000 of inventory costs .25(1,000) = $250/year or 250/365 = $.68/day.) Buffers between line stations often hold only a few minutes of output and only rarely hold as much as an hour's output. Buffers at nonprogressive stations often hold output of hours to days. Finished goods inventory is usually higher for the line, due to lack of product variety.

Supervision is more difficult for nonprogressive, because individual motivation, training, and performance problems are more obvious. There will be more paperwork for nonprogressive. Lines have line pressure, instead of supervisory pressure, on output quantity; there are lower training requirements; and quality tends to be built into the equipment. However, the effect of shocks (absent employees, machine breakdown, schedule change) is more severe in a line.

2.3 Summary
Individual assembly may seem impractical when component parts or tools are (1) too large for one workplace, (2) too many for one workplace, or

(3) too complicated for one worker to learn. However, these problems can be overcome. Problems 1 and 2 can be overcome by having the worker walk between stations. For example, at a Swedish ABB plant making motors, the worker completes motors but does the work at three workstations. He walks 20 ft between the stations once every 15 min. Since the MTM predetermined time system allocates .2 s/ft, the walking takes 4 s out of every 900 s for a penalty of 4/900 = .4%. If the job is very complex, it can be done by several operators at one station. See Box 4.1.

Even though flow lines have many disadvantages, they still are common for three reasons:

1. Some advantages of flow lines, such as lower standard direct labor cost and lower capital cost, are emphasized by decision makers.

2. Standard labor cost and capital cost are quite visible, whereas scheduling flexibility, cumulative trauma, and material handling cost often are lost in the overhead figures.

3. Many flow line disadvantages can be overcome by proper system design (primarily by design of buffers and material handling). See Chapter 5. With proper design, even batch production can be done in a flow line arrangement.

3 COMPONENTS

The **flow lines** for assembly are mostly manual and people oriented, whereas component flow lines are mostly mechanized and machine oriented. Also, in assembly lines new components are added at various stations whereas component lines generally do just additional processing to a single component. When movement and control between stations are automated, the line may be called a *transfer line* or *packaging line*. Thus, while for assembly the opposite of a flow line is the nonprogressive workstation, for components the opposite of a flow line is the job shop. A **job shop** is a group of similar stand-alone machines, each doing only a few specialized functions—for example, a group of lathes, a group of mills, a group of drills. The machines are grouped to facilitate operator movement and technical supervision since product movement between machines is not in a standard sequence. In general each machine has one worker who works only on that machine.

The key challenge of the job shop is scheduling and sequencing; for the flow line it is balance delay, buffers, and flexibility.

Between the job shop and the flow line is the *group technology cell*, which is closer to the flow line than to the job shop. Dissimilar machines are grouped to facilitate production control and setup. When the cell is highly computerized, it is sometimes called a **flexible manufacturing system (FMS)**. There is a trend to extensive use of computers in manufacturing; the systems are known by a variety of names, such as CAD/CAM (computer aided design/computer aided manufacturing) and CIM (computer integrated manufacturing).

A special case is the computer-controlled machine—the "intelligent" workstation. It may be considered a "cell of size 1." It has one or more spindles and may have computerized tool-changing capability. These are good for small to medium quantities of complex parts. They require skilled operators rather than semiskilled.

A new layout concept, the automatic storage/retrieval job shop, is now beginning to be implemented. See flexible manufacturing systems later in this chapter. The key difference is that the machines are connected by a computerized random-access material handling system; see Figure 4.2. Now there no longer is a material handling problem or a work-in-process storage problem. However, there are very big planning, control, and timing problems, as well as expensive inventories.

Machines' characteristics are different from those of people. Machines tend to be less flexible than people. (The definition of single-purpose can be given as "machine-like, robotic.") Precedence problems among work elements are more severe with machines than with people. Machines tend to be more capital-intensive; that is, it costs more to buy a machine than to train a worker. Operating costs (excluding capital costs) tend to be lower for machines than labor costs for people. In many cultures the penalty of idle equipment is less than for discharging or transferring workers. For all these reasons, in developed countries it tends to be more important to keep workers busy than to

---

BOX 4.1 *Case study: Nonprogressive assembly of cars*

In the late 1980s, Volvo built an assembly plant at Uddevalla, Sweden. It has an annual capacity of 40,000 cars per shift. Progressive assembly lines are not used; instead, the cars are assembled at six **assembly cells**; each assembly cell has 8 work areas where the 48 teams assemble complete automobiles (Krepchin, 1990).

All the parts needed (1,700 parts and subassemblies) are **kitted** and arrive on automatic guided vehicles. (There are 200 AGVs in the facility.) The car is assembled on an assembly stand that can tilt and move up and down.

Each assembly team consists of 8–10 people. Each person is responsible for thorough knowledge of 1/7 of the process and a working knowledge of 3/7. A cell has 4 cars in it at a given time, with 3 people working on each of 3 cars. There are about 30 labor hours/car.

Since the 48 teams work on 3 cars at a time, many product varieties can be assembled in a shift. In addition, if a car must be retained in the cell, it can be kept without disturbing the flow.

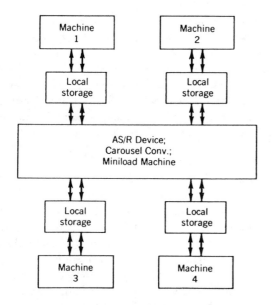

FIGURE 4.2 Automated storage/retrieval job shops use a computer-controlled transfer machine to move items from general storage to the local storage at the machine. The local storage can be quite small (e.g., 5–10 minutes production) since the delivery from general storage is so quick. The transfer machine can store not only partly processed components but also jigs and fixtures. The transfer machine might be an automated storage/retrieval system, a carousel, or a miniload machine; see Chapter 11.

keep machines busy; thus we often have more machines than workers; more places than people.

The least mechanized and specialized area is the office. Most office work is done completely at one workstation. Some degree of specialization and mechanization (word processing and duplicating groups) has begun so some offices are becoming more like a job shop. However, the production scheduling is much less complex than for a typical factory job shop and the material handling problems are much easier. See Chapter 13 on offices.

4 GROUP TECHNOLOGY

4.1 Concept
Flow line advantages are well recognized. However, many items do not have sufficient production volume to justify a flow line. The problem is how to achieve the benefits of flow lines when there is insufficient volume.

In a flow line, each part is identical. The concept of **group technology** is to form **families** of items that are almost identical. The closer the items are to identical, the more closely we can emulate a flow line and shift the fixed costs of the flow line (tooling, machines, conveyors, procedures) to the *process* rather than the part. The idea is to find common solutions to common problems. In practice, group technology is not applied to everything, because some items are simply too diverse to group.

4.2 Family Benefits
Benefits of group technology fall into two main categories: (1) product design and (2) common manufacturing.

4.21 Product Design.
Reducing the number of items made (standardization) has many benefits. Gallagher and Knight (1973) estimated that 5 to 10% of all components are duplicates or unnecessary and are used because engineers do not know all the parts ever designed by a firm over its history. Consider a simple washer. In the past, a firm's engineers may have designed hundreds of washers for many models of many products. When an engineer needs a washer for a new item, the easiest solution may be to draw another washer and put a new part number on it. However, if an existing part could have been used, this not only would save drafting time, it would eliminate all the cost of adding a new part number to the system.

One approach to variety reduction is the use of the tabulated drawing; see Figure 4.3. Note that an engineer

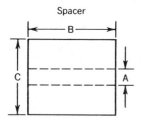

Spacer

PART NUMBER	MATERIAL	TYPE	A	B	C	USED ON
399-28	Steel	AISI 1215	.250	1.00	.500	399-77 406-84
399-29	Steel	AISI 1215	.375	1.00	.500	399-78
400-07	Alum.	2011	.375	1.00	.500	400-91
401-01	Alum.	2011	.250	1.25	.500	401-89 471-94
411-09	Alum.	2024	.375	1.25	.750	411-88 478-91

FIGURE 4.3 Tabulated drawings (generic drawings) help people keep track of similar parts.

can modify an existing design to use common parts—once there is knowledge of the common parts. For example, the newly designed washer may have a 1.25-inch outer diameter (OD). If existing parts are identical in all other respects except that they have 1.0- or 1.5-inch OD, then one of them possibly could be used.

One firm (Desai, 1981) used three categories in the justification of variety reduction:

1. *HIT.* An existing part can be used as is in place of a proposed new part.
2. *MODIFY.* An existing part can be modified slightly in place of a new part.
3. *DESIGN INFO.* The two possibilities were that the existing part could be traced to reduce drafting or that the specifications, purchase information, tolerances, etc., could aid in the design of the new part. How did previous designers treat tolerances, radii, chamfers, and undercuts? What grades and size of material were specified?

They valued *HIT* at $1,500 each, *MODIFY* at $1,400 each, and *DESIGN INFO* at $50 each. For one year, *HIT* resulted in 31% savings, *MODIFY* 57%, and *DESIGN INFO* 14%.

4.22 Manufacturing.

Manufacturing advantages fall into office and shop-floor advantages.

Comparing the similar parts—designed, planned, and costed over a period of years by a variety of people—brings out many anomalies in the processing, time standards, tooling, and cost estimates. Studying these anomalies can be beneficial to reducing cost of previous approaches as well as new approaches. What process was used for manufacture? What speeds and feeds were used on what machines? Detailed production planning can be done on a class of items when an individual item's production volume does not justify that level of analysis. Planning for a group provides a skeleton upon which the manufacturing engineer can flesh out various details and options.

Another common savings from grouping is the use of common tooling. That is, a new die or fixture need not be built, because an existing one can be used as is or can be modified (e.g., with an insert) to handle the new item also. In addition tooling can possibly be justified for a group when it cannot be justified for an individual item. For example, a die might not be justified for a production volume of 100 parts on a single item, but when the production of the total group is considered, then a die with inserts for the various parts may be justified. Each tool should have a number so that it can be entered into a tooling data base.

Group shop benefits when family members are scheduled together include setup-cost reduction and minimization of operator forgetting (i.e., reduced need for operator learning), which reduces both setup times and run times. Setup costs are reduced because the amount of change in

each setup is reduced. For example, it takes less time to change an insert in a die than to change the entire die.

Manufacturing benefits of group technology are increased when the family members are manufactured not only close in time but also close in space. Space proximity is implemented through cells, which have many benefits, including minimum paperwork, reduced material handling, and reduced inventory. For more on cells, see Section 4.4.

4.3 Family Formation

Family formation can be done at various levels. At the macro level, entire factories (focused factories) can specialize in a particular type of part or assembly. At the micro level, families can be based on similarities in the part geometry (e.g., group shafts, flat parts, gears, etc.) or in the process (castings, forgings, sheet metal parts, heat-treated parts, printed circuit boards, etc.). At the heart of the grouping (clustering, classification, pattern) process is the problem of **coding** (numerical classification).

4.31 Informal Coding.

One possibility is to group items informally. For example, a supervisor may examine all items to be manufactured in a specific week and decide to run all brass gears on Tuesday and all cast iron gears on Wednesday. However, this hit-and-miss procedure does not produce design engineering benefits or reductions in manufacturing office costs. Although it may give some of the close-in-time benefits, there are no close-in-space benefits, because no cells were formed. A formal coding system in which a code is assigned early (ideally while the item is still a "gleam in the designer's eye") will do much better.

4.32 Coding.

The ideal code contains information about the part or assembly itself and about the manufacturing process. The ideal code provides information not only to all parts of manufacturing (manufacturing engineering, industrial engineering, tool engineering, scheduling, line supervision, quality assurance, etc.) but also to purchasing, design engineering, and other areas. Using an analogy, the group technology code is the genetic code of computer integrated manufacturing. For communication excellence, this code must be very precise (Koenig, 1990):

- precise nonambiguous meaning, no double or triple definitions for the same phrases
- tightly structured and concise
- easy to use

If a code has numbers only, each digit specifies from a maximum of 10 choices. Five spaces allows for 100,000 combinations. If a code has numbers and letters, however, each symbol specifies from 36 choices (in English), providing abundant choices in a reasonable-length code. Of course, computers are used in the setup and decoding of the codes.

Codes can be chain or hierarchical. In the chain type, each digit's specific location is fixed for a particular meaning. For example, the first digit is reserved for the product type, the second digit for material, digits 3 to 6 for part geometry, etc. The advantage of a chain code is that it is easy to learn; its disadvantage is that more information requires more digits in the code, making it difficult to handle manually and with low-power computers. An example of a chain code is the Opitz system, developed in Germany. It has five digits that describe the part geometry (see Figure 4.4), and a supplementary code of four digits that give dimensions, material, initial form, and accuracy.

In hierarchical codes, each code character depends on the preceding one—a tree-type structure. A user entry sends the computer to a specific branch. Because many branches can be eliminated, the overall code can be quite short (e.g., 4–12 positions versus 32 for a chain code with the same information). The hierarchical code is difficult to learn, but the expectation is that computers will do the sorting and matching. With the decreasing cost of computers, the computation disadvantages of chain codes may become less important.

As group technology codes spread beyond machining, hybrid codes may be useful for nonmachining applications such as electronics, plastics, assembly, and so forth (Marion et al., 1986). In a hybrid code, the initial position gives the

	1st Digit		2nd Digit		3rd Digit		4th Digit	5th Digit	
	Component Class		External Shape External Shape Elements		Internal Shape Internal Shape Elements		Plane Surface Machining	Auxiliary Material and Gear Teeth	
0	Rotational components	L/D < 0.5		Smooth, no shape elements		Without through bore blind hole	No surface machining	No gear teeth	No auxiliary holes
1	Rotational components	0.5 < L/D < 3	Stepped to one end or smooth	No shape elements	Stepped to one end or smooth	No shape elements	External plane surface and/or surface curved in one direction	No gear teeth	Axial hole(s) not related by a drilling pattern
2	Rotational components	L/D < 3	Stepped to one end or smooth	With screwthread	Stepped to one end or smooth	With screwthread	External plane surfaces related to one another by graduation around a circle	No gear teeth	Axial holes related by a drilling pattern
3			Stepped to one end or smooth	With functional groove	Stepped to one end or smooth	With functional groove	External groove and/or slot	No gear teeth	Radial hole(s) related by a drilling pattern
4			Stepped to both ends (multiple increases)	No shape elements	Stepped to both ends (multiple increases)	No shape elements	External spline and/or polygon	No gear teeth	Holes axial or radial and/ or in other directions, not related
5			Stepped to both ends (multiple increases)	With screwthread	Stepped to both ends (multiple increases)	With screwthread	External plane surface and/or slot and/or spline	No gear teeth	Holes axial and/ or radial and/ or in other directions related by drilling pattern
6			Stepped to both ends (multiple increases)	With functional groove	Stepped to both ends (multiple increases)	With functional groove	Internal plane surface and/or groove	No gear teeth	Spur gear teeth
7				Functional taper		Functional taper	Internal spline and/or polygon	With gear teeth	Bevel gear teeth
8				Operating thread		Operating thread	External and internal splines and/or slot and/or groove	With gear teeth	Other gear teeth
9				Others (10 functional diameters)		Others (10 functional diameters)	Others	With gear teeth	Others

FIGURE 4.4 Geometrical code for round parts, Opitz system.

general process (machining, electronics, etc.), and then the following codes are specific for each general process. Generally, the coding is done with interactive questions between the analyst and a computer; see Figure 4.5.

To maximize use of the code, it should apply to multiple departments. In particular, the code should give characteristics not only of the item but also of how the item is processed and should include not only parts features but also production information (production equipment used, weight of item, cost of item for each processing step, current and future requirements, etc.). If the code has complete information, then higher level computer programs can use the data base for cost estimating, facilities and process planning, inventory management, work-content estimation, labor analysis, cell design, order grouping, and capacity planning. Thus, the code may be broken into two sections: the primary portion (which describes the item and doesn't change) and the secondary portion (which describes the processing and demand and which may change over time).

When using a code, Dunlap and Hirlinger (1983) recommend a layered filter approach; see Figure 4.6.

4.4 Cells
Many of the benefits of group technology come from manufacturing items close together in space—this grouping is called *cells*. This section will be divided into cell concept, cell advantages and disadvantages, and cell layout.

4.41 Cell Concept.
Manufacturing the individual items in a parts family should require almost the same steps. Therefore, all similar machines and skills are located in one area—a cell. Machines and skills include not only the machines and people themselves but also jigs, fixtures, drawings, measuring instruments, material handling equipment, and so forth. The goal is to get the advantages of mass production even though no individual family member has sufficient production volume to justify mass production.

4.42 Cell Advantages and Disadvantages.
First we will discuss advantages and then disadvantages.

Advantages. Advantages of a cell versus a job shop include use of specialization, minimum material handling, simpler production control, shorter throughput times, and lower work-in-process inventory.

Specialization in a job shop is by machine; that is, the operator runs a turret lathe, a milling machine, or other machine. In a cell, specialization is by product family, and each operator runs a mixture of machines (lathe, drill, inspection gauge). In other words, in a job shop the operator runs a variety of products on the same machine, whereas in a cell the operator runs the same product on a variety of machines. Thus, labor wage/hour should be about the same in a job shop and a cell. The cell machines, however, can be specialized to produce only what is needed

LANGUAGE: ENGLISH, FRENCH, GERMAN OR DUTCH? *english*
DIMENSION STANDARD IN MILLIMETERS OR INCHES? *mm*
DRAWING NUMBER? *abc 1*
ANSWER THE FOLLOWING QUESTIONS WHEN POSSIBLE WITH YES OR NO
   IS IT A ROTARY COMPONENT? *yes*
   LARGEST DIAMETER AND LENGTH? *196,87*
   DOES THE ROTARY FORM DEVIATE? *no*
   IS THE AXIS OF ROTATION THREADED? *no*
   MUST ANY ECCENTRIC HOLING, PLANING OR SLOTTING OPERATIONS BE DONE? *no*
   MUST THE TOP SIDES OR OUTER FORM BE TURNED? *yes*
   HAS THE OUTER FORM A SPECIAL GROOVE(S) OR CONE? *no*
   ARE THE OUTSIDE DIAMETERS INCREASING FROM BOTH ENDS? *yes*
   MUST THE INNER FORM BE TURNED? *yes*
   HAS THE INNER FORM A SPECIAL GROOVE OR CONE? *no*
   ARE THE INSIDE DIAMETERS DECREASING FROM BOTH ENDS? *no*
   LENGTH AND LARGEST LENGTH COAXIAL HOLE IN MM? *174*
   IS DIAMETER TOLERANCE 6 OR BETTER? OR ROUGHNESS LESS THAN 33 RU? *no*
   IS THE LENGTH TOLERANCE LESS THAN 0.3 MM? *yes*
   ANY FORM TOLERANCE? *no*
   MATERIAL TYPE? *st60*
[Once these questions have been answered, the computer will assign a classification number of the part. The printout is as follows:]
   DRAWING NUMBER = abc 1
   CLASSIFICATION NUMBER = 1330 4021 2104

FIGURE 4.5 Conversational coding program, MICLASS system example. (MICLASS is maintained by Organization for Industrial Research, Waltham, Mass.). The first 12 digits of the MICLASS code are universal; up to 18 additional digits can be company specific. Of the 12 digits: 1 = main shape, 2–3 = shape elements, 4 = shape element position, 5–6 = main dimensions, 7 = dimension ratio, 8 = auxiliary dimension, 9–10 = tolerance codes, and 11–12 = material codes.

on that family, so specialized machines, tools, jigs, fixtures, inspection devices, and so forth, can be used. Therefore, output/hour should be higher for the cell.

Material handling cost in the cell is minimized since material handling paperwork is zero and material handling distance is small (generally 1–5 m).

Production control is *to* the cell, not within the cell. Within the cell, the routing is fairly standardized and the timing is at the operator's discretion. In job shops, overhead often is 200% of direct labor cost; cells try to reduce that burden. The concept is to reduce micro-managing by staff personnel. Computers and paperwork encourage micro-managing by staff (a Taylorism approach), so line people need to insist that staff people leave them alone.

Shorter throughput times result from the simplified flow pattern and the eliminated paperwork. In addition, units of processed work can move to the next operation before the batch is completed.

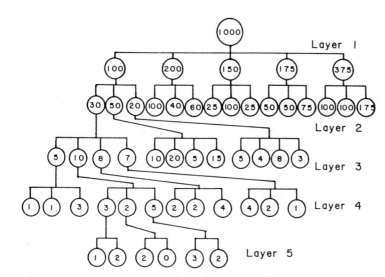

FIGURE 4.6 Layered filter approach to family formation (Dunlap and Hirlinger, 1983). Assume 1,000 parts were coded on five characteristics: material type, outer diameter, length, tooth form, and heat treatment. Assume that you decide the order of importance of the characteristics is material type, heat treatment, tooth form, outer diameter, and length. Then the "layer 1 filter" would be material type—dividing the 1,000 parts into five groups with (say) 100 of brass, 200 of stainless steel, etc. Then the "layer 2 filter" of heat treatment is used, subdividing each layer 1 group. Then the "layer 3 filter" of tooth form is used, etc. After the last filter, you know all the items that have common features and the items that are unique. If the parts families are too small, "back up" a level. To get more alternatives, repeat the process with a different priority ranking of characteristics. Copyright © Institute of Industrial Engineers, 25 Technology Park, Norcross, GA 30092. (404) 449-0460.

Work in process within the cell is small, because the sequence of machines is standardized as in line manufacture. Since the number of operators is small and the timing of the move is under operator control, the buffers between machines can be small. Low inventories within the cell, however, do not automatically lead to low total inventories, because total inventories include raw material and parts inventories before the cell and finished goods inventories after the cell. The policy of running similar parts together to maximize productivity within the cell means that due dates are ignored. That is, an item is scheduled because it is similar rather than because it is overdue. Thus, some items will be produced before they are needed and finished goods inventory will be increased. In addition, the ability to produce similar items together requires raw materials to be available before they are needed from a due-date viewpoint. Cell inventories are therefore minimized at the expense of larger inventories before and after the cell. However, cells encourage frequent small batches (due to low setup costs and short throughput times), which tends to reduce inventories before and after the cell. Thus, cell layout may or may not decrease total inventories.

Disadvantages. Disadvantages versus a job shop include lower equipment utilization, loss in flexibility, and cost of setting up and maintaining the family and cell concept.

If a cell is to be capable of all the necessary operations, it will duplicate machines located in other cells. To minimize material handling, low-cost machines (e.g., drill presses and deburring benches) may even be duplicated within the same cell. On the other hand, processing within each cell may dictate using four slow, semiautomatic machines in four different cells rather than one fast, automatic central machine. Specialized processes, such as heat treating, painting, and plating, are called *exceptional elements*. In a "pure" cell, 100% of the operations are done in the cell. In real cells, some operations are subcontracted outside the cell. Subcontracting also can occur if a specific cell's machine has insufficient capacity (a bottleneck). Kern

and Wei (1991), Sule (1991), and Boe and Cheng (1991) discuss some of the problems of minimizing intercell transfers.

Cells have less flexibility than do job shops because of the dedicated equipment, people, and procedures. Thus, new products are constrained in design and processing. If they meet the constraints, however, lower manufacturing costs may increase the potential market.

The cost of setting up and maintaining a group technology system and cell layout can be large, including the capital cost of setting up the system and buying the specialized equipment for the cells and the operating cost of analyzing the parts and the process and communicating this information to all the interested parties.

The benefits of group technology and cells are primarily in avoided costs: extra parts not designed, extra jigs not purchased, better processing for the items, larger lots made since part requirements are combined, and so forth. Unfortunately, *these avoided costs do not show up in the cost-accounting system.* What the managers do see is the out-of-pocket costs of setting up and maintaining the group technology system. Since the managers are evaluated on their short-term performance, they may resist implementing the group technology concept because reported expenses will increase and reductions will not appear in the reported costs upon which they are judged.

4.43 Cell Layout. A flow line usually is linear (although it can be L or U shaped), but cells usually are circular or U shaped. This arrangement minimizes transportation distance for operators (human or robotic) and encourages multiple machines per operator. Figure 4.7 shows an example cell. Advantages of circular or U-shaped cells include the following (Schonberger, 1983):

- *Multiple machines/operator.* Most machines are automatic or semiautomatic, resulting in considerable idle time for machine operators. In the conventional job shop, there is 1 operator for 1 machine. Group technology (with its multiple machines/operator) can substantially increase labor productivity. As shown in Figure 4.7, the second and third machine in the production sequence need not be assigned to the first operator. The additional machines also do not have to be the same type of machine. This flexibility ensures that the operator will be busy, even if cell production requirements increase or decrease.

- *Teamwork concept.* People near each other will talk to each other. Problems can be discussed immediately. The teamwork concept applies especially to quality and rework problems.
- *Minimum movement problems.* Distances for movement of people, products, and tools are minimized. Movement interference, by machines and conveyors, is minimized.

5 FLEXIBLE MANUFACTURING SYSTEMS

Flexible manufacturing systems (FMS) are automated cells and are also called *adaptive* manufacturing systems and *versatile* manufacturing systems. This section will be divided into concept and concept implementation.

5.1 Concept   The driving force behind the automation of cells is the continuing decrease in the cost of computers. It is not only that the cost of computers has been decreasing at such a rate that cost at the end of ten years is 10% of cost at the start of the ten years. In the meantime, cost of labor increases approximately 5% annually, making cost of labor at the end of ten years about 60% higher than the cost at the start of the ten years. Thus, the ratio of computer cost to labor cost changes even more rapidly in favor of computers than either of the two cost components. It is becoming a computer everywhere.

This pervasive use of computers has a number of terms: *computer-aided design* (CAD), *computer-aided process planning* (CAPP), *computer-aided manufacturing* (CAM), and *computer-integrated manufacturing* (CIM). When a manufacturing cell is automated, it is called a *flexible manufacturing system* (FMS).

5.2 Concept Implementation   In 1982 there were less than 100 metal-cutting FMSs in the world; by 1986 there were 250 (Darrow, 1987), and their number is growing rapidly. In addition, there are flexible systems for sheet metal and mechanical and electronic assembly.

The three basic machine components of the FMS are the manufacturing machines themselves, the material handling machines, and the control machines. See Figure 4.8.

The manufacturing machines (the operator) themselves often are computerized (a first-level computer). An example is a numerically controlled machine (DNC for direct numerical control or CNC for computer numerical control) such as a single-spindle, tool-changing machining

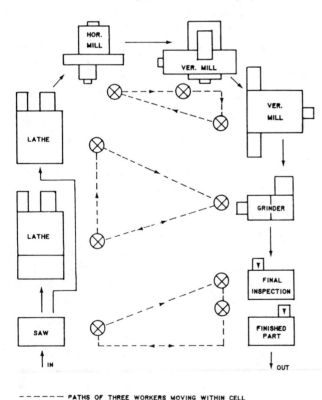

FIGURE 4.7   Example cell layout (Black, 1983). Note that multiskilled operators tend multiple machines and that the machines tended need not be in production sequence. It also is relatively easy to add or subtract workers if production requirements change. Copyright © Institute of Industrial Engineers, 25 Technology Park, Norcross, GA 30092. (404) 449-0460.

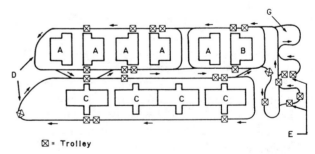

Flexible manufacturing systems (Allis-Chalmers—Kearney & Trecker)

A = Milwaukee-Matic machining centers
B = 3-Axis milling machine
C = Duplex head index units
D = Towline pallet transporting trolleys
E = Load/unload areas
G = Partial inspection

Arrows indicate direction of motion for the trolleys (boxes on towline)

FIGURE 4.8   Example flexible manufacturing system layout (Black, 1983). Copyright © Institute of Industrial Engineers, 25 Technology Park, Norcross, GA 30092. (404) 449-0460.

center. The machines have magazines of 50 or more tools; some machines even have automatic switching of magazines, which permits backup tooling (for tool failures or switching when a predetermined tool life is reached) and permits a wide variety of products to be made.

The material handling machines use humans to load parts onto fixtured pallets—each pallet having its own identification code. Then the movement can be by automatic storage/retrieval systems (see Chapter 11), conveyors with switches (see Chapter 17), automatically guided vehicles (see Chapter 18), and/or robots (see Chapter 20). The product is tracked by a variety of sensors (see Chapters 16 and 17).

The more complicated FMSs tend to have computers at three levels: the operator computer, the schedule computer, and the supervisor computer. (When one computer directs another computer, it is a **hierarchical system**.) The schedule computer tells the first-level computer which part to make in which sequence. The supervisory computer monitors production status, machine usage, and tool wear. Ideally, the computers operate in a closed-loop (feedback) mode; see Box 4.2.

Because FMSs are capital intensive, firms often use them on more than one shift. The second and third shift may have no humans (in German, *die Geisterschict*, the ghost shift). If the system fails during the ghost shift, it shuts down. Naturally, this requires a large buffer of loaded pallets in the staging area, loaded during the day shift.

What do humans do in FMS? Renault has a 7-machine FMS with transport on wire-guided carts (Dreyfack, 1982). The system machines a family of 25 varieties of 4 basic pieces. There are 6 human workers: a chief technician to handle programming and control, 2 who load and unload pallets, 2 toolsetters, and 1 electrical maintenance technician.

A cell attempts to achieve the economies of line production even though production volume is more appropriate for a job shop. An FMS tends to be less flexible and have greater output than a nonautomated cell, so it is appropriate for a relatively limited variety of standardized parts with relatively high annual volume. See Figure 4.10. The number of different parts produced in an FMS at any given time typically ranges from 4 to over 100, and the annual production rate is 40 to 2,000 (Singhal et al., 1987).

There are three FMS advantages. One is greater throughput because downtime between machines is minimized, resulting in high output for the capital cost. (Note that many of the downtime advantages may be achieved by design for manufacturability and Just-In-Time at a fraction of the capital cost of an FMS.) An interesting design choice is whether (1) to train the operator in the technology and how it operates and to permit the operator to use that knowledge or (2) to treat the operator as an unskilled material handler, leaving the technological knowledge in the technical staff. Jackson and Wall (1991) showed experimentally in a British factory that machine availability was considerably higher with the training approach rather than with operators calling the technical staff whenever they had a breakdown. "It's not so important to be able to reconnect the broken thread; what matters is to see what's likely to happen and what you have to do to prevent it snapping" (p. 1309). That is, good **reliability** (long time between failures) is more helpful than good **maintainability** (quick repairs). The concept of controlling variance at the source emphasizes Deming's point that a major impediment to workers' effective performance is the management-imposed barrier to worker discretion.

The second FMS advantage is the low direct labor cost (although the labor tends to be highly skilled). Thus, an FMS goal is minimization of machine idle time, whereas a manual cell's goal is minimization of labor idle time. One FMS problem is the tendency of engineers (who naturally love machines) to minimize the number of people in the system by replacing them with machines.

The third advantage is that FMS can be used for similar parts even if product life is short, because the capital cost is in the process, not attached to individual part numbers.

There are two FMS disadvantages. First, it needs a relatively structured environment (relatively large volumes for a few parts). Second, FMS has large capital costs, which are not only in the hardware but in the software (programming and debugging) and maintenance. (Preventive maintenance and emergency repair policies are critical to success

---

BOX 4.2 *Open- versus closed-loop systems*

Systems can be open loop (without feedback) or closed loop (with feedback). Most machine tools (including NC machines) are open loop. They become closed-loop systems with the addition of a human operator; see Figure 4.9.

For example, assume a NC drill was to drill a .500 dia in. hole at location *X*. We will just consider the dia for this example. The reference input, or set point, is .500. The operator determines the actual dia drilled and makes a comparison of actual versus desired dia. The difference between actual and desired (the error) then becomes the actuating signal. This actuating signal then becomes input to the controlled elements (also called controller; still in the human in our example), which then affects the manipulated variable (the drill dia), which inputs to the controlled system (the drill). The controlled output is the dia of the hole (hopefully .500). Disturbances may enter the system at a number of points.

In general, the operator is a "sampled data" link because data are measured only occasionally; more accurate control is obtained from continuous feedback. The feedback is negative (hopefully) as the error is reduced. The control can be proportional (small adjustment for small error, large adjustment for large error) or step (same adjustment no matter what size of error once error crosses a threshold).

Removing a person from the loop and substituting mechanisms presents many challenges in FMS.

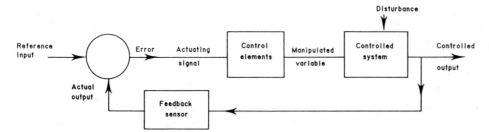

FIGURE 4.9 Closed-loop systems have feedback. In most manufacturing systems the loop is closed by a human operator.

of FMS.) A large FMS may have 25 machines and cost $30,000,000 (Meredith, 1987) although $5,000,000 to $10,000,000 is more typical. Software cost may be 50% of total system cost. Software problems are great, because many machines are not as interchangeable as assumed; rework and repair are a challenge; vendor's predictions of failure rates are not accurate; and devising interfaces between engineering, manufacturing, and business computer programs is quite difficult. An FMS is a massive agent of change, upsetting established systems, policies, procedures, and rules and affecting people throughout the firm. Utilization might be 40% after 8 months and 75% after 14 months (Gould, 1990).

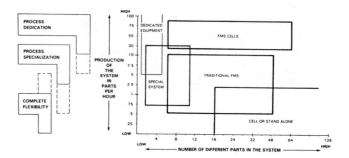

FIGURE 4.10 Flexibility versus capacity is a classic trade-off. Flexibility can be high (universal machinery, tooling, fixture, and processes), medium (some specialization in tooling and fixtures on universal machines), and low (specialized machines as well as specialized tooling and fixtures). Capacity is low for universal machinery (stand-alone machinery) and high for dedicated equipment (Heglund, 1981).

DESIGN CHECKLIST: ALTERNATIVE MACHINE ARRANGEMENTS

Assembly
  Operator to product or product to operator
  Progressive or nonprogressive
    Balance delay
    Schedule flexibility
    Shocks
    Satisfaction
    Quality
    Training
    Material handling

  Skill requirements
  Costs
  Inventory
  Supervision
Components
  Job shop, flow line, or in between
  Flexibility (variety) versus capacity (volume)
  Group technology
    Families
    Cells

REVIEW QUESTIONS

1. What is the difference between construction and manufacturing?
2. In the 3S combination, what are the three *S*s?
3. Discuss the advantages, neutral points, and disadvantages of nonprogressive assembly.
4. Briefly describe tabulated drawings.
5. What are the differences between progressive and nonprogressive assembly?
6. What is the difference between a chain and a hierarchical code?

7. Parts families give two benefits: variety reduction and common manufacturing. Discuss each.
8. Give the five advantages and three disadvantages of cells versus a job shop.
9. List the three approaches to family formation.
10. Discuss the relative costs over time of computers and labor.
11. What is the "ghost shift"?
12. What are some "avoided" costs of group technology?

13. How does releasing items for manufacture in a group technology system conflict with releasing items for manufacture in the materials management (due date) approach?

PROBLEMS AND PROJECTS

4.1 Visit a factory that does assemblies. Under what circumstances should they use progressive assembly? Does the supervisor agree with you?

4.2 Make a tabulated drawing for different types of hamburger patties at a restaurant.

4.3 Using the "layered filter" approach, devise families of menu items at a fast-food restaurant.

4.4 Have a debate between two teams concerning whether "our firm" should give special emphasis or normal attention to the use of group technology. The instructor will give information concerning the firm. There should be a written and oral presentation. The class then will vote on whether to give normal attention or special emphasis on group technology at "our firm."

REFERENCES

Black, J. Cellular manufacturing systems reduce setup time, make small lot production economical. *Industrial Engineering*, Vol. 15, 11, 36–48, Nov. 1983.

Boe, W., and Cheng, C. A close neighbour algorithm for designing cellular manufacturing systems. *Int. J. Production Research*, Vol. 29, 10, 2097–2116, 1991.

Darrow, W. An international comparison of flexible manufacturing systems technology. *Interfaces*, Vol. 17, 6, 86–91, Nov.–Dec. 1987.

Desai, D. How one firm put a group technology parts classification system into operation. *Industrial Engineering*, Vol. 13, 11, 78–86, Nov. 1981.

Dreyfack, K. Minicomputer is FMS traffic cop. *American Machinist*, 143–144, July 1982.

Dunlap, G., and Hirlinger, C. Well-planned coding, classification system offers companywide synergistic benefits. *Industrial Engineering*, Vol. 15, 11, 78–83, Nov. 1983.

Gallagher, C., and Knight, W. *Group Technology*. London: Butterworth & Co., 1973.

Gould, L. Is material handling the Achilles' heel of the FMS? *Modern Material Handling*, 56–58, July 1990.

Heglund, D. Flexible manufacturing. *Production Engineering*, Vol. 28, 5, 38–43, May 1981.

Jackson, P., and Wall, T. How does operator control enhance performance of advanced manufacturing technology? *Ergonomics*, Vol. 34, 10, 1301–1311, 1991.

Kern, G., and Wei, J. The cost of eliminating exceptional elements in group technology cell formation. *Int. J. Production Research*, Vol. 29, 8, 1535–1547, 1991.

Koenig, D. *Computer Integrated Manufacturing: Theory and Practice*. Bristol, PA: Hemisphere Publishing, 1990.

Krepchin, I. The human touch in automobile assembly. *Modern Material Handling*, 52–55, Nov. 1990.

Marion, D., Rubinovich, J., and Ham, I. Developing a group technology coding and classification scheme. *IE*, 90–97, July 1986.

Meredith, J. Implementing new manufacturing technologies: Managerial lessons over the FMS life cycle. *Interfaces*, Vol. 17, 6, 51–62, Nov.–Dec. 1987.

Schonberger, R. Plant layout becomes product-oriented with cellular, just-in-time production concepts. *Industrial Engineering*, Vol. 15, 11, 66–71, Nov. 1983.

Singhal, K., Fine, C., Meredith, J., and Suri, R. Research and models for automated manufacturing. *Interfaces*, Vol. 17, 6, 5–14, Nov.–Dec. 1987.

Sule, D. Machine capability in group technology. *Int. J. of Production Research*, 29, 9, 1909–1922, 1991.

5 FLOW LINES

CHAPTER PREVIEW

There are three types of flow lines (operation only, assembly, and order-picking). Many line problems are overcome by buffers. Buffer design is considered in detail; buffer size is considered; and balancing flow lines (equalizing the amount of work at each station) is discussed.

CHAPTER CONTENTS

1 Introduction
2 Buffer Design
3 Buffer Size
4 Balancing Flow Lines
5 Operating Flow Lines

KEY CONCEPTS

assembly line	line balancing
balanced line	operation only line
buffer	order-picking line
carriers	paired stations
decoupling	precedence diagram
flow lines	rocks in a river
Just-In-Time (JIT)	utility operator

1 INTRODUCTION

1.1 Types of Lines

Figure 5.1 shows three types of **flow lines**. In one extreme, the **operation only line**, a single component goes through a series of operations (item is "processed" or changed at stations); no additional components are added. There are operations and transportations. Examples are an engine block being machined on several machines, a steel rolling mill, a rotary index table with several "heads," and university class enrollment (visit various tables to sign up or pay for various activities).

In the other extreme, the **order-picking line**, items are accumulated together without any operation at the station. There are transportations but no operations. Examples are order picking in a warehouse and a customer obtaining food in a cafeteria.

The most common flow line, the **assembly line**, has both operations performed and items added at the station. There are both operations and transportations. Examples are product assembly (autos, TVs, clothing), packaging lines, chemical processing lines, and filling lines. A reverse version is the disassembly line, often used in food processing. Example disassembly lines are slaughter operations and grain mills.

Flow lines do not need to make a single product continuously. Figure 5.2 shows three alternatives: (1) the single product made continuously, (2) multiple products made sequentially in batches, and (3) multiple products made simultaneously. Alternative 1 might be product A made continuously. Alternative 2 might be product A made all day Monday, product B made all day Tuesday and Wednesday, product C made Thursday, etc. Alternative 3 might be one of product A made at 8:00 A.M., two of product B made at 8:07 and 8:14, one of product C made at 8:17, one of product A made at 8:21, etc.

The work (elements) of the task is divided among the line's stations. If the amount of work is equal at each station, it is a **balanced line**; if not equal, it is an unbalanced line. Depending upon the type of line, one or more elements can be done at each station. In addition, there is the possibility of the same element's being done at more than one station. A well-designed line will have

- minimum idle time at the stations
- high quality (enough time at each station for operators to complete assigned work)
- minimum capital cost (for both equipment and work-in-process)

Although it may be obvious: (1) the transport between stations need not be by conveyor, (2) the transport between stations need not be at a fixed speed or time interval, and, very important, (3) there may be storages between the operations or transportations. There may be a storage before and after each operation and transportation. This storage is known technically as a **buffer**, a bank, or float; its purpose is to decouple the line.

1.2 Decoupling

There are two primary reasons for **decoupling** (separating workstations with buffers). They will be labeled *line balancing* and *shocks and disturbances*.

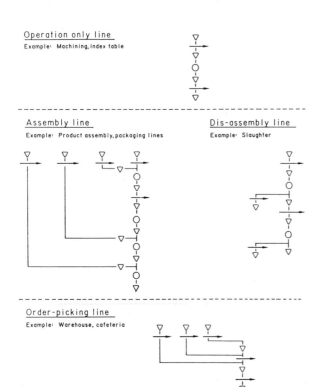

FIGURE 5.1   Flow lines can be operation only, order picking, and assembly.

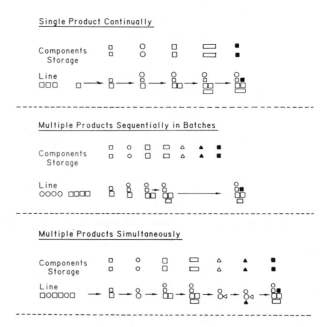

FIGURE 5.2   Flow lines can make a single product continuously, multiple products sequentially in batches, and multiple products simultaneously.

1.21 Line Balancing. The **line balancing** challenge occurs because the mean times for stations A, B, C . . . are not equal.

Assume operation A took 50 s, B took 40 s, and C took 60 s. Then, assuming there was no buffer, the line would have to index at the speed of the slowest station (the bottleneck station), station C. At stations A and B there would be idle time, called balance-delay time, the difference between the cycle time and the work time. Thus, without buffers, the line speed must be set considering the speed of the slowest station.

Figure 5.3 shows another aspect of the problem of mean times—the variation in the ability of human operators. Typically, performance among operators varies about 2 to 1—the best can produce twice what the worst can. Assuming symmetry and putting the average operator at 100%, the range is from about 67% to 133%. If the line is set at the speed of the average operator, 50% can work faster and 50% cannot keep up. (There is a temptation for those who cannot keep up to reduce safety and quality.) Thus, the speed at a station (assuming the typical balanced line where work content is equal at each station) cannot be set at the speed of an average (say 1.0) operator; it is set at the speed of a "slow" operator. Referring to Figure 5.3, if "slow" operator is defined as a 90% operator (i.e., 90% of the operators can do the job in the mean cycle time), this is 1.28σ below the mean. Considering the range of performance from 2/3 to 4/3 as 6 σ, then $(4/3 - 2/3)/6 = 1/9$. Then $-1.28(1/9) = .142$. Then $1.0 - 0.142 = .858$ for the station speed. That is, if the station could be set for an average operator instead of a slow operator, then it could be 17% faster (as $[1.0 - .858]/.858 = .17$). Thus, without buffers, the line speed must be set considering both the speed of the slowest station and the mean time of the slow operator on the slow station.

But there is a third problem: Cycle times vary.

1.22 Shocks and Disturbances. Shocks and disturbances make the cycle times vary. Figure 5.4 points out

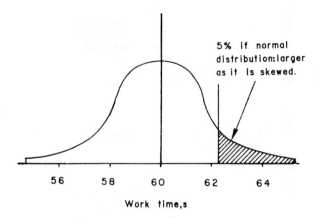

FIGURE 5.4 Distribution of times points out that only 50% of the times are done in the average time or less. If a paced line is set so that 95% of the cycles are completed before the line indexes, then the cycle time must be slower than average time. If average time = 60 s and σ = 2 s, then the line cycle time (if based on this station and assuming normality) would be 60 + 1.64(2) = 63.3 s.

the problem of distribution of operation times. Assume operation C has a mean of 60 s and σ = 2 s. To include 95% of the times, then station C, and thus the line, would operate at a cycle time of 63.3 s.

The variability of cycle times can occur for many reasons.

First, consider a station with both an infinite supply of incoming components and perfect removal of the completed unit. There may be a temporary shift in the mean because the operator normally doing operation C is absent and a substitute is doing the job. If the temporary worker took 66 s instead of 60 s, then each station on the entire line also would take 66 s, if there were no buffer. Or the time of the regular operator may vary for "normal" reasons—tools becoming dull, short breakdowns of the machine, operator's stopping to light a cigarette or sneeze, talking to the supervisor, or dropping a part. (If the operator is a machine or robot, cycle times still can vary. For example, a machine may fail if the incoming part arrives upside down.) Figure 5.5 shows typical cycle time distributions for (1) unpaced work (i.e., with sufficient buffers) and (2) paced work (usually work without buffers). The distributions for unpaced work usually are positively skewed (a few long times with a lower absolute minimum time). Paced work has a higher mean time (i.e., pace is slower) and the curve is more symmetrical.

Second, consider the station as part of a line. The station's time can vary because of inadequate supply from the preceding station (station starvation) or because the following station is not yet ready to accept a unit (station blockage). In addition, there is the problem of what to do with a defective unit ("scrap" if it can't be fixed; "rework" if it can be fixed). Time problems include when to repair the item (now or later); space and transport problems include where to put the defective component or assembly.

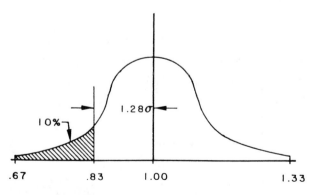

FIGURE 5.3 Inefficiency of flow lines is increased by the typical variability of workers. If individual decoupled workstations could operate at the mean speed of 1.0, then output would be about 17% faster than if the stations were arranged as a flow line.

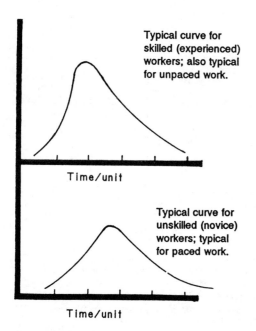

Typical curve for skilled (experienced) workers; also typical for unpaced work.

Time/unit

Typical curve for unskilled (novice) workers; typical for paced work.

Time/unit

FIGURE 5.5   Skilled workers have a lower variance and more skew as well as a lower mean. Unpaced times distributions are similar to skilled times distributions; paced times distributions are similar to unskilled times distributions.

Thus, without buffers, the line speed must be set considering (1) the speed of the slowest station, (2) the mean speed of the slowest operator on the slowest station, and (3) the slowest cycle time of the slowest operator on the slowest station.

Buffers give the flow line flexibility or tolerance. Figure 5.6 shows the flow line without buffers (the rigid system) as being similar to a train, while the flow line with buffers (the flexible system) is similar to trucks. In the "train" each rail car must travel at the same speed as every other rail car; if one rail car breaks down the entire train stops. With "trucks" each vehicle can go at its own speed; if one vehicle stops, the remaining vehicles can continue.

The total penalty of a line without buffers versus a line with buffers is difficult to determine because most lines have some buffers. However, a U. S. machine-paced auto assembly line (i.e., a line with few or no buffers) usually has a balance delay time of 8 to 15%. The U. S. auto companies

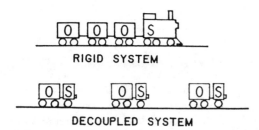

RIGID SYSTEM

DECOUPLED SYSTEM

FIGURE 5.6   Two types of flow lines are (1) flow lines without buffers, and (2) flow lines with buffers. Flow lines without buffers can be symbolized as a "train" with initial storage (S) followed by operations (O); that is, S-O-O-O. Flow lines with buffers can be symbolized as trucks; that is, S-O, S-O, S-O.

also allow an extra relief of 22 min/480 min shift (about 5%) for a machine-paced line (total of 46 min per shift). Thus, the U. S. auto companies have a built-in inefficiency of 13 to 20%, ignoring the problem of slowing the line down to the speed of the slowest station. Toyota has demonstrated that auto lines can be built with buffers. They have a buffer of three cars approximately every ten stations. Each worker has a button which can stop the line if there is a problem, such as bad quality. In U. S. auto plants there are no buffers, so only a few people are authorized to stop the line. And since if the line stops, everybody stops, they rarely stop the line. For other industries, Kilbridge (1961) gave 5 to 10% as a typical balance delay percentage. Figure 5.7 shows that balance delay percentage usually declines for longer cycle times.

Buxey (1978) reported on a number of lines in Scotland. In case 1 there were 20 stations with a cycle time of 15 s; items/station (i.e., buffer) were 1.5; the balance loss was 20%, and there was an additional loss because cycle time was determined by adding 20% to the time of the work at the most difficult station. Approximately 3% of the items were not completed when they finished the line and had to be finished at a repair station. In case 2 there were 13 stations with a cycle time of 10 s; items/station were 2; balance loss was 4%. In case 3 there were 15 stations with a cycle time of 1 h; items/station were 2. The balance loss was 6%, but, in addition, there was a utility operator, so there was an additional cost of 1/15 = 6.7%. Buxey et al., (1973) have an excellent review of line design problems.

These figures point out that although the fixed-pace line is a "manager of men," it is an inefficient manager.

See Box 5.1.

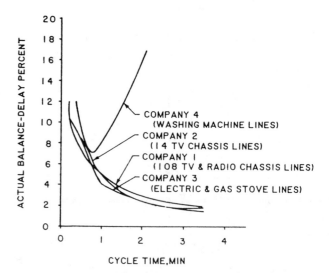

FIGURE 5.7   Balance delay time usually declines when cycle time is longer (Kilbridge and Webster, 1961). Reprinted by permission of Kilbridge and Webster, *Management Science*, Vol. 2, 1, 1961. Copyright 1961. The Institute of Management Sciences.

BOX 5.1 *Just-In-Time*

Just-In-Time (JIT) is an organizational tactic in which inventories are minimized by delivering items to the next stage of production just in time. The simplistic view of JIT is that it is a system to reduce inventory costs (i.e., have small buffers). However, inventory cost reduction is not the reason for JIT.

What is the cost of inventory, that is, buffers? A rule of thumb is that annual inventory carrying cost is about 30% of the product price. (The 30% is primarily the cost of capital, although some is cost for the building, heat, lights, etc.) Thus, a product costing $100 will cost .30($100) = $30 to store for 1 year. For simplicity, assume a factory works 250 days/year. It costs about 30%/250 production days = .125% of the product cost to store an item for 1 production day or (assuming an 8 h shift) .125%/8 = .0156%/production hour. A buffer storing $100 of material thus would have an inventory cost of $100 (.125%) = $.125 = 12.5 cents/production day or 1.57 cents/production hour. Thus, unless there are enormous inventories between stations, there is little inventory savings available.

Why did the Japanese begin JIT? They were interested in reducing *muda* (waste) and *mura* (unevenness). *Waste*, broadly defined, includes materials, capacity, labor, and time. *Unevenness*, also broadly defined, involves fluctuation in quality, arrival rates, worker skills, schedules, among others. The challenge was to develop a policy to reduce *muda* and *mura*.

Think of *muda* and *mura* as **rocks in a river**. The flow of product through the plant is the water of the river; the people of the plant are riding on a boat. How do they avoid the rocks? They can increase the depth of the river, that is, have larger inventories. But what if the water level (i.e., inventories) were reduced? Then the boat would hit the rocks, and the people in the boat would have to *remove* the rocks. Thus, to reduce waste and unevenness (*muda* and *mura*), the tactic is to reduce inventory levels so people cannot hide their problems with the "security blanket" of high inventory.

Since the concept is to remove rocks and not sink the boat, the technique is to reduce inventory until rocks (problems) appear and then increase inventory again while the rock is removed. Then the inventory is reduced again until a new rock appears, raised while it is removed, and so forth.

Examples of rocks are quality problems, equipment breakdown, and insufficient employee skills. Solutions to quality rocks might involve reducing the various components' percentage of defects. Solutions to equipment breakdown rocks might involve improving equipment

reliability and maintainability. Solutions to employee skill rocks might involve training programs in skills, such as welding, reading mechanical drawings, and use of a vernier caliper, or might involve cross-training employees (so that absenteeism has less effect when one person is not present).

In theory, management should have been working at removing the rocks all along. Inventory reduction is just a practical tactic of focusing management's attention on potential problems and forcing management to do what it should have been doing anyway.

JIT exposes manufacturing problems and changes methods and processes. It should focus on long-term solutions, not quick fixes. The time frame is months and years, not days and weeks.

Experience has shown that the JIT tactic (now sometimes called *lean production*) leads to six strategies to remove rocks:

1. little product variety (Little variety implies a focused factory with a limited number of products. However, group technology can permit greater product variety.)
2. produce only what the customer wants and only when it is wanted (Perfect scheduling often is implemented through "pull" scheduling instead of the conventional "push" scheduling. It requires close coordination with vendors; expediting no longer is feasible.)
3. high quality (These strategies often use statistical process control and employee involvement, such as Quality Circles, and need to consider vendors as well as internal operations. Reducing errors tends to lead to consistent production times.)
4. high availability (Availability requires preventive maintenance to minimize failures and skilled employees to minimize repair time.)
5. low setup times (Short setups permit small lots. Products can be varied if there is flexible equipment and workers. Small lots of varied products make it easier to match production to demand.)
6. cross-trained skilled workers (Skilled workers need to be trained. Skilled, flexible workers are essential in a high-quality, high-availability, multiple-lot, no-margins environment.)

Implementing JIT (lean production) obviously has many advantages. But not all organizations have implemented JIT, and not all those who attempted to have retained it. What do you think about lean production? If you were in charge, would you have lean production or fat production? Why?

2 BUFFER DESIGN

Figure 5.8 is a schematic of a workstation. The total workstation is composed of an operator, machine, energy and information input, energy and information output, material input, material output, input product storage, and output product storage. Buffers increase the size of input storage and output storage.

There are two buffering techniques: (1) decoupling by changing product flow and (2) decoupling by moving operators.

2.1 Decoupling by Changing Product Flow There are three subcategories: (1) buffers at or between stations, (2) buffers due to carrier design, and (3) buffers off-line.

2.11 Buffers at or Between Stations. One possibility for this type of buffer is a physical barrier on a conveyor. Figure 5.9 shows two common arrangements. In the upper figure, a piece of wood, a pipe, or piece of steel is placed across the conveyor. The pieces from the upstream workstation move along the conveyor until they hit the barrier and stop. The operator lifts them across the barrier,

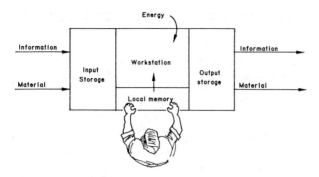

FIGURE 5.8 Schematic workstation shows that both information and material flow into input storage. Then they are transferred to processing, where they are transformed, using energy and local memory. Then the information and material go to output storage before being transferred to the next station.

works on them, and puts them back on the conveyor downstream of the dam. Rotary tables (lower figure) can be used to increase the size of the reservoir upstream of the dam. Parts stay on the rotary table and go round and round until picked up.

Buffer capacity can be increased by (1) increasing the time the item is available to the operator or (2) increasing the space within the reach of the operator. Increase time by having the operator face upstream. Arm motions forward are easier than backward, and when the object can be seen, timing can be better. If the object must approach from the operator's rear, use a rear view mirror so the operator need not turn around. Increase space within reach with a rotary table, as in Figure 5.9. A fixed-pace conveyor with items fixed to the conveyor tends to be a poor design. For a moving assembly, where access time is minimized, the components can be stored on a second conveyor positioned above the assembly conveyor or behind the operator and moving at the speed of the assembly conveyor. Figure 5.10

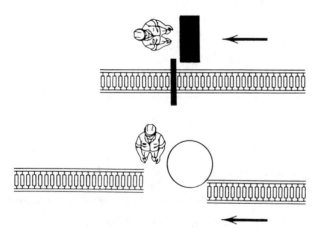

FIGURE 5.9 Physical barriers on a conveyor can dam the flow of parts, creating a reservoir or buffer. The lower figure shows a larger reservoir created by using a rotary table. If there is a "flood" and the reservoir capacity is insufficient, then the flow must be stopped upstream; this is called station blockage. The variability of the number of units in a buffer is an index of how effective the buffer is. That is, if there is no variance in the number in the buffer, there is no need for the buffer!

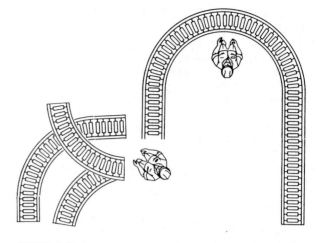

FIGURE 5.10 Increase storage space by using the cube of the space or by making a U in the line to take advantage of the operator's ability to turn. Close spacing of items on a belt and slow belt speed is preferable to wide spacing with high speed (since more items are within reach) even though average rate of arrival is the same.

shows two techniques of putting more items within the operator's reach.

The buffer can be designed for line balancing purposes as well as for shocks and disturbances—that is, the line can be unbalanced. Figure 5.11 shows a schematic pair of stations.

First we will discuss buffer input rate equal to buffer output rate. For example, assume line output is 10/h or 80/shift of 8 h; station A produces 20/h for 4 h; station B produces 10/h for 8 h. During the morning, A sends 10/h to B and puts 10/h into the bank. After 4 h, the bank has 40 units. Then B is fed from the bank at the rate of 10/h while operator A works elsewhere.

Suppose the buffer input rate is not equal to output rate. Assume, for example, that A has a rate of 20/h, B has a rate of 15/h, and desired line rate is 15/h. Then operate B at a rate of 15/h for 8 h. A produces 20/h but for only 6 h. During the 6 h, A sends 15/h to B and puts 5/h into the buffer. At the end of 6 h, the buffer holds 30 units. For the last 2 h of the shift, feed B from the buffer while A works elsewhere. Or, assume A has a rate of 20/h, B a rate of 15/h, and the desired line rate is 20/h. Then operate A for 8 h at

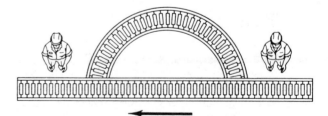

FIGURE 5.11 Buffers can be used for line balancing as well as for shock absorbers. Assume the first station produces 60 units/h for 4 h putting 30/h on the line and 30/h into the buffer (curved portion). Then the first station shuts down. For the second 4 h of the shift the second station is fed from the buffer.

20/h and have B work 8(20/15) = 10.7 h/day. (It may be easier in some cases to have a partial shift or Saturday work than to work 2.7 hours of overtime each day.)

Figure 5.12 shows the pallet on a track system to decouple operators and material handlers.

The concept so far has been to maximize time a unit is available to a station. Sometimes there are items that do not need to stop at each station; then a technique is needed that bypasses unwanted stations. Figure 5.13 shows the general concept. Figure 5.14 shows this idea applied to cafeterias, where the product is the human customers and the stations are the salads, desserts, drinks, etc.

2.12 Buffers Due to Carrier Design.

Items often are moved between workstations by **carriers**; the carrier can be a pan, box, pallet, index table, hook, or cart. If the carrier may be removed from the line, the effect can be that of a buffer. Transport between stations can be with wheeled carts pushed by hand. Heavy items (such as with tractor assembly, mounting jet engines to aircraft) can be pushed if on air-film pallets. If power is desired, it can come from overhead (power and free conveyor) or below (a towline); see Figure 5.15.

Another possibility is to keep the carrier on a conveyor but make the path omnidirectional; see Figure 5.16.

The following gives the advantages and disadvantages of multiple items/carrier. (It may be possible to obtain many of the advantages of multiple items/carrier with one item/carrier by batch processing the one-item carriers.)

Labor pickup and putdown of tools. In a high-volume operation, say a station time of .1 min, the operator picks up the tools at the start of the shift and does not put them down until break time. Thus "get" time and "put-away" time are prorated over many units so the cost/unit is small. If the station time is longer, say 1.0 min, then the operator does a number of operations and uses a variety of tools, picking up and putting down each one. However, if the items come in a carrier with multiple items, the operator can do a number of items with only one get and one put-away.

$$TTIME = (GTIME + PATIME)/N$$

where

$TTIME$ = total time/unit

$GTIME$ = get time/unit

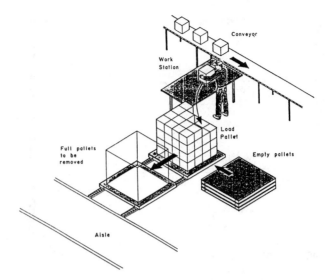

FIGURE 5.12 Pallets on tracks increase the output buffer by decoupling the operator from the material handlers. When the operator completes a pallet, it is pushed into the removal position, a new pallet is put into position, and work is resumed. Without the track, the operator must stop work when the pallet is full until the material handler moves the pallet. This tends to result in overstaffing of material handlers.

$PATIME$ = put-away time/unit

N = number of units

Assume that $GTIME + PATIME = X$. Then, for $N = 1$, $TTIME = X$; for $N = 2$, $TTIME = .5X$; for $N = 4$, $TTIME = .25X$; for $N = 8$, $TTIME = .125X$. The point of diminishing returns comes fairly rapidly. In addition, as N increases, the reach and move distances in the workstation increase, so "do" times increase. Kilbridge (1961) estimated get time + put-away time as 13% of do time for electronics assembly done on worker-paced conveyors.

Time/cycle will have a smaller variance. Although a carrier with 4 units probably will have approximately 4 times the mean do time of carriers with 1 unit, the variance of the longer cycle time probably will be less than the variance of the four short times—as long- and short-unit processing times may cancel each other on the multiple-item cycle time.

Multiple items/carrier encourages use of both hands since 2 units can be worked on at the same time. When using multiple items/carrier, make N an even number.

FIGURE 5.13 Bypassing stations not needed permits reduced work-in-process time.

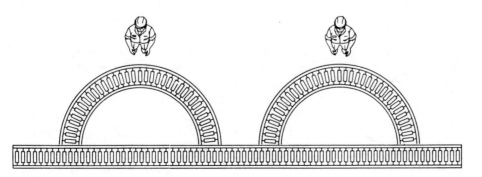

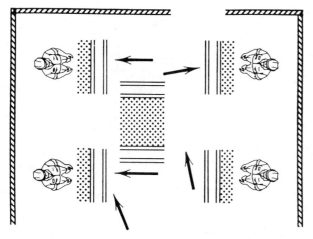

FIGURE 5.14 Scramble system cafeterias assume customers will stop at only a fraction of the available service areas. Since the goal is to minimize the customer's time waiting, the system permits customers to skip stations at which no service is desired.

Multiple items/carrier may restrict access to individual units on the carrier. For humans, this may just mean more time. For robots and machines, however, the restricted access may make the operation not feasible or the device very expensive.

Material handling. Multiple units/carrier give more units/foot of the line. For a specified distance, multiple units/carrier give more buffer. For a specified buffer size, the buffer will fit into a shorter distance for multiple units/carrier.

Multiple units/carrier require heavier carriers. The carrier will be harder to push, pull, and lift—although fewer carriers will be moved. Motors, rather than muscles, may be required for the heavier carrier.

Disposal of rejected units is more difficult for multiple units/carrier. If the defective unit remains on the carrier, take care so additional work is not done on it; be sure it does not get included into "good units." The defect also fills up buffer space. If the defect is removed from the carrier, empty space moves down the line. In addition, a carrier and means of transport must be provided for the

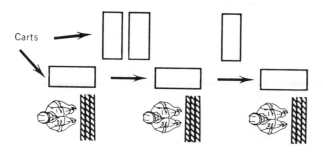

FIGURE 5.15 Removing a carrier from the line's path aids buffer creation. The carrier can be completely mobile (e.g., hand cart, air-pallet) or normally connected to a power source but able to be disconnected (power and free conveyor, towline).

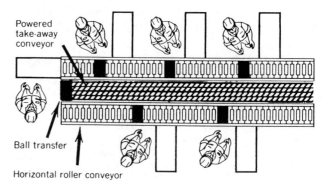

FIGURE 5.16 Multipath conveyors use ball transfers at intersections and horizontal gravity roller conveyors for accumulators. Take-away is by powered rollers. Quick modifications to the number and combination of stations permit multiple products to be built simultaneously; it is a "cart system" but with conveyors.

rejects to move them to the rework station or the scrap pile.

Equipment costs. Carrier cost/unit usually is less when there are multiple units/carrier. The cost/carrier is higher, but a smaller number of carriers are required. For example, 1 carrier holding 8 units might cost $200 while 1 carrier holding 1 unit might cost $40.

Automatic processing of units is more difficult with multiple items/carrier. One possibility is duplicate heads on one machine so that all units on the carrier are done simultaneously; another possibility is to do the units in sequence by indexing either the carrier or the head. However, unless this station is the bottleneck station, the extra heads or indexing equipment may be just extra capital expense since one head may have had sufficient capacity. If a robot is used, it may have to be a more expensive robot.

2.13 Buffers Off-Line.   On-line buffers can handle minor disturbances. Major shocks (machine breakdowns, employee absenteeism, learners, etc.) and line balancing problems may benefit from off-line buffers. Off-line can be remote either in time or in space.

Additional time to compensate for a disturbance can be obtained by overtime work, partial shifts, and working holidays on the existing equipment. For example, assume the buffer size between stations was 1 h. Then, for any shock less than 1 h of production (e.g., short machine breakdown, small difference in production between a learner and experienced worker), fill the buffer up again by having the worker work extra time at the end of the shift.

Additional space for an off-line buffer can be done several ways. Figure 5.17 shows an off-line buffer, a semi-circular conveyor. In this situation, the buffer acts as a "standby circuit." A standby circuit analysis considers: the device (in this case, the conveyor with units on it), the device that detects circuit failure (the operator), the switch that activates the standby system (the operator), and the reliability of the standby device (how good are the units on the semicircular conveyor). Naturally the off-line buffer

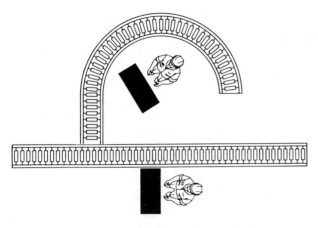

FIGURE 5.17 Off-line banks (semicircular portion) can function as a standby component in the production system.

does not have to be conveyorized. It simply may be items stored on a cart or pallet, either close to the workstation or perhaps several hundred feet away.

If a machine must be fed at a constant rate, it is more efficient for an operator to load a magazine at his own speed and the machine to work from that magazine than for a machine to feed the machine directly. This applies even if the machine cycle time from the magazine is no faster than the mean operator feed time (Corlett, 1982). See also Figure 5.12.

Off-line buffers can include processing as well as storage. See Figure 5.18. For example, assume station B was only able to produce 15 units/h when the line rate was 20 units/h. Then a duplicate of station B could be built (5 or 500 ft from the line) and operated the number of hours necessary to obtain the extra units. This technique is most useful when the capital cost of an additional workstation is relatively low. The extra workstation can be used for training purposes, or to use idle time of a worker on another job with a large machine-time component, or to use idle time of a worker whose primary job does not require 8 h/day.

So far we have decoupled by moving the product. There is another alternative (often overlooked) — move the operator.

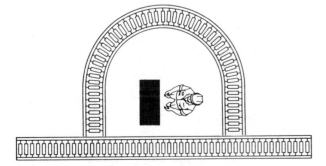

FIGURE 5.18 Standby systems can include processing as well as storage. The main advantages are better utilization of people and equipment and easier training. Disadvantages are increased material handling and scheduling problems.

2.2 Decoupling by Moving Operators

There are four subcategories: (1) utility operator, (2) help your neighbor, (3) n operators float among n stations, and (4) n operators float among more than n stations.

2.21 Utility Operator.

Figure 5.19 shows the utility operator approach. In this concept, most of the operators work at specific workstations. However, one operator is a **utility** or relief **operator**. The utility operator's assignment is to help each of the individual operators if they have trouble for a few minutes, if they want to go to the toilet, etc. In the U. S. auto industry, for example, a common situation is 1 relief worker for every 6 stations; that is, 7 people work at 6 stations with 1 always being off; the line does not stop for breaks. (Chrysler dropped the "tag relief" system to save money; it stopped the line at breaks and so reduced daily output.) In other industries the utility operator does not have formal times assigned to relieve specific operators but just helps out when and where needed.

The duties of this operator vary widely; additional work is needed for when there is no line worker taking a break. In these situations, the utility operator often is called a group leader (working supervisor); responsibilities include training new employees and making minor decisions (if the supervisor is absent) in addition to working at the various stations. Minor maintenance work or product rework are other duties.

2.22 Help Your Neighbor.

Figure 5.20 shows the help-your-neighbor approach. Each operator, by management directive, is "your brother's keeper." That is, you help your neighbor, not because you are a good person, but because it is part of your job responsibility. If management does not formally require operators to help each other out, then those helping may feel that they are "suckers." If a fellow worker gets behind, this is not a reason for everyone else to take a break until the person catches up. One approach is to divide the work at a station into thirds; the middle third is the sole responsibility of the station operator, the first third is entitled to help from the upstream operator and the last third is entitled to help from the downstream operator.

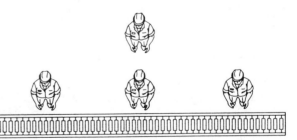

FIGURE 5.19 Utility (relief) operators move from station to station while the station operators all stay at their own station. The utility operator must know all the jobs and so often becomes the trainer. In many cases the utility operator is given minor management responsibility and is called a group leader or working supervisor.

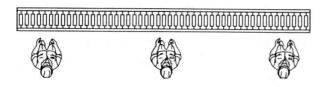

FIGURE 5.20 Help your neighbor means shared responsibility. The line shows that responsibility extends both directions from the operator.

2.23 n *Operators Float Among* n *Workstations.*

Figure 5.21 shows the *n* operators float among *n* workstations approach. Stations are not given specific times/unit; the operators are just told the total time for the entire unit. For example, instead of saying that each of the 5 stations had 1.0 min of work/unit, the operators would be told there is a total of 5 min of work. It is the operator's decision as to which operator does what job at what rate; the operators move upstream and downstream as they desire; the operators decide when to switch jobs. This type of line management tends to be efficient since buffer problems are minimized (if a buffer = 0, the operator moves to another station), group pressure to produce is high, the line runs at the average speed of the group (rather than the speed of the slowest member), everyone tends to be able to do every job (so absences or illness are less of a problem), and, since the line need not be balanced, there is no balance-delay time.

2.24 n *Operators Float Among More than* n *Stations.*

Figures 5.22 and 5.23 show two examples of the *n* operator float over more than *n* stations approach. The rationale of "more places than people" is the minimization of the total cost of the system. In the United States, labor costs tend to be higher than capital costs. Labor wages of only $5/h become labor costs of $6/h after fringe benefits are considered. At 2,000 h/yr, this means minimum labor cost is $12,000/yr. Many operators have an annual cost of over $25,000. If a workstation costs $5,000, lasts 3 years, and has a scrap value of $2,000, then annual cost is (5,000 − 2,000)/3 = $1,000/yr. Thus it is far more important to keep the $12,000/yr operator busy than to keep the $1,000/yr workstation busy.

Figure 5.22 shows the "**paired station**" approach. Two stations are built (if they are identical, it is called "double

tooling"); the operator goes back and forth between the two. This approach is quite useful when the operator has a large idle time due to machine time. This also makes the line more reliable, because if one machine fails, the other still can work; although output would not be up to full potential, it does not fall to zero.

A second variation (Figure 5.23) is the "one-worker line." The work required for a product may require more space for components, tools, and equipment than can be located conveniently at one workstation. It is not necessary to split the total task among several operators. Break the job down into, say, three stations. The operator works at station A, completing a number of units, say 25. Then the 25 units *and the operator* move to station B, completing the required operations at station B. Then the 25 units *and the operator* move to station C. Since there is only one operator, there is no balance-delay time and the stations need not have equal cycle times (i.e., the line need not be balanced). Buffer sizes are not critical.

Another version is a single-product line with, say, 12 workstations and 4 operators. See Figure 5.24. The workers work at a specific station and send products to the next station. Then they walk to another station (not necessarily adjacent) and begin work again. Different assignment policies can be used. For example, some operators can work at all 12 stations, and others (such as beginners) might be restricted to just a few.

Another version is multiple lines, each set up for a single product (say line A for product A, line B for product B, line C for product C). Then the workers work on, say, line A on Monday, line B on Tuesday and Wednesday, line C on Thursday morning, and then line A for the rest of the week. A furnace manufacturer had 9 lines for assembly of 171 models. Teams of 2 roved from one line to another, doing complete assembly of a given model. Meanwhile the setup operator converted a line for the next batch of another model. Each team specialized in a range of models so they didn't have to know all 171 models. If there was a

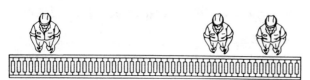

FIGURE 5.21 *n* operators among *n* stations is the full-float design. Management assigns operators to the line as a whole instead of to individual stations. Operators move from station to station as the need arises. Thus, the stations need not be balanced.

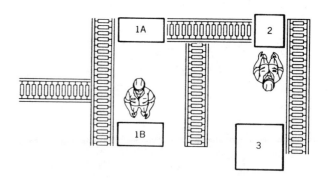

FIGURE 5.22 Paired station line for semiautomatic machines is a version of *n* operators among more than *n* stations. The two stations for an operator can be identical (see 1A and 1B) or different (see 2 and 3). The machine can either run automatically while the operator loads/unloads the other station, or can load/unload automatically while the operator processes. Thus operator utilization is high while machine utilization is low. Naturally there need not be a conveyor since material handling could be by cart or hand.

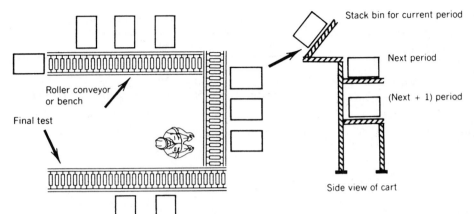

FIGURE 5.23 One-worker line uses U shape to cut worker walking. Since final test was done upon completion of each unit, the worker got quick feedback on quality, so quality problems were reduced drastically (Gargano and Stewart, 1975). The wheeled carts hold a complete period's (day or week, depending on unit) supply of parts. The next two periods' supply is on the second and third shelf; this helps keep an inventory count and pinpoints supply problems.

shortage of components for any line, they simply shifted over to another line until the supply was sufficient. Production scheduling was very flexible. This flexibility is very useful when there is insufficient demand for just one product for an entire year. Although equipment is duplicated, setup and put-away costs are minimized; maintenance can be done during normal working hours instead of at premium pay hours.

The "more places than people" concept is very useful for short runs and operators who are learning.

Letting the worker walk requires walking time, which must be added to the work time. The MTM predetermined time system allocates 5.3 TMU/foot (.2 s). Thus walking 10 ft requires 2 s. If 10 ft were required for every 20 units, then add 2/20 = 0.1 s/unit. From a physiological and comfort viewpoint, standing with occasional walking reduces venous pooling in the legs and is less fatiguing than standing without walking.

3 BUFFER SIZE

Buffer size is a tradeoff between increased capital cost of the buffer and the reduced operating cost of the line. The

number of items in the buffer should vary from zero. For example, if a buffer quantity fluctuated between limits of 10 and 20 units, then 10 units probably are excess because the buffer should fluctuate between 0 and 10. However, buffers have a psychological component; people feel safer if buffers are full (storage fills the space available) so the amount in the buffer tends to be too high. Nevertheless, sometimes the buffer should be empty; it might be empty rarely but it is very expensive to be empty *never*.

The capital cost of the buffer includes two major items: (1) the cost of the floor space and equipment for the buffer and (2) the inventory cost of keeping the items. The reduced operating costs are difficult to pinpoint since many of them are opportunity costs which don't show up in a cost accounting system. Because of a buffer, you might have a faster pace line, better quality products, greater scheduling flexibility, etc. The actual savings are difficult to determine because no one wants to build and operate an inefficient line without buffers just to prove how much money could be saved by installing buffers. (There are some special "lines" without buffers—rotary index tables; see Chapter 20.)

Some Japanese firms have emphasized minimizing work in process with their famous Kanban (Just-In-Time) system. On some dedicated automatic lines, the various stations are turned on and off to keep the buffer size from exceeding a limit (the full work system). Since there is no labor cost and the equipment capital cost is a sunk cost, they try to minimize work in process (and also floor space, which tends to be very expensive in Japan). For example, assume machine A (which produces 90 units/min) and B (which produces 60/min) are separated by a buffer with desired maximum number of 6. When there are 6 units present, machine A shuts down ("rests") for a couple of minutes. Then it runs again until the buffer is back up to 6 (Monden, 1981).

One reason for buffers is machine breakdown. Required buffer size can be decreased by (1) using more reliable equipment (longer mean time between failures, MTBF) or (2) having quicker repairs (shorter mean time to

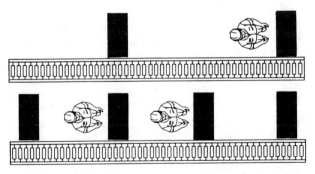

FIGURE 5.24 "More places than people" is a different way of saying *n* operators among more than *n* stations. The rationale is to keep the operator busy and not worry about machine utilization. Note that if the conveyor is eliminated from the figure, *n* operators among more than *n* stations is similar to a job shop. Flow lines are concepts; material handling and layouts follow many patterns.

repair, MTR). One way to reduce MTR is to have standby equipment. That is, failure rate and repair time can be changed by design and purchasing decisions; they are not constants.

It has been pointed out by almost everyone who has studied flow lines that short lines are better than long lines. This, in effect, is a recommendation for large buffers.

There has been some theoretical analysis of optimum buffer size. Analytical models of buffers tend to use unrealistic assumptions (such as Poisson arrival and exponential service times) because of "computational tractability" although they do not represent actual lines. Boothroyd et al. (1982), in Chapter 7 of *Automatic Assembly*, give elegant and understandable equations for optimum buffer size. Realistic solutions to real problems probably will be done with simulation rather than analytically. See Box 5.2.

4 BALANCING FLOW LINES

4.1 Standard Balancing Technique
The first question is whether you wish to balance the line. As was pointed out in the buffer design section, it is feasible, and often desirable, to use an unbalanced line.

The line balance problem has three givens: (1) a table of work elements with their associated times (see Table 5.1), (2) a **precedence diagram** showing the element precedence relationships (see Figure 5.26), and (3) required

TABLE 5.1 Elements and work times for line balancing problem. Although each element time is assumed constant, in practice each element time is a distribution.

Element	Work Time (h/unit)
1	.0333
2	.0167
3	.0117
4	.0167
5	.0250
6	.0166
7	.0200
8	.0067
9	.0333
10	.0017
	.1817

units/minute from the line. "To find" includes (1) the number of stations, (2) the number of workers at each station, and (3) the elements to be done at each station. The criterion is to minimize total idle time.

First, what is the total number to be made and in how long a time? Thus, 20,000 units could be made in 1,000 h at the rate of 20/h, 500 h at 40/h, 250 h at 80/h, or in many other combinations. Continuous production is only one of the many alternatives; Hansen and Taylor (1982), for example, discuss the conditions when a periodic shutdown is best. Assume we wish to make the 20,000 units in 1,000 h

---

BOX 5.2 Simulation

Once a flow line (or any other network of interconnected machines and buffers) is designed, one alternative is to build it, install it, and let it operate. If the design is good, then the system will be reasonably satisfactory—perhaps after a debugging period.

Fortunately, it is possible to explore environments not only by doing but perceptually. Perceptual exploration of the manufacturing environment has been greatly simplified by computer simulation.

There now are a great variety of manufacturing simulation packages available, both for mainframes and for PCs. The cost of both the programs and the computers required has dropped dramatically. The programs now available are quite user friendly. The user specifies (often from a menu)

1. nodes (i.e., machines, workstations, material handling devices)
2. the network connecting the various nodes
3. the maximum queue size before and after each node
4. the mean and distribution of arrivals to the workstation (constant, normal, log-normal distribution, etc.)
5. the service times at each node
6. the service priority at each node (usually, but not always, first come, first served)
7. the batch size to be processed

The construction of the model and its data should be put into an assumption document, which may require weeks or even months of effort. The assumption document ensures that the model is realistic; it also may highlight problems even before the model is used on the computer. The assumption document should be signed by all those involved.

The user then can play "what if," changing service times, arrival rates, batch sizes, the network, service rules, and so forth. Each set of alternatives is called a *scenario*. Since the simulation responses are often probabilistic, a user would typically run a particular scenario many times (e.g., 30) so that the output can give an estimate of best-case, worst-case, and average.

In some cases, the model is run for some time (e.g., 20 simulation shifts) and these startup data are not used; only data after steady state is achieved are used.

The program outputs typically are

- means and distributions of the queues—allowing the user to determine the mean, maximum, and minimum number of items in each queue
- total time it takes to process a batch of work
- output rates of each node
- utilization rate at each node

Typically, the output is printed in tables and histograms, but some programs have emphasized on-screen animation, using 2D or 3D machines of various colors and shapes along with dynamic changes in the flow and processing of parts. See Figure 5.25.

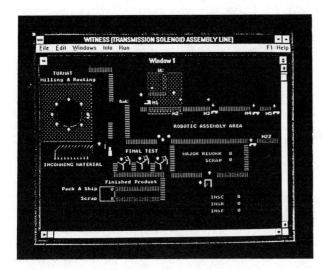

FIGURE 5.25 Status of queues can be shown graphically (i.e., animation) as simulation time changes. Color coding can indicate working or idle elements. Figure courtesy of AT&T Istel.

at the rate of 20/h. Since we are dealing with a balanced line, each station will take 1,000 h/20,000 units = .05 h/unit; cycle time = .05 h.

Second, guess an approximate number of stations by dividing total work time by cycle time: .1817 h/.05 h/station = 3.63 stations. Then use 4 stations with one operator at each.

Third, make a trial solution as in Table 5.2 and Figure 5.27. Identify each station with a cross-hatched area. Remember not to violate precedence; for example, elements 1 and 5 cannot be done at one station and element 2 at another. Then calculate idle percentage; .0183/ (4 × .05) = 9.15% in our case.

As an example of flexible thinking, consider Table 5.3. Here station 1 and 2 are combined into one superstation; the elemental time now totals .0950. Since there are 2 operators, the time available is 1.000 and the idle time is .0050 at the station and .0025 for each of the operators. So far there

TABLE 5.2 Trial solution for assembly-line balance problem with cycle time of .0500 h.

Station	No. of Operators	Element	Element Time (h)	Work Time (h)	Idle Time (h)	Cumulative Idle Time (h)
1	1	1	.0333			
		2	.0167	.0500	0	0
2	1	3	.0167			
		4	.0167			
		6	.0166	.0450	.0050	.0050
3	1	8	.0067			
		9	.0333	.0400	.0100	.0150
4	1	5	.0250			
		7	.0200			
		10	.0017	.0467	.0033	.0183

is no improvement over the solution of Table 5.2. However, note that there now is *idle time at each station*. Therefore, the amount can be reduced at all stations until one has zero idle time. Thus, the line cycle time can be reduced to .0475.

Table 5.4 shows that the new idle time is .0083 h. Expressed in percentage terms, the idle time now is .0083/(4 × .0475) = 4.4% instead of the 9.1% of Table 5.2.

With larger lines the problem complexity grows rapidly; grouping the elements into zones (which either prevent or require certain elements to be done at the same station) is one attempt at simplification. Another source of complexity is the changes in product volume; that is, 20/h in May, 24/h in June, 26/h in July, etc. Another complication is that actual lines often are multiple-product lines. First we assemble a 4-door Pontiac station wagon with a V6 engine, then a 2-door Oldsmobile hardtop with a V6 engine, then a Pontiac 4-door sedan with a V6 engine and air conditioning, etc. In addition, there are mix changes (i.e., in March we produce 70% with air conditioning and in April 80%).

The problem complexity and the need for repeated solutions have led to efforts to use computer programs. However, Ghosh and Gagnon (1989), in their extensive survey of the literature, report that very few firms (about 5%) actually used computer programs for line balancing. Part of the problem is that user-friendly programs are not available. Another problem is that computers act on input as immutable facts.

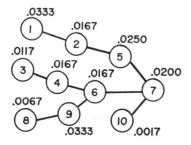

FIGURE 5.26 Precedence diagrams show the sequence required for assembly. The lines between the circles are not drawn to scale; that is, elements 4 and 9 both must be completed before element 6, but 9 could be done before or after 4. Precedence must be observed; thus elements 3, 4, and 9 could not be assigned to one station and elements 8 and 6 to another. However, 8, 9, and 10 could be done at one station.

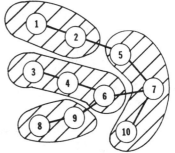

FIGURE 5.27 Solution to Table 5.2 is shown graphically.

4.2 Modifications to Standard Technique If the computer solution gives some idle time at each station, then the cycle time can be decreased until the time at one station (the bottleneck station) is zero.

Mariotti (1970) discusses some useful modifications.

First, consider element sharing. That is, operators/station need not equal 1.0. One possibility is *more* than one operator/station; some examples are 2 operators/station (operators/station = 2), 3 operators/2 stations (operators/ station = 1.5), and 4 operators/3 stations (operators/station = 1.33). This permits a cycle time that is less than an element time. For example, with 2 operators/station, each would do every other unit. (Often, combining two single stations with 1 operator each into one "super" station with 2 operators dramatically improves the balance.) See Box 5.3.

You also can have *less* than 1 operator/station by having operators walk between stations or having work done off-line (i.e., use buffers). Also it is possible in some situations to have operators from 2 adjacent stations share one or two elements. Station D does element 16 and 17 on one-half or one-third of the units and Station E does element 16 and 17 on the remaining units. Elements shared do not have to be at adjacent stations if precedence requirements are not violated.

Second, remember that cycle times are not fixed. At the start, we assumed a cycle time of .05 h (i.e., the line runs 1,000 h or 1,000/8 = 125 days). It may be more efficient to have a cycle time of .048 h (i.e., the line runs 20,000 × .048 = 960 h = 960/8 = 120 days). In addition to this balance cost, consider setup cost and inventory carrying cost. That is, what is the best combination of balance costs, setup costs, and inventory costs? The computer programs can do a quick check on many different alternatives.

Third, remember that elements often can be redefined. One possibility is to take former elements 16, 17, and 18 and eliminate element 17 by splitting it between elements 16 and 18. Or elements 16 and 17 might be combined and then split into 16a and 17a so that although the total time is

TABLE 5.3 Trial solution for assembly-line balance problem with two operators at one station and cycle time of .0500 h.

Station	No. of Operators	Element	Element Time (h)	Work Time (h)	Idle Time (h)	Cumulative Idle Time (h)
1	2	1	.0333			
		2	.0167			
		3	.0167			
		4	.0167			
		6	.0166	.0950	.0050	.0050
2	1	8	.0067			
		9	.0333	.0400	.0100	.0150
3	1	5	.0250			
		7	.0200			
		10	.0017	.0467	.0033	.0183

TABLE 5.4 Trial solution for assembly-line balance problem with two operators at one station and cycle time of .0475 h.

Station	No. of Operators	Element	Element Time (h)	Work Time (h)	Idle Time (h)	Cumulative Idle Time (h)
1	2	1	.0333			
		2	.0167			
		3	.0167			
		4	.0167			
		6	.0166	.0950	.0000	.0000
2	1	8	.0067			
		9	.0333	.0400	.0075	.0075
3	1	5	.0250			
		7	.0200			
		10	.0017	.0467	.0008	.0083

the same the relative allocation to 16 and 17 changes. Still another possibility is to split an element even further. For example, element 9 might be "Pick up screwdriver, drive 20 screws, release screwdriver" with a time of .0167 h. For balancing purposes it may be desirable to have element 9a be "Pick up screwdriver, drive 15 screws, release screwdriver" with a time of .0125 h, and element 9b be "Pick up screwdriver, drive 5 screws, release screwdriver" with a time of .0075 h. You have added extra work to the task (an extra pickup and release of the screwdriver) but may be able to cut the time of the bottleneck station and thus the time. Be careful about this technique since, when conditions change, the reason for the extra work may be forgotten and the unnecessary work retained without reason.

Fourth, interchange elements from the assembly task and the subassembly tasks. For example, a nameplate might be added at a subassembly station instead of the assembly station. An adjustment might be done on the assembly instead of the subassembly.

5 OPERATING FLOW LINES

A key consideration is whether you want to have a balanced line—that is, a line at which the work content of all stations is equal. You may not for a number of reasons.

For example, you may wish to have an easy or "loose" station for less capable operators—the boss's nephew, new workers, old workers, handicapped workers.

In theory all workers on a line are interchangeable parts. In practice, put your best operators on the toughest jobs; they can cope with problems an ordinary operator couldn't handle without slowing down the line.

Although an unbalanced line normally has less balance delay time than a balanced line (Ghosh and Gagnon, 1989), the bowl concept says that if there is unbalance, put it in the middle of the line. This can be achieved by putting either your best operators or the lowest amount of work/station there.

BOX 5.3   *Superstations*

As discussed in Section 2, designers tend to think of matching individual workers with individual stations. Decoupling can be achieved by changing the product flow or by moving the operators. However, there is another possibility—a team approach. The total line is broken into superstations and a team is assigned to the superstation rather than individuals being assigned to specific workstations.

Some reasons for a superstation are

- skill or equipment reasons (For example, all electrical work would be kept together, all welding kept together.)
- complete job concept (For example, all work on an auto dashboard may be done at one superstation, because a dashboard is a complete unit.)
- strict job classification system

Each team member would be cross-trained on all jobs within the superstation. Some advantages of superstations,

based on Toyota's experience (Monden, 1983), are

- easier to change production rate (i.e., rebalance the line)
- workers not as tired (presumably cumulative trauma is reduced by rotation because local muscle groups get a rest)
- enhanced job knowledge and responsibility of team members (encourages suggestions and changes)

A disadvantage is that balance delay will increase (Johnson, 1991). Since superstations are an additional constraint on freedom of assigning elements to workstations, it is more difficult to reduce balance delay.

If superstations and teams are used, it may be possible to split the superstation into two locations.

Thus, assembly-line design is really a multiple-criteria problem (balance delay, station cost, scheduling flexibility, buffer design, etc.), not just a problem of minimizing balance delay.

DESIGN CHECKLIST: FLOW LINES

Basic questions
    Is a flow line appropriate?
    Operation only, assembly, or order-picking line?
    Is it an unbalanced or balanced line?
    Single product continuously, multiple products sequentially, or multiple products simultaneously?
    Cost of storing one item 1 h?
    Reliability (MTBF) and maintainability (MTR) for each station?
    Capital cost/station/year? per item/year?
Material handling
    Manual or mechanized?
    Fixed or alternate paths?
    Scrap/rework disposition?

Buffer design
    Moving product or moving people?
    Items/carrier?
    On-line versus off-line?
    More places than people?
Pace
    Machine- or operator-paced?
    Balance delay percent?
    Loss due to slowest cycle of slowest operator?
    Percent of items not completed?
Balancing
    Is balancing wanted?
    Element sharing?
    Modify cycle time?

REVIEW QUESTIONS

1. Make a schematic sketch of the three types of flow lines.
2. Sketch three allocations of products to flow lines.
3. Discuss *muda* and *mura*.
4. Discuss rocks in the river and JIT.
5. Discuss the six strategies to implement JIT (lean production).
6. What is balance-delay time?
7. Discuss how the slowest operator slows down a balanced line.
8. What is station starvation? Station blocking?
9. Sketch a schematic flow line with buffers as trucks and a flow line without buffers as a train.
10. Assuming the arrival rate is the same, why is close spacing and a slow belt speed preferable to wide spacing and a higher belt speed?
11. Illustrate with numbers and sketch how a buffer can be used on an unbalanced line.
12. Give the advantages and disadvantages of multiple items/carrier.
13. How does multiple items/carrier affect labor costs?
14. Describe with sketches and numbers several ways of using an off-line buffer.
15. Why would pallets be put on tracks at an output buffer?
16. Describe the utility operator concept.

17. Describe the *n* operators over *n* workstations concept.
18. What is the rationale of "more places than people"?
19. Sketch and explain the "double-tooling workstation" concept.
20. Describe (including a sketch) a "one-worker line."

21. Discuss, using numbers, the cost of buffer storage.
22. Discuss how element sharing can improve line balance. Use numbers and a specific example including a precedence diagram.
23. Should you put your best operators at the end of the line or the middle? Why?

PROBLEMS AND PROJECTS

5.1 Visit an assembly line. Do they make single products continuously, multiple products sequentially in batches, or multiple products simultaneously? For a specific product, how much storage is there before and after each station? Which techniques are used for decoupling? What is the balance delay for the line? Any ideas? Does the supervisor agree with you?

5.2 Visit an assembly line. How many items/carrier do they use? Any ideas? Does the supervisor agree with you?

5.3 Balance the line given in Table 5.1 and Figure 5.26 for (a) 2,000 units/yr, (b) 20,000 units/yr, and (c) 200,000 units/yr. Give assumptions.

5.4 Read 3 to 5 articles on a topic related to this chapter. Summarize them in a 500-word executive summary. List the references. Don't use articles cited in the text.

REFERENCES

Boothroyd, S., Poli, C., and Murch, D. *Automatic Assembly*. New York: Marcel Dekker, 1982.

Buxey, G. Incompletion costs vs labor efficiency on the fixed item moving belt flowline. *Int. J. Production Research*, Vol. 16, 3, 233–247, 1978.

Buxey, G., Slack, N., and Wild, R. Production flow line system design—a review. *AIIE Transactions*, Vol. 5, 1, 37–48, March 1973.

Corlett, N. Design of handtools, machines, and workplaces. *Handbook of Industrial Engineering*, G. Salvendy, Ed. New York: Wiley, 1982.

Gargano, H., and Stewart, F. Material handling system is key to efficient assembly operation. *Material Handling Engineering*, Vol. 30, 4, 53–55, April 1975.

Ghosh, S., and Gagnon, R. A comprehensive literature review and analysis of the design, balancing and scheduling of assembly systems. *Int. J. Production Research*, Vol. 27, 4, 637–670, 1989.

Hansen, D. and Taylor, S. Optimal production strategies for identical production lines with minimum operable production rates. *IIE Transactions*, Vol. 14, 4, 288–295, Dec. 1982.

Johnson, R. Balancing assembly lines for teams and work groups. *Int. J. Production Research*, Vol. 29, 6, 1205–1214, 1991.

Kilbridge, M. Non-productive work as a factor in the economic division of labor. *Journal of Industrial Engineering*, Vol. 12, 3, 155–161, 1961.

Kilbridge, M. and Webster, L. The balance delay problem. *Management Science*, Vol. 2, 1, 69–84, 1961.

Mariotti, J. Four approaches to manual assembly line balancing. *Industrial Engineering*, Vol. 2, 6, 35–40, June 1970.

Monden, Y. Adaptable Kanban system helps Toyota maintain just-in-time production. *Industrial Engineering*, Vol. 13, 5, 29–46, May 1981.

Monden, Y. *Toyota Production System*. Norcross, Ga.: Industrial Engineering and Management Press, 1983.

6 LOCATION OF ITEMS: MATH MODELS

CHAPTER PREVIEW

There has been much interest in the analytical solution of locating items. Models are divided into locating one item in an existing network of customers and location of multiple items in a network of customers. A number of computer programs have been written.

CHAPTER CONTENTS

1 Location of One Item
2 Systematic Layout of Multiple Items
3 Computerized Layout of Multiple Items
4 Five Specific Programs for Multiple Items
5 Overall Evaluation of Computer-Aided Layout

KEY CONCEPTS

activity relationship diagram
construction programs
improvement programs
location of one item

relationship chart
satisfiers versus optimizers
systematic layout procedure (SLP)

1 LOCATION OF ONE ITEM

1.1 Problem

Location of one item in a network of customers is a fairly common problem; see Table 6.1. The item can be a person, a machine, or even a building. The network of customers can be people, machines, or buildings. The criterion minimized can be distance moved by people or product, amount of energy lost, or time to reach a customer.

Typically, a user is interested in finding the location that minimizes the weighted distances moved—a minimum problem (e.g., locating a copy machine in an office). Another possible objective is to minimize the maximum distance—a minimax or worst case problem (e.g., locating an ambulance so that everyone could be reached in no more than 15 minutes).

There is extensive analytical literature on planar single-facility location problems; see Chapter 4 in Francis et al. (1992) for an excellent discussion of analytical techniques.

The following material, however, is not elegant math but "brute force" calculations. With computers, and even hand calculators, the engineer can solve all reasonable alternatives in a short time, perhaps 20 minutes. Then the engineer uses the material handling cost calculations to gain insight into the location problem and, using this as one criterion (other criteria might be capital cost, maintenance cost, etc.), makes a recommended solution.

1.2 Solution

In the following example (Konz, 1970), consider (1) the item to be located is a machine tool, (2) the network of customers (circles in Figure 6.1) is the other machine tools with which the item will exchange product, and (3) the criterion to be minimized is the distance moved by the product.

In most real problems, there are only a few possible locations: the remaining space is already filled with other machines, building columns, aisles, and so forth. In the example problem, the first solution to the problem will

TABLE 6.1 Examples of locating a new item in an existing network of customers with various criteria to be minimized.

New Item	Network of Customers	Criterion Minimized
Machine tool	Machine shop	Movement of product
Tool crib	Machine shop	Walking of operators
Time clock	Factory	Walking of operators
Inspection bench	Factory	Movement of product or inspectors
Copy machine	Office	Movement of secretaries
Warehouse or store	Market	Distribution cost
Factory	Warehouses	Distribution cost
Electric substation	Motors	Power loss
Civil defense siren	City	Distance to population
IIE meeting place	Locations of IIE members	Distance traveled
Fire station	City	Time to fire

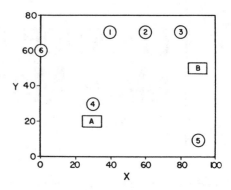

FIGURE 6.1 Should the customers (in circles) be served from location A or B? Table 6.2 gives the data.

assume that there are only two feasible locations (A and B) for the new item; they are rectangles in Figure 6.1.

Travel between a customer and A or B can be (1) straight line (e.g., conveyors), (2) rectangular (e.g., fork trucks down an aisle), or (3) measured on a map (e.g., fork trucks using one-way aisles, conveyors following aisles or connecting several machines, conveyors following nondirect paths). In real problems, travel may be a mixture of the three types.

Some customers are more important than others. Thus, the distance must be weighted. In a factory, a common index would be pallets moved/month. If the problem is location of a fire station, the weight of a customer might depend on the fire risk or the number of people occupying the site.

The operating cost of locating the new item at a specific feasible location is:

$$MVCOST = WTK\,(DIST) \qquad (1)$$

$$DIST = \sum_{k=1}^{N} (|X_{i,j} - X_k| + |Y_{i,j} - Y_k|)$$

(for rectangular)

$$DIST = \sum_{k=1}^{N} \sqrt{(X_{i,j} - X_k)^2 + (Y_{i,j} - Y_k)^2}$$

(for straight line)

TABLE 6.2 Customers 1 to 6 (of Figure 6.1) can be served either from location A ($X = 30$, $Y = 20$) or from location B ($X = 90$, $Y = 50$). Which location is better?

Customer	Coordinate X	Y	Weight or Importance	Movement Type
1	40	70	156	Straight line
2	60	70	179	Straight line
3	80	70	143	Straight line
4	30	30	296	Rectangular
5	90	10	94	Rectangular
6	0	60	225	Rectangular

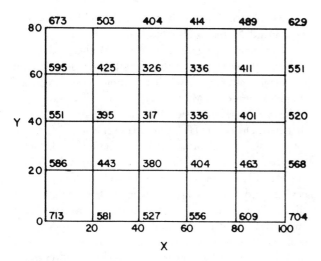

FIGURE 6.2 Evaluate your options by calculating cost at a grid of locations.

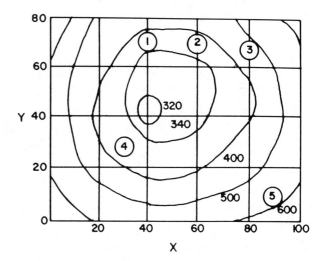

FIGURE 6.3 Contour lines are easier to understand than the numbers of Figure 6.2.

where

$MVCOST$ = index of movement cost for a feasible location

WTK = weight (importance) of the Kth customer of N customers

$DIST$ = distance moved

For the two locations given in Table 6.2, Table 6.3 shows the $MVCOST$. Movement cost at B is about 67,954/53,581 = 126% of A.

Assume you wish to know the cost at locations other than A and B for the assumption of the above problem. By calculating costs at a number of points, a contour map can be drawn; see Figures 6.2 and 6.3. This indicates that the best location is $X = 42$ and $Y = 40$ with a value of 32,000. Thus Site A is 6,000 from the minimum and Site B is 13,000 from the minimum. It may be possible to move whatever is at the minimum site to make way for the new item. Then, however, the capital cost of the move and the $MVCOST$ for the displaced item and its customers need to be considered.

The example, however, made the gross simplification that movement cost per unit distance is constant. Figure 6.4 shows the more realistic assumption when most of the

cost is loading and unloading (starting and stopping) or paperwork; the cost of moving, when "acceleration and deceleration" are omitted, is very low. More realistically,

$$MVCOST = L_k + C_k(|X_{i,j} - X_j| + |Y_{i,j} - Y_k|)$$
(for rectangular)

$$MVCOST = L_k + C_k \sqrt{(X_{i,j} - X_k)^2 + (Y_{i,j} - Y_k)^2}$$
(for straight line)

where

L_k = load + unload cost (including paperwork) per trip between the Kth customer and the feasible location

C_k = cost/unit distance (excluding L_k)

Assume for customers 1, 2, and 3 that L_k = \$.50/trip and C_k = \$.001/m; for customers 4, 5, and 6, L_k = \$1./trip and C_k = \$.002/m. Then, the load/unload + transport cost for alternative A = \$854 + 79.07 = \$933.07 while the cost of B = \$854 + 117.89 = \$971.89. Thus, B has a movement cost of 104% of A. Note that the product $(WTK) (DIST)$ (that is, the \$854) is independent of the feasible location; it just adds a constant value to each alternative.

Cost need not be expressed in terms of money. Consider locating a fire station where the customers are

TABLE 6.3 Cost of locating a new machine at locations A and B. Since WTK was pallets/month and $DIST$ was in meters, $MVCOST$ = M-pallets/month.

Customer	Weight (Pallets/Month)	Site A		Site B	
		Distance (Meters)	Cost (M-Pallets/Month)	Distance (Meters)	Cost (M-Pallets/Month)
1	156	51	7,956	54	8,424
2	179	58	10,382	36	6,444
3	143	71	10,153	22	3,146
4	296	10	2,960	80	23,680
5	94	70	6,580	40	3,760
6	225	70	15,750	100	22,500
			53,781		67,954

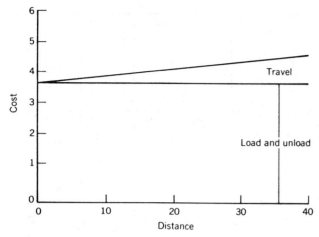

FIGURE 6.4 Material handling cost is almost independent of distance.

parts of the town and the weights are expected number of trips in a 10-year period. Then load might be 1 min to respond to a call, travel is 1.5 min/km, and unload might be 1 min; the criterion is to minimize mean time/call.

Note also that the distance cost may rise by a power of 2—the inverse square law—for such problems as location of a siren or a light.

It is a good design procedure to do a sensitivity analysis. Vary WTK and C_k and see how it affects $MVCOST$.

When making the decision of where to locate the new item, use not only the movement cost but also installation cost, capital cost, and maintenance cost. That is, "transportation cost" need not be the sole criterion. In locating the optimal restaurant for a meeting, consider food quality and price also; in locating a fire station, consider political obligations; in locating a drinking fountain, consider plumbing costs; in locating a tool crib, consider security.

2 SYSTEMATIC LAYOUT OF MULTIPLE ITEMS

In contrast to the previous section on location of one item in a network of customers, this section discusses arrangement of the entire facility.

Systematic Layout Procedure (SLP) was developed by Richard Muther and is based on extensive consulting work in plant layout (Muther, 1973; Muther and Hales, 1980). SLP can be used at the block (department) level or the detail (machine) level. The following is a concise, simplified version of Muther's approach.

STEP 1. The **relationship chart** (see Table 6.4) is the first step. Divide the facility into convenient activity areas (office, lathes, drill press, etc.). For more than 15 areas, analyze in two sections (e.g., layout of assembly departments and layout of component departments). For "closeness desired between areas," assign a letter grade: A = absolutely necessary, B = important, C = average, D = unimportant, and E = not desirable to be close. (Muther's technique uses 6 levels instead of 5.) Letters are used rather than numbers, since numbers imply more precision to the judgment than is available. Avoid too many A relationships. About 10% As, 15% Bs, 25% Cs, and 50% Ds is a good goal. Moore (1980b) confirms Muther's comment that there is a better chance of being able to satisfy high-priority relationships in the relationship diagram if the proportion of Ds (unimportant) is 50% or more. That is, you can't have everything important. Support A, B, and E relationships with a "reason for closeness." Reasons (see Table 6.5) will depend upon the problem, but common reasons are 1 = product movement, 2 = supervisory closeness, 3 = personnel movement, 4 = tool or equipment movement, 5 = noise and vibration. The reasons can be broadly divided into product flow, service, and other. The product flow relationships come from a from-to diagram (see Table 6.9 for an example). The service and other relationships come from discussion with the people working in the areas.

In addition, you may have design concepts such as

- truck docks on rear perimeter of the building
- executives on top floor of multi-floor building
- windows (view) for those with status
- cafeteria and toilets centrally located
- U-shaped flow (if shipping and receiving one department or adjacent)
- storage kept together
- utilities in a spine (lower piping costs)

TABLE 6.4 Step 1 of systematic layout is identify the desired closeness between areas with letter grades. Gives reasons for letters A, B, and E. Thus B/2 for drill press-lathes indicates a B importance for closeness for reason 2.

Area Number	Area Name	Office 1	Lathes 2	Drill Press 3	Punch Press 4	Plating 5	Shipping 6	Die Storage 7
1	Office	–						
2	Lathes	D	–					
3	Drill press	D	B/2	–				
4	Punch press	E/5	D	B/2	–			
5	Plating	D	C	D	D	–		
6	Shipping	C	D	D	C	B/2	–	
7	Die storage	D	D	D	A/4	D	B/4	–

TABLE 6.5   Typical reasons for closeness in relationship diagram.

1. Flow of material
2. Need personal contact
3. Use common equipment
4. Use common records
5. Share same personnel
6. Avoid noise and vibration
7. Supervise same personnel
8. Contact frequently
9. Service urgently
10. Minimize utility distribution cost
11. Use same utility
12. Flow of paperwork
13. Desired by management

STEP 2.   Assign floor space to each activity area, along with physical features and restrictions (see Table 6.6). Remember Moore's corollary to Parkinson's Law: "Inventories expand into whatever space is available, regardless of the need to maintain the inventory" (Moore, 1980a, p. 82). If the layout is a group of machines within an area, add space to the space for the machine alone. Consider space for the operator, for maintenance access, for movement of parts of the machine, and for local storage of product and supplies. See equation 8 in Chapter 3.

STEP 3.   Make an **activity relationship diagram** (diagram to group activities) (see Figure 6.5). First, list all the A relationships from the relationship chart, then the Bs, Cs, Ds, and Es. Then make a diagram with just the As. Then add the Bs, keeping in mind the E restrictions. Then add the Cs, then the Ds.

STEP 4.   Make a scaled layout of at least two trials from Step 3, using the areas and restrictions of Step 2 (see Figure 6.6). First, use pieces of stiff paper for each department; sketches tend to get "set in concrete" too soon. An alternative is a CAD system in which areas can be rotated and moved. Both areas and shapes of the departments can be adjusted. Some areas may be fixed in a specific location, for example, the shipping dock or the punch press department.

The reason for at least two layouts is that engineers are **satisfiers** rather than **optimizers**—they tend to stop

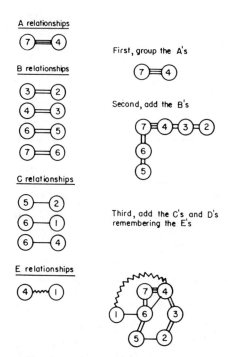

FIGURE 6.5   Step 3, the activity relationship diagram, first groups the As, Bs, Cs, and Es (left side of figure). (It may help to think of the lines as rubber bands pulling areas together. More bands make them closer together. The wavy line for the Es is a "spring" keeping them apart.) Then group all the As (right side of figure), add the Bs to the As (remembering the Es), and then add the remainder.

designing as soon as they have a solution that works. (When the decision maker is satisfied that it is not worth further effort to find something better, it is called *satisficing.*)

Note that E relationships do not have to be satisfied with distance. Walls and other barriers permit physical closeness while preventing the passage of noise, fumes, distractions, or other factors.

Note also that A relationships do not have to be satisfied with closeness. For example, if the reason for the A is communication, the communication medium may be telephones, computer lines, faxes, videos, or even pneumatic tubes, and a distance of 50 ft or 500 ft is irrelevant. If the A

TABLE 6.6   Step 2 of simplified SLP is to specify the amount of space (including about 20–30% for aisles) for each area. Give physical features and restrictions.

Area Number	Name	Desired m²	Restrictions
1	Office	50	Air conditioning
2	Lathes	40	Minimum of 10 m long
3	Drill press	40	
4	Punch press	50	Foundation
5	Plating	30	Water supply, fumes, wastes
6	Shipping	20	Outside wall
7	Die storage	50	Crane
		280	

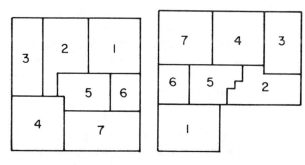

FIGURE 6.6   Step 4, scaled layout, makes several alternative arrangements. Slight modifications of some areas or of the total plant area from Step 2 may be desired (in order to keep the building shape regular). Section 4 in Chapter 9 shows that the more square the building shape, the better.

TABLE 6.7 Step 5 of simplified Systematic Layout Planning is to evaluate the alternatives. Criteria and weights depend upon specific management goals.

Criterion	Weight	Present	Alternative 1	Alternative 2
Minimum investment	6	A 24	B 18	A 24
Ease of supervision	10	C 20	C 20	B 20
Ease of operation	8	C 16	C 16	C 16
Ease of expansion and contraction	2	C 4	C 4	C 6
Total points		64	58	66
Relative merit		97%	88%	100%

is for product movement on a conveyor, again a distance of 100 to 500 ft is relatively irrelevant.

STEP 5. Evaluate the alternatives (see Table 6.7). The relevant criteria and their weights will change from situation to situation. Grade each layout (A = excellent = 4; B = good = 3; C = average = 2; D = fair = 1; and E = bad = 0) and calculate the layout's grade point (grade × weight). If there is an existing layout, include it as one alternative. Defining the best as 100%, calculate the percentage for the alternatives. Have the affected people sign off on the evaluation form. Then go back and select features from the alternatives to get an improved set of designs.

STEP 6. Detail the layout (make a working drawing, using a 1:50 scale or possibly a 1:100 scale). At this step, the department boundary lines are replaced with exterior walls (with doors), interior walls (with doors), and no walls (just department boundaries). Refine the estimate of the number of machines and operators through detailed analysis of run and setup time, production schedules, consideration of alternative staffing patterns, use of simulation, and so forth. Determine material handling and aisles and locate machines and operators. Detail utilities (electricity, water, compressed air, and communications) and service areas (offices, toilets, breakrooms, toolrooms, nurse's office, etc.).

Finally, present the alternatives to management for approval. (For techniques to improve acceptance, see Chapters 1 and 7.) After modifications, install the final design.

3 COMPUTERIZED LAYOUT OF MULTIPLE ITEMS

The problem addressed is the complete layout of an entire facility. The programs attempt to solve the same problem that was solved manually by the Systematic Layout Procedure—optimally locate the departments of a job shop or machines within a department (process layout). The five specific programs discussed below all require:

1. Departments (activities); see Table 6.8.
2. Areas of departments; see Table 6.8.
3. Relationships among departments; either from-to chart (see Table 6.9), or relationship chart (see Table 6.10).

There are two basic types of program. **Improvement programs** start with a feasible solution (possibly the present layout). The two improvement examples discussed are CRAFT and COFAD. The second type is **construction programs** (also known as "green field" or "no walls" programs since they assume starting with an empty field—a blank sheet of paper). The three construction examples discussed are CORELAP, ALDEP, and PLANET.

4 FIVE SPECIFIC PROGRAMS FOR MULTIPLE ITEMS

4.1 CRAFT Computerized Relative Allocation of Facilities (CRAFT), an improvement program, is the most used and written-about program. The goal is to minimize transportation cost. Transportation cost = (from-to matrix) (move cost matrix) (distance matrix). The program calculates the distance matrix as the rectangular distances from the department centroids.

Input required is an initial layout, a from-to matrix, and a move cost matrix, which contains the cost of handling a unit load/unit distance for each departmental pair. For example, it might cost $.02/ft to move between Department A and B, $.03/ft between A and C, etc. If these costs are not available (and they generally are not), then all move costs can be set as 1.0 and the criterion minimized is load-ft instead of dollars. Dummy departments (which have zero flow and which are fixed to a specific area) are used to (1) represent fixed facilities such as stairways, elevators, toilets, and docks; (2) represent aisles; and (3) fill building irregularities (i.e., get rectangular buildings).

Assumptions include (1) there are no "negative" relationships (departments you want to keep apart), (2) all flows start and stop at department centroids, and (3) all movement is rectangular. The cost/unit distance matrix

TABLE 6.8 Departments (functions) and area are required input in all programs.

Department Code	Function	Area (ft²)
A	Receiving	12,000
B	Milling	8,000
C	Press	6,000
D	Screw machine	12,000
E	Assemble	8,000
F	Plating	12,000
G	Shipping	12,000
		70,000

TABLE 6.9 From-to charts are required input for CRAFT, COFAD, and PLANET where material handling cost is the sole criterion. The numbers might represent weekly trips between departments.

	A	B	C	D	E	F	G
				TO			
A	–	45	15	25	10	5	
B		–		30	25	15	
C			–		5	10	
D		20		–	35		
E					–	65	35
F		5			25	–	65
G							–

assumes material handling equipment is selected before the layout, move costs are known, move costs are independent of equipment utilization, and move costs are linear with distance. (The underlying assumptions of lift-truck handling costs may not be relevant with conveyors or computer-controlled guide path equipment.)

The calculation procedure followed by the computer is (1) calculate distance matrix (from the layout) between department centroids; (2) calculate load-ft for the layout by multiplying the distance matrix by the from-to matrix; (3) try to reduce the load-ft by interchanging departments that have equal areas or common borders. If no improvement is found, the program stops. If improvement is found, the improved layout becomes the new reference layout of step 1 and the cycle repeats.

For best results, the program should be run multiple times using various initial layouts and various from-to values; that is, do a sensitivity analysis.

4.2 COFAD

COmputerized FAcilities Design (COFAD), an improvement routine, is an improved version of CRAFT; it allows more realistic calculation of material handling costs. The goal is minimum material handling cost. As with CRAFT, input is an initial layout and a from-to matrix. The move cost matrix is replaced by move cost equations for alternative fixed-path equipment (conveyors, cranes, hoists) and alternative variable-path equipment (vehicles). For example, two alternative variable-path choices (for truck, platform truck) might be entered.

$$FPMC = FPFC + \mathbf{FPVC} \qquad (2)$$
where

$$
\begin{aligned}
FPMC &= \text{fixed-path move cost, \$} \\
FPFC &= \text{fixed-path fixed cost, \$} \\
FPVC &= \text{fixed-path variable cost, \$/ft} \\
VPMC &= VPFC\,(EUT) \qquad (3) \\
&\quad + VPVC\,(MTIME)
\end{aligned}
$$
where

$$
\begin{aligned}
VPMC &= \text{variable-path move cost, \$} \\
VPFC &= \text{variable-path fixed cost, \$} \\
EUT &= \text{equipment utilization for this move, \%} \\
VPVC &= \text{variable-path variable cost, \$/h} \\
&= ANTC/(OPHY) \\
ANTC &= \text{annual total variable cost (labor, power, maintenance), \$/yr} \\
OPHY &= \text{operating hours per year, h/yr} \\
MTIME &= \text{move time (load/unload, travel time) h/yr} \\
&= \text{some function of } ML \text{ specified by the user} \\
ML &= \text{move length, ft}
\end{aligned}
$$

ML is calculated by the program; the remaining values are required input.

The calculation procedure is (1) improve initial layout (using CRAFT-like procedure); (2) determine material handling cost, assuming full equipment utilization; (3) try to improve equipment utilization; (4) if no improvement, modify layout and go to step 2. Stop when layout and handling system no longer can be improved.

As with CRAFT, best results are obtained from multiple runs using varying initial layouts, from-to values, and move costs.

4.3 CORELAP

COmputerized RElationship LAyout Planning (CORELAP), a construction routine, is a computerized version of Muther's Systematic Layout Planning. The goal is a layout with high-ranking departments close together. That is, the criterion is not the single criterion of minimum material handling cost, but the optimization of the multiple criteria of the relationship chart. Input is the department areas and the relationship chart (see Tables 6.8 and 6.10). The most common problem cited by users of computerized layout routines is lack of data, so the approximations used in the relationship diagram may be more appropriate than the exact cost approach of COFAD.

TABLE 6.10 Relationship charts are required input for CORELAP and ALDEP. The letter indicates the importance of closeness and the number the reason for the letter grade. Muther uses A = absolutely necessary, E = essential, I = important, O = ordinary, U = unimportant, and X = keep apart. Numbers indicate the criterion (e.g., 1 = material handling cost, 2 = noise, 3 = ease of supervision, etc.).

	Receiving	*Milling*	*Press*	*Sheet Metal*	*Assembly*	*Plating*	*Shipping*
Receiving	–						
Milling	E/1	–					
Press	O	U	–				
Sheet Metal	I/2	E/1	U	–			
Assembly	O	I/1	U	I/1	–		
Plating	U	U/1	O	U	A/3	–	
Shipping	U	U	U	U	I/1	I/1	–

CORELAP allows preassignment of departments to outside walls and corners, thus eliminating such absurdities as shipping docks in the plant center. The closeness relationships are translated into numbers (A = 243, E = 81, I = 27, O = 9, U = 3, and X = −729). Thus CORELAP permits undesirable closeness but assigns it a dominating value of = −729.

The calculation procedure is (1) place the department with the highest total closeness rating (sum of ratings with all other departments) in the layout center; (2) add departments with the highest relationship values first, then lower values. Do this so that total closeness rating is maximized. Continue until all departments are added.

4.4 ALDEP

Automated Layout DEsign Program (ALDEP), a construction routine, has the same input data requirements as CORELAP (with one exception) but has a philosophical difference. The ALDEP concept is that designers get in a rut, so what is needed are new alternatives from which a human can get ideas and make a final layout. Therefore, ALDEP enters departments into the layout randomly. Since department entry is random, a rerun of the program will give different answers than the first run. It does make a preliminary evaluation based on the closeness letters to aid the human. Departments (docks, elevators, aisles) can be fixed. All departments are square or rectangular so odd shape layouts are minimized.

The calculation procedure is (1) randomly select a department and place it in the upper left-hand corner of the layout; (2) add a department with an important relationship with the first department. If there is no such department, add an unimportant department. Continue until all departments are added.

4.5 PLANET

Plant Layout ANalysis Evaluation Technique (PLANET), a construction routine, gives the designer some choices. It requires the usual input of departments and areas but allows the closeness to be determined by (a) a from-to matrix (with penalty values from −9 to 99); (b) a from-to matrix with move cost/month; or (c) data that will allow the calculation of the move costs/month (parts list, monthly frequency or moves/part), sequence of departments for each part, and move cost/100 ft). No matter which of the three the designer chooses, PLANET translates it into a normalized from-to matrix. The program attempts to minimize the material handling cost, the product of (from-to matrix) (distance moved matrix). All moves are rectangular from the department centroid. Move cost is assumed to be linear with move length and independent of equipment utilization. The input requires user selection of priority of department entry. The first two departments from this list are placed in the layout center. The next departments are added considering material handling cost and department priorities until all departments are added.

4.6 Reducing the Problems in Computer-Aided Layout

Five common problems are (1) lack of confidence, (2) unrealistic department location, (3) unrealistic shape, (4) unrealistic plant shapes, and (5) unrealistic department alignment. Table 6.11 gives solutions to the problems.

Improve your confidence in the answers by running the programs multiple times. Vary the initial layout used as input. For PLANET, vary the placement priority. For CORELAP, vary the width and length of the department. For ALDEP, vary sweep width and degree of closeness. Make slight modifications of the from-to (or relationship) chart.

Unrealistic locations are not much of a problem in CRAFT, COFAD, and ALDEP because locations can be fixed. If a department shows up in an undesirable location, fix its location for the next run. CORELAP doesn't permit assignment to specific outside walls or corners. With PLANET, the lowest placement priority enters the layout last and thus is on the outside, so change the department placement priority if necessary.

Shapes might be constrained by machine type. For example, automatic screw machines might be 5 × 18 ft, so a long skinny area might be appropriate. Thus, a 10 × 10 square won't work even though total area is sufficient. For CRAFT and COFAD, take the generated layout, reshape the department as desired, and run it again with the department fixed. With ALDEP, you can either fix the area or vary sweep width. With CORELAP and PLANET, the best approach is to use the computer input as a first draft. Unrealistic plant shape (i.e., not rectangular) is not a problem in CRAFT, COFAD, or ALDEP, since they permit specifying plant shape. With CORELAP, vary the plant length-to-width ratios or fix protruding areas into a fixed location for the next run. With PLANET, just grin and bear it and remember its goal is to furnish ideas, not a final layout.

Unrealistic alignments usually are aisle placement problems. CRAFT and COFAD allow use of dummy departments (with zero flow), which can be considered aisles. ALDEP directly allows for input of aisles. With CORELAP and PLANET, the routines can't handle aisles so you must adjust the layout manually.

5 OVERALL EVALUATION OF COMPUTER-AIDED LAYOUT

As shown in Table 1.1, layout is just part of facility design. The designer also needs to consider material handling, communications, utilities, and the building.

5.1 Use

"A series of studies in the 1970s by Moore (1978) found that few industrial planners were using layout algorithms. Further, most of those who did found the results to be of limited value. . . . Contrary to popular belief, algorithms cannot produce a demonstrably best or optimum layout. At best they can provide a good solution" (Hales, 1984, p. 110).

5.2 Problems
Problems are discussed in terms of problem frequency, data base, program design and availability, and scoring of alternatives.

5.21 Problem Frequency.
In a job shop layout, the problem of arranging the entire factory or department does not often occur. Most of the layout planner's work consists of minor changes in the existing layout, handling, communications, utilities, or building. Even if a large department is to be completely rearranged, it may call for a flow line or cell arrangement, not a job shop. Therefore, planners are understandably not very interested in computer programs that might be used once in 20 years.

5.22 Data Base.
Computer algorithms call for information such as pallets/week of product A from location 1 to location 2, the transport cost/meter for lift trucks, and the like. Although the planner should have this information, it rarely is conveniently arranged in a data base. Manual approaches allow the planner to muddle on through and the analytical calculations behind the layout will not be missed.

5.23 Program Design and Availability.
Hendy (1989) points out that the conventional programs place all information concerning interacting elements in the link magnitude and that this one-dimensional approach is inadequate.

The classic programs were written in the 1960s and do not have many of the user-friendly features found in more recent programs. The classic programs generally are written for mainframes. CRAFT, COFAD, CORELAP, and ALDEP are available from SHARE Program Library, Triangle University Computation Center, Research Triangle Park, North Carolina. Khator and Moodie (1983) published PC versions of CRAFT and COFAD; MICRO-CRAFT is available from IIE, 25 Technology Park, Norcross, GA 30092; (404)449-0460.

5.24 Scoring of Alternatives.
Another reason for the lack of success is the problem of scoring. How should

TABLE 6.11 Techniques for reducing problems in computer-aided job shop layouts. (Tompkins, 1978) *Modern Material Handling*, Copyright 1978 by Cahners Publishing Company, Division of Reed Holdings, Inc.

Nature of Problem	CORELAP	ALDEP	PLANET	CRAFT	COFAD
Lack of confidence in the solution	Can test by varying the length-to-width ratios of departments and making a number of reruns of routine	Can vary sweep width and degree of closeness and make additional runs of routine	Can vary placement priorities and make several reruns of the routine	Can vary the initial layout; at least three significantly different initial layouts should be input to computer	Same as CRAFT
Unrealistic department locations	Can specify that a department be fixed to a specific outside wall or corner and rerun routine	Can move unrealistically located department to realistic location and rerun the routine	Can use placement priorities to entry sequence to layout to change location in next run of routine	Same as ALDEP	Same as ALDEP and CRAFT
Unrealistic department shapes	Although not a flexible approach, can vary length-to-width ratios and rerun routines	Can vary sweep width or fix department to area that is realistically shaped and rerun routine	Cannot correct with routine; must massage computer-generated layout manually	Can reshape department and rerun routine using layout thus modified as an initial layout	Same as CRAFT
Unrealistic plant shapes	Can vary plant length-to-width ratios or vary the filling ratio and rerun routine	Problem should not arise; you can specify plant shape at the start	Cannot correct with routine; must massage computer-generated layout manually	Same as ALDEP	Same as ALDEP and CRAFT
Unrealistic department alignment	Cannot correct with routine, must massage computer-generated layout by hand	Problem should not arise; routine has provisions for specifying aisles in original input	Same as CORELAP	Can insert departments in initial layout, align real departments in computer-generated layout, put dummies between them, and rerun routine	Same as CRAFT

good layouts and bad layouts be differentiated? In contrast to the multiple-criteria approach of step 5 of SLP, the classic computer programs use a one-criterion approach. But real-world decisions should not be based on a single criterion. For example, the planner thinks the chief executive should have a window with a view, but the computer puts the office in the center of the building to minimize travel distance, or the planner thinks the truck docks should be behind the building for esthetics, but the computer puts the docks on the street side for accessibility.

The programs tend to use material handling distance as the only criterion, which causes a number of problems. First, if a single criterion is to be used, it should be an overall criterion, such as operating cost or capital cost, not just material handling distance. Second, the programs assume that the material handling system is fixed and that the only question is the movement distance. Third, as shown in Figure 6.4, material handling cost is almost independent of distance. Fourth, the programs assume that flow is between department centroids, but, as Figure 6.7 points out, most floor-level handling uses aisles and movement is around department perimeters, not centroid to centroid. Fifth, movement distance from A to B is not necessarily the same as from B to A, due to one-way movement on conveyors, automatically guided vehicles, and so forth. Sixth, the programs tend to ignore service and support areas since they have little material handling.

In my opinion, the classic computer layout programs will remain toys for mathematicians but useless for engineers until multiple-criteria scoring systems are incorporated. The design process should return to the human designer; that is, it should use interactive models. See Malakooti and Tsurushima (1986) for comments on an interactive approach between a decision maker and a multiobjective program. Heragu and Kusiak (1988) describe an expert-system approach to layout; the concept seems promising.

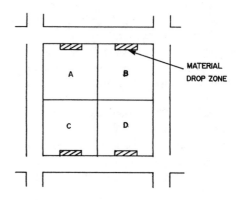

FIGURE 6.7   Movement often uses aisles, so centroid-to-centroid distances are not relevant. Note that if the aisle is one way (e.g., automated guided vehicle following a route) the distance from A to B is not equal to the distance from B to A (Francis et al., 1992).

DESIGN CHECKLIST: LOCATION OF ITEMS: MATH MODELS

Location of one item
  Who are customers
  Importance of each customer
  Travel path of each customer
  Criteria minimized
  All feasible locations considered (including moving some machines)
  Sensitivity analysis of WTK and C_k
Multiple items—systematic layout
  Relationship chart OK by users
  Not too many As and Bs

Floor space, size, shape (split and reshape)
At least two alternatives
Affected people sign off criteria and weights
Keep everyone informed during design as "no surprises" aids implementation
Modified to meet desires
Multiple items—computer layout
  Construction or improvement program
  Program available on your computer
  Multiple alternatives to check assumptions
  Computer layout translated to actual layout

REVIEW QUESTIONS

1. Give three examples of locating an item in a facility with the item, the network of customers, and the criterion minimized.

2. Sketch movement cost versus distance for some specific material handling situation.

3. What is the criterion used for evaluation in the travel chart technique? What are the evaluation criteria in systematic layout planning?

4. In SLP, there are various reasons for closeness, such as product movement, supervisory closeness, etc. In addition, there may be various design concepts. Give the seven listed in the text.

5. Give an example of the first step of the Systematic Layout Procedure.

6. In SLP, relations do not have to be satisfied with physical closeness. Give examples for an A and for an E relationship.

7. In Step 3 of SLP, what are "rubber bands" and "springs"?

8. Why aren't computerized layout programs used in practice?

PROBLEMS AND PROJECTS

6.1 For the data of Table 6.2, evaluate the cost of using location C at $X = 60$, $Y = 50$.

6.2 Make a relationship chart of the IE department at your school. Include offices, classes, labs, toilets, computers, etc. Have the department head sign off on your ratings.

6.3 Using the data of Tables 6.8 and 6.9, use one of the computer programs to obtain some layout alternatives.

6.4 Write a 500-word executive summary of a computerized layout program not reported on in this chapter.

REFERENCES

Francis, R., McGinnis, L., and White, J. *Facility Layout and Location: An Analytical Approach.* Englewood Cliffs, N.J.: Prentice Hall, 1992.

Hales, H. *Computer-Aided Facilities Planning.* New York: Marcel Dekker, 1984.

Hendy, K. A model for human-machine-human interaction in workspace layout problems. *Human Factors*, Vol. 31, 5, 593–610, 1989.

Heragu, S., and Kusiak, A. Knowledge based system for machine layout. *Proceedings 1988 International Industrial Engineering Conference.* Atlanta, GA: Institute of Industrial Engineers, 159–164, 1988.

Khator, S., and Moodie, C. A microcomputer program to assist in plant layout. *Industrial Engineering*, Vol. 15, 3, 20–23, March 1983.

Konz, S. Where does one more machine go? *Industrial Engineering*, Vol. 2, 5, 18–21, 1970.

Malakooti, B., and Tsurushima, A. Some experiments with computer aided facility layout selection. *Proceedings of the 1986 International Industrial Engineering Conference.* Atlanta, GA: Institute of Industrial Engineers, 124–129, 1986.

Moore, J. Computer methods in facilities layout. *Industrial Engineering*, Vol. 12, 9, 82–93, Sept. 1980.

Muther, R. *Systematic Layout Planning (SLP)*, 2nd ed. Boston: Cahners Books, 1973.

Muther, R., and Hales, L. *Systematic Planning of Industrial Facilities (SPIF)*, Vol. I and II. Kansas City, Mo.: Management and Industrial Research Publications, 1980.

Tompkins, J. How to massage the computer output (5). *Modern Material Handling*, Vol. 33, 9, 102–107, Sept. 1978.

7 | PRESENTATION OF LAYOUTS

CHAPTER PREVIEW

A proposed facility design has three components: a data base, a written proposal, and an oral presentation. The data base is divided into two sections (graphics and nongraphics). Recommendations are given for improving written proposals and oral presentations.

CHAPTER CONTENTS

1 Data Base
2 Written Proposal
3 Oral Presentation

KEY CONCEPTS

briefings
computer-aided design (CAD)
data base

layers (overlays)
milestones
plot plan

A proposed facility design normally has three components: a data base, a written report, and an oral presentation.

1 DATA BASE

The **data base** will be discussed in two sections: the graphics data base and the nongraphics data base. The most advanced CAD systems have a single data base for both kinds of data. When the data base is not in a single source, the data should be well integrated (which may be difficult).

1.1 Graphics Data Base

Graphics data bases will be divided into two dimensional (2D) and three dimensional (3D).

1.11 Two Dimensional.

Layout drawings are a subdivision of **computer-aided design (CAD)**. Two popular commercial programs are AutoCAD and CADKEY. The user searches a series of menus, using a mouse to select features. Common features include drawing lines (straight or curved, various widths, solid or dashed), cross-hatching (different patterns), precision input (locating an item at a specific *X, Y* coordinate), dimensioning (determining distances between locations, calculating areas if desired), and use of standard symbols. Among the useful editing features are the abilities to pattern (create an image once and then duplicate it), to rotate an image (e.g., when there are 10 identical desks), to move the image (e.g., move a lathe from one part of the shop to another), to mirror a pattern (e.g., left- or right-opening door), to delete a feature, to scale a feature up or down in size (zoom), to add text (various sizes and fonts), and to layer.

Layering is an important CAD characteristic (Malde and Bafna, 1986). Programs can create over 100 layers, which can be printed superimposed on each other. Layers can show the layout at various levels (building only, machines located, electrical grid, water grid, etc.). Each department can also be placed on a separate layer, permitting flexibility in printing. For example, alternate versions of each department can be printed on different layers, allowing four different versions of the office to be shown by printing out layers 4, 5, 6, and 7. When changes are made to those four layers, the overall layout can be shown by picking the desired layers.

Zooming allows the user to change the scale on the screen (showing the entire plant, the shipping dock, or just the area around a single machine). If you want to know the scale, you must include a dimension or use a grid scale when setting up the program (see Figure 7.1).

Other available programs interact with AutoCAD. They can have preprogrammed standard American National Standard Institute (ANSI) symbols for air, water, and steam; standard machines and office furniture; and standard architectural features such as doors, windows, and bathroom details.

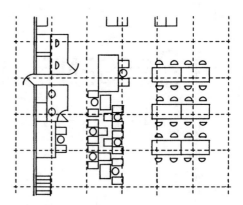

FIGURE 7.1   A grid scale, with a known grid such as 10 ft × 10 ft, is drawn on one layer. Then, when any other layer is printed, the grid scale is superimposed on it (Malde and Bafna, 1986). The grid lines may be 10 mm apart for a view of the entire plant and 100 mm apart for a view of a specific machine.

When desired, a printed copy can be made through either a local printer or (using a modem) a remote printer. Complex printers (called *plotters*) can print multiple colors (usually 4 maximum).

Until about 1985, most layouts were presented on polyester (Mylar) sheets with printed grid lines every 1/4 inch. For larger layouts, a popular scale was 1/8 in = 1 ft, but for most layouts the scale was 1/4 in = 1 ft. Now, however, most layouts are made on computers, and the engineer has a choice of scales. See Figure 7.1. Printers and plotters have six paper sizes: A (8.5 × 11 inch), B (11 × 17), C (17 × 22), D (22 × 34), E (34 × 44), and F (44 × 68). Small printers associated with PCs may only have capability for A and B sizes.

The first drawing should be the **plot plan**, which shows the building on the site as well as features such as driveways, sidewalks, parking lots, security fences, and utilities (water, sewage, fire hydrants, gas, electric, telephone). Since a building typically takes up only a small percentage of the site, the plot plan drawing usually has a different scale than the building drawing.

The next drawing is of the empty building. Give information such as exterior walls, interior free (non-loadbearing) walls, load-bearing walls, windows, doors, stairwells, elevator towers, docks, washrooms (including fixtures), auditorium and cafeteria walls, columns, column centerlines and numbering, title block with organization logo, key, and north arrow. Identify departments and areas with words or abbreviations rather than numbers on a list; show department boundaries with dashed lines so they are not mistaken for interior walls. Differentiate outside and inside walls. Indicate ceiling heights and permissible floor loadings. A pattern of light grid lines is useful.

Next you should have **layers** (also called *overlays*) for utilities. It is convenient to have one layer for each function—electrical, plumbing, HVAC (heating, ventilating, and air conditioning), conveyors, and pits and trenches—but it is common to have several functions on one overlay.

Main lines should be shown thicker than branch lines. For example, 2-inch water lines are shown thicker than ½-inch lines. Distinguish 440 V electrical lines from 220 and 110; distinguish storm sewers from sanitary sewers. The overlay output gives the empty building plus the utility of interest.

A good approach is to have multiple layers showing flow, starting at receiving and ending at shipping. For example, use 10 layers for the top 10 products and 1 layer for the total of all products. When you print it, super-impose the desired flow combination onto either a block layout (building layout showing departments but not machines) or a detailed layout (layout showing machines).

Finally, you will have a layer of the building with equipment (but without the detail of utilities). Show the operator (a 24" × 36" rectangle) for each machine. Scales smaller than 1:50 (i.e., 1" = 50") tend to be too small for equipment layouts.

1.12 Three Dimensional.
As of 1994, most layouts are presented as two-dimensional plan views. However, some programs can make isometric (3D) views. Since they are less abstract, they are easier to understand, especially for nontechnical users, such as management and sales. The continuing decline in computer costs should increase the utilization of these programs.

The 3D programs aid in interference detection. The most complex of these programs allow the viewer's eye to be at any *X, Y, Z* coordinate. This permits the user to walk through the layout and see it as it would appear. These renderings are popular with architectural presentations because they allow the client to see far views, close-ups of details, views when entering the space, views with different amounts and locations of lighting fixtures, solids modeling, color fill and shading, and so forth.

For some designs, it may be useful to construct a 3D mock-up of a workstation. Typically, mock-ups are constructed of cardboard (i.e., non–load bearing), but occasionally they are wood. Elaborate mock-ups with functional controls and displays are called *simulators*.

1.2 Nongeometric Data
Relational data bases should interface with the graphics output; they store data about specific machines (such as manufacturer, size, height, power requirements, part numbers made on the machine, group technology classifications, output rate, etc.). The data bases then can be passed to engineering analysis programs, such as duct sizing, electrical load analysis, queuing simulators, and so forth, for engineering and economic calculations. For example, structural routines do calculations (load analysis, code checking, etc.) on structural members. HVAC routines do calculations on ducts, dampers, fittings, fans, and the like. Piping routines can determine types of pipes, valves, and connections as well as do calculations. Electrical routines calculate cables, conduit, loads, and so forth. Available routines also can interface graphics programs with bills of material and group technology data.

2 WRITTEN PROPOSAL

Eventually the project will come to the official decision-making stage, in which a formal written proposal is needed. The goal is acceptance of the proposal's recommendation and implementation of the recommendation. Thus, the proposal needs to convey convincing information. Information itself is not sufficient; the proposal must convince decision makers to accept the recommendation. Their decision will convert a recommendation into action—a deed.

Before writing the report, you need a plan—an outline. You need to answer the following: What is the purpose of the report? Who will read it? What do they not know that the report will tell them?

Some decision makers tend to avoid risk; others emphasize economic gain. You need to know your decision maker. Generally, decision makers are interested primarily in the economic aspects of the proposal rather than the technical aspects. Since they probably will accept the proposal as technically sound, avoid technical overkill. Remember that the decision maker will review many proposals and not all will be accepted. Why should yours be accepted?

Decision makers are busy; therefore the report should present the proposal concisely. The proposal may also be in depth, but decision makers are unlikely to go through the details. (They hired you to do the detail work.) They will want details to be available, should they choose to inspect them. A short report, with details in appendices, may be the solution.

The following is a good general format with a logical structure obvious to both the reader and the writer. On the cover sheet, put the project title, the date, your name, and the word *RECOMMENDATION*, followed by 50–100 words giving the recommendations. At the top of the second page, put the word *PROBLEM*, followed by a 10- to 50-word statement of the problem. Next put the word *ANALYSIS*, followed by the analysis. This section may take 3–10 pages. (Reduce the material by putting the detail in appendices. The appendices should be preceded by a table of appendices, and each appendix should be identified by a letter, with pages numbered in each appendix; for example, B14 identifies page 14 in appendix B.) Finally, put the word *CONCLUSIONS* and give them briefly. Some decision makers also like an *EXECUTIVE SUMMARY*, which summarizes the entire project, including recommendations, in about 400 words.

Determine the desired writing style and use it. Some people want an informal, active voice style. ("We measured background noise levels" or even "I measured background noise levels.") Others want a formal, passive style. ("Background noise levels were measured.")

Proposals should include the following economic information: annual cost of the present and proposed methods, capital cost of the proposed method (e.g., new equipment, installation costs, training costs), estimated project life, and expected savings. On simple proposals, an annual return on investment is sufficient. However, expensive, long-range projects often require a cash flow analysis. State your assumptions specifically (product life = 3 more years at 8,000 units/yr, no change in product design, the change will not affect sales of spare parts, quality is not affected). Good proposals also have a schedule of **milestones** (1 May—proposal accepted; 15 May—change schedule OK by facility manager; 15 June—all machines moved; 20 June—production back to normal).

Proposals should go through several drafts, so be sure to schedule enough time for multiple drafts. Consider handwritten material as notes rather than as a draft. Too many errors of spelling, composition, and structure lurk in handwritten material to call it a draft. Plan to type the report at least twice for a small project and several times for an important project. The report must not have any spelling, grammatical, or typographical errors if it is to have a reasonable chance of acceptance. Insist that the word processing operator use spelling checking programs. In general, decision makers cannot evaluate the technical merits of proposals because they are not experts. They can, however, judge typing, spelling, and grammar, and if these are poor, they will consider the technical material to be poor also.

Many word processing programs now have grammar subroutines. They check the use of active versus passive voice, agreement of subject and verb, trite sayings, and redundant words; they may also suggest alternative adjectives and so forth. Unless you are a skilled writer, you should use a grammar program routinely—whether you take the program's advice or not.

Konz (1990) provides eight guidelines for good table design and discusses features of good and poor graphs.

3 ORAL PRESENTATION

Oral presentations are usually accompanied by visual aids. Although many speakers pay more attention to their clothes than to the quality of their visual aids, audiences focus on and remember the visual aids.

The first choice to make concerning visual aids is the medium. Transparencies or slides are used most often, but occasionally videos are the choice. See Table 7.1 for comments on good transparencies and slides.

Video is effective at showing existing operations. Use a remote control when presenting. For voiced videos, dub the voice after filming rather than while filming. Unless professionally made, videos should be used sparingly because audiences will compare them with other, more professional videos they have seen.

Slides produce a better visual image than do transparencies. Slides also have better options with color, for both photographs and text. Although the slide has the highest visual quality, it also takes the longest to prepare and costs the most. In addition, because slides are more formal, the speaker often takes more care with them. Transparencies are more likely to have poorly organized material presented with inferior images. If you do use transparencies, dark blue text on a clear background produces a good image, with little loss in legibility (Konz et al., 1988). Multicolor transparencies can be made with computer graphics.

An alternative to a transparency is the flip chart, a large paper pad to use with a marker. Advantages are that flip charts can be used in any room (e.g., a restaurant meeting room) and that the series of sheets are like frames that can be prepared in advance or during the presentation. Two

TABLE 7.1 Effective overhead and slide presentations. An important characteristic of slides and transparencies is that they force you to organize your presentation.

Overheads

Stay within $7\frac{1}{2} \times 9\frac{1}{2}$ format, since projector platen is not $8\frac{1}{2} \times 11$.

Use color to "outline" the talk as well as for figures and overlays.

Organize your presentation so it has a beginning, middle, and end. Tell what you will tell them, tell them, tell them what you told them.

Use high-contrast originals; be sure the transparency is easily readable from the farthest viewer position; check by projection, not hand-held viewing.

Use, for graphs, grids or graph paper under paper originals to get proper scales and relationships.

Keep projector on your left if you point with your left hand, on your right if you point with your right. Point to the transparency with a pointer or pen, not your finger. Face the audience, not the screen.

Have material prewritten on the film. Handwriting speed is about .4 words/s, speaking is 2 to 4 words/s, and reading is 3 to 9 words/s.

Mount frame borders on transparencies to serve as a notecard—to the audience you appear to be speaking without notes.

Slides

Don't put too much onto one slide. Three guidelines are: 1 slide/min of presentation, 20 words maximum per slide (6–7 words/line; 5 lines; 3 vertical columns), and maximum of 9 double-spaced lines high and 54 elite (45 pica) characters wide.

Use color, not black and white. Make color slides from black and white text by adding a yellow or *light* pastel overlay to the slide. Reverse slides (light letters on dark background) can be made in color from black and white text.

Keep information/slide to one idea; no more than 3 curves/graph.

Make material *easily* readable from the farthest viewing position. Graphs and tables that are satisfactory in print need to be simplified and lines emphasized for slides. Leave space—at least the height of a capital letter—between lines of text.

Use duplicate slides rather than backing up during a presentation.

Practice your talk. Do it *early* so you can make changes in the slides.

disadvantages are the lack of magnification and the difficulty of duplication.

Generally, oral presentations (**briefings**) should be relatively short, 15–30 min. Additional detail can be provided in answers to specific questions or through the written report, but do not assume the people present for the oral presentation have read the written report.

Visual aids and lighting are inadequate in many meeting rooms. If possible, practice your talk ahead of time in the room in which you will make the presentation. Check power cord lengths, light switch locations, projection distances, viewing distances (your material should be legible from the back of the room), sight angles, and so forth. For important presentations, rehearse not only the talk but also the question and answer period (as politicians do before appearing at a news conference).

The presentation itself should have a logical structure that the audience can follow. An effective sequence goes from *what* to *why* to *how*. Describe the implementation plan, and conclude with a summary (e.g., table of cost savings, space requirements, etc.). If you don't know the answer to a question say, "I don't know, but I will find out."

After the presentation, you may wish to summarize the consensus or decision on a transparency, a flip chart, or the blackboard. Anyone who disagrees with your written interpretation can give immediate feedback.

The meeting should include those who can contribute, those with divergent views, those who can make the decisions, and those who will carry out the decisions.

DESIGN CHECKLIST: PRESENTATION OF LAYOUTS

Title block should include:
  Name of project
  Name of designer
  Date
  Scale
  North arrow
Key to special symbols; key in logical order (Don't depend too much on key; labels generally are better.)

Lettering readable from same direction; different sizes of lettering used
Aisles indicated
Areas indicated
Machines labeled
Operators indicated
Product flow indicated

REVIEW QUESTIONS

1. Give two uses of layering in CAD.
2. Give the paper sizes for A and B size plotters.
3. What is the most popular layout scale in the United States?
4. What are the advantages and disadvantages of 3D over 2D models?
5. What should be on a plot plan drawing?
6. What are some 3D mock-up alternatives for a workstation?
7. Give the recommended format for a written report.
8. Write the same information in informal active voice and formal passive voice.
9. Why should reports not have any spelling or grammar errors?
10. Discuss the use of slides versus transparencies for a presentation.
11. Why should material on overhead transparencies stay within 7.5 × 9.5 inches?
12. When using a transparency projector, which hand should you use when pointing to the transparency on the projector? Why?
13. Give the relative speed of handwriting, speaking, and reading.

PROBLEMS AND PROJECTS

7.1 Observe a speaker at a meeting or conference. What did the speaker do to make the presentation effective? What else could have been done?

7.2 Write a 500-word executive summary of an available computer program for layout (architectural) graphics.

REFERENCES

Konz, S. *Work Design: Industrial Ergonomics*, 3rd ed. Scottsdale, Ariz.: Publishing Horizons, 265–271, 1990.

Konz, S., Jackson, R., and Verschelden, M. Legible and attractive transparencies. *Proceedings of the International Ergonomics Association 1988*, London: Taylor and Francis, 1988.

Malde, A., and Bafna, K. Facilities design using a CAD system. *Proceedings of the 1986 International Industrial Engineering Conference*, Atlanta, GA: Institute of Industrial Engineers, 118–123, 1986.

PART

II

BUILDING SHELL AND SPECIALIZED AREAS

8 | FACILITY LOCATION

CHAPTER PREVIEW

Facility location should be planned carefully, since it is expensive to change later. With communication and transportation changes, distance has become less important, so more locations (even in other countries) are feasible. Within the United States, the trend has been to locate in the Sunbelt and away from central cities; rural and suburban locations are popular. The site needs to be satisfactory.

CHAPTER CONTENTS

1 Introduction
2 Specification of Objectives
3 Location Within a Geographic Area
4 Location Within a Region
5 Physical Considerations of Site

KEY CONCEPTS

amenities
back office
commuting radius
demographic reversal
distance effect
expansion strategy

front office
incubator
industrial park
spine
sunbelt effect

1 INTRODUCTION

The following briefly covers a field of great theoretical interest. For the economic geography viewpoint, see Lonsdale and Seyler (1980). For an operations research viewpoint, see Yoon and Hwang (1985a, 1985b) and Love et al. (1988).

Facility location is like selecting a spouse. Although a change may be possible, it may be both expensive and unpleasant. Make a good choice the first time.

2 SPECIFICATION OF OBJECTIVES

The product and quantity must be described. How many workers of each type will be needed? What wastes will be generated? How much space will be required? What utilities will be needed? See Systematic Layout Planning in Chapter 6.

The criteria to evaluate alternatives vary greatly and differ for factories, offices, and warehouses (Hales, 1977). In addition to listing criteria, the relative weight of each criterion should be given.

Weight	Criterion
5	Distribution cost
6	Low labor costs
3	Water supply
2	Material supply cost
2	Adequate labor supply
3	Electrical cost and supply
3	Local taxation
3	Community facilities
4	Transportation adequacy
3	Suitable climate
2	Freedom from regulation
10	Location CEO likes

Note that many location decisions are made for noneconomic reasons, such as locating the plant in the town where the company president was born. Tong (1979) surveyed foreign firms that located factories in the United States. Of the 32 reasons cited for the choice of location, Table 8.1 gives the five that were most important on an overall basis. Ady (1992) said that the main factors in locating a BMW plant in South Carolina were financing, tax incentives, infrastructure, and (especially) the state's intense pre-employment worker training program.

Figure 8.1 shows that wage differentials in different regions of the country have narrowed due to the mobility

TABLE 8.1   Foreign firms said the following factors were the most important when locating their plant in the United States (Tong, 1979).

Rank	Factor
1	Availability of transportation services
2	Labor attitudes
3	Ample space for future expansion
4	Nearness to markets within the United States
5	Availability of suitable plant sites

of labor and capital. There are still substantial differentials, however. See Table 8.2.

The facility location decision uses five considerations: (1) location within a geographic area, (2) location within a region, (3) location within a site, (4) site selection, and (5) physical consideration of the site.

3 LOCATION WITHIN A GEOGRAPHIC AREA

In addition to cost factors, two effects to consider are the decreasing distance effect and the Sunbelt effect.

3.1 Decreasing Distance Effect   The **distance effect** has decreased because of long-term changes in transportation and communications technology.

3.11 Transportation.   The changes in transportation are seen most easily when long time spans are considered. For example, consider the amount of time needed to ship a product from a factory in Kansas City to New York City in 1850, 1900, 1950, and today. Or consider the time from Osaka, Japan, to Kansas City. Even the shipping time for

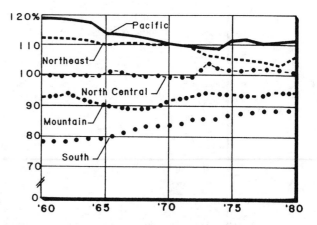

FIGURE 8.1   Regional income/capita has converged since 1960. The Northeast is New England and the Middle Atlantic states; North Central is Ohio west to the Dakotas; South is Maryland south to Florida and west to Texas; Mountain is from Arizona north to Montana; Pacific is California, Oregon, Washington, Alaska, and Hawaii (Malabre, 1982).

TABLE 8.2 Evaluation of 50 cities for business (Huey, 1991). After the top 10, the remaining cities are in alphabetical order.

Rank City	Pop. (Mil.)	Office Lease Rate ($/sq ft/yr)	Tax Rate Max. % Corp. Income	Tax Rate Max. % Per. Income	Av. Annual Salaries ($) Mfg.	Av. Annual Salaries ($) Clerical	Rank Access to Quality Labor	Rank Pro-Business Attitude
1. Atlanta	2.9	24	6	6	27,000	19,900	17	2
2. Dallas	4.0	18	*	none	29,300	19,500	4	13
3. Pittsburgh	2.0	20	12	6	30,000	18,000	4	8
4. Kansas City	1.6	19	6	7	28,100	18,100	1	10
5. Nashville	1.0	22	6	6	24,300	18,300	7	4
6. Salt Lake	1.1	18	5	7	24,300	16,700	2	13
7. Charlotte	1.2	19	7	8	21,900	18,500	14	1
8. Orlando	1.1	21	6	none	25,400	17,300	36	10
9. Austin	0.8	16	*	none	29,500	17,000	24	13
10. Phoenix	2.2	20	9	7	27,700	17,800	7	21
Albany	0.9	17	9	8	27,300	21,500	7	28
Baltimore	2.4	25	7	8	28,800	18,800	24	36
Birmingham	0.9	17	5	6	23,100	16,500	48	34
Boston	3.1	29	10	6	30,700	20,300	4	40
Buffalo	1.0	20	9	8	28,200	16,300	30	48
Chicago	6.1	34	7	3	29,200	19,700	3	30
Cincinnati	1.5	22	11	9	30,000	17,700	18	31
Cleveland	1.8	20	11	9	30,400	18,100	18	13
Columbus	1.4	21	11	9	28,800	17,600	30	50
Dayton	1.0	18	11	9	32,300	17,400	36	8
Denver	1.6	13	5	5	30,800	19,100	39	31
Detroit	4.4	17	4	8	36,800	19,500	45	46
Greensboro	0.9	17	7	8	22,300	18,500	24	13
Hartford	0.8	26	12	5	30,500	19,600	45	50
Houston	3.3	16	*	none	31,300	19,300	14	8
Indianapolis	1.3	20	8	4	29,400	17,300	30	4
Jacksonville	0.9	20	6	none	23,700	18,300	35	13
Los Angeles	9.0	28	9	10	27,500	22,200	24	40
Louisville	1.0	16	10	8	28,500	16,600	39	4
Memphis	1.0	14	6	6	24,000	17,300	49	21
Miami/Ft. L.	3.2	30	6	none	19,200	18,400	39	21
Milwaukee	1.4	23	8	7	28,500	17,900	18	34
Minn./St. Paul	2.5	27	10	9	32,000	18,200	18	45
New Orleans	1.2	15	8	6	27,000	17,900	50	45
New York City	8.6	39	18	12	27,700	22,500	7	42
Norfolk	1.4	20	6	6	24,800	15,900	30	25
Oklahoma City	1.0	17	6	7	25,900	16,300	14	28
Philadelphia	4.9	24	12	8	28,100	18,600	24	46
Portland	1.2	20	7	10	26,600	18,600	18	36
Richmond	.9	21	6	6	30,800	16,900	36	25
Rochester, N.Y.	1.0	19	9	8	32,500	21,500	7	13
Sacramento	1.5	27	9	9	26,700	21,500	7	36
St. Louis	2.5	25	6	7	31,300	17,400	18	21
San Antonio	1.3	13	*	none	19,500	16,600	30	13
San Diego	2.6	23	9	9	28,100	19,400	43	42
San Francisco	5.2	24	9	11	29,900	22,800	24	49
Seattle	2.0	22	none	none	32,400	18,700	7	4
Tampa	2.1	22	5	none	23,000	16,500	43	25
Washington, D.C.	4.0	27	11	10	30,300	20,400	42	44
West Palm Beach	.9	18	6	none	33,100	19,900	45	2

* Texas imposes a franchise tax based on net worth and profits.

relatively short distances (250 miles) within the United States has decreased.

Improvements have been made in ocean shipping (ship size, power, speed of loading and unloading), in land shipping (trucks, being more flexible, have replaced railroads in most applications), and especially in air transportation (people and products). Passenger air travel is on commercial and noncommercial planes (charter or company owned). Air transport of products (both international and domestic) is extensive. Various services (Federal Express, UPS, etc.) deliver small packages anywhere in the United States by 10:30 A.M. if picked up by 5:00 P.M.; they also deliver overseas. Improved shipping greatly reduces the need for local warehouses and permits centralized location of specialty items. It also means offices can be located anywhere.

3.12 Communication. An alternative to transportation is communication. Communication costs have declined even more rapidly than transportation costs. Again, consider the time necessary to communicate between Kansas City and New York City in 1850, 1900, 1950, and today. Communication through the mail has improved greatly with the improved transportation system, but even more impressive are advances in communication over wires—first the telegraph and then the telephone. The phone system has improved in many ways: an increase in the number of people with phones; local, national, and then international call direct dialing; satellite communications; computers to reduce billing costs, permitting Wide Area Telephone Service (WATS) lines; 800 and 900 phone numbers; fax messages; electronic mail; computer data transfer; and other new technology. Table 21.11 shows the relative cost of phone calls and letters over the years. Communication is becoming less and less expensive relative to transportation.

For facility location, the centralization/decentralization decision is affected by all these technological changes. **Back office** operations, such as accounting, finance, and engineering, can be more centralized. **Front office** facilities (customer contact points) can be at more locations since they need not be self-contained. For example, a bank can have automatic teller machines at several locations connected by phone lines to a single computer. Your credit card can be read at many locations (even overseas) and, within 15 s, a central computer can issue authorization for your transaction and print a receipt with your name.

The net result of the decreasing costs of transportation and communication is that facilities can serve larger markets. Thus, a factory in Osaka can furnish stereos to the world rather than just to part of Japan; a factory in Kansas City can furnish greeting cards to the world, not just Missouri. A larger market permits more output per facility and more specialization, because each factory need not produce all models of the product line. For example, some processes now are split between factories in different countries rather than being done at a single plant.

In addition, factories need not be close to either their suppliers or to their customers. For example, a factory in Houston may get its components from Singapore and ship products all over the world. Automobile components made in Taiwan may be assembled in Mexico for sale in the United States. Iron ore from Australia is sent to Japan to become steel, which is sold in Thailand for components to be assembled in France.

As business has gone from local to regional to multinational, communication barriers within firms also have been reduced by special communication networks (e.g., WATS lines and computer tie lines) and standard business procedures and objectives. A German firm, for example, may produce products in 10 different countries and sell products in 100.

3.2 Sunbelt Effect   The **Sunbelt effect** is the U.S. trend for new facilities to go south and west (to the Sunbelt and away from the Frostbelt). See Figure 8.2. Hot climates formerly had less economic development and lower labor wages than did cool areas. The development of air conditioning for homes, factories, and autos reduced the unpleasantness of warmer climates, encouraging manufacturing and offices to move to the warmer portions of the United States, and allowing moves anywhere in the world. A book can be printed in Columbus, Cologne, Coventry, or Calcutta. A device can be designed in Dallas, Dacca, Damascus, or Dublin.

4 LOCATION WITHIN A REGION

4.1 City, Suburb, Rural   In the United States, a historic shift in the location of manufacturing plants began after World War II—the movement of factories from cities to rural areas (defined as counties or groups of counties that do not contain at least one city of 50,000). For example, in the period 1960–1977, rural areas accounted for over 50% of all new industrial jobs; rural areas gained 300,000 jobs while metropolitan areas lost 1,000,000 (Lonsdale and Seyler, 1980). In Pittsburgh, the last blast furnace closed in 1980, and manufacturing jobs dropped from 50% of the area's total in 1953 to 20% in 1985. In New York City, 40% of the employed worked in manufacturing or construction in the 1950s, but fewer than 20% did in 1980 (Frieden, 1990). Most of the rural gain was in the South, although some was in the Midwest and West.

There has been a **demographic reversal** in the developed countries (United States, Western Europe, Japan). The flow of people is from the cities to the suburbs and rural areas, reversing a trend of at least 100 years (Vining, 1982). In the Third World, cities are still growing rapidly.

Office employment began to move from cities to suburbs and smaller cities, but developers continued to construct new office buildings. From 1960 to 1984, the 30 largest U.S. metropolitan areas added as much new

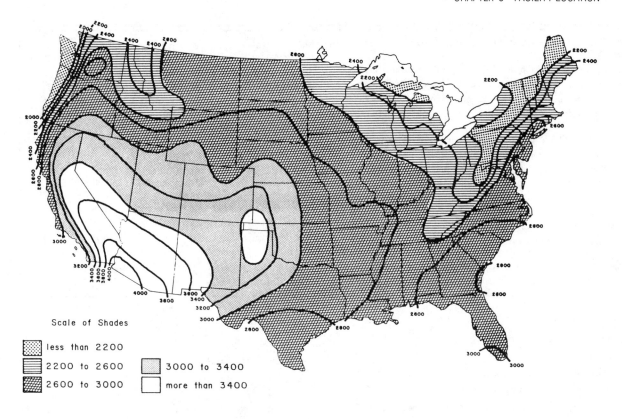

Scale of Shades

▨ less than 2200

▦ 2200 to 2600   ▨ 3000 to 3400

▩ 2600 to 3000   □ more than 3400

Average Annual Amount of Sunshine, In Hours

FIGURE 8.2   Hours of sunshine/year encourages movement to the Sunbelt.

downtown office space as they had accumulated in all the years up to 1960—some 1,300 buildings altogether, with as much office space as 250 Empire State Buildings (Frieden, 1990). Continued construction and decreasing jobs resulted in high office vacancy rates (over 20%) in many cities.

Manufacturing plants in developed countries are subject to the rural pull factors of lower wages and less unionization and the urban push factor of social and environmental deterioration of the large cities. Moves are possible because of

- improved mobility of capital
- changes in industrial organization (especially the development of branch plants for mature products)
- decreasing distance effects (transportation and communication changes)

The automobile has increased the **commuting radius** of the worker. People generally want to live within 1 hour commuting time of work, with many making 30 minutes their maximum. By foot, 1 h is about 3 miles; by bus, about 20 miles (due to stops); by automobile, about 50 miles. A 50-mile circle around a plant (or even a 25-mile circle), includes a large potential labor pool, even in rural areas. Of course, if wages are good, people will migrate to the jobs from other parts of the country.

Trucks and the interstate highway system speed the movement of materials; air travel (scheduled airlines and company planes) facilitate executive travel. The telephone minimizes the distance effect of conversation, and fax and courier services minimize the distance effect for mail. TV (especially when supplemented by cable or satellite dishes) brings mass entertainment to rural areas. These factors have led more organizations to locate in nonurban areas.

Manufacturing firms most likely to locate in rural areas are those with mature products. Mature products made in satellite plants using standardized procedures call for relatively unskilled, low-priced labor rather than scarce engineering, managerial, and marketing talent. (If this talent must be on-site, locate the facility in a suburb.) The rural area does not need to have skilled labor for high-priced jobs, because most of those jobs will be filled by incumbents (from the original plant site) or by emigrants to the area, not by people presently living in the rural area.

4.2 Site Selection   New construction is not the only alternative. Offices, especially, often use previously occupied space. See Box 8.1 for a case study of an office relocation.

BOX 8.1   Case study of a national headquarters' relocation

A national organization decided to move the headquarters from the central city of a large East Coast city. The reasons given for making the decision to move were high operating costs and a desire to be centrally located in the United States.

After a preliminary screening, the choice came down to Chicago, Kansas City, and Dallas. Key factors in selecting these candidates were superior airline service and accessibility, good meeting facilities and lodging, a central U.S. location, adequate base of business services for the firm as well as cultural and recreational opportunities for the employees, considerable operating cost savings for the firm and lower cost of living for the employees who relocated.

The majority of the one-time costs were personnel costs (personnel relocation and replacement), with only about 25% of

the cost being for the move and management costs. Failure of employees to move depended on salary (lower-income employees tended to stay), age (younger stayed), length of service (short-term stayed), and sex/marital status (married females stayed, married men moved). Operating cost savings were from lower labor costs, reduced travel expense, and reduced rent, utilities, and parking.

The three candidate cities then presented several proposed sites in each city. Other "sweeteners" (such as reduced mortgage interest rates for employees and special hotel discounts for the organization) were offered; these sweeteners seem to be getting larger in value as competition for jobs increases. The city chosen was not the lowest cost alternative, since other considerations tipped the balance. The move was implemented.

In selecting a specific site, consider

1. location (including zoning)
2. site features (dimensions and shape; distance from highways, airports, residential areas, topography and soil conditions, zoning restrictions)
3. utilities (power, water, drains, waste disposal, fire and police protection)
4. costs (land, financing, taxes, construction costs)
5. intangibles (neighborhood, community)

See Box 8.2 for pitfalls of site selection. If the site has been used, environmental contamination is possible. Decontamination can be very expensive (in addition to unpredictable legal expenses). Consider the following (Betz, 1992):

- hazardous materials on site
- underground storage facilities
- transformers (PCBs)

- building and equipment insulation
- soils
- groundwater and streams
- residue on all building and equipment surfaces

With a few exceptions, the number of employees per site no longer reaches 10,000 or more. Greater employee productivity and the development of branch plants, aided by improved transportation and communication, have resulted in plants with 300–500 people. If more output is needed, another plant is built elsewhere. One justification for the 300–500 size is that the manager can know everyone personally. On the other hand, Black and Decker set up four "plants within a plant" in their 1,500-person Hampstead plant. The work force was divided into four teams, one for each area. Each of the four areas had a different color scheme and different bulletin boards.

BOX 8.2   Pitfalls in finding new sites (Hales, 1977)

1. *Failure to specify what is needed.* The first step is to make a specific list of needs (objectives, factors, criteria). Visiting sites and contacting people before this is done is a waste of time.
2. *Misunderstanding cost relationship.* Objectives need to be chosen and ranked so unimportant variables don't get too much attention.
3. *Misunderstanding taxation.* Get precise information on state income taxes, payroll taxes, and unemployment compensation taxes. Consider state plus local taxes.
4. *Overestimation of labor supply.* The commuting pattern defines the total labor market. Break this total into those that have the skills you require. Other firms may compete for the same people.
5. *Improper estimation of labor cost.* Wages might be based on type of industry or based on wages for the area regardless of industry. Add in fringe benefits costs.
6. *Failure to identify local land-use and growth patterns.* Incompatible use risks future restrictions. Environmental

impact requirements vary greatly from site to site. Location in an overindustrialized area may lead to increased competition for labor; location in an underindustrialized area may mean insufficient supply of skilled labor.

7. *Purchasing an inadequate site.* A site can be too small due to a number of reasons, including yard space requirements, failing to anticipate growth, and poor plant arrangement on the site. The soil should be tested.
8. *Underestimating adequacy of support services.* Many firms contract out specialized services such as plating, equipment repair, advertising, legal work, etc. Are these services available?
9. *Overlooking quality of life.* Quality of life is very important in obtaining skilled labor, technical personnel, and managers.
10. *Looking for a specific site too soon.* The region should be selected first. Looking at specific sites too soon wastes time and increases the chances of land values rising before negotiations are opened.

The total space in a site can be divided into required and vacant:

$$S = R - V$$

where

S	=	site space, sq ft
R	=	required space, sq ft
	=	$U + Y + P + RGS$
U	=	under-roof area (will be less than floor space if mezzanines or multiple floors are used)
Y	=	yard storage space, sq ft
P	=	parking space, sq ft
RGS	=	roads, greenery, setback space, sq ft
V	=	vacant space (may not all be usable)

When evaluating specific sites, be sure that V is sufficient for expansions and desired **amenities** (attractive features). For example, a firm in Georgia had its new headquarters building straddle a man-made lake. The lake is aesthetically pleasing and also has a functional use for holding excess storm water, which is filtered before discharge to a nearby river. In addition, fish are stocked to control insects.

Land cost is relatively small compared with building and equipment costs. Assume a 50-acre site can be purchased for $6,000/acre—a total of $300,000. Assume there are 200 employees, each having 500 sq ft, for a total of 100,000 sq ft. If the plant costs $40/sq ft, then the plant would cost $4,000,000, before any equipment was installed. Thus, land cost is a small portion of the total cost.

There are enormous administrative-involvement and purchase-service costs when acquiring land, demolishing adjacent property, or abandoning a property as too small. Thus, a good insurance policy is to buy more land for the site than is immediately required. The building itself may occupy less than 10% of the site. In the above example, the 50-acre site has 50 × 43,560 = 2,178,000 sq ft. A building of 100,000 sq ft would occupy only 100,000/2,178,000 = 4% of the site.

A popular alternative is location at an **industrial park**. Developers (either individuals or cities) improve a site with roads, utilities, drains, and the like, and then sell portions to firms for their plants. A firm can thereby avoid the problems of developing the site. Industrial parks tend to be convenient to transportation facilities, and where utility and zoning requirements are met.

Some industrial parks subdivide space in a building and lease from 1,000 to 25,000 sq ft to small firms. A variation of this is an **"incubator"** facility in which new small firms share specialized services (infrastructure), such as shipping docks, mailrooms, telecommunications, toilets, HVAC, secretarial services, conference rooms, and so forth. Sometimes this incubator facility is a former factory, warehouse or office building which has been subdivided for the small firms. Another group of firms using this subdivided space are established firms serving specialized niche markets, such as precision grinding or locksmith work.

Some governmental agencies are attempting to lure businesses back to central cities by creating enterprise zones, which typically involve some form of tax incentive.

5 PHYSICAL CONSIDERATIONS OF SITE

The engineer must consider five physical factors for any site (Muther and Hales, 1980). (Ideally these should be considered in the final round of site selection rather than after a specific site is selected.) The five factors (see Table 8.3 for an amplification) are

1. layout (building on site)
2. handling (external to building)
3. communication (external to building)
4. utility (access and egress)
5. building (program and site characteristics)

TABLE 8.3 Physical considerations of the site (Adapted from Muther and Hales, 1980).

Layout
- Size
- Site shape
- Property lines, rights-of-way, easements, etc.
- Types of neighbors (residential, commercial, industrial)
- Hazards from fire, flood, etc.

Handling
- Location versus suppliers and customers
- Access to trucks, rail, airports, etc.
- Access for employees and visitors
- Fire protection access
- Trash and waste handling and disposal

Communications
- Telephone lines (including computer lines)
- Mail and parcel service

Utilities
- Capacity of water (potable, process, fire-protection), electricity, gas, etc.
- Disposal/discharge of sanitary sewage, wastes, scrap, storm water drainage, etc.
- Tie-in points for utility services
- Utility easements for pipelines or power lines

Building
- Legal ownership
- Zoning, building codes
- Flood conditions, site drainage
- Soil characteristics (water table, subsoil)
- Direction of prevailing winds, angles of sun

Design and locate the building on the site to minimize the costs of future changes over a 25- to 50-year planning horizon. The layout should show the present site with adequate space for access roads, parking, dock traffic, and protection from future street widening.

In addition, show the direction and extent of proposed future changes. Figure 8.3 shows four **expansion strategies:** mirror image, centralized expansion, decentralized expansion, and spine expansion. In mirror image, the facility expands symmetrically to the periphery. In centralized expansion, the facility grows outward nonsymmetrically.

In decentralized expansion, the facility grows inward and outward to connect separated areas. In the **spine** system, the building spine functions as a human spine—an integrated connector of the circulatory and nervous systems. The spine (which can be multilevel) contains the central electricity, compressed air, heating, plumbing, telephones, and other systems, as well as service facilities (maintenance, office, personnel, material handling). In expansion, the spine is relatively unaffected because growth takes place in the "limbs."

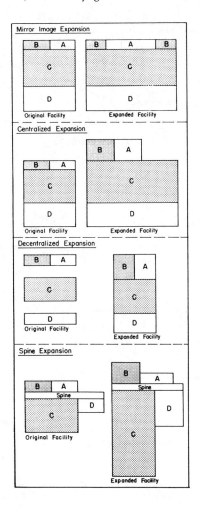

FIGURE 8.3 Expansion strategies can be divided into mirror image, centralized expansion (center out), decentralized expansion (periphery in), and spine (Tompkins, 1980). Copyright © Institute of Industrial Engineers, 25 Technology Park, Norcross, GA 30092. (404) 449-0460.

DESIGN CHECKLIST: FACILITY LOCATION

Location within a geographic area
    Objectives specified
    Objectives weighed
    Distance effect
    Climate effect
Location within a region
    City, suburb, or rural
    Labor supply

    Labor cost
    Transportation
Site
    New or used facility
    Land (type, amount)
    Expansion need
    Expansion strategy

REVIEW QUESTIONS

1. Why is facility location like selecting a spouse?
2. Discuss the decreasing distance effect of transportation on location of facilities.
3. Discuss the changing costs of communication on location of facilities.
4. Discuss the relative costs of a building and its equipment versus the cost of land for the site.
5. Describe an incubator facility.
6. Discuss the Sunbelt effect.
7. Give the advantages of locating a factory or office in a central city, in a suburb, and in a rural location.
8. Discuss the four expansion strategies given in Figure 8.3.

PROBLEMS AND PROJECTS

8.1 Visit the local Chamber of Commerce (or other agency trying to recruit for your community). What do they see as the strong points and weak points of your community? Is there a particular type of firm they are trying to encourage to move?
8.2 For a local facility, determine (a) site area, (b) building area. Are they happy with their present ratio of building area/site area?
8.3 Have a debate between two teams concerning whether to locate a branch plant in town X or town Y. The instructor will furnish details on the firm's wants. There should be a written and oral presentation. The class then will vote on where the plant should be located.
8.4 Have a debate between two teams concerning whether "our firm" should centralize or decentralize operations. The instructor will furnish details on the firm's operations. There should be a written and oral presentation. The class then will vote on whether to centralize or decentralize.

REFERENCES

Ady, R. Why BMW cruised into Spartenburg. *Wall Street Journal*, July 6, 1992.

Betz, G. Environmental responsibilities in plant closing. *Plant Engineering*, pp. 58–60, Feb. 6, 1992.

Frieden, B. American business still wants to go downtown. *Wall Street Journal*, Jan. 16, 1990.

Hales, L. Site selection. Chapter 5 in *Management Handbook for Plant Engineers*, E. Lewis, Ed. New York: McGraw-Hill, 1977.

Huey, J. The best cities for business. *Fortune*, pp. 52–84, Nov. 4, 1991.

Lonsdale, R., and Seyler, H. *Nonmetropolitan Industrialization*. New York: Halstead Press, 1980.

Love, R., Morris, J., and Wesolowsky, G. *Facility Location: Models and Methods*, Amsterdam: North Holland, 1988.

Malabre, A. Income differences between regions narrow. *Wall Street Journal*, p. 48, March 4, 1982.

Muther, R., and Hales, L. *Systematic Planning of Industrial Facilities (SPIF)*, Vols. I and II. Kansas City, Mo.: Management and Industrial Research Publications, 1980.

Tompkins, J. Modularity and flexibility: Dealing with future shock in facilities design. *Industrial Engineering*, Vol. 12, 9, 78–81, 1980.

Tong, H. *Plant Location Decisions of Foreign Manufacturing Investors*. Ann Arbor, Mich.: University Microfilms International, 1979.

Vining, D. Migration between the core and the periphery. *Scientific American*, Vol. 247, 44–53, Dec. 1982.

Yoon, K., and Hwang, C. L. Manufacturing plant location analysis by multiple attribute decision making: Part I—Single plant strategy. *Int. J. Production Research*, Vol. 23, 2, 345–359, 1985a.

Yoon, K., and Hwang, C. L. Manufacturing plant location analysis by multiple attribute decision making: Part II—Multi-plant strategy and plant relocation. *Int. J. Production Research*, Vol. 23, 2, 361–370, 1985b.

CHAPTER

9 BUILDING DETAILS

CHAPTER PREVIEW

Any design is a collection of details. This chapter discusses details concerning the building and grounds.

CHAPTER CONTENTS

1 Foundations and Floors
2 Windows
3 Roofs
4 Building Shape and Orientation
5 Fire Protection
6 Parking

KEY CONCEPTS

atrium	emergency organization	mezzanines
building shape	encapsulated storage	slips and falls
classes of fires	isolation joints	sprinklers
compartmentation	isolation techniques	stall layout
control joints	low bay (versus high bay)	surface area/volume

1 FOUNDATIONS AND FLOORS

The functions of a floor on grade, that is, on the ground, are to transmit the loads to the soil beneath the floor and to provide a smooth, easily cleaned and maintained bearing surface.

1.1 Load Transmission   Common design practice is to design for a live load (load in addition to structure weight) of 75 lb/ft² for light manufacturing and 125 lb/ft² for heavy manufacturing and storage (Klein, 1982). The floor should be uniformly supported by the soil beneath it. Thus, the soil should be uniformly compacted or a granular 4 in. thick sub-base used. Even with a uniform sub-base, the floor probably will settle differently from the abutting wall and column foundations; thus the floor must be isolated from them. These **isolation joints**, sometimes called *expansion joints*, must permit both vertical and horizontal movement (Gustaferro, 1980). See Figure 9.1.

Other types of joints, **control joints**, divide a large floor area into relatively small rectangular (preferably square) panels. They accommodate shrinkage along predetermined paths so cracks are not random but are straight line—and thus easier to seal. The reason for steel reinforcement in concrete is that it minimizes the widths of random cracks; thus the steel should be near the slab top rather than the bottom. After the cause of a crack (shrinkage, settlement, or overload) is eliminated, the cracks can be welded. Epoxy can weld cracks as narrow as .01 in.; the repaired area will be as strong as the original concrete. Metzger (1983) says to avoid high-strength epoxy (stronger than concrete) as it just causes new cracks. For most commercial floors on grade, slab thickness is between 5 and 9 in. Often, control joints are saw cuts of 20 to 25% of slab thickness. Joints should be as narrow as possible (say by a 1/8 in. wide diamond saw blade) to reduce the volume of

filler material needed and reduce surface area exposed to vehicle wheel damage. Joints should always be filled. Since the filler material is a compromise between firmness for protection against wheels and flexibility for expansion, install the filler as late in the construction schedule as possible. Then the building is under controlled temperature, most of the shrinkage and moisture loss has already occurred, and filler flexibility is not as critical.

Floors on the upper levels of multi-story buildings and on mezzanines are supported by columns and beams; see Figure 9.2. The farther a load is from a column, the higher the stress and the greater the vibration; in practice this means the worst location for a heavy load is midway between columns. Midway between columns thus is a good location for an aisle, if floor loading is a potential problem. For heavily loaded racks, use an oversize base plate under the rack uprights. If there is clear height for a potential mezzanine, provide footings for it, since the incremental cost is nominal at this stage but expensive later. Minimize interior load-bearing walls, since they inhibit material flow as well as later building modifications.

In multi-story buildings, passage of cables and wires from one floor to another is a problem, especially once the building is completed. Plan ahead by putting a number of holes through the ceiling/floor and cap those not immediately needed.

I-beam columns support upper floors and mezzanines and also support the roof (in some types of construction). Smaller I beams may be 4–6 inches across the flanges and 6–8 inches along the I. I beams for multi-story buildings might be as large as 24 × 30 in. The spacing between columns in single-story buildings depends upon the roof

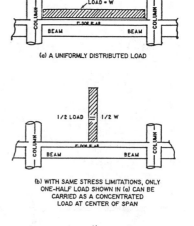

(a) A UNIFORMLY DISTRIBUTED LOAD

(b) WITH SAME STRESS LIMITATIONS, ONLY ONE-HALF LOAD SHOWN IN (a) CAN BE CARRIED AS A CONCENTRATED LOAD AT CENTER OF SPAN

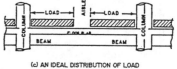

(c) AN IDEAL DISTRIBUTION OF LOAD

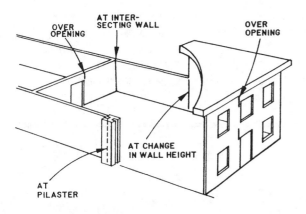

FIGURE 9.1   Isolation (expansion) joints separate structural elements (wall, columns, and foundations) from the floor; the joints allow these elements to move independently.

FIGURE 9.2   Aisles should be midway between columns on upper floors if floor loading is a problem (McElroy, 1988).

type. A gable roof might have a 50 ft clear span (distance between columns) while a triangular frame roof might have a 170 ft span. Complex designs can be used for large spans (such as hangers).

1.2 Machinery Mounting/Isolation

When machinery is installed it needs to be stable and level and not transmit or receive vibration and shocks. Shock and vibration sources include in-plant traffic, overhead cranes, presses, machines making heavy or interrupted cuts, and highway or road traffic. Resonance can make problems worse; a typical concrete industrial floor resonates at 20–24 Hz (Whittaker, 1990).

There are a number of different **isolation techniques** (Marinello, 1983; see also Figure 23.6). Isolation pads (inertial blocks), placed under the machine legs or base, can serve the majority of installations; see Figure 9.3. Leveling mounts (leveling screws or leveling wedges) combine leveling with vibration control. They do not require a special foundation so the machine can be moved relatively easily. Steel spring isolators are good for low-frequency, high-amplitude vibration. Air springs have low transmissibility coefficients, and load capability is varied by changing the air pressure. Pneumatic systems add a servo-control so air pressure is adjusted automatically to any change in load.

To hold equipment down, use foundation bolts and grout or use isolation material. Foundation bolt design sometimes must consider not only the typical vertical force (due to weight) but also the overturning moments, horizontal forces, and torques. Block foundations (nearly cubic structures) anchor machinery with large overturning moments (such as jib cranes). Grout is a hardenable adhesive material placed between a machine and the floor to ensure positive contact. Avoid ordinary cement-mortar mix grouts for reciprocating machinery.

1.3 Concrete Surface Characteristics

The American Concrete Institute (ACI) standard 302 recommends a class 4 floor for foot and pneumatic wheeled traffic, class 5 for foot and wheels (abrasive wear), and class 6 for foot and hard-wheeled vehicles (severe abrasion). Heavy trucks and carts with small, hard wheels cause problems; guided vehicles that follow exactly the same path quickly make a rut on a nonhardened floor. Vacuum dewatered concrete is 2 to 2.5 times more wear resistant than normal concrete with the same surface finishing. The choice is initial capital cost or repeated maintenance cost.

Metallic aggregates can be applied as a topping (¼ to 2 inches thick) over fresh or hardened concrete; they have 4 to 8 times the abrasion resistance of plain concrete (Wells and Lupyan, 1991). A polymer topping is a common solution to chemical and chloride (i.e., salt) attack of concrete.

Surface flatness (ACI-301) has three tolerances: Class A (true plane within ⅛ in. in 10 ft), Class B (¼ in. in 10 ft), and Class C (¼ in. in 2 ft). Class A probably is most appropriate for aisles used by high-lift trucks. A ¼ in. floor difference can translate into a 2 in. deflection on the top of a 40 ft truck mast, causing difficult positioning of blades into pallets. Note that the floor under the racks could be Class B or even Class C. Class C is not desirable for machine shops.

If liquids are used or stored in an area, there will be spills. Install drains and slope the floor appropriately.

Sealing concrete floors has several advantages. It makes the floor easier to clean and reduces damage from spilled chemicals. Sealing also reduces dust and thus the number of air changes and thus energy costs. Painting the floor a light color substantially increases the lighting coefficient of utilization so less energy is needed and lighting is more even. See Table 9.1. Psychologically, it is easier to get employees to use good housekeeping on a sealed, painted floor than on an unsealed floor. Painting concrete floors is difficult because of air pockets in the concrete surface and laitance (incompletely hydrated cement scum which floats up and clings loosely to the surface).

To reduce problems, clean and acid etch and brush before painting (Riders, 1989). Then use special concrete paint. A low-maintenance alternative is to clear-seal but not paint the floor.

Concrete can be made antislip by adding silicon carbide or aluminum oxide to the installation mix; abrasive-reinforced coating can be applied as paint.

1.4 Tile and Carpet

Slips and falls on the same level are a very serious problem. In general, reduce falls with

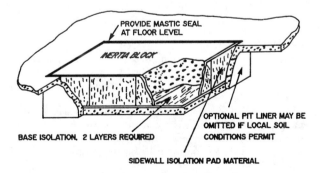

FIGURE 9.3 Isolation pads (inertial blocks) consist of an excavation (lined with a suitable vibration isolation material) filled with concrete (Whittaker, 1990). All air, water, and electrical connections to the machine should have flexible links.

TABLE 9.1 Light reflected (footcandles) from concrete panels when illuminated by a single 650 W source at 3,200 K at 20 ft (Ernst, 1976) (1 footcandle = 10.8 lux).

Panel	Reflected Light (Footcandles)
Uncoated clean concrete	30
Uncoated dirty concrete	7.4
Coated with clear urethane	30
Coated with red urethane	30
Coated with yellow urethane	60
Coated with white urethane	120

good housekeeping, proper floor materials, shoe materials, and floor finishes. See Box 9.1.

Good housekeeping requires keeping the floor clean of foreign materials, which generally lower slip resistance. Never use an oil mop to dust a waxed floor.

Floor materials and shoe materials are related. Dry leather shoe soles are the least slip-resistant of all shoe materials; if a floor has satisfactory slip resistance for leather soles, it will be acceptable for almost any other material. Neoprene gives much better slip resistance; see Table 9.2. Table 9.2 points out that some floor materials are more slip-resistant than others. Antislip surfacing comes in different types, widths, and colors. Use it for ramps and stairs and for areas that might become wet. Mats and runners are needed especially where there is water or oil, such as at outside doors and around machines with oil or coolant. Floor finishes, although not as important as the floor material, should be selected for their nonslip characteristics as well as their protective characteristics.

TABLE 9.2 Coefficients of friction.

Floors	*Coefficient of Friction*
Soft rubber pad	1.0
End grain wood (not slippery when soiled)	.8
Rough-finish concrete	.7
Working-decorative	
Dry	.65
Soiled	.5
Steel	.4
Shoes on Clean Floors (Soiled Floor .2 less)	
Soles	
Rubber-cork	.8
USA-USAF standard	.75
Rubber-crepe	.7
Neoprene	.6
Leather	.5
Heels	
Neoprene	.7
Nylon	.65

BOX 9.1   *Slips and falls*

Of compensable injuries in the United States, 20% are related to slip and/or fall incidents (Szymusiac and Ryan, 1982). In Sweden, slips and falls in industrial accidents account for 11% of all occupational injuries (Strandberg, 1983).

Reduce injuries with proper floor design, proper shoe design, and good housekeeping.

Proper Floor Design

Many people are injured when entering and leaving vehicles with poorly designed access systems, especially construction and mining equipment (Albin, 1988). See Chapter 10 for comments on ramps, stairs, and ladders.

People fall into holes or off platforms. Use a standard railing on platforms and around permanent holes. OSHA defines a standard railing as one with a smooth top rail 42 inches above the floor and an intermediate rail at about 21 inches. A standard railing must be able to withstand a 200-lb force at any point on the top rail.

A toeboard (4 inches high and with a clearance between the toeboard and floor of less than 1/4 inch) prevents the foot slipping over the edge and also prevents objects being pushed over the edge.

For temporary holes, a warning barrier may be sufficient (Marshall, 1982). One possibility is a sawhorse or other freestanding object; another is a yellow tape placed to prevent access to the danger.

Because walking is an automatic activity that people normally don't think about, a small change in height or friction is very dangerous. A permanent small change in height (e.g., one-step stairs) may not be noticed and can lead to falls or tripping (depending on the movement direction). Therefore, use ramps instead of steps. If a step must be used, add a handrail and change the floor color at the step. A temporary small change in height (usually an object on the floor causing a tripping hazard, although it could be a temporary hole) calls for housekeeping and maintenance.

A high-friction floor reduces falls. Some materials are better than others; for example, rugs and mats are better than concrete, tile, and steel (see Table 9.2). If a mat is used, it should have beveled edges to reduce tripping hazards; if water or oil is present, the mat should have drain holes. In the food industry, mats have to be cleaned periodically. For standing comfort, a mat should have an intermediate compressibility—not too hard or too soft (Konz et al., 1990). A high coefficient of friction is needed especially for nonlevel or oily or wet floors. Wet floors near outside doors are common because people track in moisture.

Floors (e.g., steel gratings and stairs) often have abrasive materials added (either to the material itself) or as traction strips. If strips are used, the gap between each tape should not exceed about 1 inch (Strandberg, 1983). Strips can be safety yellow to call attention to the hazard.

Steel and concrete floors also can be textured (with sandblasting or chemicals) to increase surface roughness.

Proper Shoe Design

See Table 9.2 for friction of shoe materials.

In normal walking, the heel hits the ground first and the foot rolls forward onto the ball of the foot for the thrust-off. A heel slip is dangerous because one loses balance and falls backward. Thus, on slippery surfaces, try to walk on the balls of your feet; then if you slip, you will most likely go forward and will be less likely to lose your balance and fall.

Good Housekeeping

Materials on the floor (litter, oil, water, screws, dirt, etc.) act as lubricants; Table 9.2 shows coefficients of friction are about .2 lower on dirty floors than clean floors. Large materials can be tripping hazards. Prevention is the best solution, so fix leaky sources of water or oil, reduce oil misting, and provide sufficient containers for trash. Maintenance workers should not add lubricants, such as in oil mopping a waxed floor or using slippery waxes on wood or tile floors.

Housekeeping is especially important in areas such as offices because people expect these areas to be clean and do not pay attention to the floor. People walk carefully through a construction site because they expect the floor to be poor.

Just as tire ads emphasize the value of tread designs that place more "rubber on the road," some shoe shapes are better than others. Deck shoes have a continuous sole without a heel arch and give the best contact; boots and women's high heels give little contact area and thus are the most dangerous. An intensive French study reported (1) the heel should be eliminated (i.e., use deck shoes), and (2) the tread design should have many "braking edges," acting equally in longitudinal and transverse directions. The best tread had small flexible rectangular cleats with drainage canals that facilitated movement of liquid from under the sole. Strandberg (1983) emphasized the importance of the slip resistance of the heel's rear edge.

Covering an unheated concrete floor with carpet will save about 1% in heating of the building (Hager, 1977).

Advantages of carpet (versus tile) are noise reduction, lowered maintenance, more comfort for standing and walking, a higher coefficient of friction, and better appearance; disadvantages are higher capital cost and difficulty of cleaning, such as for oil, grease, or cigarette burns. Carpet tiles with "forever alive" glue backing can be used in these severe environments.

2   WINDOWS

To have or not to have windows is a conflict between esthetics and a view versus energy conservation.

2.1 Window Advantages/Disadvantages   The one advantage of windows is that they provide a view. Window disadvantages (compared to solid walls) are numerous.

1. They cost more in both capital cost and operating cost (heat entry in summer, cold entry in winter, washing, and repair costs). A single-pane glass window has a U value of about 1 BTU/h-ft^2 °F versus .5 for double pane, .4 for triple pane, and .2 for a normal wall. Thus, a single-pane window transmits 5 times as much as a wall.

2. They are a source of glare. Many windows are almost permanently covered with shades, drapes, and curtains. Glare is a special problem in computer areas (monitors) and conference rooms (visual aids). Windows can also restrict arrangements of nearby workstations.

3. They are not a practical source of illumination in factories because the light is too variable, depending upon time of day, season, and weather, and because the light decreases by distance squared from the window. Work close to the window gets too much light, and work far away gets too little. If you depend upon sunlight through windows for illumination, equipment and workstations must be within about 20 ft (6 m) of a window, making maximum building width about 40 ft (12 m). Artificial illumination permits varied building size, shape, and layout.

4. They may admit air. In industrial buildings, exterior windows are not desirable sources of ventilation. The air volume passing through a window is too variable, because of wind velocity and direction, and is difficult to use in hot, cold, or wet weather. The air admitted and released is also not controlled, resulting in low velocities at locations far from windows. In addition, pollutants (e.g., dust, pollen, insects) may be admitted or if pollutants are being discharged, they are not controlled (e.g., passed through a filter). In comparison, mechanical ventilation is relatively cheap and easily controlled.

5. They pass noise and distractions, out to in and in to out.

Some window characteristics can be good or bad, depending on the situation.

1. They decrease privacy for people on both sides of the window.

2. If openable, they allow air passage. If the heating, ventilating, and air conditioning (HVAC) system fails, either locally or totally, windows may be useful. Openable windows may give people a feeling of personal control, but for normal operation of a HVAC system, open windows decrease system control and generally increase system costs.

Window advantages are:

1. They permit a view.

2.2 Recommendations   To decrease window disadvantages, maximize the view by using windows with a long horizontal axis and a short vertical axis. Keep the sill low (.5 m above the floor) so that a seated worker can see the view. Ideally, the horizon should bisect the window and the view is of nature (e.g., water, foliage) rather than man-made objects. To maximize the number of viewers, make windows public instead of private. Place windows in break rooms and at the ends of corridors. Use interior windows between windowed perimeter rooms and the interior core so that people in the core can see through both sets of windows.

Minimize window surface area. Reduce the area of existing windows with opaque insulated panels.

Use double or triple panes to reduce heat (and noise) transfer. Glazing may be different on the south and west side than the east and north.

Reduce transfer of solar heat and glare. Permanent treatments include window and door design (especially on the west and south sides), recessed windows, films on windows, and windows on enclosed courtyards (atriums). Deciduous trees shade in summer and admit light in winter. Interior treatments include blinds and curtains. If the blind or opaque curtain is never opened, board up the window. The drape should be close (.8" (2 cm)) to the window; a spacing of 5" (12 cm) with a sill permits convective rollover as the air moves around the edges and base of the drape (Horridge et al., 1983).

Decide whether the window will be permanently sealed, openable with a key, or openable anytime. Since openable windows leak air (even when closed), consider making only a few (say 1/3) of the windows openable. Another option is a large sealed window and a small openable window.

If using skylights, make them vertical rather than horizontal (reducing dirt and breakage), plastic rather than wired glass (better light transmission and less breakage), translucent rather than transparent (less glare), and openable (supplementing mechanical ventilation).

3 ROOFS

3.1 General Considerations

A roofing system must meet five design requirements (Johnson, 1991):

1. weather resistance
2. external fire resistance (fire from sparks)
3. internal fire resistance
4. wind uplift resistance
5. thermal performance

3.2 Roof Construction

A roof generally consists of a roof deck covered by the roof. About 50% of decks are metal, 25% wood, and 25% concrete. The roof itself is (1) a built-up roof (BUR) (33%); (2) an elastomeric single-ply roofing, of which ethylene-propylene-diene monomer (EPDM) is the most popular (33%); and (3) a modified bitumen, metal, or other single-ply formulation (33%) (Steiner, 1989).

The conventional BUR has a base sheet, a layer of saturated felt, and a layer of asphalt and gravel over fiberglass insulation. If insulated, the roof insulation can be above or below the deck. Basic specifications for a 20-year roof over a nailable deck call for two nailed 15-lb felts under three solidly mopped (covered with asphalt or tar pitch) 15-lb felts (Built-Up Roofing, 1991). Because the membrane is manufactured at the job site, its quality depends upon the roofing contractor.

Unless coated with fire-resistant mastic, spray-on foams may present a fire hazard.

The roof needs sufficient strength to support loads placed on it. In addition to the weight of the roof itself, there may be equipment placed on the roof and cranes and hoists suspended from it. Snow may weigh 10–15 lb/ft³; if it has thawed and refrozen, it may weigh 40–60 lb/ft³ (Hoover, 1981). The snow problem is critical in valleys between sloping portions of a roof or where drifts form behind obstructions.

If a roof must be replaced, there are three options: reroofing, retrofit, and recovering (Johnson, 1991). The reroofing process strips the existing roof system to the deck and installs new insulation, vapor barriers, and membrane. The retrofit process places a new roof over the existing roof (usually improving the slope and the insulation). The recovering process places new roofing material over the existing material. A recovered roof may not be insurable. Both recovering and retrofit add to roof weight.

3.3 Roof Leaks

Most industrial roofs are flat or have a low slope. The problem with flat roofs is that they are not really flat. Low spots develop and permit ponding. The pond of water stresses the roof underneath it and eventually (perhaps in several years) it begins to leak. The problem is more severe in climates with a freeze-thaw cycle. The National Building code now requires a minimum slope of ¼ inch/foot when more than 25% of an existing roof is replaced within a 12-month period (Peterson, 1991).

Many leaks occur where items (ventilators, chimneys, etc.) project through a roof. Do periodic inspection and maintenance before and after winter and just after a rain (when it is easier to detect the problems). Another leak problem is due to water falling continuously on the roof and eroding it away. Examples are runoff from upper to lower sections of multi-level roofs and leaking water from cooling tanks (Hoover, 1984).

Rainwater from the roof should be piped (either inside or outside the building) to a disposal area, typically a storm sewer line. Conduct water away from the building walls by sloping walks, pavement, and lawns to drain away from the building.

3.4 Roof Ventilators

Roof ventilators commonly supply or exhaust air from factories. Intake air may not be clean because it may come through other roof ventilators. Consider the location of the various ventilators as well as the wind directions. Shields may be necessary. If intake air is for air conditioners, it should not come from the roof because that air is preheated. Have the intake piped from a cool location such as a grassy area on the side of the building.

Exhaust air may contain moisture or compounds that damage the roof; the problem is more severe in hooded ventilators (which exhaust the air downward toward the roof).

The roof receives radiant energy from the sun. To add heat, make it a dark color; to reduce heat, make it a light color. Roofs can be cooled with roof spray cooling, which may reduce interior temperatures 10–15°F (Campbell Soup, 1989).

4 BUILDING SHAPE AND ORIENTATION

4.1 Perimeter/Area

The area of a building is enclosed by walls (perimeter). **Building shape** affects perimeter (skin) size, and the larger the perimeter, the more heat transfer to and from the environment (heating and cooling) and the larger the cost for initial construction. On the other hand, a longer wall allows more windows, so more people have a potential view. In general, as perimeter length increases, the internal distance for transportation increases.

Figure 9.4 shows, for a building of 10,000 units of area, how the perimeter varies with building shape. (The ranking of the shapes stays the same for small or large areas, but the perimeter/area ratio increases for small areas.)

A circle has the minimum perimeter. Unfortunately, curved walls are difficult to build.

The square is next and is easy to construct.

The perimeter of a rectangle (for the same area) depends upon its shape. A "shoebox" shape has less perimeter than a "finger," which has less than a "hair," which has less than a "leaf." In design of ducts, the length to width ratio of a rectangle is called the *aspect* ratio. A rectangular building with a large aspect ratio has a short distance between walls (the roof span), reducing need for roof support columns, but also producing a space that is more difficult to use for equipment arrangement. The space between columns is called a *bay*.

Next come letter shapes. The 10,000 units of area have been divided into 5 squares of 2,000 each. A **T** has 12 wall sections of 44.7 or a total perimeter of 536.4. If the two blocks of the leg of the **T** are rearranged to form an **L** (either 4 and 1 or 3 and 2), there are still 12 sections of 44.7 = 536.4 for the perimeter. The blocks can be rearranged into the shape of **C**, **U**, or **H** and the perimeter does not change—if the shape of the 5 sections does not change. If the shape of the section becomes rectangular instead of square, then the perimeter goes up. The internal transportation cost of the different shapes may also differ.

The shape of the letter **O** is interesting. When five 32.1 × 64.2 rectangles form a letter **O**, the perimeter is 449.6. There is an inside perimeter of 192.6 for a total perimeter of 642. But if the inside of the **O** building is covered by a transparent roof, the courtyard (or **atrium**) walls have minimal energy exchange with the environment and only a small loss in perimeter for windows. Heating and cooling of the atrium generally is free due to the minimal energy exchange. The interior space of the atrium can be a commons for a cafeteria, recreational space, or other general use. In addition, the central area provides an opportunity for a view of a man-made scene (e.g., a swimming pool, garden, etc.).

The ratio of perimeter to area has applications beyond building design; for example, a circular parachute falls more slowly than a rectangular one. For pipes and ducts, a circular shape not only has less energy exchange with the environment but also less frictional loss of the fluid or gas flowing through the pipe or duct (thus requiring smaller motors for pumps or fans). However, ventilating and heating ducts sometimes are made rectangular to fit into a small plenum or thin wall.

4.2 Surface Area/Volume   The preceding two-dimension relation can be extended to three dimensions: the relation of **surface area/volume**. The sphere minimizes surface area in relation to volume. The sphere is not generally used for buildings occupied by people but is popular for storage of liquids and gases because it minimizes material cost and energy exchange with the environment. A popular compromise is a cylinder (e.g., for storage of pressurized gas). A dome minimizes material use for volume enclosed and strongly resists exterior pressure (although weaker for internal pressure); thus it finds applications in military bunkers and igloos.

The square becomes the cube. This efficient enclosure is fine for cartons and boxes, but for buildings it may be inefficient because transportation (of people and objects) between floors is difficult. Some warehouses, however, are cubes.

The shape generally used for industrial buildings is the square or rectangle with a wall height of about 15 feet (**low bay**) or about 30 feet (**high bay**). This shape also produces an efficient cavity ratio for illumination. Note that the air space in high bay buildings can be used for conveyors, utilities (air, water, electricity), or other necessities, so it is not wasted. In addition, partial second floors (**mezzanines**) can be used for people (offices, cafeterias, toilets) or for storage. They do not interfere with product flow on the main floor and allow the service facility to be close to customers on the main floor. Mezzanines can be enclosed, with their own heating, cooling, and ventilation, and can be part of the original design or be added later.

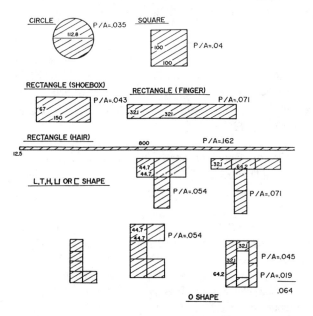

FIGURE 9.4   Effect of shape on the perimeter/area ratio. All the figures have an area of 10,000. The circle minimizes the perimeter. Squares are next and then rectangles. Next is the shape which can be identified as L, T, U, or C (depending on the arrangement). The O shape has an interior and exterior perimeter; if the center of the O is covered by a roof, the central area (atrium) walls have little energy exchange with the environment.

Office buildings in urban areas often are designed to minimize the "footprint" and use the free air rights. However, as building height increases, additional weight must be supported, and so the pressure on the foundation and structural members increases. Stronger supports use more floor space for each floor; the cost and space requirements of elevators and utility lines also increases with height. Another problem is the wind loading on the side of building. (Wind loading and osmotic pressure—ability of water to rise—limit tree height.) Thus, even if capital cost/cubic meter of the building were the only criterion, buildings would still be limited in height. However, other problems exist (employee and visitor congestion in parking, access/egress, etc.). Since in new offices most communication between departments will be by computer, phone, and mail, there probably will be few new skyscrapers built.

Cargo ship design also considers surface area/volume. As the length of a ship increases, the cargo volume increases by the cube, but the area under water increases by the square. The required propulsion force (and thus engine size) is determined by area under water times friction. Thus, the engine size of large cargo-volume ships does not increase proportionally.

Biological applications of surface area/volume are called the *law of similitude*; Haldane (1928) called it "On Being the Right Size." If a 2 m man became a 20 m giant, the cross section of the leg bone would increase by the square, but the weight on the bone would increase by the cube. Thus, the giant would break his leg when he walked (if he stayed the same shape). Thus heavy animals, such as elephants, have thick legs. A change in size requires a change in shape.

People living in cold climates tend to be "spherical" (Roberts, 1975). In the tropics, however, a radiator shape would be better than a boiler shape, so long arms and legs are good. The surface area of elephants, for example, is increased by their ears (which act as fins). Gloves, because of the separate fingers, have more surface area/volume than mittens and thus lose more heat.

4.3 Building Orientation

Building orientation, especially on small lots, may depend upon the lot orientation. You must consider which side of the building is to be the front, access from roads and railroads, and clearance requirements. Also consider possible future building expansion.

The thermal load from the sun (good in winter, bad in summer) may be a factor in orienting the building. In the northern hemisphere, the sun is in the southern sky and heat is maximized on the southern and western faces of the building. Generally, cooling costs more than heating, and so heat load is reduced by minimizing the western side of the building. The low western sun also causes more glare through windows than does the higher southern sun.

Wind load generally does not affect building orientation but may affect placement and protection of doors and may influence window insulation decisions.

5 FIRE PROTECTION

5.1 Safety Concepts

Each facility should have an **emergency organization**. The duties of the team should be (1) to identify the hazards, (2) to equip the facility, (3) to develop a map, (4) to plan to communicate, and (5) to maintain the program.

Depending upon the facility, there may be emergencies such as fires, tornadoes, floods, and explosions. In tall and large buildings, for example, the primary hazard to life is smoke and toxic gases, not the fire itself. This is due to new materials for construction and furnishings and to construction techniques such as sealed windows, large central air-conditioning systems, and large open work areas. The general emergency goal should be life safety of the building occupants; the secondary goal is to save as much of the building as possible.

The facility should have emergency equipment and people. Emergency equipment includes fire extinguishers, first aid kits, etc. The people should be trained. Since an emergency can occur at any time, people on all shifts should be assigned specific key duties for times of emergency.

A map aids the emergency team in one of its primary duties—traffic management of the outgoing employees and incoming emergency personnel. A detailed emergency map should pinpoint emergency equipment locations and emergency exits and escape routes. Outside firefighters will benefit from a map which identifies main power switches, hydrant and standpipe locations, and location of hazardous and flammable material storage. An annual visit from the local fire department as well as periodic practice alerts are good ideas.

Communication usually is by telephone. Post emergency numbers at each phone. Make a written list of whom to contact in an emergency. The person who will be in charge during an emergency should be identified. There should be an authorized person to handle all communication with the press.

5.2 Design Considerations

This section is divided into building construction, portable fire extinguishers, sprinklers, sensors, and fire sources.

5.21 Building Construction.

Compartmentation reduces the spread of fire and smoke in a building. Compartments are formed by vertical and horizontal barriers such as fire walls and fire doors, horizontal barriers in racks, and closures in ducts. Smoke is more difficult to control than fire. Therefore, ventilation fans should shut off in a fire, *and* the duct should close. Shafts (e.g., for elevators and stairs) are a problem because they concentrate smoke through the chimney effect.

If flammable and combustible liquids are received in drums, a drum storage room is necessary. See the National

Fire Protection Association (NFPA) 30 (Flammable and Liquid Code) and NFPA 251 (Fire Tests of Building Construction and Materials). The NFPA address is Batterymarch Park, Quincy, MA 02269; (617) 770-3000. Flammable liquids storage rooms (Rekus, 1989) should have fire-rated construction; liquid-tight wall-floor joints; a listed fire door that is normally closed; a noncombustible 4-inch raised door sill to prevent liquid flow; ventilation to remove vapors; and Class 1, Division 2 electrical wiring.

If the room is used solely for storage, natural ventilation may be used. However, if flammable liquids are dispensed there, the room must have mechanical ventilation with an air-flow interlock which activates an audible alarm in case of ventilation failure. Ventilation rate must be at least 1 CFM (cubic ft/minute)/sq ft of floor area but not less than 150 CFM. Drawing flammable liquids from drums into a safety can should be either through (1) a safety faucet (which, however, leaves about 2 inches of liquid in the drum) or (2) a safety pump. Liquid transfer causes static charges. Ground these by using a bonded wire to connect the drum and the receiving container.

Reduce the effect of explosions by venting and blow-out panels; see NFPA 68 (Venting of Deflagrations) and NFPA 69 (Explosion Prevention Systems).

5.22 Portable Fire Extinguishers.

There are four **classes of fires**: A (combustible), B (flammable), C (energized electrical), and D (combustible metals). (See Figure 9.5.) Portable extinguishers are recommended even where sprinklers are used, because frequently they can put out the fire before the sprinklers are triggered. (Sprinklers put out a fire but often cause water damage.) Call the fire department before or at the time portable extinguishers are used.

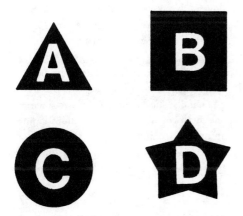

FIGURE 9.5 Class A extinguishers use a quench-cooling effect on wood, paper, etc.; if the triangle is colored, it should be green. Class B extinguishers exclude oxygen or interrupt flame for flammable liquids such as oil and grease; if the square is colored, it should be red. Class C extinguishers have nonconducting compounds for fighting electrical fire; if the circle is colored, it should be blue. Class D extinguishers are for combustible metals such as magnesium and zinc; if the star is colored, it should be yellow. NFPA Standard 10 (1978) says markings on the extinguishers should be legible from 3 ft but wall markings should be legible from 25 ft.

Locate extinguishers close to hazards but not where they would be cut off from a fire. Thus, place them near exits and outside of small rooms or spaces with flammables (inside might become inaccessible). The location should be conspicuous and not blocked by machines or stored material. A common technique is to paint a large red area on the building column or wall where the extinguisher is located. OSHA requires heavy extinguishers (> 40 lb) to have their tops less than 42 inches from the floor; lighter extinguishers ($\leq$ 40 lb) should have their tops less than 60 inches from the floor. All material handling vehicles should have an extinguisher, because they can get to the fire quickly.

The number of extinguishers depends upon the type of fire and the relative hazard. Light hazard includes offices and schools; ordinary hazard includes most factories, warehouses, and stores; extra hazard includes woodworking and paper and plastic storage. As a general rule, maximum travel distance to a class A extinguisher should be 75 ft and to a class B extinguisher, 30–50 ft. NFPA 10 (Portable Fire Extinguishers) has more detail.

5.23 Sprinklers.

Sprinklers are essentially free because their cost is offset by reduced fire insurance premiums. There are four main types (Stein et al., 1986):

- **wet pipe** (most common): Water is always standing under pressure in all pipes and mains.
- **dry pipe** (used in unheated areas): Pipes are filled with compressed air (or nitrogen) until the opening of a sprinkler head permits water flow. The system requires an air compressor, a heated main control valve housing, and pitching of all piping (to allow drainage after use).
- **preaction** (similar to dry type): Water is admitted to the pipes *before* any sprinkler head has opened. This early alarm (of water filling the pipes) may permit extinguishing the fire manually before the sprinklers open and subject the building contents to water damage.
- **deluge:** *All* sprinkler heads go off at once. Use where there is rapid fire spread (e.g., flammable liquid fires).

Figure 9.6 shows a typical wet-pipe system. The system components are (1) water supply (city water main + secondary supplies such as storage tanks), (2) valves and piping leading to the sprinklers, (3) Siamese connection for fire department hoses, and (4) alarm and communication systems. The alarm may sound at the fire department as well as on site and may trigger automatic shutdown of processes and conveyors, start and stop ventilation systems, and close fire doors and tank covers. Be sure to install drains for removing the water from activated sprinklers.

Sprinklers can be on ceilings, in storage racks, or both. Sprinkler requirements depend primarily on three factors: product combustibility (e.g., steel, paper, plastic), the

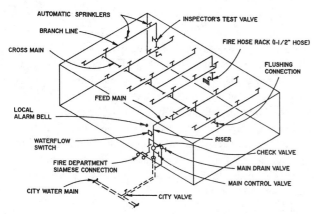

FIGURE 9.6 Wet-pipe sprinkler systems are the most common. A tree and branch system is shown. The riser can connect to the trunk at the middle or one end; the branches can diverge from the trunk from one side (comb) or two sides (tree). A grid layout (a cross main at front and rear connected by branches) also is popular.

packaging (e.g., wooden pallets, paperboard cartons, plastic packaging), and storage size and density (e.g., 8-ft versus 4-ft aisles, 12-ft versus 25-ft height, 8-ft rack depth versus 40-ft rack depth). In the NFPA tables, more sprinklers are required when aisle width is less than 8 ft, because narrow aisles act as a flue and also allow the fire to cross the aisle more easily. For **encapsulated storage** (material covered by a plastic film), increase sprinkler density 25% because the fire can burn while the plastic shields it from sprinkler and hose water. If there are multiple levels of sprinklers, the lower ones need to be shielded so they are not cooled from above, inhibiting their operation.

The amount of water necessary can be found in three standards: NFPA 13 (Installation of Sprinkler Systems), NFPA 231 (Indoor General Storage), and NFPA 231C (Rack Storage of Materials). If water pressure is inadequate, use a booster pump. If insufficient water is available from public mains, consider a pond or tank. Underground storage tanks generally need not be heated.

5.24 Fire Detection.
Fires can be detected automatically from smoke, heat, or flames. See Table 9.3 and NFPA 72E (Automatic Fire Detection). Another option is a waterflow detection device located in the sprinkler riser. Once the fire is detected, the detector (1) initiates an alarm, (2) activates an extinguishment system, or (3) does both 1 and 2.

If manual pull stations are used in an industrial plant, pull boxes should be located within 50 ft of any equipment that has moderate to serious fire potential (NFPA Standard 101, Life Safety Code).

The alarm can be communicated (1) to a central plant location that is staffed 24 h/day, 7 days/week (e.g., a guard station), which then notifies the fire department; (2) to the fire department; or (3) to an answering service, which then notifies the fire department. See NFPA 72A (Local Protective Signaling Systems), 72B (Auxiliary Protective Signaling

Systems), 72C (Proprietary Protective Signaling Systems), and 72D (Proprietary Protective Signaling Systems).

5.25 Fire Sources.
Fires require oxygen, a source of ignition, and combustible materials.

Maintenance activities may be ignition sources. For joining and separating, use mechanical fastenings and saws and cutting wheels; forbid welding, cutting, soldering, and brazing in storage areas or near paint booths or flammable storage areas. If nonmechanical means are used, require a permit from the safety department and take special precautions during their use. A smoking ban is a good idea, but employees must be given a place to smoke or they will ignore the ban.

Reduce combustible materials by storing spare pallets outside, away from the building. If pallets are stored inside, the pile should be under sprinklers and less than 6 ft high. Minimize the amount of plastics (e.g., a 4-h supply of plastic components) and remove trash frequently. Foam protective packaging, which burns rapidly and has an extremely high heat release, is quite dangerous. Flammable liquids should be stored in safety cabinets or a flammable liquid storage room, not near ignition sources. Flammable materials at the workbench should be in safety cans.

6 PARKING

6.1 Construction
Parking will be assumed to be self-park in a lot, as opposed to attendant park and a garage. With good planning, the parking lot will not need to be moved if the facility is expanded. The lot should be hard-surfaced. Salt applied to melt snow corrodes concrete (as well as vehicles). Use protective coatings on the lot surface. For drainage, use a minimum grade of 1% for asphalt (0.5% for concrete); maximum grades are 3% longitudinal to the stalls, to 5% for cross slopes and aisles. Parking lots should slope away from buildings. Bricks with holes oriented vertically permit grass to grow in the lot; this gives good drainage and esthetics but is poor for snow removal. Lighting generally should be 10–20 lux; since color is not critical, use high- or low-pressure sodium lamps for economy. Minimize glare and spillover onto adjacent property. Often the parking lot is semiconcealed from the street by a ridge of lawn 2 to 3 ft higher than the pavement; low bushes also are used.

Table 9.4 gives the advantages and disadvantages of separating adjacent cars with barriers. If barriers are not used, paint white lines 3 to 4 in. wide to indicate the stall. Double lines between stalls tend to cause drivers to center the car better within the stall.

6.2 Layout
There are five criteria for **stall layout**:

- maximize ease of parking (search pattern, stall entry and exit)

TABLE 9.3 Detector selection guide (Fire Detectors, 1991). Many factors influence the selection of fire detectors. This guide summarizes the major considerations. It is not all-inclusive, and each case must be viewed individually for overriding circumstances.

Type	Applications	Response	Limits/Advantages	Cost
Smoke				
Photoelectric	Offices, computer rooms; projected beam type used in open areas and to protect high rack storage	Early warning to smoldering fires, sometimes in seconds	Must be used indoors where smoke can be contained; not adversely affected by wind	Moderate
Ionization	Offices, computer rooms, combustibles	Early warning to fast-flaming fires, sometimes in seconds	Adversely affected by wind; should be used indoors	Moderate
Heat				
Fixed temperatures	Large areas where life safety is not paramount; to protect heat-generating equipment	Responds when a predetermined temperature is reached, usually in minutes	Use usually limited to indoor applications; has a very low false alarm rate; a simple reliable device	Low
Rate of rise; rate compensated	Large areas where life safety is not paramount	Responds to a specified temperature rise or a selected protection level; usually faster than a fixed-temperature detector	Should be used indoors; may be affected by space heaters; low false alarm rate; suitable for corrosive environments	Low
Flame				
Infrared	Hazardous processes; explosion suppression; ducts or other dark, enclosed areas; aircraft hangars	Rapid response in milliseconds to infrared radiation generated by fire	Indoor use; may be affected by temperature, other infrared sources; explosion-proof housings available	High
Ultraviolet	Hazardous processes; explosion suppression; fuel loading; aircraft hangars	Rapid response in milliseconds to ultraviolet radiation generated by fire	May be used indoors or out; explosion-proof housings available; may be blinded by oil film, thick smoke; sensitive to arc welding	High

- maximize number of stalls
- minimize accidents
- maximize ease of vehicle circulation in lot

TABLE 9.4 Advantages and disadvantages of curbs (bumpers, barriers) within parking lots.

Advantages of curbs, bumpers, barriers

    Define stalls better
    Protect vehicles
    Reduce aisle encroachment
    Define stalls when snow covers lines

Disadvantages

    Increase capital cost
    Make snow removal difficult
    Catch debris and litter
    Interfere with pedestrian traffic

- maximize ease of pedestrian circulation in lot

Note that parking need not be in one lot. Having one lot simplifies security by centralizing the area and by reducing the number of permitted building exits. Multiple lots reduce employee walking (assuming they can leave the building through multiple exits), reduce traffic congestion, and permit better use of irregular sites.

The street entrance and exit should be over 50 ft from intersections; single-lane entrances should be over 14 ft wide, single-lane exits over 10 ft wide; combined entrance-exits should be over 26 ft wide.

Aisles running the long dimension of the lot generally give better search patterns and larger number of stalls. One-way versus two-way aisles and stall angle depend upon the tradeoff of ease of parking versus number of stalls. Figure 9.7 and Table 9.5 give recommended dimensions for various angles. Stall width (dimension D) can be squeezed to 8.5 ft (with greater damage to car doors and

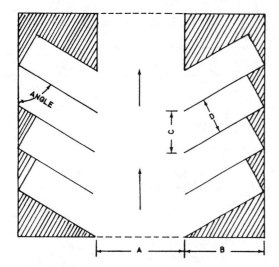

FIGURE 9.7   Dimensions for Table 9.5. Aisles are one-way except for 90° stall angle in which they are two-way.

TABLE 9.5   Recommended dimensions (feet) for parking stalls shown in Figure 9.7 (McElroy, 1988).

Stall Angle (degree)	Alternative	Dimensions			
		A	B	C	D
0	1	12	9.0	23	9.0
	2	12	9.5	23	9.5
45	1	13	19.8	12.7	9.0
	2	13	20.2	13.5	9.5
60	1	18	21.0	10.5	9.0
	2	18	21.2	11.0	9.5
90	1	24	20	9.0	9.0
	2	24	20	9.5	9.5

fenders) if it is essential to maximize the number of stalls. Dimensions A, B, C, and D can be reduced about 1 ft for compact cars (15 ft long). However, it is difficult to administer parking in a lot with different size stalls, so use just one size stall unless space is at a premium.

A lot will use 240 to 300 ft²/car, depending on the stall angle and stall-aisle layout. For minimum space/car and minimum driver convenience, use two-way aisles and a 90° angle and have drivers back into stalls; backing in permits aisles to be 4 ft narrower. For medium space/car and convenience, use two-way aisles and a 90° angle and drive forward into stalls. For maximum space/car and maximum driver convenience, use one-way aisles and angled parking. (For odd-shaped lots a combination of 90° and angled may be best; confine the deviation to one row or end.) Generally, the cars should be bumper to bumper. The herringbone pattern saves space, but since car bumpers match fenders of the opposite car, there is considerable risk of car damage, even if wheel stops are used. If wheel stops are used, assume a 30 in. overhang in front.

Although reserved parking stalls can be used for status symbols, benefit for car pooling, or convenience of handicapped or selected people, stall utilization is better for unreserved stalls. However, some stalls should be reserved for visitors and service vehicles.

In urban areas, where security and nonemployee parking may be a problem, consider fencing and magnetic-card–actuated gates. A sticker on the bumper of employee vehicles helps identify nonemployee vehicles.

6.3 Number of Spaces   Most of the space will be used by employees (plus a few for visitors). With multiple shifts, use the total of two shifts, because one shift arrives before the previous one leaves.

$$SPACES = VISITORS + SF\,(VPE)\,(EMPLOYEES)$$

where

$SPACES$ = number of parking spaces in lot

$VISITORS$ = number of spaces for visitors

SF = safety factor

VPE = vehicles/employee (depends on car pooling arrangements and public transportation, e.g., .98 for a rural site and .5 for an urban site)

$EMPLOYEES$ = number of employees

Stall utilization is better if no parking spaces are reserved. However, reserve a few spaces for visitors, service vehicles, and the handicapped. The Americans with Disabilities Act requirements (28 CFR, part 36) state that a handicapped parking space must be at least 96 in. wide and the curb ramp (maximum slope of 1:12) must be at least 36 in. wide (exclusive of flared sides). Reserve space also for those coming and going on company business. In addition, reserved spaces can be used to reward behavior. For example, reserve a space for the employee of the month with a changeable nameplate. Or reserve spaces for those in car pools or participating in community activities.

Consider small stalls for motorcycles and bicycles if any employees use them.

DESIGN CHECKLIST: BUILDING DETAILS

Foundations and floors
    Load transmission
    Expansion joints

Isolation
Flatness
Concrete, tile, or carpet

Windows
    Cost
    Glare
    Ventilation
    Noise
    Privacy
    View
    Disadvantages overcome?
Roofs
    Type of roof
    Leak prevention
    Ventilators
Building shape & orientation
    Perimeter/area
    Surface area/volume

    Orientation on site
Fire protection
    Emergency organization
    Compartments?
    Flammables
    Portable extinguishers
    Sprinklers
    Fire detection
    Fire sources
Parking
    Surface
    Layout
    Lighting
    Security
    Number of spaces

REVIEW QUESTIONS

1. What is the typical size of an I beam to support the roof in a single-story building?
2. List five different techniques of isolating machines from the floor.
3. Discuss the advantages and disadvantages of windows versus solid walls.
4. Why shouldn't air-conditioner intakes use air from the roof?
5. Discuss, for a 10,000 sq ft building, the ratio of perimeter to area for different building shapes.
6. What are typical low and high bay heights?
7. What shape container minimizes material cost and energy exchanges with the environment?
8. Why do elephants have large ears?
9. Discuss thermal load from the sun as it affects orientation of a building on a lot.
10. Briefly discuss compartmentation for fire design in buildings.
11. Briefly discuss flammable-liquid drum storage rooms.
12. For small rooms with flammables, should the fire extinguisher be inside or outside? Why?
13. Why are sprinklers essentially free?
14. Briefly describe the four main types of sprinkler systems.
15. Briefly discuss the three options in communicating a fire alarm.
16. What three things are required for a fire?
17. Give the formula for required number of parking spaces.
18. Give the three basic approaches for reducing slip and fall injuries.

PROBLEMS AND PROJECTS

9.1 Observe the foundation and interface of machines to the floor for a specific department in a factory. Discuss your thoughts with the department supervisor.
9.2 Observe the windows in 10 offices. How does each conform to the recommendations of this book?
9.3 For a specific building, calculate the perimeter to area and the surface area to volume. Discuss.

9.4 For a specific factory, evaluate its fire protection features. Discuss your conclusions with the safety director. What does the director think of your ideas?

9.5 Design a parking lot for 100 cars. Give assumptions. Justify your design.

REFERENCES

Albin, T. Relative contribution of behavior to slip and fall accidents in mining maintenance. *Proc. of Human Factors Society*, Santa Monica, CA: Human Factors and Ergonomics Society, pp. 511–514, 1988.

Built-up roofing. *Plant Engineering*, pp. 182–186, July 18, 1991.

Campbell soup cools down with roof sprinklers. *IE*, p. 42, Sept. 1989.

Ernst, D. How to increase light levels without increasing lighting. *Material Handling Engineering*, Vol. 31, 5, 56–59, May 1976.

Fire detectors. *Plant Engineering*, pp. 196–198, July 18, 1991.

Gustaferro, A. How to plan and specify concrete floors on grade. *Plant Engineering*, Vol. 34, 23, 73–78.

Hager, N. Energy conservation and floor covering materials. *ASHRAE Journal*, Vol. 19, 9, 34–39, Sept. 1977.

Haldane, J. Possible worlds, 1928. Also reprinted in *The World of Mathematics*, J. Newman, Ed., pp. 952–957, New York: Simon and Schuster, 1956.

Hoover, S. How to prevent snow load damage. *Plant Engineering*, Vol. 35, 25, 81–82, Dec. 10, 1981.

Hoover, S. Inspecting plant roofs. *Plant Engineering*, pp. 57–61, May 24, 1984.

Horridge, P., Woodson, E., Khan, S., and Tock, R. Thermal optical comparisons of accepted interior window treatments. *ASHRAE Journal*, Vol. 25, 2, 45–49, Feb. 1983.

Johnson, D. Preparing to choose a new plant roof. *Plant Engineering*, pp. 30–35, May 16, 1991.

Klein, G. How to determine floor load capacity. *Plant Engineering*, Vol. 36, 5, 69–72, May 27, 1982.

Konz, S. *Work Design: Industrial Ergonomics*, 3rd ed. Scottsdale, AZ: Publishing Horizons, 1990.

Konz, S., Bandla, V., Rys, M., and Sambasivan, J. Standing on concrete vs. floor mats. *Advances in Industrial Ergonomics and Safety II*, B. Das, Ed. London: Taylor and Francis, 1990.

Marinello, R. Vibration isolation mounts. *Plant Engineering*, Vol. 37, 4, 36–40, 1983.

Marshall, G. *Safety Engineering*. Monterey, Calif.: Brooks/Cole, 1982.

McElroy, F., Ed., *Accident Prevention Manual—Engineering and Technology*. Chicago: National Safety Council, 1988.

Metzger, S. How to prevent failures of industrial floors. *Plant Engineering*, Vol. 37, 27, 57–60, Dec. 22, 1983.

Peterson, W. Selecting tapered roofing insulation. *Plant Engineering*, pp. 117–118, March 21, 1991.

Rekus, J. Finishing touch. *Occupational Health and Safety*, pp. 16–19, 48, Feb. 1989.

Riders, Z. Preparing concrete and steel surfaces for painting. *Plant Engineering*, pp. 42–44, April 27, 1989.

Roberts, D. Population differences in dimensions. Chapter 2 of *Ethnic Variables in Human Factors Engineering*, A. Chapanis, Ed. Baltimore, Md.: Johns Hopkins University Press, 1975.

Stein, B., Reynolds, J., and McGuiness, W. *Mechanical and Electrical Equipment for Buildings*, 7th ed., New York: Wiley, 1986.

Steiner, V. Plant roofing: Real world problems and perceptions. *Plant Engineering*, pp. 94–100, March 23, 1989.

Strandberg, L. Ergonomics applied to slipping accidents. Chapter 14 in *Ergonomics of Workstation Design*, T. Kvalseth, Ed. London: Buttersworths, 1983.

Szymusiac, S., and Ryan, J. Prevention of slip and fall injuries. *Professional Safety*, p. 11, June 1982.

Wells, R., and Lupyan, D. Materials for preventing concrete abrasion damage. *Plant Engineering*, pp. 47–49, August 8, 1991.

Whittaker, W. Machinery mounts isolate shock and vibration. *Plant Engineering*, pp. 66–69, Dec. 13, 1990.

CHAPTER PREVIEW

Aisle width is a compromise between wideness (for ease of movement) and narrowness (for conservation of floor space). Traffic type, vehicle type, and layout also affect aisle width. Changing levels with ramps, stairs, ladders, or elevators is discussed, as are various types of doors. Security begins with controlling access to a space.

CHAPTER CONTENTS

1 Aisles
2 Change-in-Level
3 Doors
4 Security

KEY CONCEPTS

Americans with Disabilities Act
   (ADA)
backbone and branch aisles
counterbalanced truck
doors

industrial espionage
Life Safety Code®
lines of defense
narrow aisles
path of egress

ramps
reach truck
security (for computers)
stairs
straddle truck

1 AISLES

1.1 General   Aisle and corridor width depends upon the type of use, the frequency of use, and (with vehicles) the travel speed permitted. In general, they should be as narrow as possible since they use floor space. For example, for Durland Hall at Kansas State University, the second floor, which has a mix of faculty offices and classrooms and thus more traffic, has an 8 ft wide corridor. The third-floor corridor, where there are faculty offices but no classrooms, is 6 ft wide. The 6 ft corridor permits the offices on the third floor to be 2 ft longer than on the second floor; thus each office has $2 \times 10 = 20$ more ft².

In general, use a "**backbone aisle** and **branch** (cross, intersecting) **aisles**. Branch aisles should be located opposite door openings in exterior walls. Aisles serving only one side—such as against walls—waste space. Aisles against columns may be good if the space along the column line is not usable; however, consider column-line space for conveyors and storage. Cross aisles trade off loss of floor space against reduced travel distance. For safety, avoid blind intersections.

Flat or convex mirrors mounted at intersections give reverse situations—you look to the left to see what is coming from the right. See Figure 10.1. An alternative is a spherical mirror mounted at the center of the intersection; its disadvantage is that it needs to be mounted high to avoid being an obstruction, and thus people have to look up.

Aisles in offices, designed for people rather than vehicles, need not be straight. With free-form or landscaped offices, straight lines for the branch aisles are deliberately avoided; see Chapter 13. Manufacturing or storage aisles, however, which have vehicular traffic, should be straight—no curves or jogs. Don't obstruct vision at corners through placement of machines or product. If vision is obstructed (such as by a wall), place mirrors so drivers and pedestrians can see around the corner. Aisle boundaries should be marked with white or yellow paint 3 in. wide. Don't obstruct aisles with temporary storage of product or

people. Don't permit pallets to be placed in the aisle. Don't put columns in the aisle. If a person works on the edge of the aisle, or even stops temporarily such as to get a drink of water, there should be a barrier between the person and the aisle; see Figure 10.2.

1.2 Mixed (Vehicle and People) Aisles   The general-purpose main aisle used in manufacturing to accommodate both human traffic and sit-down counterbalanced lift trucks is 12 ft wide. Since a typical counterbalanced sit-down lift truck is less than 4 ft wide (usually 38–44 in.), this allows plenty of room for two-way traffic and people. For through traffic, reduce this width if the vehicle traffic is one-way, if narrower vehicles are used, or if it is acceptable and possible for people to step out of the aisle to permit a vehicle to pass. See Table 10.1.

Obstructions into an aisle that are out of the line of sight (especially below the thigh) are bumped into more frequently; the frequency of these collisions can be reduced if the object is more conspicuous (e.g., use yellow paint). Doorknobs protruding into corridors tend to snag clothing. Recess the doors or the door grip.

Two-way aisles may be converted to one-way width if a bypass of 10 to 15 ft is located every 100–200 ft along the aisle; however, it is very difficult to enforce one-way traffic. If stacking is required from the aisle rather than just through traffic, then vehicle type becomes the dominant factor in aisle width.

1.3 Vehicle-Only Aisles   If the aisle is in a storage area and not for general through traffic, then narrower aisles make more space available for storage. **Narrow aisles** are 5 to 10 ft wide versus the conventional width of 12 ft. If

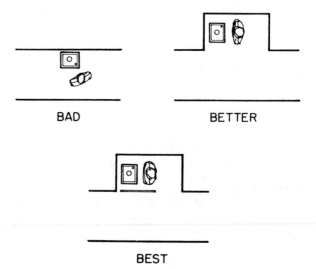

BAD   **BETTER**

BEST

FIGURE 10.2 Drinking fountains and other obstructions should not be in the aisles. The **Americans with Disabilities Act (ADA)** Requirements (28 CFR, part 36, *Federal Register*, July 26, 1991) has specific requirements on fountain design; for example, spouts must be no more than 36 in. above the floor, the spout must be at least 4 in. high (to allow insertion of a cup or glass), and so forth.

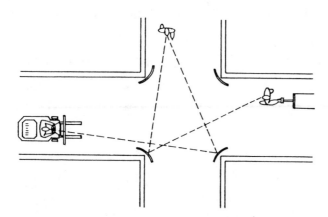

FIGURE 10.1 Corner mirrors reduce traffic collisions.

TABLE 10.1 Recommended one-way aisle widths (Thompkins and White, 1984). For two-way aisles, use the sum of flow in each direction.

Aisle Width (ft)	Type of Flow
	Personnel
3	Personnel
6	Door opening into aisle from 1 side
8	Door opening into aisle from 2 sides
	Equipment
5	Manual platform truck
6	Narrow aisle truck
9	Forklift—1 ton lift
10	Forklift—2 ton lift
11	Forklift—3 ton lift
12	Tractor

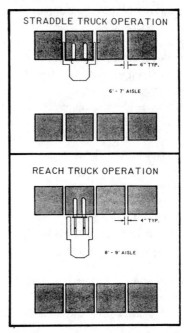

FIGURE 10.3   Straddle versus reach truck comparison (McGaffigan, 1984). The straddle truck straddles its loads and thus needs about 6 in. clearance between stringers for its 4 in. wide straddle arms. If a single wing pallet with 2 in. overlap is used, the stringers can remain 6 in. apart, but the deckboards (and load) are only 2 in. apart; that is, the area of the pallet holding the load is 4 in. wider than the stringers. The reach truck pantograph extends (24 in. for single deep pallets, 42 in. for pallets stored 2 deep) while the truck stays in the aisle. Clearance between loads (stringers) can be 4 in.

the storage aisle is less than 8 ft wide, the National Fire Protection Assoc. (NFPA) requires special sprinkler systems. For information, contact NFPA, Batterymarch Park, Quincy, MA 02269, (617) 770-3000.

Perhaps the first possibility in reducing aisle width is to use a stand-up rider truck instead of a sit-down rider truck. The stand-up version can operate in an aisle .5 to 2 ft narrower, because it is shorter and thus has a smaller turning radius. Walkie trucks save even more in aisle space—but with a penalty in driver fatigue and lower vehicle speed. Most narrow aisle equipment requires the driver to stand or walk.

The next decision is type of vehicle. The primary alternatives are a counterbalanced truck, a **straddle** (outrigger) **truck**, a **reach** (extend) **truck**, a turret or swingmast truck, a sideloader truck, and a stock-picking truck. See Chapter 18 for more detail (figures, costs) on vehicles.

A **counterbalanced truck**—a general-purpose vehicle —counterbalances the load (which is in front of the front wheels) with a battery or engine over the rear wheels. See Figure 18.5. Thus the truck is long and heavy. The heavy battery (engine), however, gives a lot of power and speed— important for high lifting. The truck's large wheels give it good clearance, which is helpful on dock ramps. Because it has no straddles, it can pick up closely spaced loads and various size loads. A 3,000 lb capacity walkie is about 6 ft long and, with a 4 ft load, needs 10 ft aisles to right-angle stack.

Two kinds of narrow-aisle trucks are straddle (see Figure 18.5) and reach (see Figure 18.8). The operator stands or walks. Figure 10.3 shows how they operate; Table 10.2 gives advantages and disadvantages of straddle versus reach trucks.

A turret truck has the load on a turret in front of the front wheels. Thus, although the truck tends to be quite long to counterbalance the weight of the load plus turret in front of the wheels, the truck need not turn to load/unload, and aisles can be 6 to 7 ft. A swingmast truck pivots the entire mast. The truck need not turn and there is no turret to balance, so the truck can be shorter; aisles can be 5 to 6

ft. Since aisle width does not affect truck length, the operator can sit, stand, or walk without affecting aisle width.

A side loader carries the load (generally long items such as pipe) along the side of the truck. Since the truck does not need to turn to load/unload, aisle width can be truck width plus load width plus about 12 in. clearance, if the truck is manually guided. For a 4 ft load and 4 ft truck, this would be 9 ft. Since aisle width does not affect truck length, the operator can sit, stand, or walk without affecting aisle width.

A stock-picking truck has a stockpicker ride on a platform which moves up and down as the truck moves forward and back. The truck does not turn while in the

TABLE 10.2   Comparison of advantages and disadvantages of straddle trucks versus reach trucks. The key advantage for the reach truck is the ability to handle different size pallets.

Advantages of Straddle Trucks

Narrower aisle

Operate faster (no reach mechanism)

Less maintenance (no reach mechanism)

Lower capital cost

Disadvantages of Straddle Trucks

Single size pallet only (in practice)

6 in. between loads (versus 4 in. for reach)

Upper loads in rack need to be aligned with load on floor

Difficult to use in bulk storage

aisle. Aisle width can be truck width plus about one foot clearance on each side.

Aisle width can be reduced to 4 in. clearance on either side of the truck for turret, swingmast, and stock-picking trucks by means of automatically guiding the vehicle on a wire or rail while it is in the aisle.

Larger trucks tend to need wider aisles. Assume a facility needs only a 3,000 lb capacity truck except for a few times per year, when it needs 9,000 lb. Buying just the large truck not only gives high truck operating costs but requires wide aisles everywhere. It is better to buy a 3,000 lb truck and keep the aisles narrower. For the large loads, rent a big truck when needed, or buy a used one; keep a special wide-aisle area for heavy-load storage.

If vehicles and carts use a corridor, have a bumper panel along the walls, corner protectors, and a protective panel on doors.

Box 10.1 indicates the importance of machine arrangement for width of manufacturing aisles. The arrangement and machine orientation have considerable influence on aisle width.

1.4 People-Only Aisles and Corridors

For very light traffic, such as traffic only at shift change time, people can walk between workstations and no aisle is necessary. Corridors are aisles with walls. Since people cannot step out of corridors to avoid oncoming traffic as they can with aisles, corridors should be wider than aisles. Table 10.3 gives some recommended corridor widths. The *Life Safety Code®** specifies 44 in. as the minimum corridor or aisle width in a path of egress. See Box 10.2. Peteroy (1982) says shoulder width is the key factor. For one person, he gives 30 in. as minimum corridor width. For two people, he gives 36 in. if one turns sideways, 44 in. as possible, but 56 in. as minimum for comfort. Rodgers (1983) gives 54 in. as the minimum for two persons passing and 72 in. for three persons abreast; if handcarts are used, there should be 10 in. clearance between the cart and a wall or cart going the other way.

*Life Safety Code® and 101® are registered trademarks of the National Fire Protection Association, Inc., Quincy, MA 02269.

TABLE 10.3 Recommended corridor widths (inches) for people-only traffic.

Number of People	Situation	Minimun	Better
1	Avoid touching equipment or hitting switches	20	24
2	Passing one person standing with back against wall	30	36
3	All three walking abreast in same direction	60	72

Clark and Corlett (1984) give 13 in. as a minimum (moving sideways), 20–30 in. for feeder aisles, 30–36.5 in. for two people passing. Catwalks should be a minimum of 14.5 in. wide for the walking surface but 25 in. wide at the shoulders. It really is the amount of personal space that is the determining factor (see Chapter 13). RTKL Associates (1977) give 60 in. as the minimum width sidewalk so two pedestrians can pass each other without unreasonable evasive maneuvering. They have developed calculating procedures for more congested situations.

2 CHANGE-IN-LEVEL

Moving between levels may be for regular travel or for access to equipment platforms. In general, ramps have about a 5° rise, stairs from 15° to 50°, and ladders 75° to 90°.

2.1 Ramps

If there is no restriction in length, **ramps** are an alternative. Ramp angle should be about 5°; Vaughan (1981) gives 4.9° as the maximum angle for the handicapped, 5.7° as the maximum for the public. For new construction, the ADA requirements (28 CFR, part 36) mandate a maximum slope of 1:12, a minimum rise for any run of 30 in., and a landing length of at least 60 in. Do not put steps on a low-angle slope as people often fall—especially when descending. If handcarts or trucks are to be pushed up the ramp, put a level landing large enough to hold the vehicle at least every 9 ft change in elevation. Ramps should have nonskid surfaces. For ramps used for hand trucking, lay a nonskid strip in the center of each lane. Vehicle ramps exposed to rain or snow should have a 24 in. strip of abrasive metal plate in the track of each wheel; attach it to the concrete with countersunk holes and flat-head expansion screws. This precaution will permit shipping and receiving departments to operate in bad weather. Ramps should have handrails and, when used by heavy vehicles, substantial curbs. See Table 12.5 for ramp angles for vehicles.

2.2 Stairs

Stairs permit much shorter lengths than ramps, an especially important point for inside a building.

Irvine et al. (1990) had people use stairs of systematically varied riser heights and tread lengths. They concluded riser heights should remain between 6 and 8 in.; tread depths should remain between 10 and 13 in. The ADA requirements, however, do not permit treads less than 11 in. A narrow tread supports only the ball of the foot; this is more tiring than if the whole foot is supported. Very long treads (over 16 in.) disrupt walking and cause falls.

Table 10.6 gives recommended riser heights (vertical height of a single step) and goings (horizontal distance from the nosing edge of a step to the nosing edge of the next step) for comfort and safety. Riser heights over 7.33 in. and goings less than 11 in. are not recommended. (Tread depth is the horizontal distance of a tread.)

BOX 10.1 *Machine arrangement patterns*

When orienting machines, consider maintenance access as well as operator convenience. Four arrangements are given (Reed, 1961):

1. *Parallel.* The main axis of the machine is parallel to the axis of the adjacent aisle. See Figure 10.4. Generally there is an aisle, two rows of machines, and then another aisle.
2. *Acute angle.* The main axis of the machine is at an acute angle to the axis of the adjacent aisle. See Figure 10.5. Input to the machines can be from a one-way aisle and output from another one-way aisle. The operation is protected from vehicle traffic. Since the vehicles need not turn

90° and since aisles can be one-way, aisle width may be reduced. Since machines normally are longer than they are wide, the overall percentage of aisle space to total space tends to be lower.

3. *C shaped.* Generally the open side of the C is on the aisle. See Figure 10.6. The operator tends a number of machines.
4. *Free-form.* There is no fixed relation between the machine and the aisle. See Figure 10.7. Machines are arranged to facilitate the transfer of product from one machine to the succeeding machine.

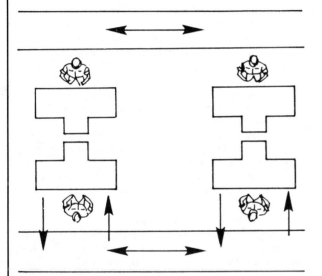

FIGURE 10.4   Parallel axes for aisles and machines are common. Input and output are from the same aisle, and generally there can be two machines between aisles. Input (for right-handed people) should be to the operator's right and output from the left, since reach and grasp are more difficult than dispose. Since the truck needs a 90° turn, the aisle can be two-way.

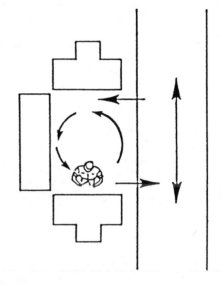

FIGURE 10.6   C-shaped arrangements permit one person to tend multiple machines. The aisle can be one-way or two-way.

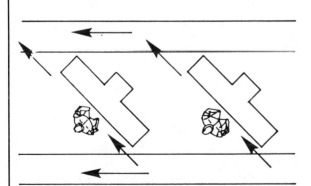

FIGURE 10.5   Acute-angle orientation is especially advantageous for long, narrow machines as the truck need not make a 90° turn in the one-way aisle. If one of the machine orientations is reversed, one operator can tend two machines.

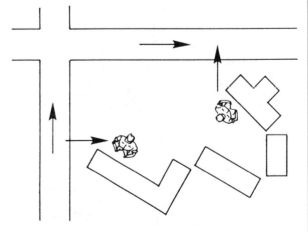

FIGURE 10.7   Free-form arrangements emphasize ease of transfer of product from machine to machine within the cell. The aisle can be one-way or two-way.

BOX 10.2   Egress (fire safety)

General

The **Life Safety Code®** (1991) is based on the concept that people in a fire may not think clearly due to stress and inhalation of smoke and fumes. Thus, egress design must be simple.

The Code varies for different types of occupancies and for new and existing buildings ("grandfathered").

The **path of egress** has three portions: the exit access, the exit, and the exit discharge.

The **exit access** is the portion of the route from any given point in the building to the exit, including not only aisles and corridors but also paths from workstations to the aisles.

The exit is where the inside of the building is separated from the outside. In many cases this is just a door, but it also can be a stairwell (including doors and stairs). An exit passageway provides a protected path of travel within a building. An exit passageway can be used when locating stairs on an outside wall is impractical. It is a horizontal, protected passageway leading to the top of the stairs or from the bottom of the stairs. It qualifies as an exit for measuring travel distance.

The exit discharge is the path from the exit to the "public way"—a place entirely safe from the building.

Egress Capacity

Occupant load is based on the maximum number of occupants at any one time, for any occasion, including unusual circumstances. It is not based on normal occupancy. For offices and industrial space, use 100 ft^2 per person for areas within the interior confines of the building. For exits that serve more than one floor, calculate the load for each floor individually—provided capacity is not reduced in the direction of exit travel.

Stair size is determined by the occupancy of each floor, based on staging. That is, by the time the occupants of the fourth floor reach the third floor, the third floor occupants are already gone.

Flow rate is based on a flow time of about 3.5 min: that is, in a fire people will flow past a specific point for 3.5 min. Table 10.4 gives egress capability per person. For example, for a 32-in. door under normal use, capacity in 3.5 min is 32/0.2 = 160 people. If the door opens on a stair 54 in. wide, the stair has a capacity of 54/0.3 = 180 people. Thus the capacity of the system is that of the smallest component—160.

Egress Components

Egress components include corridors, doors, stairs, and fire escape stairs.

Corridors. The minimum ceiling height in the egress path is 229 cm (90 in.). Any projection from the ceiling must clear the floor by at least 203 cm (80 in.). These dimensions apply also to stairs. Width should be calculated for the occupant load. The minimum corridor width is 112 cm (44 in.) in industrial occupancies.

Doors. Although **doors** restrict access by unauthorized people and minimize environmental disturbances (weather, drafts, noise) from adjacent areas, the Life Safety Code is concerned primarily with passage of fire and smoke.

Doors can be non–fire rated doors, smoke-stop doors, or tested fire doors. Note that none of these doors functions for fire and smoke if left open. Doors should therefore not be secured open because that practice facilitates the spread of fire and fumes.

Doors in a path of egress (see Figure 10.8) should

- be hinged and swing in the egress direction (if the occupant load is over 50 or is from a high-hazard area)
- have no latch or a simple, easy-to-operate latch
- not narrow the egress path
- be side-hinged or the pivoted swinging type
- accommodate the flow of traffic (i.e., have enough capacity)
- require little force to open (quantified as 67 N (15 lb) to open the latch, 133 N (30 lb) to set the door in motion, and 67 N (15 lb) to open the door to the minimum required width)
- use locks that do not require use of a key from the inside (A two-step door, such as knob and independent slide bolt, is not acceptable.)
- be recognizable as doors (Doors cannot be covered with hangings or drapes, have mirrors on them, or blend in with the wall so as to be undetectable.)

Stairs. Table 10.5 gives the code values for new stairs; tread depth and riser height can be different for existing stairs. Note that carpeting on stairs may reduce the effective tread depth. Many accidents result from nonuniform stairs. Nonuniform stairs can be due to construction errors or design (e.g., spiral stairs or stairs with inconsistent tread lengths).

Stair treads should have a high coefficient of friction. Thus, outside stairs need to be sloped to shed water. Carpeted stairs not only improve friction but give a softer impact surface if a fall occurs.

Visual clues are important not only for the presence of stairs but also for the distinction of individual steps. Short stairs (one or two steps) are especially dangerous because people do not notice them. Replace them with a ramp, or, if they must be used, change the color of the floor where the stairs are, change the wall paint, and use a handrail (to serve primarily as a visual notice of the presence of stairs).

TABLE 10.4   Egress capacity requirements (adapted from *Life Safety Code®*, 1991).

Type	Use	Inches/Person	Centimeters/Person
Levels and	Normal	0.2	0.5
Class A ramps[a]	High hazard[b]	0.4	1.0
Stairs	Normal	0.3	0.8
	High hazard[b]	0.7	1.8

[a]Class A ramps have a minimum width of 44 in. (112 cm), maximum slope of 1 in 10, and a maximum height between landings of 12 ft (3.7m).
[b]High hazard: Areas with highly combustible, highly flammable, or explosive products or products that produce poisonous gases or fumes. Also includes areas with fine particles or dust subject to explosion or spontaneous combustion.

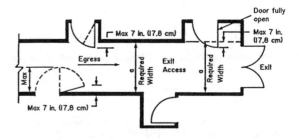

FIGURE 10.8   Door swing into a corridor. The best arrangement for clear passage in an exit access corridor uses doors that swing 180°.

(continued)

BOX 10.2   Continued

All stairs and ramps should have handrails. The height of the top of the rail (to the top of the tread) should be from 86 to 97 cm (34 to 38 in.) with values at the upper end of the range preferred. Handrails should permit a power grip, and so the outside diameter of the circular cross section should be 3.2 to 5 cm (1.25 to 2 in.).

Guardrails should be at least 107 cm (42 in.) high. The concept is to keep them above the hip pivot point.

TABLE 10.5   Stair dimensions (cm) from the *Life Safety Code*® (1991).

Dimension	New Stairs	Existing Stairs Class A	Existing Stairs Class B
Minimum width clear of all obstructions*			
Occupant load of all floors served, over 50	112	112	112
Occupant load of all floors served, less than 50	91	91	91
Riser height, maximum	17.8	19.1	20.3
Riser height, minimum	10.2	—	—
Tread depth, minimum	27.9	25.4	22.9
Headroom, minimum	203	203	203
Height between landings, maximum	370	370	370
Door opening immediately onto stairs (without landing at least width of door) permitted	No	No	No

*Except projections not exceeding 8.9 cm (3.5 in) at and below handrail height at each side.

Fire Escape Stairs. The goal of fire safety can best be served by the proper internal design of means of egress and the phasing out of fire escape stairs as new buildings replace old ones. However, it is possible to design outside stairs that meet the code.

Exit Number and Arrangement

The minimum number of exits is two (three if there are over 500 occupants, four if there are over 1,000 occupants). The minimum is based on a floor-by-floor consideration, not an accumulation of floors.

The multiple exits should be not only separate but remote from each other. *Remote* is quantified as a ratio to the diagonal of the area. Exits must be separated by 50% of the diagonal (except for areas with sprinklers, where 33% is permitted). There shall be no dead-end pockets, hallways, corridors, passageways, or courts. Maximum travel distances vary from 50 to 400 ft, depending on the hazard and the use of sprinklers. See the code for specific values.

Illumination on the floor shall be at least 10 lux for exit access, exits, and exit discharges.

Reprinted with permission from NFPA 101-1991, *Life Safety Code*®, Copyright © 1991 National Fire Protection Association, Quincy, MA 02269. This reprinted material is not the complete and official position of the National Fire Protection Association, on the referenced subject which is represented only by the standard in its entirety.

When the tread is less than 9 in., the ANSI and OSHA standard 1910.24, Fixed Industrial Stairs, requires open risers (no obstruction below the nosing). However, open risers do increase the fire hazard and the ADA does not permit them.

Dimensions of adjoining risers and treads must be constant (dimensional differences between the largest and smallest in a flight should not exceed 3/16 in.) and regular. Therefore, avoid spiral stairs.

ANSI also specifies that each tread and the top landing should have a nose which extends .5 to 1 in. beyond the face of the lower riser. Templer (1992) recommends that overhangs should be 11/16 in. or less and that the nosing overhang should be backward sloping rather than abrupt (formed when the tread projects beyond the riser).

External stairs should have a wash (slope of less than 1:60) to permit water to drain. Water freezing on treads of outside stairs is a serious problem. In addition to drainage, consider preventing the water from getting to the stairs.

A flight of stairs should have 3 or more risers; in many codes the maximum number of risers in a flight is 18 (Templer, 1992). Very short flights (1 or 2 steps) are especially dangerous because people don't notice them. If very short flights are needed, use handrails, color, and lighting to emphasize their location.

Landings provide a resting place, form turning zones, and, in the event of a fall, break the force of the fall. The minimum landing width must be equal to that of the widest flight of steps that reaches it. The minimum clear depth must equal the widest stair run or 48 in., whichever is larger (Templer, 1992). Intermediate landings have no change in direction; a quarter landing changes direction by 90°; a half landing changes direction by 180°.

The procedure for designing straight-flight stairs (Templer, 1992) is as follows:

1. Determine floor-to-floor height (H). Select acceptable riser (R) and going (G) dimensions. If space is limited, choose maximum R and minimum G.

2. Determine the number of risers (n): $n = H/R$. Set n equal to the nearest whole number and adjust the riser dimension.

3. Determine the number of treads (t): $t = n - 1$

4. Determine run length (L): $L = G(t)$

To minimize impact injuries, stairs, handrails, and balustrading should be free from projecting elements, sharp edges, and corners.

TABLE 10.6   Recommended riser and goings (in inches) for stairs (Templer, 1992).

Riser	Goings						
7.2	11						
7	11						
6.5	11	11.5	12	12.5			
6	11	11.5	12	12.5	13	13.5	14
5.5	11	11.5	12	12.5	13		
5	11	11.5	12				
4.6	11						

When visibility is poor, stair accidents increase. In addition to no or low light, excessive light (outdoor sun on white stone) also can be a problem because of difficulty in differentiating between riser and tread. Rogers (1983) suggested the following:

- Provide color contrast between the tread nose and the rest of the tread.

- Have matte rather than high-gloss finishes on the steps.

- Avoid carpet patterns that confuse depth perception (e.g., narrow strips with strong contrast).

- Have the handrail color contrast with that of the wall and the stair.

See Box 10.2 for more comments on stairs.

Maki et al. (1985) recommend a minimum handrail height of .91 m (35.8 in.) and a maximum of .97 m (38.2 in.). ADA requirements (28 CFR, part 36) state that the top of the handrail gripping surface must be between 34 and 38 in. above the stair nosing.

Armstrong et al. (1978) recommend a 4.6 in. wall clearance with 2.2 in. minimum. ADA requirements (28 CFR, part 36) state that the clear space between the handrail and the wall shall be 1.5 in.

The handrail (a grab bar) should be small enough to be gripped with a power grip (i.e., thumb lock on fingers) —the fingers must reach completely around the rail. Armstrong et al. recommend a 4.4 to 5.2 in. handrail circumference.

Scaffolds and work platforms (e.g., for stock picking or replenishment or maintenance) should have a waist-high guardrail.

Minimum stair width is 22 in. RTKL Associates (1977) recommend a minimum width of 48 in. as two pedestrians can pass comfortably on the stairway. Vertical clearance should be at least 84 in. above each tread's nosing. Design load should be 5 times the normal live load with a minimum of 1,000 lb. Use 12 ft as the maximum vertical distance between landings. A landing should be at least 30 in. long, measured in the direction of travel. Another recommendation is that minimum landing length equal stair width.

Steps are difficult for the handicapped and for cart movement. Elevators, ramps, and curb cuts are required in many situations.

2.3 Ladders

Figure 10.9 gives the recommended dimensions for stair ladders (slope 50–75°) and fixed ladders. Ladder angle should not exceed 90°. Fixed ladders (generally 90°) should have a cage or ladder safety device if they are more than 20 ft in height. They should extend 3.5 ft above the landing. MIL-STD-1472B and OSHA .25–.27 have more details. Provide a foot scraper and a mat at the top and bottom so users can remove slippery materials from their shoes. Ladder rungs can have abrasive material on the rungs.

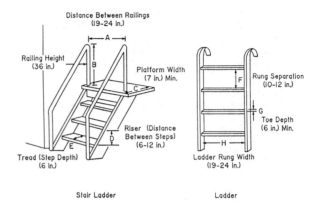

FIGURE 10.9 Recommended dimensions for stair ladders and ladders (Rodgers, 1983; Woodson, 1981).

The ideal angle for a portable ladder is about 75°— greater increases the danger of the ladder falling backward and less increases the danger of the base slipping (Marshall, 1982). This angle can be approximated by having the user stand erect with toes against the ladder and arms parallel to the ground and comfortably grasping a rung. For 95% of the male population this gives an angle between 70.7° and 71.7°; for 95% of females the angle is between 73.8° and 74.4°. Figure 10.10 is a pictogram which could be used for instruction.

Portable step stools and stairs with retractable casters (which become stable as soon as a person stands on them) are easy to move about the workplace (Rodgers, 1983).

2.4 Elevators

For low-rise buildings, hydraulic power for elevators is most common. The best location for the power unit (machine room) is on the lowest floor adjacent to the elevator shaft; remote location is also possible. Typical machine room size for a one-car installation is 7.8 × 5.5 ft by 8 ft high; sound isolation is recommended; heating and ventilation are required. Elevator doors can be center opening (one door moves left and one right—best for quick entry and exit), two-speed sliding (both doors move in the same direction, one sliding behind the other—gives

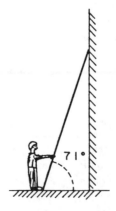

FIGURE 10.10 Place a ladder by putting feet against ladder beams and grasping a rung with arms parallel to the ground (Irvine, 1978). Figure adapted from *Proceedings of the Human Factors Society*, 1978, p. 586. Copyright 1978, by the Human Factors Society, Inc., and reproduced by permission.

maximum opening width), or single sliding (one door moving either right or left—cheapest and slowest). Be sure the door opening is wide enough (42 in.) for handicapped use. ADA requirements (28 CFR, part 36) for door opening width are 36 in. minimum; there are other requirements for control locations, car-to-floor tolerance (less than .5 in.), call button size and location, illumination, and so forth. Allow 4 ft² per person for passengers waiting for the elevator. Allow 1.9 ft² per person (mixed gender occupancy) within the elevator (Templer, 1992).

If the elevator is used for freight, protect the walls and corners with bumper strips.

3 DOORS

The first question is whether the door is necessary. Determine its purpose and how it affects security, privacy, temperature control, noise control, fire control, and insect control. Determine why the door must be closed (i.e., what environments are separated?) and why it must be opened (i.e., who and what passes through the door?). Determine the relative priority of opening versus closing.

The *Life Safety Code®* recommendation is a maximum of 150 ft between a person and an exit for buildings that have no high-hazard occupancy and that are completely protected by automatic sprinklers; 100 ft is maximum for unsprinklered buildings, and 75 ft for high-hazard buildings.

For exterior doors, consider a canopy for sun and rain protection; in some climates a wind screen is desirable to protect against the direct blast of prevailing winds.

Given its necessity, the next question is the type of door. Most doors, especially for people, are hinged at the side and are solid. Doors with louvers generally are undesirable since the louver permits passage of noise, as well as smoke in case of a fire. Sliding and folding doors have a tendency to jam, so they should never be used for emergency exits.

Considerable energy can be lost through air exchange when a door is open to the exterior.

$$AVOL = DOORA \cdot AVEL \cdot TIME \qquad (1)$$

where

$AVOL$ = air volume, ft³

$DOORA$ = door opening area, ft²

$AVEL$ = air velocity, ft/s

$TIME$ = time door is open, s

For example, if $DOORA$ = 10, $AVEL$ = 14.7 (i.e., 10 miles/h), and $TIME$ = 6 s, then $AVOL$ = 20 (14.7)(6) = 800 ft³ of air. ASHRAE (1977) gives 900 ft³/person for a single-bank door, 550 for a door with a vestibule, and 60 for a revolving door.

Revolving doors generally are not recommended. To reduce interior climate loss, use an airlock (doors on either side of a 6 to 10 ft vestibule).

For large numbers of users in a cold climate, a revolving door can be used, but it should be flanked on either side by a conventional door (for fire safety and handicapped access).

Swinging doors (opening both ways) are more hazardous than doors that have a closer and check and that open only one way. If swinging doors are used for people, they should (1) operate in pairs, in and out; (2) be separated by a door post; (3) have windows or openings for traffic visibility; and (4) have the hinges "in" to avoid interference. If vehicles use the door, mount hinges out, with no center post.

Industrial (vehicle) doors can be subdivided into horizontal acting, up-down acting, and air curtains and strips. See Figures 10.11, 10.12, and 10.13. Controls can be manual or automatic. A simple, effective control is a pull-cord switch on each side of the door: Pull one cord to open the door, the other to close it. Automatic opening and closing can be by radio or proximity wire loops. A photoelectric beam and a timer are satisfactory for one-way traffic but tend to keep the door open too long for two-way traffic.

Door height for people is not adequate for lift trucks. Check the minimum mast height.

For safety, doors should open *out* of buildings and corridors. In production areas, be sure the door is wide and high enough to permit vehicles and equipment to pass through. Door sills are not desirable. If sound or climate control requires sealing the lower gap, use a flexible strip on the door bottom. Doors of completely transparent glass are dangerous because people walk into them. Mark them, making some portion opaque or translucent rather than transparent, or use a pattern to avoid accidents. On the other hand, a completely opaque door may not always be best if collisions are to be avoided. To preserve appearance, use stainless steel push panels when the door is pushed open by carts or by hands. Is a doorknob necessary? They

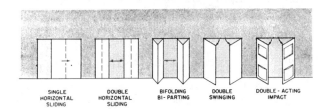

SINGLE HORIZONTAL SLIDING   DOUBLE HORIZONTAL SLIDING   BIFOLDING BI-PARTING   DOUBLE SWINGING   DOUBLE-ACTING IMPACT

FIGURE 10.11 Horizontal doors always provide full opening height (Industrial Doors, 1978). All but the impact doors use electric or air-powered actuators. The doors can be insulated and fire rated. The single opens only half as fast as the double but requires clear wall space on only one side. The double is center opening, thus keeping vehicles away from the jambs. The bifolding, biparting door requires no wall space and opens even faster than the double horizontal; however, the folded panels project from the wall. The lightweight double swinging door is fastest of all to open and requires no wall space; however, the panels project from the wall. The nonpowered impact door has low capital and operating cost. Windows in the swinging and impact doors reduce accidents. Strong winds may make opening difficult or prevent complete closing.

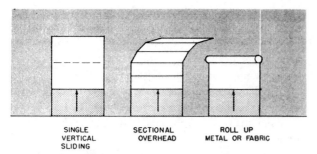

SINGLE VERTICAL SLIDING   SECTIONAL OVERHEAD   ROLL UP METAL OR FABRIC

FIGURE 10.12   Up-down doors can be manually or power operated (Industrial Doors, 1978). For manual operation, use springs or counterweights. The low capital cost single-panel door requires ample headroom; sectional and roll-up doors can use low headroom. Both the sectional and single panel can be better insulated than the roll-up. Because of low maintenance, steel doors are more popular than wood. An uninsulated steel door can have a U factor of 1.0 BTU/hr-s ft²-F. Insulated doors have U factors of .14 to .20. Air infiltration (leakage) around the door is another large source of heat loss. Reduce infiltration with bottom seals, perimeter seals, weatherstripping joint sections, double glazing of windows, and, especially, powered opening and closing. Paybacks of 1 to 2 years are reported.

are difficult to use when the hands are loaded. Can the knob be replaced with a lever or a push plate, at least on one side? Need the door be lockable? If it is lockable, can a person unlock the door and open it with one hand, for example, with a briefcase in the other hand?

For small areas such as private offices, locate the door in the corner so it swings through 90° of arc. For larger areas (holding over three people), put the door in the center of the wall (180° swing). Allow at least 6 in. of space between the moving door edge and any fixed object (such

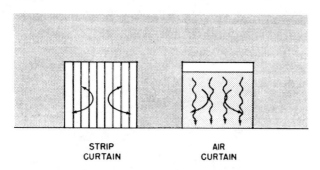

STRIP CURTAIN   AIR CURTAIN

FIGURE 10.13   *Air curtains*, invisible doors, project an air stream through which people, vehicles, or product (on a conveyor) pass without interference; they provide a barrier against wind, dust, heat, cold, and insects. They can reduce heat loss from an oven traversed by a conveyor. In addition to saving energy in the oven, oven temperatures are more consistent (improving quality) and fewer fumes enter the workspace. Air curtains require continuous power while operating but can be turned on and off. They are often used as auxiliary doors with other doors. *Strip doors* are made of overlapping strips of plastic (West, 1988). The thickness of the plastic affects primarily wind resistance and infiltration. Strip doors provide two advantages: The door closes immediately, and only the required width is used (a 4-ft wide truck going through an 8-ft wide door moves only 4 ft of strips). In addition to providing a partial barrier against wind, heat, cold (freezers and air conditioning), and insects, they reduce noise. When containing welding stations, strips offer transparency and accessibility and are light enough to be portable. Strips are useful on refrigerated trailers. Strips with a rib reduce the amount of contact, and thus abrasion, from passing traffic.

as file cabinets) to minimize accidents. For doors opening from a corridor, the distance from the hinge side of the door to the wall should be 3 in., although 0 is feasible; the distance from the knob side of the door to a perpendicular wall should be 12 in., although 1 in. is feasible.

Office doors are typical obstructions for moving equipment. An office door 32 in. wide or wider is best. A 30 in. door causes some trouble for passage of 30 in. wide desks, especially if the door doesn't pull clear of the opening. A 28 in. wide door is big trouble.

4   SECURITY

4.1   General
Security begins with controlling entry and access to a space. There are three ways to identify whether a person may or may not enter; by what the person has (key or badge), by something the person knows (combination, password), or by something the person is (photo of face, voice, signature, fingerprints). It usually is wrong to assume that a potential intruder denied access by the system will not try to force the system or seek to enter some other way.

It may be desirable to be able to seal off certain work areas (night work, weekend work, and confidential material) from the remainder of the building. Be sure to provide toilet facilities. A unisex toilet is sufficient unless there is a large number of people.

Security for a facility should have three **lines of defense**: perimeter (the property edge), area (the building), and object (individual room or cabinet). Because human labor is very expensive—especially if needed 24 hours/day, 7 days/week—the tendency has been to use devices more than people for security.

The perimeter can be protected by fences, walls, and gates that lock as well as by various sensors such as closed circuit TV (CCTV). CCTV can be a simple system with fixed cameras, a switch, and one monitor. It can be upgraded to include a videotape recorder, a time/date generator for all scenes, pan and tilt cameras, zoom capability, and two-way voice communication. Low-light-level cameras may eliminate the need for supplementary lighting. In general, it is better for alarms to sound remotely at some central location rather than at the point of entry. A zone indicator pinpoints the alarm location. Baker and Lyons (1978), in their detailed paper, give many specific lighting recommendations for security.

The building needs to be secured for any opening over 96 in.² that is less than 18 ft above ground. Thus, not only doors need to be secured but also windows, basement grates, ventilation equipment openings, and skylights. Receiving and shipping areas are the most vulnerable to theft (Jaspan, 1981). Dual accountability (having two people check shipments) works well to prevent collusion between employees and truck drivers. Pedestrians should be kept away from the docks. Trash collection areas and trash vehicles are notorious for smuggling out goods. Night access to trash areas should be restricted, since it becomes a convenient pickup point for something stolen during the day.

Objects protected are cashier offices, research laboratories, classified materials, and computer facilities. Interior spaces, as well as the perimeter, can be protected by ultrasonic motion detectors, microwave instrusion sensors, and passive infrared sensors. There is a false-alarm problem with most of these devices. Most of this third line is protection against employees, especially from "white-collar crime" such as embezzlement or industrial espionage.

Industrial espionage is the theft, usually on a continuing and systematic basis, of confidential and sensitive information. "Right to privacy" also is a consideration for personnel records. Mailrooms, administrative support/filing areas, and word-processing areas are targets. Menkus (1981) suggests the following: (1) Lock cabinets; (2) label drawers with numbers rather than words (strangers have a hard time finding things); (3) shred excess output; and (4) disconnect electric power at night at a power panel and lock the panel. To prevent theft of data sent over phone lines, the data can be coded. The Data Encryption Standard (DES) has a 56 bit key programmed into the machines on each end of the line; the key is changed periodically.

Computers usually can be accessed from many terminals—often over telephone lines. This has led to many **security** problems. Control techniques include the following:

- Restrict access to terminals. Use lockable keyboards. Program the computer not to accept certain types of transactions outside of specified hours.

- Use passwords. Avoid easily identified passwords such as the operator's first name. Use symbols (e.g., ?) and numbers as well as letters because more alternatives make the code more difficult to break. Intruders may not even try symbols when attempting to break the code. Password codes should be of variable length beyond the minimum. The password can be multiple level. For example, after the operator enters the initial password, the computer asks a random personal question from a prearranged list. (What is your spouse's first name? What is the dog's name? What town were you born in?) The password should be changed periodically. Different passwords should be used for different types of transactions. The computer should print out a daily list, by terminal, of all invalid passwords attempted.

- If the computer is accessible by telephone, the telephone numbers should be secret and changed periodically. In another option, the remote individual dials the computer, the computer records the password, and then the call is returned to the authorized remote site.

4.2 Locks and Keys   Doors can be held closed simply with a latch-set button (passage set) that does not lock. A privacy lock is locked on the inside by a push button and opened by turning the knob; from the outside it is unlocked by a key. A classroom lock is always free on the inside; on the outside a key is needed to lock or unlock. A storeroom lock is similar to a classroom lock, but the door locks automatically when it is closed.

The choice of a lock is a compromise between convenience and security. Remove as much trim (knobs, pulls, handles, cylinders) as possible from the exterior exits to make forced entry difficult. The hinge side of the door is the low-security side because, given enough time, the hinges can be removed. Many door locks (and, of course, windows) can be quickly defeated with a crowbar or hammer.

For security, minimize the number of access points to and from the perimeter, area, or object. They can be supplemented with emergency exits equipped with a panic bar deadbolt which sounds an alarm when opened.

In addition to individual (change) keys, master keys generally are given to specified individuals in each area, such as maintenance, offices, and production. A grand master key opens all locks in all areas. Some locks should not be master-keyed. Two examples are employee tool boxes and maintenance lockout locks (thus no one can start equipment when maintenance is working on it because the maintenance employee has the only key). Padlocks and shackles are rarely picked; generally the shackle is snipped or sawed. Use large-diameter shackles of hardened steel with short or shrouded shackles.

Mechanical keys in a cylinder have been complemented by electronic keys, in particular by a marriage between magnetic cards (which act as a key) and computers. The card, the size of a credit card, has a magnetic code sandwiched between the two exterior surfaces (there also are optical-coded, passive-electronic, active-electronic, and capacitance-coded cards). The card surfaces can be printed or embossed with the cardholder's name, organization logo, or personal picture. The advantages of the card—even without a computer in a stand-alone operation—include the picture in combination with the key (so humans can check if the proper person has the key), the difficulty of counterfeiting or duplicating the card, and the difficulty of picking the lock that is controlled by the card reader.

An important advance in card access was the marriage with the computer. When the card is inserted in the reader, the information is sent to a computer for a decision on whether to open the lock. Some systems operate 24 hours/day and others only outside normal working hours. The computer advantages are control, monitoring, and documentation.

The computer knows whether a card is to open all doors or just specific ones. In addition, a card can be valid only during certain time periods. Thus, the decision to open can be based on whether the card is proper not only for the door but also for the time of day or week. Elevators in office buildings can be made inoperative without the proper card and then programmed to go only to the floor authorized for that card. Cards also can be used to actuate copy machines and gasoline pumps, with automatic billing! Readers can be programmed to "eat" the card (not give it back) if it is invalid or to actuate alarms. The computer also is useful in eliminating "passback" (e.g., a person using a card to gain access to a parking lot and then passing back the card to the driver of the next car). The easiest solution is to have the computer deny the use of the same card twice within a specified time period. The computer also can keep track of who is in the building. Example applications are keeping track of which doctors are in the hospital and who is where in an office building at night, in case of emergencies such as fires. The cards also can replace punching in and out on time clocks. Documentation is a powerful tool, as the daily printed record of all users of a card (whether valid or not) is a deterrent to thieves.

If a card is lost or stolen, the present procedure is not to reassign a new card number but to assign a new issue number of the same card number. The computer accepts only the latest issue number.

In very high-security situations, such as power plants or computer centers, the card can be combined with a keyboard requiring the keying of a personal code of 4 or 5 digits. (Banks use this system for cash withdrawals at remote terminals.) Thus, even the use of the proper card at the proper time is not sufficient.

Systems that depend on physical traits, such as fingerprints or voice, are the most difficult systems to defeat.

Whether a security system depends on what a person has, knows, or is, no system is perfect. In particular, consider error rates (false rejection of authorized people and acceptance of imposters), effect of power loss, sabotage and vandalism resistance, maintenance, and ease of modifying the system.

DESIGN CHECKLIST: AISLES AND SECURITY

Aisles
    Vehicle, vehicle + people, or people only
    One-way or two-way
    Type of vehicle using aisle
    Location versus doors
    Straight for nonoffice locations
    No blind intersections
Change-in-level
    Ramps, stairs, ladders, or elevators
    Ramp angle, nonskid surface, handrails
    Stair angle, tread, railing
    Step visibility
    Ladder angle

    Elevator type, door type, door width
Doors
    Purpose, type, size, height of each door
    Traffic pattern, direction of opening
Security
    Threat of physical theft, espionage, embezzlement
    Perimeter security
    Area security (receiving and shipping, computers, offices)
    Object security (money, documents, computer terminals)
    Types of locks
    Types of keys

REVIEW QUESTIONS

1. Cross aisles trade off what two things?
2. Aisles in manufacturing should be straight, but aisles in offices need not be. Why?
3. What is the storage aisle width below which special sprinkler systems are required?
4. Discuss how different truck types affect aisle width.
5. Why does a stand-up truck require a narrower storage aisle than a sit-down truck?
6. Sideloading trucks need narrower aisles than counterbalanced trucks. Why?
7. Why, for people-only traffic, can aisles be narrower than corridors?
8. Design interior stairs between two levels 10 ft apart.
9. Should doors open into or out of a corridor?
10. How wide should an office door be?
11. Where are the three lines of defense against theft?
12. What are the differences among a privacy lock, a classroom lock, and a storeroom lock?
13. Give three advantages of magnetic card keys when a computer is not used.
14. A magnetic card key plus a computer gives advantages of control, monitoring, and documentation. Give an example of each of the three.
15. A person wanting to enter a space can be identified in what three ways?
16. Give seven recommended password approaches.

PROBLEMS AND PROJECTS

10.1 Design stairs between two floors 14 ft apart. Justify your design.

10.2 Measure the width of the aisles in a factory. Any ideas? Does the supervisor agree with your ideas?

10.3 Evaluate the security of a factory, office, or warehouse. Any ideas? Does the supervisor agree with your ideas?

10.4 Read 3 to 5 articles on a topic related to this chapter. Summarize them in a 500-word executive summary. List the references. Don't use articles cited in the text.

REFERENCES

Armstrong, T., Chaffin, D., Miodonski, R., Strobbe, T., and Boydstuw, L. An ergonomic basis for recommendations pertaining to specific sections of OSHA standard, 29 CFR part 1910, subpart D—walking and working surfaces. Washington, D.C.: U.S. Dept. of Labor, 1978.

ASHRAE. *Handbook of Fundamentals*, p. 21.11, Atlanta: ASHRAE, 1977.

Baker, J., and Lyons, S. Lighting for the security of premises. *Lighting Research and Technology*, Vol. 10, 1, 10–18, 1978.

Clark, T., and Corlett, E. N. *The Ergonomics of Workspaces and Machines: A Design Manual*. London: Taylor and Francis, 1984.

Hohman, J. Plastic strip and sheet barriers. *Plant Engineering*, Vol. 34, 16, 71–73, Aug. 7, 1980.

Industrial doors. *Modern Material Handling*, Vol. 33, 3, 94–103, March 1978.

Irvine, CA. A human factors approach to slippery floors, slippery shoes and ladder designs. *Proceedings of 22nd Annual Meeting of the Human Factors Society*, Santa Monica, Calif.: Human Factors and Ergonomics Society, pp. 583–587, 1978.

Irvine, C., Snook, S., and Sparshatt, J. Stairway risers and treads: Acceptable and preferred dimensions. *Applied Ergonomics*, Vol. 21, 3, 215–225, 1990.

Jaspan, N. Warehouse theft—It can be stopped. *Modern Materials Handling*, Vol. 36, 10, 58–61, July 6, 1981.

Maki, B., Bartlett, S., and Fernie, G. Effect of stairway pitch on optimal handrail height. *Human Factors*, Vol. 27, 3, 355–359, 1985.

Marshall, G. *Safety Engineering*. Monterey, Calif.: Brooks/Cole, 1982.

McGaffigan, J. Facts about narrow-aisle lift trucks. *Plant Engineering*, Vol. 38, 4, 60, Feb. 24, 1984.

Menkus, B. A practical approach to office security. *Administrative Management*, Vol. 42, 6, 31–33, 92, June 1981.

NFPA. *NFPA 101 Life Safety Code® 1991*, NFPA, 1 Batterymarch Park, Quincy, MA 02269.

Peteroy, F. Studies of human motor ability yield space planning solutions. *Contract*, Vol. 24, 8, 91–95, Aug. 1982.

Reed, R. *Plant Layout*. Homewood, Ill.: Irwin, 1961.

Rodgers, S., Ed. *Ergonomic Design for People at Work*. Belmont, Calif.: Lifetime Learning Pub., 1983.

RTKL Associates. "Pedestrian Planning Procedure Manual." Baltimore: RTKL Associates, 1977.

Templer, J. *The Staircase: Studies of Hazards, Falls, and Safer Design*. Cambridge, Mass.: MIT Press, 1992.

Thompkins, J., and White, J. *Facilities Planning*. New York: Wiley, 1984.

Vaughan, P. Vertical movement. In *Architectural Graphic Standards*, 7th ed., C. Ramsey and H. Sleeper, Eds., p. 7. New York: Wiley, 1981.

West, F. Systematic inplant door selection. *Plant Engineering*, pp. 70–72, Sept. 22, 1988.

Woodson, W. *Human Factors Design Handbook*, New York: McGraw-Hill, 1981.

CHAPTER PREVIEW

Much of a facility's space is required for storage, not for operating equipment. Storage can be divided into three types: moving the order picker to the package, moving the package to the picker, and eliminating the picker (mechanization). Organization of items in storage can reduce space requirements as well as improve picking efficiency.

CHAPTER CONTENTS

1 What Is Stored
2 Equipment
3 Organization of Items in Storage Units

KEY CONCEPTS

accumulation conveyors	kitting	racks, drive through
aisle captive	last in, first out (LIFO)	racks, flow
automatic storage retrieval system (AS/RS)	miniload	racks, single depth
	order picking	rack-supported structure
bulk storage	package to picker	random storage
cantilevered shelves	Pareto principle	replenishment aisle
carousel	picker to package	stock-keeping unit (SKU)
case system (versus split case)	picking aisle	storage categories
first in, first out (FIFO)	racks, double depth	storage/retrieval (SR) machine
high-density storage	racks, drive in	

1 WHAT IS STORED

If you look at a scale layout of an existing factory, there seems to be far too much room for the small number of machines. Why is the building so big? Storage! And even storage areas often have 40% or more of their space for aisles, offices, and pick-up and drop areas. Table 11.1 shows eight **storage categories**. Often forgotten is enough space for supplies and rework, especially unreported rework. (Much rework never enters the formal paperwork system. After all, why should anyone report mistakes? It just gets supervisors angry. People just fix the units and put them back into the system. Thus, don't depend on the paperwork trail for the real volume of activity.)

How much storage space is required for product? The following formula can be used for rough approximations.

$$SPACE = ICU * NEST * UNITL * STORE * DAYS$$

where

$SPACE$ = ft³ of space required per part number (e.g., cans of 3.2 light beer)

ICU = ft³ of space required per individual item (e.g., for one can of beer) *Note:* many items are not flat, rectangular, rigid, or regular.

$NEST$ = multiplier to adjust for a package of items (a six-pack or case of beer)

$UNITL$ = multiplier to adjust for a unit load (e.g., a pallet of cases of beer)

$STORE$ = multiplier to adjust for utilization of the unit load in the building (e.g., pallets occupying only 60% of the storage space)

$DAYS$ = multiplier to adjust for the number of days of inventory desired to be stored

Item weight also should be recorded for rack, floor, and equipment load calculations. Finally, record quantity received per shipment and quantity issued per requisition. A good rule is that a small percentage (e.g., 20%) of the part numbers use a large percentage (e.g., 80%) of the space. Generally only a sample of 500 or so parts are examined in detail with the total extrapolated from the sample. Some cartons must be stored upright because the product might be damaged in a different orientation. Other cartons have built-in skids (blocks of wood, either two way or four way) on one face; this also limits possible storage orientations. There are complex interactions among carton size, pallet size and loading pattern, and storage size as you trade off storage utilization, transportation utilization, and flexibility.

One way to reduce the amount of storage space required is to reduce the procurement lead time. Some possibilities (Lauer, 1966) are as follows:

TABLE 11.1 Storage can be divided into eight categories.

Abbre-viation	Type of Storage	Examples
RM	Raw material awaiting manufacturing	Steel sheets
PP	Finished parts awaiting use in manufacturing	Castings, resistors, purchased subassemblies
SU	Supplies and records: maintenance supplies, office supplies, packaging supplies, tools, manufacturing supplies	Drill bits, welding rod, paint, floor wax, light bulbs, cardboard boxes, shipping labels, personnel records, catalogs, invoices
EQ	Equipment: auxiliary equipment, containers, material handling equipment, unused manufacturing equipment	Extra lathe, floor sweeper, fork truck, extra conveyor hooks, pallets
WP	Manufactured items between operations 1. Between fabrication and assembly 2. Between operations	Completed clamp Clamp after forming but before drilling
SR	Scrap and rework 1. Scrap material or units	Lathe chips, torn paper, units that can't be fixed, vendor packaging
	2. Rework	Units that are repaired
PE	Personal equipment	Coats, lunches
FG	Finished goods	Completed units in package—ready for shipment

Check to see if the lead time is realistic (the estimate may be 5 or 10 years old).

Have the vendor ship to you on consignment. What you don't use, you send back. This doesn't save you space, but the vendor absorbs the carrying cost (although possibly increasing the selling price to cover the cost).

Vendor stocks the items in a nearby warehouse; this cuts transport time.

Vendor accepts telephone orders for standard items; this eliminates paperwork processing time.

Vendor partially processes item or stocks long-lead components; this reduces manufacturing time.

The above are preliminary steps toward Just-In-Time (Kanban) production, where the concept is to deliver the item just in time.

With the increasing computerization of inventory information in retail outlets and improved telecommunications, many retailers now communicate sales of each item directly to the manufacturer's factory (rather than to the warehouse). The sale is entered automatically into the production schedule and the shipping schedule. The shipment goes to the retailer directly or, if it stops at the warehouse, it is only for switching rather than for holding (Drucker, 1992).

2 EQUIPMENT

Storage has a variety of tasks:

- Stocking (putting a **stock-keeping unit [SKU]** into a storage location), also called *replenishing*.
- Storage (letting the SKU remain).
- **Order picking** (removing SKU from storage).
- Packing (putting the SKU into a new container).
- **Kitting** (collecting related items into a kit, generally for assembly by others).

Note that storage is a challenge in manufacturing as well as in distribution. Although provision should be made in all storage systems for the 3 Rs (rejects, repacks, and reassignments), design has become more critical as automation and throughput rate increase.

See Table 11.2. The variety of storage units, as with material handling equipment, is very large. The following will cover just the highlights. Equipment will be divided into that which moves the picker to the package, that which moves the package to the picker, and that which eliminates the picker.

2.1 Picker to the Package

When moving the **picker to the package**, there are five general alternatives: floor storage, static shelving and racks, drawers, flow racks, and storage/retrieval machines.

2.11 Floor Storage.

The least efficient approach is to store random items one deep on the floor. This not only does not use the cube, but tends to cause excess material handling to get at material stored behind something else, and, because of its unsystematic nature, tends to make it difficult to find items.

A better approach is **bulk** (block) **storage**. See Figure 11.1. Pallets, cargotainers, drums, etc., are stacked three to five units high without any aisles. Stacking frames also can be used. This method uses the cube fairly well, but it takes a lot of work to get the items from the back or the bottom of the pile. If all the items are identical and **last-in, first-out (LIFO)** is not a problem, bulk storage can give quite dense storage (**first in, first out (FIFO)** is not feasible with bulk storage). Stackability may be a problem (especially above 2 high) and the bottom cartons may be crushed; a palletboard between cartons may help distribute the load. If a series of items are bulk stacked next to each other, note that each part number needs an aisle dimension of the width of the lift truck (say 40 in.). That is, if product A = 30 in. wide, B = 40 in., and C = 60 in., then use 40 in. for A and B and 60 in. for C. If 6 in. is added to the pallet for access, then A gets 40 in., B gets 46 in. and C gets 66 in.

TABLE 11.2   Characteristics of very narrow aisle transportation deposit/retrieval vehicles for racks (Mulcahy, 1991). See also Chapter 18.

Vehicle Type	Aisle Width (ft)[1]	Transfer Aisle Width (ft)	Stacking Height, (ft)	1991 Cost ($1,000)[2]	Operator's Position	Guidance	Floor Type	Movement Activities	Power	Transactions Per Hour[3]	Fork Direction	Height of 1st level, (inches)
Sideloader, man down	5.5–7	11	30	70–75	stand up, floor	rail	F-100	storage	battery	12–15	side	6–12
Sideloader, man up	10–10.5	11	30	75–80	man up	rail	F-100	storage	battery	13–16	side	6–12
Sideloader, man down, counterbalanced	5.5	12	30	65–70	man down	wire optical	F-100	storage, transport	battery	10–13	front, side	0
Sideloader, man up, outrigger	5.7	12–13	20–21	40–45	man up	rail/wire optical	normal	storage	battery	10–12	side	14–16
Sideloader, man down, counterbalanced	5.7	16–17	40	85–90	man down	wire optical	F-100	storage	battery	18–20	side	0
Sideloader, double deep	10–10.5	11	30	80–85	stand up, floor	rail	F-100	storage	battery, DC	10–13	side	6–12
Man up, counterbalanced	5.5	16–17	40	85–90	man up	wire optical	F-100	storage	battery	18–20	side	0
Rising cab, auxiliary mast	5.5	16–17	40	90–95	man up	wire optical	F-100	storage	battery	19–21	side	0
Man up, rigid mast	5.5	27–28	60	100	man up	rail	F-100	storage	battery	13–15	side	16
Storage/retrieval, manual controlled	4.5–5	25–40	60	125	man up	rail	F-100	storage	DC	18–20	side	34
Storage/retrieval, automated	4.5–5	25–40	80	300	—	rail	F-100	storage	DC	20–25	side	34
Stacking vehicle, automated guiding	6.3	10–12	15	50–60	—	wire optical	normal	storage, transport	battery	10–12	front, side	0
Order selector, high rider, counterbalanced	4.3	10–12	20	20–21	man up	rail/wire optical	normal	storage, order sel.	battery	—	front	0
Order selector, high rider, straddle	4.3	10–12	30	25–27	man up	rail/wire optical	normal	storage, order sel.	battery	—	front	0

[1]Based on a 48 × 40 inch pallet with 2,000 lbs
[2]Includes battery and charger
[3]Based on a warehouse aisle of "normal" length

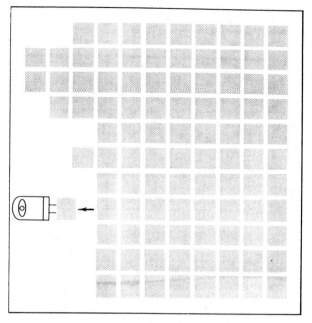

FIGURE 11.1 Bulk or block storage has high density but low accessibility (Which System, 1981). Stacking at a 90° angle to the aisle (as shown) maximizes density. Stacking at a 45° angle to the aisle reduces density but permits narrower aisles. The situation is similar to parking lots for cars.

2.12 Static Shelving and Racks. Static shelving and racks are low in capital and maintenance costs. See Table 11.3. They are generally a LIFO system and are best suited to very low activity. A rack (a strong shelf) generally holds pallets; heavier loads require stronger racks. Both shelves and racks use the cube but tend to use considerable floor space as there is one aisle per two storage rows and items take up a small percentage of the shelf cube. See Figures 11.2 through 11.6. Double or triple mezzanines

TABLE 11.3 Storage rack summary (Storage, 1991).

Type	Installed Cost ($/pallet position)	Application Notes
Rack		
Single row	30–40	Load accessible from 2 sides unless back-to-back
Two deep	38–53	Truck with reach mechanism
Drive in	45–65	LIFO; for identical, nonspoiling-goods
Gravity flow	175–225	FIFO; high-density storage
Systems (Cost does not include vehicle, controls, or software)		
High rise	35–40	Stacking height to 40 ft; truck under rail or wire guidance
High rise	35–40	Storage/retrieval machine with stacking heights to 75 ft; exacting tolerances on racks and floors; for large-volume, high-throughput applications
High density	90–140	Car in lane; high-rise configurations; good use of cube with dense storage of like items

can extend shelf height for manual picking from 7 ft to 14–20 ft.

Whereas the rack configuration determines the number of aisles used, the type of truck determines the aisle width and stacking height. Counterbalanced rider trucks require an aisle of about 12 ft for right-angle stacking, stand-up models about 10–11 ft. Reach trucks need about 9 ft. Straddle trucks need about 7–8 ft. Turret trucks, side-loaders, and automatic storage/retrieval trucks need only 6–8 ft. See Chapter 10 for more on aisles and Chapter 18 for more on vehicles.

Box 11.1 shows the effect of vehicle type on storage design. (See Chapter 10 for the effect on aisles.) Generally, pickers walk along shelves and racks. For long distances, a cart or vehicle should be used and, when entire pallets are moved, a lift truck is used. Stockers tend to use vehicles more as loads are heavier. Consider conveyors at the end of the aisle.

To reduce the cloak of manufacturer's claims and to minimize liability, specify that shelves and racks conform to the Rack Manufacturers Institute (RMI) specification "Minimum Engineering Standards for Steel Storage Racks," later adopted by the American National Standard Institute as ANSI Standard MH 16.1. See Table 11.7. For more about racks and shelving, order the "Industrial Steel Storage Racks Manual" (Z-10) from the Material Handling Alliance, 8720 Red Oak Boulevard, Charlotte, NC 28210; (704) 522-8644.

For information on fire protection, see Chapter 9.

Figure 11.8 shows a composite shelf module with some common options. Shelves tend to be labor intensive with poor stock control; see Figure 11.9. Shelves are commonly made of steel, either solid or perforated; for racks, steel frames with plywood or expanded mesh steel decks are common. About 75% of shelving is 36 in. wide, either 18 or 24 in. deep, with a height of 84 or 87 in. See Table 11.7 for loads. Reinforcement of the shelf front is not so much for the product load as to reduce shelf damage from people stepping on shelves as they try to see what is on the top shelf. The vertical posts can be of different shapes, such as L, T, or tubular. L and T shapes are popular but may snag hands or pallets when items are removed; in addition to causing injury, the post may be damaged. Posts decrease the nominal shelf width, and so a 36 in. wide item cannot be stored on a 36 in. wide shelf. For easier front access and storage of longer items, the front posts can be removed and the self supported only from below—**cantilevered shelves**. See Figures 11.10 through 11.14.

See Chapter 9 for floor requirements. Selective pallet racks do not, in general, have to be anchored to the floor, but drive-in, drive-through, and stacker racks should be anchored.

A typical rack has a frame on each end plus two beams for each shelf. However, if there are multiple racks end to end, they can share frames, and so each rack after the first one needs only one frame. (The beams, not the frames,

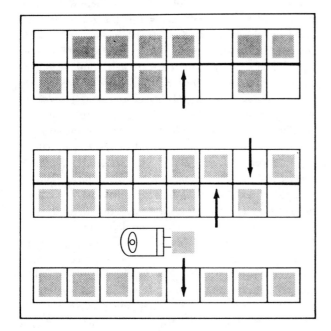

FIGURE 11.2   **Single-depth racks** have better accessibility (versus block storage) but at a cost of increased aisle space (Which System, 1981).

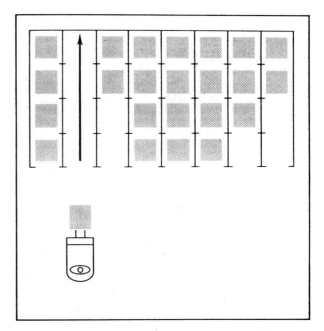

FIGURE 11.4   **Drive-in racks** have small proportions of aisles, but access to items is only from one side; thus it gives last-in, first-out storage (Which System, 1981).

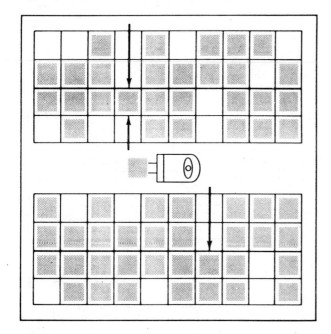

FIGURE 11.3   **Double-depth racks** reduce aisle space (versus single-depth racks). Reach trucks are needed to get the rear item; accessibility is worse than for single-depth racks (Which System, 1981).

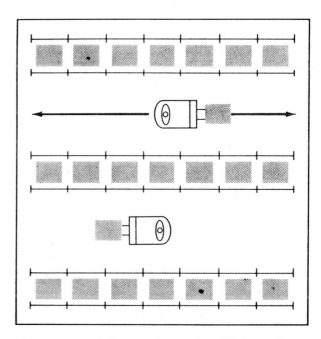

FIGURE 11.5   **Drive-through racks** are similar to drive-in racks, but the lane is open at both ends; this can be either LIFO or FIFO. The lane also can be used to store two different part numbers (Which System, 1981).

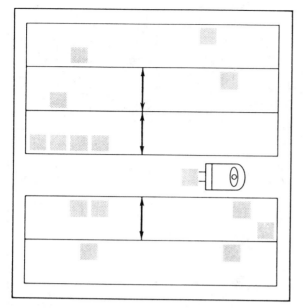

FIGURE 11.6 Mobile racks are moved together for storage and opened for an access aisle. When racks are moved toward each other (as shown), several sections may have to be moved for access. An alternative is to move the desired rack parallel to the rack axis (horizontal in the figure). This minimizes rack movement but requires a wide aisle space. Accessibility is better for mobile racks than for block storage, but the racks have a capital cost (Which System, 1981).

usually are the weak part of the rack, so there still is sufficient strength.) In general, allow 2 in. horizontal clearance between a load and the rack structure and 4 in. between loads. (For high racks when the operator is at floor level, the clearance may need to be 5 to 10 in.). Thus a bay holding two 48 in. wide loads would have 2 × 48 = 96 in. for the loads plus 12 in. for clearances, or 108 in. For ease of locating the pallet on the support, frame depth should be 4 to 6 in. less than the stringer length. Thus, if the stringers are 48 in. long, the supports would be 42 in. apart. (If the pallet or load projects into the aisle, this reduces the usable aisle space. That is, a 96 in. aisle with 2 in. projections on each side gives 92 in. of usable space). The height-to-depth ratio of the rack should not exceed 5–6; that is, the distance to the top of the uppermost load should not be more than 5 to 6 times the frame depth. (The ratio can be greater if the racks are attached to a wall and the floor.) Where sprinklers are to be installed in racks, allow for the necessary sprinkler clearance. The rack structure also needs to be anchored to protect the sprinklers and sprinkler piping. When two rows of racks are back to back, use ties to maintain alignment and stability. For fire, no longitudinal flue space (back-to-back clearance) is necessary if the double row is less than 25 ft high; above 25 ft, a minimum of a 6 in. flue is required (Rack Storage, 1980). If the product is susceptible to water

---

BOX 11.1   Effect of vehicle type on warehouse design

Table 11.4 gives characteristics of four different types of vehicles with example rack and building costs. The example is based on Egland's (1982) article, but with costs increased 30%. Each rack slot is assumed to be 4 × 4 ft, with no projection into the aisle from the pallet or load. For the example, a 180 × 190 ft building (32,400 ft²) is assumed with a 14 ft access aisle, leaving a 180 × 176 storage area. The 14 ft aisle is put on the end of the building rather than in the center to cut the number of aisle entrances in half. Aisle entry slows cycle time of guided vehicles. In many buildings, posts and other obstructions limit layout flexibility. A post can go through a rack (with loss of that space) but should not be in the aisles!

Figure 11.7 shows the layout using the counterbalanced

truck. The counterbalanced truck serves racks 48 in. deep on either side of a 144 in. aisle. In addition, assume a 6 in. flue between racks; allocate 3 in. to each side of each module. Table 11.5 shows that initially there is 246 in. planned. But

TABLE 11.4 Characteristics of various vehicles for use in warehouses. Assume a building capital cost of $40/ft² for 5 storage levels and $50/ft² for 10 storage levels. The higher cost for 10 levels is due to both additional height and a more level floor. (At a height of 40 ft, a ¼ in. difference in floor height on two sides of an 8 ft aisle is ¼ (40/8) = 1.25 in.)

Vehicle	Typical Aisle Width (Ft)	Typical Rack Height (Levels)	1982 Cost ($/Slot)
Counter-balanced	12	5	40
Narrow aisle, 25 in. reach	8	5	40
Narrow aisle, 50 in. reach (double deep)	8	5	36
Narrow aisle, turret	5.7	10	45

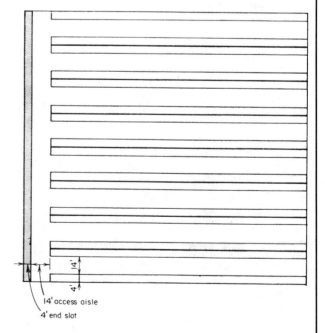

14' access aisle
4' end slot

FIGURE 11.7 Counterbalanced trucks use a 12 ft aisle with one rack on each side; the space has 733 slots/level.

(continued)

BOX 11.1 *Continued*

the 2,160 in. width holds only 8 modules, so there is 24 in. extra available per module. This could be used for more flue space or wider aisles. Along the 14 ft aisle there is a single row of slots; 180/4 = 45 slots. Thus, each module is 190 – 4 – 14 = 172 ft long and holds 172/4 = 43 slots on each side of the aisle, or 86 total. The 8 modules hold 8 × 86 = 688 slots. Each floor has 688 module slots + 45 end slots = 733 slots.

Table 11.5 repeats the calculations for the narrow aisle with 25 in. reach, the narrow aisle with 50 in. reach (2 deep storage), and a narrow aisle turret truck.

Table 11.6 shows the cost of the four alternatives. For the

5-level building, the total cost is about the same, but the number of slots and thus the cost/slot varies considerably, showing why narrow aisle equipment is so popular in warehouses. The comparison for the turret truck between 5 and 10 levels shows why stacking to higher and higher levels has become popular. If all clearances are minimum, however, vehicle speeds will be low and damage high—even with the maximum number of spaces. That is, consider the productivity as well as the number of spaces.

The main floor need not be at the bottom rack level. That is, storage can be below grade as well as above grade.

TABLE 11.5   Slots for various aisle and truck alternatives for a 180 × 190 ft building.

Truck Deep	Counter-balanced Single 144 in.		Narrow-25 in. Single 96 in.		Narrow-50 in. Double 96 in.		Turret Single 68 in.	
Aisle (nominal)	Initial	Final	Initial	Final	Initial	Final	Initial	Final
Flue	3	3	3	3	3	3	3	3
Pallet	48	48	48	48	48	48	96	96
Aisle	96	96	144	144	68	68	96	96
Pallet	48	48	48	48	48	48	96	96
Flue	3	3	3	3	3	3	3	3
Spare	0	18	0	24	0	10	0	14.6
Total	198	216	246	270	170	180	294	308.6
Bldg. width		2160		2160		2160		2160
No. of modules		10		8		12		7
Pallets/modules		86		86		86		172
Module slots		860		688		1032		1204
End slots		45		45		45		45
Total slots/floor		905		733		1077		1249

TABLE 11.6   Warehouse capital cost for various alternatives.

Alternative		Building Cost ($)	Rack Cost ($)	Total Cost ($)	Number of Slots	Total Cost ($/slot)
Levels	Vehicle					
5	Counterbalanced	1,778,000	205,900	1,583,900	3,665	541
5	Narrow aisle (25 in. reach)	1,778,000	251,700	2,029,700	4,525	449
5	Narrow aisle (50 in. reach)	1,778,000	288,300	2,066,300	6,245	331
5	Narrow aisle turret	1,778,000	297,400	2,075,400	5,385	385
10	Narrow aisle turret	2,223,000	544,400	2,767,400	10,770	257

damage, be sure the lowest row is on pallets at least 4 in. above the floor. Allowing a 12 to 16 in. clearance between

TABLE 11.7   Permitted evenly distributed loads (lb) on a 36 in. long by 24 in. deep shelf (Shelving, 1974). Table courtesy of Material Handling Engineering.

Shelf Class	Load (lb)	Reinforced
1	750	no
2	750	front
3	900	front, rear
4	1650	front, rear, middle

back-to-back rows permits using 36 × 42 in. or 40 × 48 in. pallets to be stored flush to the aisle, using either orientation as the longer loads project into the gap.

Undamaged racks and shelves rarely fail. The key word is undamaged. Damage usually is caused by vehicle impact. To reduce abuse by vehicles, use double posts, reinforced front posts, concrete or steel protectors about the post bottom, and steel and concrete guides at the ends and entrances of aisles to prevent trucks from cutting the corner too sharply. For sprinklers within racks, build guards. Rail- or wire-guided vehicles in the aisles not only prevent rack damage but permit narrower aisles. Wire

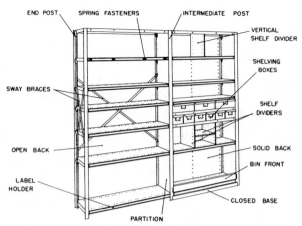

FIGURE 11.8   Shelves have many options (Shelving, 1974). The sway braces generally are not needed if the sides or backs are solid. Usually the vertical distance between shelves can be adjusted. Shelves can be modified to hold drawers and can be divided horizontally or vertically with partitions. Bin fronts prevent items from falling off the shelves. Label holders give location and stock number. Perforated shelving permits passage of air, which reduces dust accumulation, and of light, which makes it easier to see items in the shelf back. Figure courtesy of Material Handling Engineering.

guidance can be used also for data transmission of stock-picking and inventory information. Since picking and stocking are done from the same aisle, you must mesh the two activities or else stock when you aren't picking.

2.13 Drawers.   A drawer is a shelf that moves in and out, has four walls, and is contained. Drawers can be cabinet mounted or shelf mounted. For cabinets, normal suspension systems, with rollers and slides, have a load capacity of about 200 lb. Full-suspension systems have a capacity of 400 to 600 lb and achieve 105 to 110% withdrawal through use of a telescoping slide and carriage (Quinlan, 1980).

Storage of air is a primary disadvantage of shelf storage since most items fill only a small portion of the available cube. Drawer storage of small (shoebox size or less) items gives high-density storage (i.e., low air storage). See Figure 11.15. One firm stores 19,000 high-activity parts in 750 ft² of storage space by stacking modular drawers 20 ft high. Using a man-ride picking machine and batch picking,

FIGURE 11.9   Shelves tend to be labor intensive (cost and possible injury) with poor stock control.

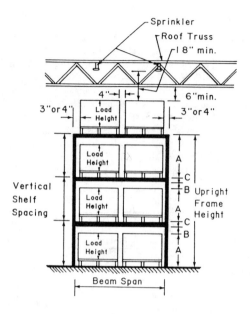

FIGURE 11.10   Racks hold pallet loads. Usually there are 3 to 8 levels. A = load height including pallet (usually 5 in.). B = clearance between load top and beam; 4 in. minimum on the floor level but 6 in. for higher levels, as the visibility of the driver is poor and the truck mast deflects, and so the load is not parallel to the floor. Beams usually can be located at 2 in. intervals on the column. C = beam height (3–4 in. is typical). Upright frame heights are available in 2 ft increments from 8 to 30 ft. Typical frame depths are 28, 34, 42, 48, and 60 in. Beam span ranges from 36 to 180 in., with 54, 90, 96, 108, 120, and 144 being popular. (The longer spans allow pallet racks to compete with cantilever racks.)

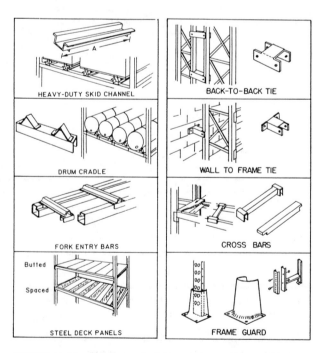

FIGURE 11.11   Rack accessories improve rack strengths or usefulness. 1) Skid channels and drum cradles permit use of drums and skids. 2) Fork entry bars permit lift trucks to move nonpalletized loads. 3) Deck panels permit smaller size loads. 4) Ties aid rigidity. 5) Cross bars aid rigidity and permit smaller size loads. 6) Guards protect against truck impact.

FIGURE 11.12   Cantilevered beams (A frames, "fir trees") hold long, rigid items such as bar stock, pipes, and lumber. The arms can be tilted or horizontal. Cantilevered shelves hold nonrigid long items. The shelves and arms can extend from one or both sides.

they pick 120 parts/hour. Drawers get these more address-able storage locations/ft^3 (often over 50% more than for shelves) by better use of the horizontal dimension and of the vertical dimension. In the horizontal dimension, drawers usually are subdivided by inserts, giving more locations/ft^2. These subdivisions facilitate having a place for everything and everything in its place. A number of drawers also can be put in the space of one shelf. As the drawer moves out, vertical clearance when removing an item is not a problem as it may be in shelf storage. Another drawer advantage is that when the drawer is pulled out, the picker can see every part. Thus, there is little access penalty between a front and rear location, as there is with shelves, and drawers need not be last-in, first-out systems as shelves tend to be.

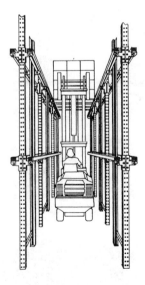

FIGURE 11.13   Drive in and drive through are for high density at the cost of restricted access. If the pallet is two-way, the pallet is a wing type and the shelves on each column support the wings. If a four-way is used, enter the forks perpendicular to the stringers and let the stringers rest on the shelves. The shelves must be far enough apart to let the truck through but not too far to support the pallet. See Figure 11.14.

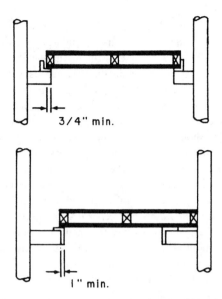

FIGURE 11.14   Gap between pallet guides is pallet width— 3/4 in. With no guides, keep the distance from the column to the opposite rail $\leq$ pallet width – 1 in.

here is an engineered analysis to the problem:

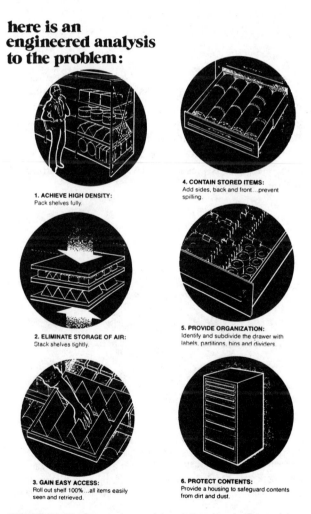

1. ACHIEVE HIGH DENSITY:
Pack shelves fully.

2. ELIMINATE STORAGE OF AIR:
Stack shelves tightly.

3. GAIN EASY ACCESS:
Roll out shelf 100%...all items easily seen and retrieved.

4. CONTAIN STORED ITEMS:
Add sides, back and front...prevent spilling.

5. PROVIDE ORGANIZATION:
Identify and subdivide the drawer with labels, partitions, bins and dividers.

6. PROTECT CONTENTS:
Provide a housing to safeguard contents from dirt and dust.

FIGURE 11.15   Drawers store much less air than do racks. Two modular drawers (each 59 in. high by 28 in. deep by 30 in. wide) hold the same tools as on five standard shelving units (each 87 in. high by 18 in. deep by 36 in. wide). (Photo courtesy of Stanley Vidmar.)

Since a drawer has four walls, that is, it is contained, there are two additional benefits—cleanliness and security. Dust and dirt are less of a problem than with shelves. Security is better because individual drawers as well as the stockroom door can be locked; security can be selective, with more for some drawers and less for others. There should not be any area near storage where items can be hidden, nor should locker areas be close to storage areas. It should be mentioned that not all security is for thieves; most is for raiders. Raiders take tools, equipment, and parts for use without completing the paperwork, resulting in items not being available to others when needed and in overstocking when the stockkeeper orders duplicate items to bring the actual inventory up to specified inventory.

Since items in drawers are stored compactly and are unlikely to be missing, walking by stockers and pickers should be reduced, allowing them to service more customers per day.

Safety may be a problem, as the cabinet may tip forward when the drawer is fully extended; this is a common problem with file cabinets. A simple solution is to attach the cabinet to the floor or another cabinet.

See Chapter 14 for design of tool cribs and stockrooms.

2.14 Flow Racks.

Figure 11.16 shows a flow rack. A rule of thumb for **flow racks** is that the stock-keeping units (SKU) should hold 3 to 5 days' supply. Thus, if a single container is not emptied in 3 to 5 days, it should be on a static shelf. If it is emptied in less than 3 to 5 days, it belongs in a flow rack (Gravity Flow Racks, 1980). If more than one pallet is picked per day, use a more automated system.

There are pallet systems, case systems, and split-case systems. In pallet systems, the entire pallet is removed by a lift truck. Because of the weight of the pallet plus load and the relatively high center of gravity, braking devices are recommended for runways over 10 ft long. Load isolators to keep the second pallet from exerting pressure on the first pallet can also be used. Guide rails hold the pallet on the lane center line. In **case systems** the case is removed; in **split-case systems**, the container is opened and only a part of the contents is removed for any one order. Cases—generally cardboard boxes or items on a pallet—are loaded from a stocking aisle onto a roller conveyor. Cases move by gravity to the end of the queue of that product. As items are removed from the picking aisle side, the cases automatically advance down the grade. The last case stops at the case stop, which can be a simple piece of steel or a set of wheels. Wheels aid case picking, since the picker just lifts the front of the case over the wheels and it rolls out, being pushed by the full ones behind it. Remember to provide for take-away of empty boxes, pallets, and packaging material in the split-case systems.

There are two frame options; see Figure 11.17. The rectangular (vertical) frame is satisfactory for full cases but there isn't much space above the case so it is difficult for split-case picking. For split-case picking, use a "layback" design, which will expose more of the first-position case. When case size varies, use a design that lays back the front face about 10°. When cases on all tracks are the same size, then a layback plus a 14° to 18° tilt on the shelf front exposes even more of the case top for picking. A "knuckled" front has the most tilt. Layback frames require slightly more space between levels than vertical frames.

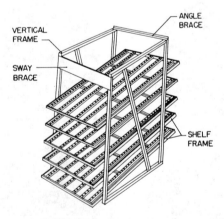

FIGURE 11.16 Gravity-flow racks automatically advance the stock. In most applications they are loaded from the rear from a **replenishment** (deposit) **aisle** and unloaded from the front from a **picking** (withdrawal) **aisle**; stock rotation is FIFO. An alternative is a push-back rack, in which only the front aisle is used for both picking and replenishment; stock rotation is LIFO.

A typical frame is 8 ft high and 4–6 ft wide; standard lengths are 5, 7.5, 10, 12, 15, and 20 ft. Vertical height between shelves is adjustable, as is shelf angle (3° is common). An impact shelf at the entry point prolongs roller life at the entry point.

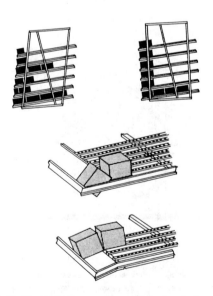

FIGURE 11.17 Vertical frames with straight shelves on flow racks require minimum space but are only for case picking (because of restricted access to the case top). For split-case picking, use layback models with straight, tilted, or knuckled fronts. (Sketch courtesy of Buckhorn Material Handling Group.)

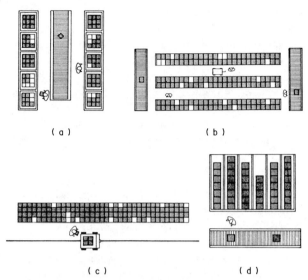

FIGURE 11.18 Picker moved manually to the pallet is shown four ways: (a) an in-aisle conveyor (use for fast-moving cases); (b) a nonpowered collection cart and a take-away conveyor at the end of the aisle (use for slow-moving cases); (c) a powered pallet truck or tow train facing bulk storage, flow racks, or drawers (an alternative design for b and d); and (d) a conveyor facing drawers or floor racks (use for fast-moving small parts) (Manual Orderpicking, 1983).

Flow racks combine automatic stock movement at no cost, FIFO procedures, separate aisles for pickers and stockers, minimum frontage per item (i.e., minimum walking distance for pickers), and multiple rows of storage/aisle. One aisle is for picking and one for stocking. With lane depth properly related to item activity, one stocker can keep up with many pickers.

Figure 11.18 gives some layouts for pickers moving manually to the part.

2.15 Storage/Retrieval (S/R) Machines.
In the four previous options, the picker generally walked to the product. With the **storage/retrieval machine (S/R)**, the picker rides a platform which goes among racks 30–70 ft high. Note that S/R machines can store more than raw materials and finished goods. They are used also for supplies, work in process, and tooling fixtures. The platform can be part of a rubber-tired or rail-mounted vehicle (above 30–40 ft); the man-aboard crane runs on a single floor rail at the bottom and is supported by an upper guide beam at the top. The platform also can be attached to a truck mast (up to 30–40 ft). The truck may be aisle-captive (operate only in one aisle). **Aisle-captive trucks** also can be equipped to move between aisles on transfer cars so that one truck can pick several aisles in sequence. Noncaptive trucks generally use wire guidance while in the aisles to maximize speed and minimize rack damage.

In the low-energy warehouse, the lighting, heating, and cooling are installed only locally on the picking platform—using only minimal light, heat, and cooling (fans reduce thermal stratification) in the building.

The platform moves the picker not only horizontally but, more important, vertically to increase the feasible storage heights. Naturally, safety belts are required. Equip the platform with shelves or multiple levels of roller conveyors to aid batch picking. In the more automated versions, a computer controls the platform movement, including diagonal movement, and the picker just picks. For high-activity operations, take-away conveyors in the racks eliminate the need to move the platform to the end of the aisle to offload. The picker can pick either cases or split cases; the stocker loads pallet loads using either a stacker crane or a high-lift truck; both picker and stocker use the same aisle.

Figure 11.19 gives some layouts for pickers moved by machine to the part.

For any of the picker-to-product systems, consider a picking list of pressure-sensitive labels instead of a printed list. The picker peels the labels from the paper and attaches them to the orders. At the end of the work period, leftover labels indicate out-of-stock products. Another concept is forms with redundant information: The same information is bar coded (for scanners), keypunched (for data processing equipment), and printed (for humans). Electronically assisted picking eliminates picking documents (reduces errors) and cuts picking time. An indicator lamp, a counter display, and a button are mounted next to each picking location. The computer turns on the lights for the items to be picked and displays the number wanted. The picker notifies the computer the item has been picked by pushing the button, which also turns off the light.

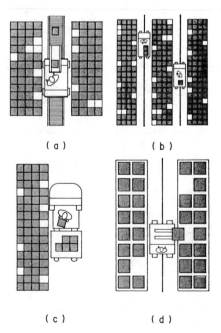

FIGURE 11.19 Picker moved by machine to the part is shown four ways: (a) a pick car system with in-aisle conveyor; (b) a man-ride machine that is aisle-captive (follows a guide path and is used for high storage); (c) an order-picking truck (noncaptive system lifting the operator to about 30 ft); and (d) a hybrid truck (follows a guide path within the aisle but can move between aisles) (Manual Orderpicking, 1983).

To improve stockroom accuracy to the point where, for example, 95% of the part numbers have actual counts within acceptable accuracy limits, use cycle counting. Cycle counting requires checking actual inventory against paper inventory every week or so for several months for a sample of, for example, 100 part numbers. If the two numbers do not match, the short time interval usually allows determination of the reason. The technique is used to determine reasons for errors and to correct procedures rather than update all the records.

2.2 Package to the Picker

There are three main approaches to bringing the **package to the picker**: miniload, carousel, and conveyor. Figure 11.20 gives some layouts for part to the picker. A **miniload** system (automated drawers) brings a bin. A **carousel** rotates the entire rack until the bin is in reach of the picker. In both approaches, just as the picker stays at one location, the stocker stays at one location. These locations can be well lighted and have good access to other equipment such as scales and bags. The product area need not be lighted or air-conditioned, although air flow should be maintained to prevent excess humidity from becoming a problem. A subtle advantage of bringing the package to the picker is that the systems require discipline—they force you to be organized. Take-away is easier to design and supervision is easier when the picker is in one place. Pilferage is reduced because the picker is working in the open. In general, build enough picking and packing stations for your peak loads; on less busy times of the month assign operators to a smaller number of stations.

The third method of bringing the package to the picker is storage on a moving conveyor—often with overhead storage. The picker often is an operator, that is, does something to the product. See Chapter 17 on conveyors for some options.

2.21 Miniload.

The miniload (automated drawers) is a miniature **automated storage/retrieval system (AS/RS)**, a computer-controlled system for pallet storage. Instead of pallets, it stores bins holding 100 to 500 lb, each generally holding several different types of parts. Most often the loads are sent to a picking buffer area and then sent back into storage rather than being sent directly to shipping. See Figure 11.21. Control systems can range from pushbutton to full computer control. Light bins are removed with a magnetic grab that attaches to the end of the bin and pulls it onto the lift carriage. Heavier bins are engaged with a device that makes mechanical contact with the end of the bin. By bringing parts to the picker, a miniload machine eliminates the nonproductive work of walking, searching, bending, reaching, and climbing. Miniloads tend to have the storage efficiencies of drawers rather than racks—they store less air than racks. Because they are modular, many can be installed in 10 days rather than the 10 months or more needed for an AS/RS. They also can be moved within the plant if so desired.

Miniloads usually are keyed. Only the operator has a key; only the operator can remove items from the totally enclosed miniload. This gives good inventory control.

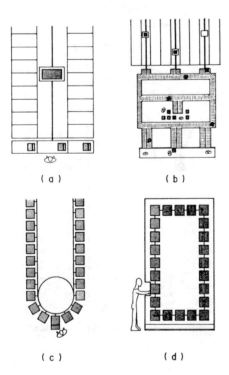

(a)   (b)

(c)   (d)

FIGURE 11.20 Part moved to the picker is shown four ways: (a) a miniload bringing individual bins to the operator's picking station (have multiple stations per operator so the operator doesn't wait for the machine); (b) a unit load AS/RS bringing pallets to the operator's picking station; (c) a horizontal-travel carousel; and (d) a vertical-travel carousel (Manual Orderpicking, 1983).

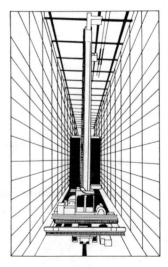

FIGURE 11.21 Miniloads (automated drawers) combine high-density storage with moving of the part to the picker in a drawer, which then returns to storage. In contrast, AS/RS moves a complete pallet of product; generally there is no picking from the pallet.

FIGURE 11.22   Carousels move the rack to the picker; they tend to have lower capital costs than miniload systems. In the most advanced models, each height (row) moves independently. Courtesy of White Conveyors.

2.22 Carousels. A carousel loop consists of a series of floor-supported bins driven by a powered chain around a long, narrow loop. Carousels have a capital cost about 50% lower than miniload units. See Figure 11.22. They do not have the stock control potential against pilferage of miniloads, but they have most of the advantages of miniload systems. Carousels are especially useful for small parts in buildings with low ceilings. They are excellent for kitting operations, since they present a number of trays to the operator at one time.

Generally, miniload and carousels will give less than 1 pick/min/machine (Kemp, 1982). Carousels may or may not be computerized. One operator usually controls one or more carousels, so they should be mounted side by side or end to end to minimize walking. Carousels can be multilevel with the use of mezzanines. Install adjacent carousels with an 18 in. space between the outside of the loops so stock pickers can walk between units to pick if a carousel is down.

Carousels also can be located vertically. For example, units at Texas Instruments holding 114 totes have a 5.5 × 10 ft footprint and a 12.5 ft height (Vertical Carousels, 1991). In an application at Hayworth, the units have a 6 × 11 ft footprint and are 23 ft high; typical shelf spacing is 16 in., although some shelves for small parts have 8 in. spacing. Picking lists are entered in the carousel's computer and

the carousel automatically controls shelf movements to reduce picking time. When the proper shelf is positioned, an indicator lights up in front of the appropriate carton on the shelf (Dunn, 1986).

2.23 Conveyors. See Chapter 17.

2.3 No Picker   In the first section, we discussed bringing the picker to the package. In the second section, the package was brought to the picker. In this section the picker has been eliminated.

2.31 Automatic Storage/Retrieval System. AS/RS is a computer-controlled system. The first computer-controlled warehouse was in Detroit in 1956 (Nagy, 1979). In AS/RS, the computer tells the machine to take a pallet out of a specific location and the machine does so. The pallet then generally is sent by conveyor (at the end of the lane) to **accumulation conveyors**. See Figure 11.23. When the accumulation conveyor has the entire order, a barrier is released and the total order is moved to shipping. A variation is to have a variety of machines picking different parts of the order; each pallet is sent to a different accumulation conveyor. Then, when the order is completely picked, barriers are released and the items from each accumulation conveyor are sent simultaneously to shipping. With proper

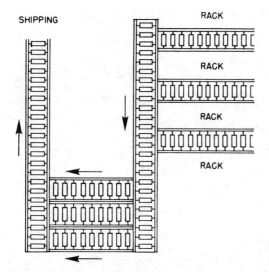

FIGURE 11.23 Accumulation lanes (sortation lanes) can be operated two ways. In method 1, all items for order 1 are sent to lane 1, items for order 2 to lane 2, etc. Then when an order is complete, the gate is released and all items are sent to shipping. In method 2, line 1 of order 1 is sent to lane 1, line 2 to lane 2, line 3 to lane 3, etc. Line 1 of order 2 is sent to lane 1, line 2 of order 2 to lane 2, etc. When an order is complete, all the gates open simultaneously and one item is let through to the shipping conveyor. Generally there are 10 to 20 accumulation lanes. Method 1 is useful for batch picking. Method 2 allows the line sequence in the order to be maintained to shipping even with multiple pickers.

control, multiple orders can be stored in each accumulation conveyor, as the items for each order will be in the proper sequence. Another alternative is for the AS/RS to bring the pallet to a picking station buffer. After the desired case or split case is picked, the AS/RS machine puts the pallet back into storage. With randomized storage, this may not be the same location that it came from.

The general tendency is to run the AS/RS not with one computer but with several. Under this distributed processing, the host (administrative) computer supervises a number of minicomputers. The advantages are flexibility and reliability.

A recent development is to put AS/RS in manufacturing to use not only its storage capabilities but its transfer capabilities. It can process not only parts but also tools. The result is an automated job shop, because the AS/RS holds not only finished goods but also work in process and even production tools and jigs.

Advantages of AS/RS include fewer material handlers and order pickers, fewer office support staff, lower utility cost, and more efficient use of storage space; the disadvantages are high capital and maintenance costs. In manufacturing, another disadvantage is that simple rack storage is easy to modify if there is a need to expand the space for production machines; with AS/RS the space is difficult to change. In addition to these hard savings, there are two types of savings that are intangible because the dollar benefit is difficult to predict. The first is time lag. The time to process the paperwork for a transaction is reduced from

days to minutes. The second is reduced human errors. The fixed storage locations of conventional storage occasionally lack sufficient space for a given product; when a product is stored temporarily somewhere else it gets lost. Humans tend to make errors in recording and transcribing numbers; so the more forms they need to use, the more chances for error. Another source of error is that people don't like to follow the paperwork rules of removing stock from stockrooms; neither do they wish to admit they damaged or lost inventory—so the restricted accessibility of AS/RS helps. Accessibility is restricted by system height, by random storage, and by allowing only machines (not people) into the storage areas. Use a STO versus FLO analysis (storage versus throughput) to decide approximately whether a static rack, SR machine, or AS/RS system is best. See Figure 11.24. From this approximation, fine tune your analysis and pick your vendor.

If you build an AS/RS, there are many advantages to building it as a **rack-supported structure** (RSS) rather than putting it in a free-standing building (FSB) (Quinlan, 1981):

An RSS will require about 10% less steel for racks and structure.

An RSS foundation will be smaller.

An RSS building is considered equipment with a single-purpose use, and most of it can be depreciated in 10 years (versus 25 to 30 for an FSB).

An RSS building gets an investment tax credit since it is a "machine."

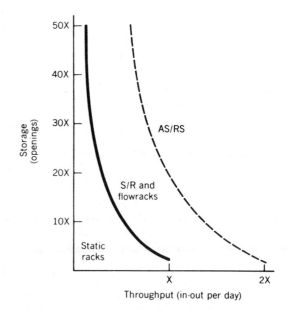

FIGURE 11.24 STO versus FLO analyses aid decision making on the proper type of system (Koenig, 1980). Use approximate numbers to get the general picture, and then get detailed estimates to select your vendor. Copyright © Institute of Industrial Engineers, 25 Technology Park, Norcross, GA 30092; (404) 449-0460.

2.32 Automatic Item Picking. The common AS/RS machine picks an entire pallet. There are a small number of item pickers. One type, an automated vending machine, holds individual items in a magazine. The computer triggers release mechanisms which cause the item to drop onto a conveyor. The other type has several hundred picking lanes and a movable picking head. Automatic item picking is used only for small high-activity products with hundreds of orders/day and a relatively small number of lines/order (e.g., 25). Table 11.8 summarizes unit storage systems.

3 ORGANIZATION OF ITEMS IN STORAGE UNITS

There are a number of considerations in organizing storage units: location identification, FIFO versus LIFO, space utilization, distance moved, and container. The decision will depend upon the relative importance of tight material control, inventory accuracy, space savings, labor and energy savings, and capital cost. See Table 11.8.

3.1 Location Identification

Identify the *XYZ* coordinate of each storage location with an alphanumeric code. First identify the rack with a letter (two letters if there are over 26 racks). Then identify the position in the rack with a number. Start at one wall and keep numbers aligned for all racks, possibly skipping numbers in some racks if necessary. Finally, identify the vertical dimension with a letter—A is on the floor. Thus, code M62C means rack M, 62 spaces from one wall and the 3rd level up.

3.2 FIFO Versus LIFO

First in, first out (FIFO) versus last in, first out (LIFO) generally is resolved in favor of FIFO. (Even though your tax policy is LIFO, your actual inventories do not have to be LIFO.) A related problem is keeping a lot together in storage. For shelves, FIFO tends to require stock storage one row deep. Drawers and flow racks have no problems with FIFO. Static racks and shelves, if one deep, present little problem except remembering which SKU has the oldest unit. If units are stored in bulk storage or double deep in racks, then extra material handling is required to move the back unit to the front after the front unit is picked, or to move the front

TABLE 11.8 Unit storage system comparison (Material Handling: Storage, 1991).

	Block Storage	Tier Rack & Stacking Frames	Standard Pallet Rack	Double Deep or Two deep	Bridge Across	Drive-in	Drive-through	Mobile Rack	Gravity Flow	Push Back	Car-in Lane	Cantilever
Cost/unit load, dollars[1]	—	—	100–150	100–150	50	200	225	300	250	225	300	150–200
Utilization factor, %	60	60	85	80	85	66	66	85	90	66	66	85
Building footprint	Small	Small	Large	Medium	None	Small	Small	Small	Small	Medium	Small	Large
Effectiveness of space use	Very good	Very good	Fair	Good	Good	Very good	Very good	Excellent	Excellent	Good	Excellent	Fair
Accessibility to unit loads	Poor	Poor	Excellent	Fair	Excellent	Poor	Poor	Good	Fair	Fair	Poor	Excellent
Probability of damage	High	Low	Low	Low	Low	Medium	Medium	Low	Low	Medium	Low	Low
Sprinkler requirement[2]	Ceiling	Ceiling	Ceiling, in rack	Ceiling, in rack	Ceiling	Ceiling, in rack	Ceiling, in rack	Ceiling	Ceiling, in rack	Ceiling, in rack	Ceiling, in rack	Ceiling, in rack
Unit load stability	Poor	Fair	Good	Good	Good	Good	Good	Good	Good	Good	Good	Good
Rotation of unit loads	LIFO	LIFO	FIFO	LIFO	FIFO	LIFO	FIFO[3] LIFO[4]	FIFO	FIFO	LIFO	FIFO[3] LIFO[4]	FIFO
Number of aisles	Few	Few	Many	Medium	None	Few	Few	Few	Few	Few	Few	High
Security	Poor	Poor	Good	Good	Good	Good	Good	Good	Good	Good	Excellent	Good
Unit loads deep/ opening	8–10	8–10	1	2	1	8–10	8–10	1	3–20	3	10–20	1
Storage density	High	High	Low	Medium	Varies	High	High	High	High	Good	High	Low
SKU openings/aisle	1	1	2	2	2	1/vertical lane	1/vertical lane	2	2	1/lane opening	1/lane opening	1–3
Throughput volume	High	Medium	High	Medium	Medium	Medium	Medium	Low	Low	Medium	Low	Medium
Stacking height, ft	20–25	20–25	20–40[5]	20–40	20–25	20–30	20–30	20–30	20–30	20–30	40–80	20–40
Pick positions	1 high	1–2 high	1–2 high	1–2 high	Not feasible	1 high	1 high	1–2 high	1–2 high	1 high	Not feasible	1–2 high
Types of pallets	All	Captive	All	All[6]	All	Varies	Varies	Varies	Captive	Captive	Captive	None
Retrieval equipment[7]	W,N	W,N	W,N, VNA	N	W,N	W,N	W,N	W,N	W,N	W,N	VNA	VNA, sideload
Number of unit loads/ SKU	High	High	Variable	2 or more	High	High	High	Varies	High	2 or more	High	Varies

[1]Approximate installed cost.
[2]Always check local codes.
[3]Single aisle.
[4]Two aisles.

[5]Depends on vehicle lift height.
[6]With small pallet boards, bottom level is raised.
[7]W = wide aisle counterbalanced; N = narrow aisle straddle or sideload; VNA = very narrow aisle hybrid or AS/RS.

NOTE: This table is intended only as a general guide.

unit to get the back unit. This problem may be reduced by storing the units in two or more areas. First the stock is picked from area 1 while all new stock is put in area 2. Then, when area 1 is empty, area 2 is picked while new stock is put in area 1. However, the technique does waste space. The worst FIFO problem occurs with bulk storage, in which the lack of access of the items in the rear tends to give an actual LIFO system even if management desires a FIFO system. See Box 11.2 for comments on order picking.

3.3 Space Utilization

"Use the cube" has been advocated for many years. Even more efficient than using three dimensions is effectively using the fourth dimension, time, by using **random storage**. In dedicated storage, a space is reserved for an item whether it is present or not, so the space tends to be empty a substantial portion of the time. In dedicated storage, area A is reserved for product 1, B for product 2, etc. Every part has a home. However, the area reserved must equal the maximum amount needed for the product, but the amount stored probably will be closer to the average. (A closer estimate might be reserve stock + order quantity/2; see economic lot size calculations.) In random storage, the items are stored in a location determined randomly. (Actually it is not random. It is just that the location is determined only at the time when the new item is to be stocked; the resulting locations, however, tend to be located randomly within the total storage.) Random

storage tends to fill a much higher percentage of the total SKUs—90% versus 50% for dedicated storage. This, in turn, means a much smaller total stock space is required. The disadvantage is keeping track where the stock is; in practice, only the computer knows where the stock is. The dedicated system assumes it is better to preassign space than to keep detailed records; computers have changed the relative cost of detailed record keeping, so you can "buy" space at the cost of detailed record keeping. A compromise is random storage within a limited area—the picker needs to do some searching but more space is saved than with dedicated storage. When humans pick orders from random storage, the computer prepares the picking list in a sequence to maximize ease of picking.

Another obvious, but often neglected utilization concept, **high-density storage**, is to store big items in big spaces and small items in small spaces (but not small items in big spaces). Inventory needs to be categorized for drawer, shelf, rack, etc. See Table 11.1. In automated systems, the incoming shipment often goes through a sizing station (measuring load height and pallet overhang) so that the shipment can be placed in the optimum size space.

Reserve (backup) stock generally is placed close to the stock being picked. Common locations are the top racks or racks in an aisle immediately behind the picking racks. See Figure 11.25 for a well-designed system showing a mixture of types of storage.

BOX 11.2 *Order picking*

Order picking is removing material from storage for customer orders. The customer can be internal or external. Seven guidelines (adapted from Frazelle, 1990a; 1990b) are as follows:

1. Eliminate and combine operations when possible. Picker work elements are traveling, extracting, searching, documenting, reaching, sorting, counting, and packing. For example, can a picker sort material directly into a shipping container, thereby combining picking, sorting, and packing?

2. Assign the most popular items to the most accessible locations. Using a Pareto analysis, store popular items at waist height near the front of the storage area. Just 15% of the SKUs may account for 80% of the picks.

3. Store correlated items together. Certain items may be ordered together (orders are correlated), thus reducing travel time. Examples are parts in a repair kit, left and right brackets, and items and their related screws and bolts.

4. Separate slow and fast movers. One alternative is to store fast moving items in a forward picking area and slow movers in a reserve area. Another alternative is to store all SKUs in the forward area but in relatively small quantities for most items. Additional SKUs are in the reserve area, and when the forward area runs low, it is replenished from the reserve stock.

5. Improve pick rates. Travel time can be reduced by dense storage. Two examples are high-density storage (big spaces for big items, and small spaces for small items) and use of mezzanines. When using mezzanines, pick all the items on a specific floor before going to the next level. Travel time can

be reduced if orders for multiple customers are picked on the same picking tour, but the items also need to be sorted. This can be done by the picker (sorting while picking, perhaps using a "pigeon hole" cart) or later (usually with automated equipment on a conveyor, which scans a bar code applied by the picker). Travel time also can be reduced by zone picking, in which a picker picks only the items within a zone of the warehouse. Later, the items need to be sorted and merged into the customer's total order. If the picker has to wait for a machine to furnish an item (e.g., carousel or miniload), multiple machines/picker are useful.

6. Presequence the picking list. A computer can select the picking sequence, eliminating picker decision making and allowing use of complex strategies tailored to the specific list being picked. One simple strategy is to take advantage of the horizontal speed (faster than vertical) of man-aboard vehicles. To minimize vertical travel, on the outbound leg of the picking route, pick all items on the lower half of the rack; on the return leg, pick all high items.

7. Minimize search time and errors. Errors can be caused by poor document ergonomics (small letters, poor contrast, long item codes, poor spacing, etc.) or by poor bin identification and poor identification of items in the bin. One alternative is a display light at the bin of each item to be picked (indicating location and quantity). When the picker turns it off, the computer notes that the item is picked. Weighing stations in the conveyor to the shipping dock also can weigh the picking container as it passes, compare predicted weight with actual weight, and shunt aside nonconforming containers.

Stock availability is another challenge. In redundant storage, items are stored in several aisles of an AS/RS so that if the machine fails on one aisle, the item can be picked from another aisle. Another technique is to put critical items low so that they can be picked by humans if the machine fails.

3.4 Distance Moved
Distance moved by the picker (man or machine) is an important factor. In general, lift trucks should not move more than 100 to 200 ft; beyond that, use driverless carriers. When humans are picking, the optimum picking zone, to minimize stooping and stretching, is from the hips to the shoulder; that is the area for the "fast movers." For humans picking while walking, don't store any items above the top of the head, and then only on the front of the shelf. Most machines used while picking—either manned trucks or cranes or unmanned AS/RS and miniload machines—travel 5 to 25 times faster horizontally than vertically. Thus, to minimize vertical travel, pick the entire length at a low height and then make another pass at a higher height. This leads to the concept of picking zones in which a picker doesn't pick an entire order but only the

parts in one zone. (This does require later sorting of picked items and merging items from several pickers to get the total order.) Another technique is to store items as close to the end of the aisle as possible to minimize travel distance; using the **Pareto principle**, 20% of the parts have 80% of the activity.

Some items (paperwork kept for possible audits, spare parts for your equipment, spare parts for customers, etc.) may have very low activity, once/month or even once/year. These items should be in a long-term storage area rather than with more active inventory. Offices especially are cluttered with long-term storage items which should be moved or discarded.

When the picking list is computerized, the computer can sequence the picking order to minimize travel. Store items together that are likely to be picked together, such as left and right brackets or other mating parts, or, in manufacturing, parts of the same subassembly. On the other hand, don't put look-alikes (different sizes of the same product; different products in the same color and shape carton) into the same drawer or bin, as the picker may pick the wrong item.

(1) Trash Conveyor
(2) Transport Conveyor
(3) Staging Conveyors

FIGURE 11.25 Combinations of storage systems are best for most applications. In this design, gravity flow shelves are used for the majority of items. Static shelves are used for slow movers and goods that, because of size or shape, do not lend themselves to gravity flow shelves. Pallets are used for fast-moving items; the pallets are on pallet flow shelves for full-case picking. Reserve pallet storage (for fast movers) is kept on top of the gravity flow rack. Reserve pallet storage (for slow movers) is in the rear. (Courtesy of Kingston-Warren Corp.)

Safety may require storage of some items together (or apart). For example, water-reacting materials such as calcium, phosphorus, potassium, and sodium salts should be in a special area. Water-reacting materials should be identified with a capital *W* with a horizontal bar across it.

Outdoor storage often is floor storage because space utilization is not critical. There may be a "floor" of pavement, gravel, or dirt. Naturally, outdoor storage requires items that can withstand rain, snow, insects, the sun, etc. Tents and inflatable buildings provide minimum protection at minimum cost.

3.5 Container   The container in which the item is received also may be used for storage and shipping. Then picking, counting, sorting, and the like are reduced, although the space cube may not be used as efficiently. It is

the total storage cost you are trying to minimize, not just the cost of storing an item on a specific shelf. That is, using one container throughout may be better, although somewhat more inefficient in storage than using one container for the vendor to ship in, another to transport to storage, another to store in, another to send to shipping, and another to send to the customer.

Table 11.9 summarizes storage principles.

TABLE 11.9   Principles of storage.

1. Reduce the number of aisles in the storage space.
2. Reduce the width of the storage aisles.
3. Increase stacking height of stored items.
4. Consider random storage.
5. Use high-density storage.
6. Minimize picking cost.
7. Secure the storage area.

DESIGN CHECKLIST: STORAGE

What is stored
    Eight types of storage
    Amount of each type
    Can space be reduced?
Storage equipment
    Picker to package
        Floor or outdoor storage
        Shelves, racks, mezzanines
        Vehicle size and type
        Size of aisles
        Rack type and accessories
        Size of racks
        Drawers
        Flow racks
        S/R machines

Package to picker
    Miniload
    Carousel
    Conveyors
No picker
    AS/RS
    Accumulation lanes
Storage organization
    Location identification
    LIFO versus FIFO
    Space utilization
        Random versus dedicated
        High-density storage
Picker movement
One container throughout?

REVIEW QUESTIONS

1. List the eight types of storage and a typical item in each.
2. What does SKU stand for?
3. What is kitting?
4. Does static shelving tend to be a LIFO or a FIFO system?
5. Why should the front edge of a shelf be reinforced?
6. When should a cantilevered shelf be used instead of a standard shelf?
7. List four ways of protecting shelves from vehicle damage.
8. Why do drawers get more addressable locations/ft³ than shelves?
9. What is the raider problem?
10. Why might a flow rack have a stop of wheels instead of a bar?

11. What are the two frame options in flow racks?
12. Give the seven order picking guidelines.
13. What is an aisle-captive machine?
14. What are the three approaches to bringing the package to the picker?
15. Sketch a system of an AS/RS, a take-away conveyor, three accumulation conveyors, and a conveyor to shipping.
16. Give the advantages and disadvantages of random versus dedicated storage.
17. Do picking machines travel faster horizontally or vertically?
18. Do racks or flow racks require more aisle space/amount of storage?
19. What is the fourth dimension? How can it be applied to storage?
20. List the seven principles of storage.

PROBLEMS AND PROJECTS

11.1 List, using the 8 categories of Table 11.1, all the items stored in the IE Department of your school. (To reduce time/student, make this a team project.) List each item name, quantity stored when it was observed, location, approximate square feet of storage space, and approximate cubic feet of storage space.

11.2 Design a warehouse with 5,000 slots, each holding a 42 × 48 in. pallet. Include your slot identification plan. Justify your design.

11.3 Design a selective pallet rack to hold one hundred 42 × 48 in. pallets. Give assumptions, including material handling technique. Use a manufacturer's catalog to specify all necessary purchasing information. Justify your design.

11.4 Design a cantilevered rack to hold 50 pieces of lumber (2 in. × 4 in. × 12 ft long). (An alternative is 25 carpet rolls, each 18 in. in diameter and 12 ft long.) Give assumptions, including material handling technique. Use a manufacturer's catalog to specify all necessary purchasing information. Justify your design.

REFERENCES

Drucker, P. The economy's power shift. *Wall Street Journal,* Sept. 24, 1992.

Dunn, R. Vertical storage carousels improve small parts handling. *Plant Engineering,* pp. 68–70, Sept. 11, 1986.

Egland, A. Analyzing the usefulness of specialized lift truck features. *Industrial Engineering,* Vol. 14, 3, 100–105, March 1982.

Frazelle, E. Orderpicking: Part 1. *Modern Material Handling,* Vol. 45, 10, 61–63, Oct. 1990a.

Frazelle, E. Orderpicking: Part 3. *Modern Material Handling,* Vol. 45, 12, 58–60, Dec. 1990b.

Gravity flow racks—Faster picking in less space. *Modern Material Handling,* Vol. 35, 8, 59–73 (Ref. 21), Aug. 1980.

Kemp, J. What to consider in choosing a system for automating small parts picking. *Industrial Engineering,* Vol. 14, 3, 68–75, March 1982.

Koenig, J. FLO/STO and GO/FLO analyses aid in decision. *Industrial Engineering,* Vol. 12, 9, 28–32, 1980.

Lauer, S. Six good ways to shorten your procurement lead time. *Factory,* pp. 88–89, Sept. 1966.

Manual orderpicking. *Modern Material Handling,* Vol. 38, 2, 53–61, Jan. 20, 1983.

Material handling: Storage. *Plant Engineering,* pp. 167–168, July 18, 1991.

Mulcahy, D. Unit handling and storage vehicles. *Plant Engineering,* pp. 56–57, May 2, 1991.

Nagy, A. It does you good to look back now and then. *Material Handling Engineering,* Vol. 34, 8, 120–122, Aug. 1979.

Quinlan, J. The ins and outs of drawer storage. *Material Handling Engineering,* Vol. 35, 2, 60–65, Feb. 1980.

Quinlan, J. How to build a case for a rack supported structure. *Material Handling Engineering,* Vol. 36, 8, 48–54, Aug. 1981.

Rack Storage of Materials, NFPA 231C. Quincy, Mass.: National Fire Protection Association, 1980.

Shelving (MHE Ref. 1). *Material Handling Engineering,* Vol. 29, 6, June 1974.

Storage. *Plant Engineering,* p. 153, July 18, 1991.

Storage of Rubber Tires, NFPA 231D. Quincy, Mass.: National Fire Protection Association, 1980.

Vertical carousels control material flow at TI. *Material Handling Engineering,* Vol. 62–64, May 1991.

Which system gives you the best cube utilization? *Modern Material Handling,* Vol. 36, 8, 78–83, June 5, 1981.

12 SHIPPING AND RECEIVING

CHAPTER PREVIEW

The first decisions are the number of docks and where they should be located in the building. Then design the dock approaches. Finally, design the dock itself (including matching dock and vehicle height) and dock accessories.

CHAPTER CONTENTS

1 How Many Docks?
2 Where in the Building?
3 Dock Approaches
4 Dock Structure and Equipment

KEY CONCEPTS

dock levelers   staging area
ICC bar   truck leveler
marshalling area

1 HOW MANY DOCKS?

All materials and supplies come through the receiving area; all products leave through shipping. Inadequate receiving and shipping facilities can choke production.

Consider whether two special-purpose areas (one for receiving and one for shipping) will be better than one general-purpose area (shipping and receiving combined). Combining them will reduce capital cost and give better equipment utilization as well as possible labor flexibility. In general, if the areas are not combined, they at least should be adjacent. Shipping areas are more adaptable to mechanization than receiving areas because of the more uniform packages in shipping. Rail docks generally are separate from truck docks.

Determine the number of pallets inbound and outbound for each dock each day of the month. Note that not all pallets or trucks will be full for either inbound or outbound shipments. You may request vendor shipment on a specific day, but usually you have no control on the specific hour the truck will arrive. Unless the shipment is to a remote warehouse, it is unrealistic to expect to ship Monday's production on Monday night and Tuesday's production on Tuesday, and so on, because customer desire for product-model mix and quantity will differ from production. Thus, shipments come from finished goods storage rather than from the production line. Remember the Pareto principle: 20% of the products are 80% of the workload.

After making a schedule for arrivals and departures of raw materials, purchased components, and finished goods for each day of a typical month, you need to consider miscellaneous shipments of supplies and also scrap and waste (which may require special handling).

You will discover that some days have little work and others are quite busy. In addition, some shipments will be handled with lift trucks and others manually. Some will be a full truckload and others less. Staffing the docks is therefore a challenge and makes a good simulation problem. Docks could be classified as "available," "truck awaiting processing," "truck being processed," and "truck already processed." To minimize the number of docks and truck idle time, reduce the time a truck occupies a dock but is not being processed. However, to minimize the staff idle time, have a supply of trucks waiting to be unloaded. That is, the best design depends upon the criteria.

Docks can be general purpose or special purpose. For example, a dock might be designed to receive supplies only from local vendors (often odd sizes and shapes and usually requiring manual handling), delivered in pickup trucks and vans.

The trend to Just-In-Time (JIT) shipments has increased the frequency of shipments and decreased the amount per shipment. Often the JIT shipment does not go to long-term storage within the plant but goes relatively quickly from the dock to the using department. This decreases material flow from the dock to long-term storage and from long-term storage to the using departments, but increases the flow from the dock directly to the departments. Thus, the dock versus long-term storage arrangement is less important, but the dock versus using departments arrangement has become more important.

Docks are relatively permanent, and future expansion should therefore be considered when they are first constructed (e.g., knock-out panels for doors and knock-out pits for dock levelers).

As a rough estimate, fork trucks can (1) load or unload a full trailer from or to the adjacent dock staging area in 45 min, (2) load or unload a full railcar from or to the adjacent dock staging area in 100 min, and (3) move pallets between a dock staging area and storage in low-bay floor stacking or storage racks (either direction) at the rate of 20 pallets/h. For hand stacking of cases, a team of two can fill a trailer in 120 min and a railcar in 240 min, assuming the cases were delivered on pallets by fork trucks or extendable conveyors (Stallard, 1974). Also see Box 19.1.

$$NDOCKS = SF * TRCKPDA * AVH/AVHA$$

where

$NDOCKS$	=	number of docks of each type
SF	=	safety factor
$TRCKPDA$	=	average number of trucks/day
AVH	=	average hours to load/unload a truck
$AVHA$	=	average hours available/day

The SF depends upon many factors, including distribution of arrival rates (within day, week, month, year), distribution in load/unload times, size of the marshalling area, and desired maximum waiting time in the marshalling area. The more the dock is specialized (between shipping/receiving and by vehicle size), the larger the SF should be.

The equation is best investigated with a simulation program so that the effect of various assumptions can be examined.

See Chapter 16 for more on unit loads and containers.

2 WHERE IN THE BUILDING?

In addition to one versus two areas, consider the relation of the dock to transportation facilities and your desired flow pattern within the plant. See Figure 12.1. For the sake of energy conservation and dock worker comfort, locate the dock on the side opposite the prevailing winds. Generally the southeast corner would avoid the north winds of winter and the southwest winds of summer. Since docks are not pretty, esthetics (and some zoning rules) may dictate a dock shielded from the view of bypassers. Since docks are expensive to move, consider how the facility will expand.

The entire dock area may be sealed off from the remainder of the facility for security and energy reasons.

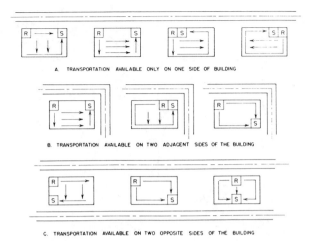

A. TRANSPORTATION AVAILABLE ONLY ON ONE SIDE OF BUILDING

B. TRANSPORTATION AVAILABLE ON TWO ADJACENT SIDES OF THE BUILDING

C. TRANSPORTATION AVAILABLE ON TWO OPPOSITE SIDES OF THE BUILDING

FIGURE 12.1 Docks for receiving and shipping can be separate or combined. One advantage of combined docks is reduced security problems. The dominant factor may be access to transportation and truck staging areas, especially on existing buildings on crowded lots.

Be sure to provide the appropriate size of quick-acting doors with locks. Dock seals and shelters restrict access through the truck–building gap. Many firms wish to restrict driver access to the plant; in these cases, toilets, waiting areas for drivers, and a telephone (even if it is a pay phone) are needed. They can be placed on mezzanines. Other uses for shipping dock mezzanines are for storage of packing materials and office supplies, warehouse offices, and lunch rooms.

Receiving departments have four functions: validation of receipts, handling for inspection, interim storage, and moving material to storage or production.

Table 12.1 gives recommended distances between the leveler and the staging area.

Staging areas are needed for both incoming and outgoing shipments. For more on storage, see Chapter 11. A good goal is a staging area equal to a trailer length (Peterson, 1981). Typically, when a truck or railcar is unloaded, the material goes not directly to stock but just to a staging area where quality control and inventory control information is processed. The load may go from the dock into a separate CBOSS area (count, back order, sample, select). In count, the vendor count is verified; in back order, any back-ordered stock is sent directly to the customer; in sample, items are removed for inspection; and in select, the specific storage location is selected. One important decision is the size of the unit load, since there generally are several different size storage locations. Many organizations weigh the incoming shipment and check it against invoice weight as a first screen. Then the load is moved to its location within the plant. The staging area should be separated from the plant itself by a physical barrier to provide inventory control and security. This buffer area decouples the dock and truck from in-plant storage but tends to require double handling. One alternative is to move directly from the truck to in-plant storage. Although this reduces double handling, it ties up the dock and truck while the goods are being checked, counted, inspected, and processed. Another alternative is to move directly to inplant storage without inspection and inspect later. This reduces dock congestion and permits inspection at nonpeak times but requires double handling and may permit noninspected items to slip through.

Goods from local delivery vehicles tend to be odd sizes and shapes, often badly packed. Special processing (both dock and buffer area) may be desirable.

Dock office needs vary, but the minimum would be a standup desk, a telephone, and a terminal connected to the computer network. See the section on shop offices in Chapter 13.

3 DOCK APPROACHES

3.1 Truck Docks
Dock and dock approach designs need to consider the entire site. Specifically, consider truck access to the docks from the street, marshalling areas, employee parking, outdoor storage, and future expansion.

Table 12.2 gives recommended site road and gate dimensions. A **marshalling area** where trucks can wait before going to the docks permits fewer docks but increases truck waiting time. Trucks in the marshalling area should not interfere with trucks maneuvering into or from

TABLE 12.1 Minimum maneuvering distance between the back of the dock leveler and the beginning of the staging area (Thompkins, 1982). Copyright © Institute of Industrial Engineers, 25 Technology Park, Norcross, GA 30092. (404) 449-0460.

Equipment Used	Distance (ft)
None (manual)	5
Hand truck	
Two wheel	6
Four wheel	8
Hand lift (jack)	8
Narrow aisle truck	10
Lift truck	12
Tow tractor	14

TABLE 12.2 Recommended dock staging dimensions (Thompkins, 1982). Copyright © Institute of Industrial Engineers, 25 Technology Park, Norcross, GA 30092. (404) 449-0460.

Item	Dimension (ft)
Served road width	
One-way traffic	12
Two-way	24
Pedestrian walk (physically separated) along road	4
Gate openings, vehicles only	
One-way traffic	16
Two-way	28
Gate openings, vehicles + pedestrians	
One-way	22
Two-way	34
Right-angle intersections, minimum radius	50

the docks. If incoming or departing trucks have heavy loads, consider heavy-duty pavement in the dock area, marshalling area and on-site roads.

Sharp turns are difficult. If possible, avoid 90° turns in approaches to the plant gates, on roads within the site, and at the docks. Drivers prefer to back up using the left side mirror. Visibility when going forward also is better to the left.

Table 12.3 gives truck dimensions. Lane width normally should be 12 ft; 10 ft is feasible (with increased maneuvering time and accidents), and 14 ft should be used for very busy docks. Paint lane lines on the pavement.

Lane length depends on truck length, lane width, and whether the driver is looking out the left or right window. Some rules are

LANE LENGTH (left window) = *VEHICLE LENGTH* + 40 ft

LANE LENGTH (right window) = *VEHICLE LENGTH* + 100 ft

LANE LENGTH = 2 (*LENGTH OF LONGEST VEHICLE*)

LANE LENGTH = SF (*VEHICLE LENGTH*)
where
SF = 2.1 for 10 ft wide lanes
2.05 for 12 ft wide lanes
2.02 for 14 ft wide lanes

Figure 12.2 shows four design alternatives. In the usual design (alternative 1) the truck is outdoors and perpendicular to the dock. Sometimes a canopy is placed over the trucks for partial protection from sun and rain. Alternative 2 shows the trucks inside. The advantages are better weather protection and the capability to unload from the side door or with overhead cranes and monorail hoists suspended from the roof. The disadvantage is the use of expensive interior space. Alternative 3 (sawtooth docks) reduces exterior maneuvering space, as does alternative 4 (finger dock). Alternative 4 has the advantage of only one door opening into the building but the disadvantage of congestion on the finger itself and at the door.

The dock and the floor of the production-storage area should be at the same height. Unless the ground slopes, the dock approach lane will be below grade. If the dock

TABLE 12.3 Typical truck dimensions (Loading Docks, 1991).

Vehicle Type	Overall Length (ft)	Width (in.)	Height (ft)	Truck Bed Height (in.)
Straight truck	15–30	96	11–12	36–48
Semi, city	30–35	96	11–13	44–48
Semi, road	55–70	96–102	12–14	48–52
Flatbed	55–70	96–102	—	48–60
Refrigerated	40–55	96–102	12–14	50–60
Container	55–70	96	12–14	56–62
High cube	55–70	102	13–14	36–42

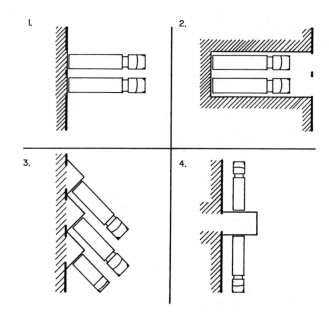

FIGURE 12.2 Four dock alternatives are (1) trucks perpendicular to the building and outside, (2) trucks perpendicular to the building and inside, (3) trucks at an angle to the building and outside, and (4) trucks parallel to the building and outside. (Adapted from Muther, 1985.)

approach lane slopes toward the dock, vehicle control is difficult and the top of the tilted truck may hit the building. If the drive must slope, keep the grade less than 10%. Another design is a level segment (e.g., for 10 ft) before sloping. Consider how rainwater will drain and how snow will be removed and where it will be put.

A separate personnel entrance to the dock is common. If none exists, install a steel ladder with a handrail and nonslip steps to give safe dock access.

3.2 Rail Docks   Boxcars typically are 102 in. wide, 108 in. high and 40 or 50 ft long. Docks for rail cars present different problems than do truck docks. Maneuvering areas and provisions for drivers are not needed. Loading and unloading cycles are longer. If rail cars will have heavy materials on flat or gondola cars, consider using a bridge or gantry crane over the dock. As with truck docks, inside rail docks eliminate weather and security problems.

Provide clearances between the car and obstructions. Required clearances vary by specific railroad and state. In general, vertical clearance is 22 ft above the rail top; horizontal clearance is 8 ft from the track centerline; and dock height is 48 in. above the rail top. Before construction, submit all plans for approval to the railroad serving the facility.

4 DOCK STRUCTURE AND EQUIPMENT

4.1 Docks   Dock heights tend to be about 48 in. A good dock height is 50 in. At one time there was an OSHA regulation requiring guardrails for docks 48 in. or higher, so many people made the docks 47.5 in. The regulation no longer is in effect, so any height can be used now. Good practice is to have the dock slightly below truck bed level so runaway loads do not trap people in the truck; 24% of all industrial accidents occur on the dock (Dock Location Principles, 1973).

Docks should be on 11, 12, or 14 ft centers. The dock area should provide sufficient maneuvering space so that one lift truck doesn't have to wait until another gets out of the way. Dock areas also should have communication facilities so that inventory information can be updated as quickly as possible.

4.2 Dock to Truck Adjustment

Truck bed heights range from 36 to 58 in. above ground. Because of springs and tires, they will vary as much as 8 in. in height from fully loaded to empty. Thus, there generally will be a difference in level between the truck bed and the dock. The difference can be reduced by dock specialization. That is, some docks can be for semis and some for high cube vehicles. The disadvantage, of course, is that all vehicles cannot use all docks. There are three solutions: dock levelers, truck levelers, and dock lifts. See Figure 12.3 and Table 12.4.

Dock levelers—portable, mechanical, or hydraulic—create a ramp. There is a noticeable shift from manually and mechanically powered units to hydraulically powered units.

See Table 12.5 for permitted ramp angles. Typical range is plus or minus 12 in.; thus, a 50 in. dock can accommodate truck beds from 38 to 62 in. Be sure the leveler has sufficient width so that the lift truck can enter and leave the truck in a straight line. Maneuvering while entering or leaving takes more time. Most trailers' inside width is 90 in. (some can be 92 in., depending on the door hinging); dock leveler width probably should be 78 to 84 in. Below 78 in.

DOCK LEVELER   TRUCK LEVELER

LIFT TABLE

FIGURE 12.3 Dock levelers, truck levelers, and lift tables compensate for height differences. Dock levelers make a ramp on the dock. Truck levelers make the ramp under the truck wheels. Lift tables have advantages of level transfer and a wide range of vertical adjustment. Lift tables also are used for in-plant handling.

TABLE 12.4 Application guide for dock equipment.

Application Factor	Recommended Method		
	Dock Leveler	*Dock Lift*	*Truck Leveler*
High volume loading	x		
No dock		x	
Low dock (under 28 in.)		x	
Low dock (28–36 in.)		x	x
Lift truck access to grade level		x	
High dock (50–59 in.)		x	x
Low to moderate use		x	
Wide variation in truck heights		x	x
Narrow dock apron		x	x
Restricted truck area		x	
Unstable loads		x	
Level handling requirement		x	x

presents lift truck–pallet alignment problems; above 84 in. makes it difficult to align the trucks to the dock. Dock leveler capacity should be twice the loaded lift truck to allow for dynamic loads. To calculate leveler length, solve

$$LEVL = \frac{MAXDI}{\sin(MAXA)} \qquad (1)$$

where

$LEVL$ = leveler length, in.

$MAXDI$ = maximum difference between dock and truck bed, in.

$MAXA$ = maximum leveler angle, degree (see Table 12.5)

Truck levelers have a higher capital cost than dock levelers. Their wider range permits a wide range of trucks to be handled. Because of their adjustment range, the resulting ramp angles may be closer to zero.

Lift tables, a low–shipping-volume alternative, have a strong advantage of level handling, which may be convenient or even necessary for a product. Lift tables, which have an even wider vertical range than truck lifts, also can be used when there is no dock (i.e., load from ground level) or to move items from a dock to the ground. They are very handy for station wagons, vans, and pickup trucks. Pit-mounted dock lifts have lower capital costs than surface-mounted units, but surface-mounted lifts can be moved and are handy for leased facilities. Since lift tables use some of the truck driveway, be sure there is sufficient driveway left.

The most common dock lift table has a 5,000 lb capacity. To calculate capacity required, assume a hand pallet

TABLE 12.5 Recommended maximum grades for ramps.

Vehicle	Maximum Grade (Percent)
Power-operated hand truck	3
Powered platform truck	7
Low-lift pallet-skid truck	10
Electric fork truck	10
Gasoline fork truck	15

truck has 50% of its load on each axle while a lift truck has 80% on the front axle. Thus a 5,000 lb load on a pallet truck has a roll-on load of 5,000 × .5 = 2,500 lb. A 6,000 lb lift truck with a 4,000 lb load has a roll-on load of 10,000 × .8 = 8,000 lb. A heavy-duty lift table can accept 75% of its rated capacity over the platform ends (50% for medium duty) and 50% over the sides. Thus, when loading over the table sides, a 10,000 lb truck with load has an 8,000 lb roll-on load and, at the 50% factor, a 16,000 lb capacity table is required.

If forklifts will move between the dock and the driveway, provide either a concrete ramp or a lift table.

4.3 Dock Accessories   Protect the dock itself from impact by bumper pads. A 1 in. pad will decrease impact force to 10% at 4 miles/hour. Concrete-filled pipes (6 or 8 in., painted yellow) will protect the dock and equipment from impact by trucks and lift trucks. Wheel chocks, tethered to the dock face with a chain, prevent trucks from moving accidentally from the dock and thus causing an unexpected gap for the lift truck.

To reduce "submarining" of vehicles hitting the rear of a truck, the Interstate Commerce Commission requires most trucks to have an **ICC bar** on the rear of the truck. When the truck docks, a restraint device hooks the bar (which is 12–30 in. above the road) and a light indicates that the hook is engaged. Thus, the truck cannot move away and the unloading vehicle cannot drop off the ramp.

Mount dock loading lights on arms to illuminate truck interiors; point lights into the lower or upper corners to reduce glare. For long trucks and night operations, lights on the lift truck masts are good. Fans may be needed to ventilate truck interiors during the summer; radiant heaters on the dock are useful in the winter.

Conveyors, either mobile or retractable, are very useful for hand loading or unloading of trucks and railcars. Some trucks have roller conveyors built into their floors. Fixed conveyors, unless flush with the floor, inhibit cross traffic. Floor-level conveyors, however, cause back problems when people bend over to pick items off them; product often is damaged when people throw items onto them.

A good dock door is 9 ft wide and 10 ft high; if it needs to accommodate super vans, it should be 12 ft high. A narrow door presents unloading problems because trucks rarely align perfectly with the dock. Doors can be actuated by pressure pads, pull cords, or photoelectric cells; they should have security locks. Powered doors (e.g., opening rate of 5 ft/s and closing rate of 1.5 ft/s) reduce air exchange with the environment.

To protect the building climate, use dock seals and shelters to close the gap between the building and the truck. See Figures 12.4 and 12.5. They will seal out light, so supplementary light is needed in the truck. They also seal out fumes, insects, and people (who might enter the building through the truck–building gap if there were no seal).

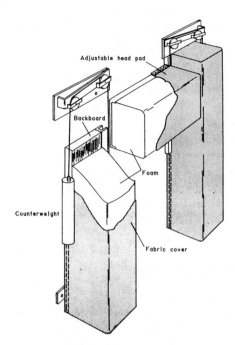

FIGURE 12.4   Dock seals are like pillows ringing the doorway (Seals and Shelters, 1983). They cost about $400 per door and cut heat losses about 90%. Foam seals (shown in sketch) generally are polyurethane with a protective cover; provide extra protection at the wear points (seal faces that contact the trailer and inside faces of side panels, which are brushed by lift trucks and moving cargo). Foam seals have no moving parts. If the drive slopes, have side pads taper 1 in. from top to bottom for every 1% of slope (to even out pressure on pad). Air seals (not shown) are deflated against the building when not in use. After a trailer backs into position, the operator trips a switch to activate a blower, which presses the seal against the trailer. They tend to be more universal (fit more size vehicles) than are foam seals.

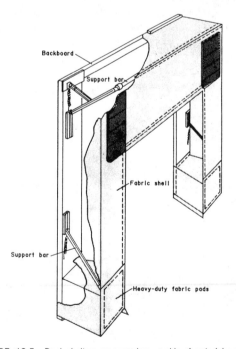

FIGURE 12.5   Dock shelters are canopies—a skin of material on a frame (Seals and Shelters, 1983). They cost about $900/door and cut heat loss about 75%. Shelters tend to come in one standard size which will seal any doorway up to 10 ft high by 10.5 ft wide. They can service a wide variety of vehicles and do not restrict access to the vehicle.

Seals and shelters tend to have a payback of less than two years. Consider them even for rented or leased facilities. Impact doors and flexible strip doors are used for temporary closing of the gap. They are pushed aside when the lift truck goes through.

An interesting approach is the dock-mounted loader. Rather than loading items directly into a truck, items are brought to a prestaging area and then assembled on a roller near the dock. When the truck arrives, the entire load is powered into the truck in one motion. The primary advantages are access to the roller load from all sides (resulting in better cube utilization of the truck, lower loading labor costs, better inventory count control) and reduced truck loading time (giving less truck waiting time and requiring one dock instead of four). The disadvantage is the equipment capital cost; however, one company reported a seven-month payback, using reduction in direct labor costs only.

Many injuries occur when cutting steel and plastic bands, straps, and wires. Use a cutting tool (not hammer or crowbar to pry), cut square (diagonal cuts have sharper edges), wear gloves and goggles, and hold the strap with the closest hand (if the strap springs, it will go away from the face).

DESIGN CHECKLIST: SHIPPING AND RECEIVING

1. What is shipped? Received? For each dock, what is minimum weight (volume)/load? Mean? Maximum? What records are modified?
2. Where to? Where from? Dock location within a building? Flow pattern within dock area/within plant?
3. How many docks? Some docks specialized? Expansion (contraction) of shipping requirements?
4. Why and when is load shipped (received)? Can shipment (receipt) be made sooner (later) in shift? For each dock, what is minimum, mean, and maximum volume per shift (week) (month)? Who decides frequency? Effect of more storage before shipping (after receiving) by ourselves or vendor (customer)?
5. Who will use docks? Full or part time? Job classifications? Access restrictions for employees (non-employees)?

REVIEW QUESTIONS

1. Discuss determining the number of docks.
2. List at least four factors to be considered in determining the number of docks.
3. Give the four dock design alternatives given in Figure 12.2.
4. Give some desirable characteristics of the truck marshalling area.
5. Discuss some dock design features to increase security from theft.
6. Why shouldn't the truck lane slope toward the dock?
7. What are the three approaches to matching dock level and truck bed level?
8. Discuss use of fixed conveyors in truck loading and unloading.
9. What are two uses of an ICC bar?
10. What is a dock-mounted loader?

PROBLEMS AND PROJECTS

12.1 Evaluate the security at an existing dock. What does the supervisor think of your ideas?

12.2 Evaluate the staging area for an existing dock. What does the supervisor think of your ideas?

12.3 Design the approach area for an existing dock. Is there anything different you would design if starting with an empty field? What does the supervisor think of your ideas?

12.4 Design a dock structure and equipment for an existing facility, but assume you were starting with an empty field. Justify the dock design, dock-to-truck alternative recommended, and dock accessories recommended.

REFERENCES

Dock location principles. *Industrial Engineering*, Vol. 5, 10, 41–43, Oct. 1973.

Loading docks. *Plant Engineering*, Vol. 47, July 18, 1991.

Muther, R. Relationship between material handling and plant layout. Chapter 2 in *Material Handling Handbook*, R. Kulwiec, Ed. New York: Wiley-Interscience, 1985.

Peterson, N. Designing the plant loading dock. *Plant Engineering*, Vol. 35, 11, 118–120, June 11, 1981.

Seals and shelters—Saving energy at the dock. *Modern Material Handling*, Vol. 38, 4, 64–67, March 7, 1983.

Stallard, D. How to make ballpark estimates in systems planning. *Material Handling Engineering*, Vol. 29, 2, 51–53, Feb. 1974.

Thompkins, J. Plant layout. Chapter 10.2 in *Handbook of Industrial Engineering*, G. Salvendy, Ed. New York: Wiley, 1982.

CHAPTER PREVIEW

Offices process information (from papers, computers, telephones, people); factories process things. Esthetics and decor are important in offices. The open plan has become the conventional arrangement of the computerized office, and specialized areas such as conference rooms are receiving more attention.

CHAPTER CONTENTS

1 How Office Layout Differs from Factory Layout
2 Types of Office Arrangements
3 Information Processing and the Office
4 Furniture
5 Special Areas

KEY CONCEPTS

conventional arrangement	open-plan office	training room
electronic mail (EMail)	personal space	VDT (VDU)
fax	shop offices	virtual office
landscape office	social environment	visual privacy
local area network	sociofugal (versus sociopedal)	visual variety
modem	teleconferencing	voice mail
nonterritorial office	territory	WATS lines

1 HOW OFFICE LAYOUT DIFFERS FROM FACTORY LAYOUT

Offices differ from factories in three main ways: the product, the physical environment, and the social environment.

A factory produces things. These things are moved with conveyors and lift trucks; factory utilities include gas, water, compressed air, waste disposal, and large amounts of power, as well as telephones and computer networks. A layout criterion is minimization of transportation cost.

An office produces information—subdivided into physical items (paperwork, including computer input and output), electronic files, and oral communication (face to face and telephone). The paper is moved by hand, foot, and, occasionally, mail carts; utilities are telephones and computers and a small amount of electrical power and waste disposal. If the information is moved electronically or orally, nothing is actually moved. Office layout criteria, although hard to quantify, are minimization of communication cost and maximization of employee productivity. *Administrative Management*, in selecting the "Office of the Year," uses the following criteria:

suitability (operational effectiveness) of space allocation, work and traffic patterns, accommodation for required equipment, and energy conservation

flexibility to permit efficient change and growth

habitability through features designed to heighten human efficiency, such as lighting, sound and climate conditioning, decor, and various employee facilities

advancement of the administrative profession through state-of-the-art improvements and other innovations in office planning and systems designs

The second difference, the physical environment, tends to be better for the office than the factory. Office lighting tends to be slightly better than the factory, primarily due to tasks that usually are visually more difficult. Toxicology and ventilation problems are minimal within the office since there rarely are pollution sources. Acoustics tend to be better in the office since noise, which interferes with oral communication, may be distracting to those doing paperwork. Climate control tends to be considerably better, because of lack of large heat sources within the space, high worker density/ft³, and the historical concept that white-collar workers should get better treatment than blue-collar workers. This leads to the third difference between office and factory—the **social environment**.

Historically, office workers have been much more concerned with status and esthetics than factory workers. A key consideration in many office layouts is the question of who will get the best window location. To show their status, executives expect preferred locations and larger amounts of space. Rank expects more privacy and more plush physical surroundings. In the factory, esthetics are

unlikely to be a concern, but office personnel expect carpets, color-coordinated furniture, artwork, and even plants. "Artificial plants are totally unacceptable" (Pile, 1978, p. 118). Offices are designed to be tasteful and to "reflect the organization's approach to business dealings."

2 TYPES OF OFFICE ARRANGEMENTS

Figures 13.1, 13.2, and 13.3 show examples of the conventional, landscape, and open-plan arrangements.

2.1 Conventional

The typical office building of the 1890s, when large organizations were unusual and office staffs were small, had a plan much like that of a hotel: small rooms lined up along a window wall, with an access corridor running the length of the building (Pile, 1978). There was little concept of the arrangement of the rooms in relation to each other except that high-ranking people got preferred locations. By the 1950s, the technological revolutions of artificial light and air conditioning reduced the importance of the window. The **conventional arrangement** now had a large central open area of desks in regular rows (a "bullpen") with private offices along the perimeter window walls. Because of the recognition of the effects of distraction and the desires of the workers for some privacy, the central area now often was subdivided by rectangular partitions 4 to 6 ft high. High-ranking people still had enclosed rooms along the perimeter—although sometimes, because of the large size of buildings, without a window. With the implementation of the computer, repetitive clerical work now is done by computers instead of people, so the size of the central space has shrunk to hold tens of people instead of hundreds.

2.2 Office Landscape (Burolandscaft)

In 1958 Eberhard and Wolfgang Schnelle in the Hamburg suburb of Quickborn formed the Quickborner team and advocated a drastically different office arrangement, the **landscape**

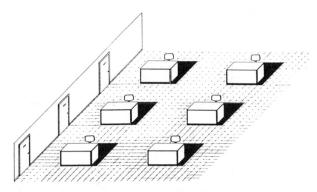

FIGURE 13.1 Conventional arrangement of an office. Characteristics include private offices for higher ranks, no partitions between desks, no plants, straight lines, desks only.

FIGURE 13.2 Landscape arrangement of an office. Characteristics include no private offices, some partitions due to furniture, plants, no straight lines, desks, plus some storage units.

office. They had two revolutionary concepts—equality and absence of straight lines.

By *equality* they meant status and rank symbols of offices. "The officers should be down in with the troops." Private offices were abolished. They reluctantly permitted the officers more plush furniture, but additional space had to be a function of need, not of rank. There were to be no permanent interior walls. When a task required complete privacy, the individuals were to move (for those few minutes) to a special privacy area. Visual privacy at the normal workstation was attempted by curved partitions blocking lines of sight; auditory privacy was attempted by acoustical treatments and was dependent upon office hum. There was a feeling that demands for privacy really cover a desire to hide from contacts and from work.

The abolition of straight lines (so beloved by planners) was equally shocking. Straightforward aisles and corridors now were replaced by crooked paths through the forest—they emphasized considerable use of plants as well as meandering aisles.

The rationale of the abolition of private offices was that fixed partitions restrict the constant rearrangement of offices. The rationale for the random desk arrangement was

that it was actually a functional arrangement based on minimization of communication cost. Neither rationale stands up to detailed scrutiny.

Partitions need not be torn out or moved when people are rearranged, especially if there is some spare space and if the partitioned areas are of unequal size. In addition, the amount of rapid organizational change typical of Germany in the 1950s is not common to all cultures, countries, or decades. Groups of desks can be arranged to minimize communication costs while still retaining rectangular aisles.

2.3 Open Plan   In 1964, Herman Miller Inc. introduced the concept of the "action office"; this now is known as the **open-plan office**. The key concept was that the needs of each workstation varied. Standard desks, chairs, and files were replaced by units that combined work surfaces, storage, and seating (or standing) in endlessly adjustable components and configurations. Panels, standing free on feet, carried worktops, shelves, and storage. Files could be embedded in work tables or hung on panels. Storage also was divided into three categories: workstation, departmental, and long-term (in a different part of the building). Users could sit or stand (Pile, 1978, p. 32).

The individualized workstation tends to have a systematic arrangement rather than a rectangular landscape one. That is, although there still are main aisles, there also are "residential neighborhoods" with byways and dead-end streets to minimize through traffic. The designer needs to consider the problem a casual visitor to the office might have in finding people. Some strategies are to color-code various departments (e.g., green for accounting, yellow and brown for engineering, blue for purchasing), coloring the corridor carpet brown (contrasting with colors of carpets in work areas), and having partitions between functional areas higher than partitions within an area.

Private offices still are used, but not for the low-ranking officers. A typical office space might have 25% of the space for private offices and 75% for individual workstations. Figure 13.4 shows four typical linkages of open-plan workstations.

Visual variety is emphasized, with color-coordinated furniture replacing battleship gray and industrial green. Using different colors to define different areas adds visual variety for the occupants and aids visitor traffic. If employees do not have a window, the wall is often covered with artwork or picture wallpaper. Windows should be shared by putting them at the end of corridors, in areas with low partitions, near elevator banks, or in stairwells. Atriums are popular for cafeteria, lounge, and corridor areas. Considerable effort is devoted to **visual privacy** through placement of partitions (usually curved), interrupted lines of sight, opaque glass partitions rather than clear, partition heights of 62 to 80 in., and placement of plants.

In acoustics, the ratio of the signal (or message) to noise (or background) is important. Acoustical treatment

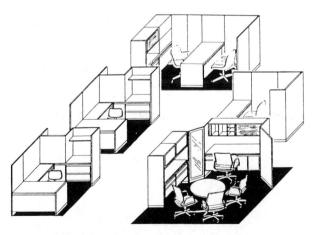

FIGURE 13.3 Open-plan arrangement of an office. Characteristics include a few private offices, extensive use of partitions, plants, straight and curved lines, work surfaces, and storage units in a wide variety of sizes and shapes.

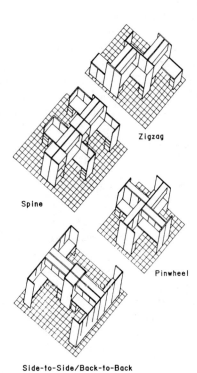

Zigzag

Spine

Pinwheel

Side-to-Side/Back-to-Back

FIGURE 13.4   Typical workstation linkages are zigzag, spine, pinwheel, and side-to-side/back-to-back (Sachs, 1981; excerpted from *Administrative Management,* October 1981, by Geyer-McAllister Pub., New York.)

of partitions and furniture makes them absorb noise and thus makes the signal more noticeable. However, it may be that the signal is really a distraction (e.g., conversation from an adjacent desk). Then a technique is to mask the signal by *increasing* the background noise. A common approach is to increase air distribution noise. Another technique is to buy special "white noise" generators. The goal is to have the ambient noise level about 45 dbA. It may also be possible to reduce the distracting signal. For example, an unanswered telephone is very disturbing in open-plan areas. A central secretary to answer all calls and then buzz specific extensions will reduce the annoyance, as well as ensure that all calls are answered.

Privacy in the open plan is much better than in the bullpen of the conventional office or in the office landscape

but not as good as in private offices. The concept of individualized workstations has led to more functional as well as more aesthetic workstations for the vast majority of office workers.

Changes will continue (see Box 13.1). One possible direction is the **nonterritorial office** in which workers move from place to place, spending the day at three or four workstations of their choosing. Another approach is the "hot" workstation. This nonterritorial station is used by employees who are in the office at infrequent intervals. Examples might be sales or service personnel, part-time employees, or people working at several locations. The goal is to avoid tying up floor space for only occasional use.

Another possibility is the **virtual office**. In the virtual office, the worker can be almost anywhere at any time. The worker has a portable terminal or personal computer and connects to the physical office via modems and the telephone network. For example, a sportswriter can use a portable terminal holding 20,000 words. The terminal can have any notes or background information stored in the terminal or transmitted by telephone from the main office to the terminal. After the story is written on the scene, it is transmitted through a modem to the central computer; then the newsroom calls it up for editing and sends it to the composing room for electronic typesetting. Thus, the job no longer is tied to flow of paper across a specific desk; it is tied directly to the office worker. Although some enjoy the home office, others miss the social interaction and sense of community of the conventional office.

2.4 Comments on Open Plan Versus Conventional

Open plan and modular furniture have a number of important advantages over conventional offices, that is, combinations of private offices and bullpens: (1) more people/ft^2, (2) less rearrangement cost, (3) less energy cost, and (4) possibly more productivity. (Note that the alternative to open plan for most people is the bullpen, not the private office.)

More people/ft^2 comes from a number of factors. In conventional offices, vertical space often is wasted since little is placed higher than desk height. Walls take some space in conventional offices but the primary problem is that walls tend to make offices into fixed areas which are

---

BOX 13.1   *Impact of the electronic office (Wilson, 1983)*

The proportion of office workstations with electronic equipment will become quite high.

Power requirements in offices will be higher than in present offices; this power will generate heat as well as require more power, especially "clean" power.

Equipment tends to take up considerable space, so space/workstation may increase 50 to 100%.

Since information flow is by telephone and electronics rather than face to face, decentralization has more potential.

Small, dispersed locations are feasible replacements for large central offices. In particular, offices need not be in central cities.

Office space will need to accommodate the power requirements in the office as well as cables for computers. This requirement will hinder use of buildings that do not permit zoning of air conditioning and that restrict horizontal and vertical ducts for cables.

TABLE 13.1   Space recommendations for university faculty in conventional private offices (Bareither and Schillinger, 1968).

Number of Full-Time Faculty (Including Dept. Head)	Space/ Faculty (ft²)	Comment
1–5	120	Add 15 ft²/faculty (15–75 ft²) for larger office for head.
6–15	120	Add 15 ft²/faculty (90–225 ft²) for larger office for head and a reception area. Add 200 ft² for a conference room for 8 to 10.
16–25	120	Add 15 ft²/faculty (250–625 ft²) for larger office for head, a reception area, and a conference room for 8 to 10. For a conference room for 12 to 15, add 50 ft² more.

not easily varied. Furniture location often is limited by location of utility outlets, windows, and doors. Bareither and Schillinger (1968) recommend the values given in Table 13.1 for university faculty in conventional offices. In the open plan, workstation size, components, configuration, and work surface can be determined by an analysis of functional needs. See Tables 13.2 and 13.3. One open plan allocated 164 ft² for each manager's workstation, 137 for each supervisor's, 55 for sales, 54 for engineers, 42 for technical support, and 36 for clerical. (In both open-plan and conventional offices, it is generally a mistake to give people of equal rank a different size office; that is, space requirements are set by prestige as well as function.) The open-plan fixtures often combine functions, such as a wall-mounted drafting table which folds flat against the wall when not in use and then doubles as a poster board. The door of a clothes wardrobe doubles as a chalkboard.

Less rearrangement cost is a result of the modularity of the furniture and the elimination of moving walls and permanent utilities. Electrical connections are plug-in and do not require an electrician; most workstations can be assembled or disassembled with screwdrivers and allen wrenches. Carpet tiles with "forever alive" glue on the backs simplify both carpet movement and maintenance. Compared with moving a permanent wall in a conventional office, change costs in open plan are 1% to 5%. A hidden cost of a conventional office is that, since change costs are so high, desirable changes will not be made.

Three reasons exist for reduced energy cost. First, since fewer ft² are used/person, total energy costs are lower. (This assumes the office space is rented or other uses are available for the excess space in a conventional office.) Second, there are lighting economies. Conventional offices use ceiling fixtures. Because of the walls, lighting patterns from each fixture are interrupted and large, efficient fixtures cannot be used everywhere. In general, lighting costs will be about 20% lower in the open plan. Open plan offers another possibility—task lighting. The modular fixtures can include lights mounted directly above the work surfaces. Thus, if 1,000 lux is desired on a work surface, the ceiling illumination could furnish 200 lux while the relatively small bulb only 2 ft away can furnish the remaining 800. An open plan with task lighting may give lighting savings of 50% over conventional offices. Third, there are air conditioning economies, primarily due to the lower lighting loads. Most offices have relatively large "metabolisms" (many people and considerable heat-generating equipment). This, combined with the common desire to maintain close temperature control, means that air conditioning is much more of a problem than heating.

More productivity in the open plan is difficult to prove. One advantage of the modular furniture is that it can be adjusted in height to the nearest inch so as to fit the individual and the task. The modular workstations have been well thought out, with attention to left- versus right-handed workers, placement of files, types of drawer

TABLE 13.2   Office space requirements Courtesy Steelcase Inc., Grand Rapids, MI 49501.

Need	Support[1]	Professional[2]	Supervisory[3]	Managerial[4]	Executive[5]
Functional					
Primary work surface, sq in.	1,125–1,800	1,125–2,450	1,500–2,100	2,100–2,450	2,450
Secondary work surface, sq in.	1,200	2,000	2,000	3,000	3,000
Display	Small pin up	Large pin up + easel	Large pin up + easel	Large pin up + easel + proj. screen	Large pin up + easel + proj. screen
Filing, in.	Up to 190	50–225	50–250	Up to 200	Up to 250
General storage, in.	Up to 100	50–500	50–250	125–300	100–500
Conference					
No. of guests	0	Up to 2	Up to 3	4	6
Privacy	Low	Medium	Medium	High	High

[1]Clerk, clerk-typist, machine operator, mail messenger, secretary, receptionist, drafter.
[2]Scientist, engineer, accountant, programmer, sales, customer service, architect, lawyer, purchasing, marketing, promotion.
[3]Group leader.
[4]Department head, function director.
[5]Director, officer, division head.

TABLE 13.3 Office space requirements (Wrennall, 1992). Net space excludes main aisles, conference rooms, toilets, computer rooms, and supply stores. Gross space includes these areas. Both net and gross space exclude cafeterias and mechanical rooms.

Space		(sq ft)	Job
Net	**Gross**		
35–75	180		Clerical
65–120			Secretaries
60–120			Professionals
100–200			Managers
150–600	320		Executive (senior)

contents, etc. Thus, when modular is compared with a conventional desk, it is very difficult not to have some improvement. Another advantage (especially versus bullpen arrangement) is the careful attention to minimizing visual distractions. The problem with visual distractions is not that people see but that when they see they feel required to talk. When they talk, they talk about children, ball games, etc., and only occasionally about work. Naturally there is a conflict between a supervisor's desire to observe people easily and the wish to reduce distractions with barriers. With the exception of receptionists, people should not face doors or aisles, since this encourages non–job-related communication with bypassers. A disadvantage of the open plan versus the private office is the problem of auditory distractions. Use of carpets, acoustical treatment of the walls and panels, and acoustical hoods over typewriters and printers are needed to eliminate excess noise. However, the noise cannot be too low either. Some firms even use a low level of electronic white noise to mask conversations. More aesthetic white noise is water splashing from fountains and background music.

Cost of office space varies widely, depending primarily on the cost of land and availability of competing office space. See Table 8.2 for the rental cost of office space in various cities in 1991.

3 INFORMATION PROCESSING AND THE OFFICE

Offices process four different "items":

1. paper messages (e.g., reports, letters)
2. computer messages (generally without a "hard copy" input and output)
3. telephone messages
4. in-person messages (from department coworkers, organization coworkers, and visitors)

3.1 Paper   Although the paperless office is a goal, most offices still generate reams of paper. Very little of it,

however, now is handwritten; almost all of it is from forms or from printers and copiers. With the computerization of the office, even the common typewriter is disappearing. In general, electronic machines (computers, laser printers) are quieter than the machines they replaced (typewriters, impact printers). Copy machines (which became common in the 1960s) have improved in quality and capability. Common capabilities now include multiple sizes of paper, sorting, enlarging, and reducing. Some devices now transmit information from paper directly into electronic memories. Three examples are bar code readers (transfer a bar code into a symbolic message in memory), optical and magnetic character readers (translate ZIP codes, check numbers, etc., into memory) and page scanners (transfer text and graphics into memory).

Paper records typically are kept in temporary storage (on the work surface), short-term storage (file cabinets at the workstation) and long-term storage (file cabinets elsewhere). See Box 13.2. Space for file drawers to pull out is necessary, but the drawers can block low-traffic aisles temporarily. Rotary files combine dense storage with easy access. Files may need to have restricted access: See Chapter 10 concerning security.

Depending on what is processed in the office, storage space may be needed for books, catalogs, magazines, computer disks, visual aids (transparencies and slides), microfilm, and microfiche. If microfilm or microfiche is used, a reader is necessary.

Paper messages can be internal mail within the department, internal within the organization, and external. External mail once came only from the Post Office, but now courier services are common. Messages can be delivered directly to each individual or, more commonly, to a mailbox area that individuals visit periodically. A workstation for sorting mail to individual boxes may be needed.

3.2 Computers   The word processor has replaced the typewriter in most applications. Word processors have three advantages: quick document production, especially for repetitive documents; fast, flexible document revision; and low error rates.

Originally, word processors were "stand-alone" machines; communication was between an operator and the machine. Gradually they have become part of a network, which can be at several levels. Typically, the first network (a **local area network**) connects only a few machines in a single office. For example, two word processors and a laser printer might be connected. Gradually, other machines in the same office are linked together by dedicated lines. Then other offices and areas in the same building or adjacent buildings become part of the larger local area network. Finally, a message can be sent electronically from any location to any other location, and upon arrival, it can be read from a screen or printed. See communication systems in Chapter 21.

BOX 13.2 *Tips on filing paper records*

When attaching papers, use staples, not paper clips. Paper clips come loose and attach paper to the wrong item. In addition they add bulk to the file.

When removing an item from a file, pull the folder only part way out so it won't have to be refiled and thus possibly be misplaced.

When you take an item out of a file, write the file name on the item so it is placed back in the correct file.

When refiling, put the item in the front of the folder, because the most recent items are wanted most often.

Color coding helps reduce misfiled folders (Konz and Koe,

1969). There are a number of different systems. Generally the second letter is coded. For example, using the letter B, orange files are Ba–Bd, yellow Be–Bh, green Bi–Bn, blue Bo–Bq, and violet Br–Bz. Color coding also can be used for files by activity—e.g., all appointments are in blue, all purchase requisitions in green, all personnel matters in red. Or, for a word processor, all files for person A are green, for person B red, for person C yellow. Report binders and computer tape labels are other items that can be color-coded. Use color for rough area location and a high-density code (such as a name) for discrete location.

3.3 Telephones   For over 100 years, the telephone has been used for oral communication. In the 1960s, long-distance calls within a country became direct dial; in the 1980s, international calls also became direct dial. The reduction of human labor in phone companies (in both call processing and billing) permitted dramatic reductions in phone rates. See Table 21.11. It also has permitted **WATS** (wide area telephone service) **lines**. An outgoing WATS line typically costs 20–30 cents/minute for a phone call anywhere in the United States. With incoming WATS lines (800 numbers), the firm receiving the call pays for the call.

When the person called is not available, the phone is either not answered or is answered by a secretary, who takes a message. This leads to "telephone tag," in which each person tries to catch the other. One solution is an answering machine. A "super" answering machine is called **voice mail**. Another alternative is a pager (beeper).

Teleconferencing (multiple people on the same phone line) has become more common with the introduction of speaker phones (multiple people on the same phone) and conference calls (multiple phones on the same call).

In the late 1980s, cellular phones became widely available. The key characteristic of a cellular phone is that it transmits by radio instead of by telephone wire, resulting in "office" calls that originate from a car, truck, boat, or anywhere.

Phone lines can transmit nonoral messages as well. One approach is **electronic mail** (EMail). The sender keys in a message and the address of another computer on the network. The computer then sends the message in a burst to the other computer, using telephone lines. The message is stored in the reader of the receiver's computer under the addressee's file. Since the time on the phone line is quite short, many organizations do not charge employees for use of the phone line, and so EMail is "free."

Another way of using the phone lines for data is a **fax**. The fax machine scans the input document and sends a photographic reproduction to another fax machine. The speed of transmission is limited by the slower of the two machines (sender and receiver), but 15–30 s/page is a typical transmission rate. If the cost of a WATS line is

$.30/minute for a call anywhere in the United States, then the cost of the fax is less than the cost of postage. Fax machines connected to a computer can send the same message to multiple locations or redial if the receiver's line is busy.

Yet another way to send data on phone lines is through a **modem**. The modem converts the digital information to an analog signal for transmission over the phone lines; it is reconverted to digital at the receiving end.

It also is possible to send a video signal (i.e., a picture) over phone lines. For full motion, transmission requires considerable bandwidth, but the feasibility and cost of transmitting still and compressed motion video (picture phones) over standard T1 (telephone) lines have improved dramatically in the last few years.

3.4 In Person   Computer communications, EMail, voice mail, faxes, cellular phones, and other technological revolutions have reduced the need for face-to-face contact. Office work no longer is done just in the office, and communication is replacing transportation. However, offices should still provide for face-to-face contact. The first information need is for the visitor to be able to find the desired people. Because the receptionist function has been automated in many places, giving directions requires good signs, color-coded carpets, and other aids. The need for a conference area at each individual workstation is reduced and shared use of conference space becomes more feasible.

4 FURNITURE

4.1 Arrangements   Furniture in the office is akin to machines in the factory. Spatial arrangements in the factory emphasize material handling, but in the office their primary function may be to support social structures. Position, distance, and symbolic decoration define "territory." See Box 13.3. A private secretary/receptionist once was common and served as a status symbol; the mechanization of the office and high labor costs have reduced this convenience, and so status tends to be displayed by the office furnishings and space.

BOX 13.3 *Territoriality and personal space*

One design consideration is territoriality and personal space. *Territory* is a visible area in a stationary location; *personal space* is an invisible boundary which moves with the person. Figure 13.5 gives personal space zones, using Hall's (1976) terms of intimate, personal, social, and public zones. The concept of territoriality comes from studies of animals who "mark out" their territory; personal space can be considered the analogous term for temporary rather than long-term occupancy. Thus people might mark out their work area with family pictures, art objects, and a calendar; in the cafeteria they may put their coat on the opposite chair so no other person sits at the same table. People feel more at ease on their turf—less at ease in yours. Generally designers should encourage personalization of long-term space, since it costs the organization nothing and satisfies the occupant. Plants and wall posters may lead to positive feelings by visitors, although decor has a greater impact on females than males (Campbell, 1979). Designers may or may not want to decide whether a space is to be sociofugal (leading away from social activity) or sociopedal (leading toward social activity) (Bennett, 1977); the best space can be either, at the user's option. An example of sociofugul design is the typical airport lounge; its purpose is to drive travelers and visitors to the concessions (Sommer, 1974).

The primary variable concerning space is ability to communicate; this communication can be subdivided into visual and auditory.

Visual communication is encouraged by being close, by eye contact, and by lack of barriers; it is discouraged by being far apart, by lack of eye contact, and by barriers. Thus to encourage social activity between people in an office, keep the workstations close, have the people face each other, and minimize barriers. (Barriers can range from none, to a low coffee table, to a desk, to a clear glass partition, to an opaque partition.) A clear glass partition defines a territory while still permitting visual contact. To discourage social activity between people in an office, keep the workstations far apart, have the people's backs to each other, and maximize barriers. In factory jobs with large amounts of "automatic" psychomotor activity, visual contact should be encouraged to reduce boredom.

Auditory communication, which is less directional than visual communication, is encouraged by being close and by having a quiet background; it is discouraged by being far apart and by having a noisy background. Thus to encourage social activity between people in an office, keep the workstations close and the environment quiet. To discourage social activity, keep the workstations apart and the environment noisy.

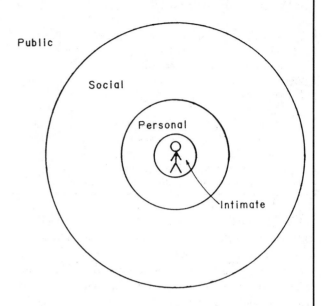

FIGURE 13.5 Personal space zones (This material was written by Osborne and Heath (1979) and appeared in *Applied Ergonomics*, Vol. 10, 2, 99–103, published by Butterworth Scientific Ltd, Guildford, Surrey, U.K.) Each of the four zones has a close and far phase. Exact dimensions of each zone vary with the situation. Females tolerate closer zones than males; other nationalities may prefer different dimension zones than Americans. Dean, Willis, and Hewitt (1979) pointed out that interaction distance depends on the "ranks" of the two people—a superior has the option of a more formal (more distant) or more intimate (less distant) interaction whereas the subordinate usually must initiate a formal interaction.

Intimate—close (0–6 in.): Other person is overwhelming; actual physical contact; taboo in many cultures.
Intimate—far (6–18 in.): less contact but still increased visual awareness.
Personal—close (18–30 in.): for "well-known" friends.
Personal—far (30–48 in.): begins at arms length; area for normal social contact.
Social—close (4–7 ft): for people who work together; casual social gatherings.
Social—far (7–12 ft): formal business.
Public—close (12–25 ft): well outside social involvement.
Public—far (over 25 ft).

The arrangement conveys information (1) about the occupant and (2) about how the occupant wants others to behave in the space. The arrangement describes the occupant's role and status in the organization.

Figure 13.6 shows six arrangements of a one-person office. The key room feature is the door; most people prefer to face the door. Although this may lead to distractions, it also implies a readiness for interaction. The chairs should be arranged so that the occupant can choose a formal or informal arrangement—perhaps even depending on the visitor. If there is a window, it should be behind or to the side of the desk, not in front. (See windows in Chapter 9.)

Privacy in a work environment does not necessarily mean being alone but more the ability to control social interaction with others. Privacy can be aided by controlling telephone calls, by acoustical control, by visual control, and by controlling who can enter the area (a locked door or a receptionist barrier).

4.2 VDU Workstations

In the 1970s, computer input and output began to appear within the office itself rather than just from the service window at the computer center. The data are displayed on a cathode ray tube—formerly abbreviated as CRT but now generally as **VDU** (video display unit) or **VDT** (video display terminal). In addition to the screen, there usually is a keyboard for input; there may or may not be a printer for output. The input material (source documentation) generally is on paper (previously typed material, handwritten material, computer printouts) but also can be audio (oral information, either in person or over the telephone). VDU tasks include (1) data entry (document intensive—the operator routinely enters data

from documents), (2) data acquisition (screen intensive—information from a form is matched with information from the computer), (3) word processing (document and screen intensive—attention focuses on both) (4) interactive user using screen occasionally (e.g., travel agent or CAD/CAM user), and (5) programming (similar to interactive).

There has been considerable investigation on the optimum design of the VDU workstation. Three key items in the workstation are the screen, the keyboard, and the document holder. There are two key items for the user: the eyes and the hands. How should these five be arranged?

Start with a seated person. This gives a location for the eyes and the hands. However, since people vary in size and shape (1) the hand location in relation to the floor will vary, (2) the eye location in relation to the floor will vary, and (3) the relation between the eye and the hand will vary. Although the above statements seem obvious, they need to be said because they lead to a key design principle: *The workstation elements must be adjustable.* If the workstation elements are not adjustable, the workstation will not be optimum.

Next locate the screen in relation to the eye. (In a document-intensive task, put the document, not the screen, in the favored position.) A natural, relaxed position of the head is with the eyes gazing slightly downward—20° is a commonly accepted value. The screen (document) should be placed along this line with its face perpendicular to the line of sight. The distance from the eyes depends on the size and quality of the displayed material. A number of studies have shown that the optimum size of alphanumeric characters on VDUs is approximately 20–30 min of arc. Large characters can be farther away, smaller ones closer. Considering many tradeoffs, the consensus seems to be to locate the screen 20 to 25 in. from the eyes.

Next locate the source documents. Since the same considerations apply as with the screen, ideally it should be where the screen is. (Although typed copy tends to be more legible than characters on a VDU, the copy could be computer printouts, photocopied material, or handwritten material and could be more difficult to read than VDU characters.) To minimize the need for visual accommodation when the user is shifting the eyes back and forth between the VDU and the document, both should be the same distance from the eyes. Where should the document be placed? Below the screen is impractical since it would block visual and manual access to the keyboard. Above the screen is satisfactory if the eyes can use the 20° for the document and a lower angle (say, 45°) for occasional glances at the screen. However, the head can move side to side more easily than up and down (aided by swiveling motion of the chair). If the document and display are placed side by side, then which goes where depends on what the user looks at more. If the document is used more, place the document ahead and the display at its right or left (operator's choice). If the display is used more, place the display straight ahead and the document on the operator's

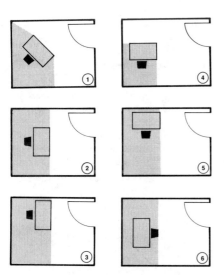

FIGURE 13.6 Doors are the key room features in a one-person office. Desks can be used as barriers and to define spaces formally. The shaded area is the private zone—normally the occupant's territory—and the remainder is the public zone. Arrangement 1 has the highest degree of zone definition and arrangement 6 the least (Joiner, 1971).

preferred hand side, to avoid cross-body reaches when the operator is changing or marking the document.

Finally locate the keyboard in relation to both the hands and the eyes. (Many graphic input devices, such as light pens, mouse, and trackballs, have become available to supplement keyboards.) Although VDU operators spend considerable time looking at the keyboard (because the extra function keys are used as well as the alphanumeric keys, and thus the keys should be at the same distance from the eyes as the screen and the documents), the hands are the dominant location consideration. The consensus is that the forearm should be approximately horizontal and the hands should keystroke at about elbow height; some recommend hands slightly above the elbow. The keyboard should be directly ahead of the person so no twisting is needed. The third coordinate is determined by forearm length and shoulder position (assuming the 90° angle between the upper and lower arm). The person should be able to sit with a slight slouch, not a ramrod straight posture. An adjustable height chair with a high backrest and adjustable backrest inclination is therefore the best.

Finally, the workstation should have some space for work-in-process storage of documents (to be worked on as well as completed), supplies (pencils, disks), and personal items (purse, family pictures). For VDUs used by engineers and executives, the input and output are less standardized. Additional work surfaces are needed to hold drawings, reports, documents, and other material for interactive work with the computer.

See VDU workstations in Chapter 22 for comments on lighting.

5 SPECIAL AREAS

5.1 Conference, Meeting, and Training Rooms

5.11 Room Use. Decide whether the room will be single purpose or multiple purpose. A multi-purpose room might also be used as a cafeteria or for social interaction. In that case, windows should be considered if there is a view. The windows should be treated with two light control devices—a sun screen and an opaque drape or shade. The sun screen will reduce glare while the windows are uncovered. The drape or shade (not venetian blinds, which tend to work poorly) should be used when presentations are being made. If the room is used only for meetings using audiovisual presentations, it should not have windows (McVey, 1988; 1991). (Many of the following recommendations are taken from McVey's two articles, which will not be cited specifically. McVey also made many improvements on a draft of this section.)

Since refreshments often are served, provide a permanent space for serving them or at least space for a cart. More elaborate conference rooms will have a nearby galley, pantry, or coffee nook. Rest rooms also should be

conveniently located. If people from outside the building use the room, provide a place for coats.

5.12 Dimensions. Another basic decision is the room shape and size. See Tables 13.4 and 13.5. The typical **training room**, which holds about 30 people, is rectangular and has a length-to-width ratio of 1.2–1.33/1. Seats begin at a distance of 2 *H* (*H* = vertical size of projected image) and extend to 6 *H*; optimum is at 4 *H*. Seats should be within an imaginary line projected at 60° from each end of the display. Circular seating arrangements are acceptable, but avoid circular rooms—they tend to have acoustic problems.

For legible displays and accurate viewing, the screen's vertical height should be 1/6 of maximum viewing distance as measured from the far side of the display to the furthest off-axis viewer. For relatively unobstructed and comfortable viewing, the bottom of the projection should be about seated eye height (assumed to be 48 in. above the finished floor). The screen should extend upward a distance equal to 1/6 of the maximum viewing distance. Add at least 6 in. for trim above the screen. Thus,

$$MCH = 54 + MVD/6$$

where

MCH = minimum ceiling height, in.

MVD = maximum viewing distance, in.

Note that maximum viewing distance may be less than room length.

In rooms where MCH is not possible throughout the room, legible display and good viewing still are possible

TABLE 13.4 Space requirements for conference rooms (McVey, 1991). The space allocations will provide 7–8 ft at the front of the room for a conference leader, a worktable, and a Vugraph and 2–3 ft at the rear for AV equipment.

Style	Space/Conferee (sq ft)	Seating
Auditorium	12–15	Flat floor open space with stackable chairs (23–26" centers; 36–42" rows)
	13–14	Room with risers and fixed tables with seats on pedestals (26–28" centers; 48" rows)
	12–13	Room with risers and fixed tablet arm chairs (24–26" centers; 42" rows)
Classroom	15–18	Fixed tablet arm chairs (26–28" centers; 42" rows)
		Movable tablet arm chairs (28–30" centers; 42–48" rows)
	20–22	18" × 28" table area (48" rows)
	22–24	24" × 32" table area (54" rows)
	28–33	24" × 36" table area (60" rows)
U-shaped	38–42	24" × 30" table area
	45–50	24" × 36" table area
Circular	45–50	24" × 30" table area
	50–57	24" × 36" table area

Note: With computers at each seat location, use 30" deep × 36" wide table area or 24" deep × 36" wide table area with keyboard drawers.

TABLE 13.5 Recommended aisle widths for conference rooms (McVey, 1991).

Width (inches)		
Minimum	Preferred	Situation
30		Between table rows; pedestal seats
36	42–48	Between table rows; castered conference chairs
36	42–48	Multiple side aisles
48		Single (center or side) aisle
54	60	Perimeter aisle (U- and O-shaped seating)
60		Space between last row of tables and rear wall (slide and movie projectors located in adjacent projection room)
84		Space between last row of tables and rear wall (with slide and movie projectors located in room)

because only the front portion of the room needs to be at *MCH*. The raised ceiling section (soffit) should be as wide as the projection screen (minimum); the entire width of the display surface (marker boards, etc.) is preferable. The soffit should extend 13 ft toward the back of the room. In addition to improving viewer sight lines, the soffit can also accommodate the recessed installation of a TV projector and supplementary lighting.

If the projection screen will display computer-generated alphanumerics, its vertical height should be at least *MVD*/6 and its width at least *MVD*/4.5. If screen width is *MVD*/3, multi-image display is possible. A screen of these dimensions will permit comfortable, clear viewing of the following: horizontal format Vugraph transparencies or LCD (liquid crystal display) panels, standard video projection, 16-mm film, and dual horizontal or vertical slides. In a typical 32 ft long conference room where *MVD* = 30 ft, such a multi-purpose screen would be 60 in. high and 120 in. wide.

5.13 Equipment.

The primary display will be the projection screen. Screens should be permanently mounted (rather than using portable, tripod-mounted screens). Screens can be front projection or rear projection. Front projection screens require less space for projection equipment and provide the best image quality. The depth of a rear projection room can be *2 H* (2 times screen vertical height) if front surface mirrors and short focal-length lenses are used.

For front projection, use the standard matte screen for general applications, wide-angle viewing, and low cost; the ultramatte screen will give higher image luminance while still providing wide-angle viewing, but at a slightly higher price. For high image luminance in narrow-angle viewing situations and where room illuminance levels can be kept very low, use beaded screens. For rear projection, use low-gain (1.2–1.7) gray screens for wide-angle viewing, and moderate gain (1.8–3.0) for more narrow-angle viewing

situations. High-gain rear-projection screens (> 3.0) tend to be unsatisfactory due to excessive hot spots.

Other displays include the following:

1. Enamel or plastic-coated steel marker boards can be used with various color pens that permit simple erasure. Magnets can hold items on the boards.

2. Flip charts (large paper pads to use with markers) should be on wheeled stands. Disadvantages are lack of magnification and difficulty of duplication.

3. TV monitors can be wall mounted, on carts, or at each seat location. If wall mounted, one 25" monitor for every 25 people can accommodate standard videotapes. For displaying computer-generated alphanumerics, larger monitors or more of the small units are necessary. If monitors are at each seat location (e.g., when using "groupware"), keyboards should also be at each seat, along with a keyboard and monitor at the leader location. VCRs should have a remote control, including a freeze-frame feature.

4. Hooks mounted around the room perimeter permit hanging various other displays.

For convenience, there are other considerations: In some conference rooms, a telephone and a fax machine may be useful; there should be sufficient floor and wall electrical outlets, suitably arranged to minimize tripping hazards on the cords; a wall-mounted clock should be in the rear of the room; and a lockable storage area should be provided for equipment, supplies, and presentation materials (slides, videotapes, transparencies, and handouts).

5.14 Furniture.

Furniture decisions are based on how the room will be used—for meetings, for classes, or both. A meeting format implies interaction among all people in the room, whereas a class format implies interaction primarily between the instructor and the students, with little interaction among the students.

Conference tables are used as writing surfaces and as storage locations. Each person should have a clear table width of 30 to 40 in.; a depth of 18 in. is sufficient for writing, but 24 in. is better if items are also stored on the table. Thus, if people sit on both sides, the table should be 36 to 48 in. across. Table surfaces are usually 28.75 to 30.5 in. above the floor; 29.5 in. is the most popular. The lower height promotes leaning forward and verbal interaction; the higher height promotes sitting back and less verbal involvement. Table supports should not restrict seating or leg movement. Use pedestals, trestle bases, C leg, or inverted T legs rather than legs along the edges.

The conventional rectangular table results in poor eye contact among people at the table; the problem magnifies with increasing length. If you have only long, rectangular tables, put them in a V shape, and put the group leader at the end opposite the point. Arranging the table in a U is not as good as a V for visual contact, although it permits more people to be in the room and promotes social interaction. If multiple rectangular tables are used, eye contact tends to be better when the short axis of the table points toward the leader or the screen. Figure 13.7 shows two improved table shapes—the boat and the circle. Do you want to encourage confrontation or cooperation? The rectangular shape ("negotiation rectangle") encourages people to take sides, whereas the circle has no sides. The circle also has no head position, which is useful to emphasize equality. However, even a round table may be a barrier. Hendrick (1979) found that people sitting in a circle with no intervening table came to a "more accurate" decision than when seated in other arrangements, including sitting around a round table.

Chair seat height and backrest adjustment are not as critical as with office chairs. The user may be sitting back, rather than using the table surface intensively, and uses a specific chair for only a short time, rather than 2,000 h/year. If the decision is made to purchase chairs with adjustable height seats, then the adjustment needs to be convenient (e.g., pneumatic adjustment while sitting in the chair).

Consider the following in your purchasing decision:

- Chairs should have casters and should swivel and tilt so that the person can make minor adjustments in

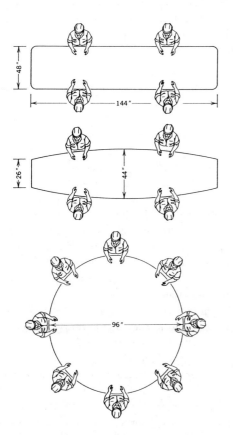

FIGURE 13.7 Table shape influences eye contact as well as acting as a barrier.

posture to reduce fatigue from statically loaded muscles. A 5-star base is recommended for stability.

- Chairs should have padding covered by a fabric. (Padding increases comfort; fabric breathes.) The seat should have a waterfall front.

- Chairs should have short, padded armrests. Without armrests, the arms would have to be supported by the individual or the table, which would restrict posture). *Short* is defined as not extending over the front 7 in. of the seat. Short arms minimize interference with the table.

In elaborate conference rooms, provide chairs along the walls for aides and observers (who are not part of the group at the table).

5.15 Illumination.   Certain aspects of illumination are fixed. General room illumination is usually fluorescent. Surfaces should be matte, and reflectances should be

- 70–90% for ceilings
- 40–60% for walls
- 30–50% for floors and desktops
- 20% for green chalkboards
- 10% for black or gray chalkboards

Parabolic (eggcrate) fixtures reduce glare and produce primarily direct illumination, with relatively little indirect illumination (which tends to wash out AV displays). Consider using T8 (32 Watt) fluorescent lamps, with color rendition of 80 or more, in combination with electronic ballasts for attractive and energy-efficient lighting. Fluorescent lamps can be dimmed but require special circuits, ballasts, and controls.

Room decor should be restrained, with bold colors, such as blues and reds, restricted to artwork and the rear and side walls.

Visual requirements vary with the task. Room occupants should be able to choose normal lighting, supplementary lighting for the speaker, reduced lighting for slides, and so forth. Therefore, provide multiple controls. Some should be discrete (on/off switches) and some continuous (rheostats for lamp control). The lighting in the display area, in particular, should be independent of the remainder of the room, and have its own switches and dimmers. In addition, supplementary light should be available for non-electronic displays such as panel boards, flipcharts, and podiums. Typical solutions are "wall washer" fluorescent luminaries and track lighting (which can be aimed).

For conversation activities, illumination should be 300 to 500 lux, with higher levels on flip charts and panel boards. Low levels are required for many visual aids, and providing adequate light for note taking during their use requires careful lighting design. In general, lighting levels can be higher with rear projection than with front projection.

5.16 Auditory Environment. Kryter (1985) recommends an ambient noise level of < 43 dbA for conference rooms for 20 people (15 ft table), < 45 dbA for meeting rooms with sound amplification, and < 48 dbA for conferences at a 7 ft table. McVey recommends background noise criteria maximums of ≤ NC35 for general activities and ≤ NC20 for conference rooms where audio recording is a regular feature.

To achieve these levels, the walls should reduce noise intrusion from adjacent space by at least 45 db. Use walls with slab-to-slab construction and with a 50+ sound transmission class. The air diffusers should be high volume and low velocity (< 300 ft/min). Light fixtures, ballasts, and AV equipment should be quiet rated. Carpets reduce maintenance cost as well as reducing noise levels. These noise control features also will reduce conference room noise disturbing people outside the room.

5.2 Shop Offices

Manufacturing facilities often are noisy. Supervisors and clerical personnel often work in prefabricated **shop offices** (4 walls and a roof), which are available from various vendors. Figure 13.8 shows a typical layout. Generally, several windows allow people in the office to monitor shop activity.

5.3 Reception Area

The security needs of the facility influence the reception area design. Will visitors need to sign in before entering? Will a badge or pass be issued? Will the visitor need an escort? Are conference facilities needed in the reception area so that visitors will not enter the facility? Will the receptionist be a member of the security staff or the office staff?

The number of visitors and their access to personnel also influence the design. If there are many visitors, the receptionist job becomes full time. With few visitors, other jobs need to be provided. The typical solution places the office adjacent to the reception area and uses an office worker as a part-time receptionist. The receptionist function also can be handled remotely through a telephone in the reception area. The visitor dials the necessary numbers, and the person answering the phone unlocks the door remotely. Occasionally, for reasons of prestige or security, the firm wants a full-time receptionist in the reception area, even if there are few visitors, leading to considerable idle time for the receptionist. A receptionist stationed in the reception area needs a workstation which provides some privacy for writing and keying.

The reception area should have a telephone; many firms have an internal facility phone and a pay phone. There should be convenient access to a toilet inside the facility or (depending on security) a toilet in the reception area.

Choice of furniture depends on whether the reception area will be simply a temporary waiting area or whether small meetings will be held there. People using the area may or may not want to talk to each other. Figure 13.9 shows an arrangement that encourages conversation; an L-shape arrangement of chairs (with sufficient personal space) permits a person to encourage or discourage eye contact (and thus conversation). In general, seating should be individual chairs. Three-piece sofas hold only two people because no one wants to sit in the middle.

The reception area provides a first impression of the firm, which should be that the firm is frugal, business like, first class, or other positive image. Factors that influence the image are the room decor and the absence or presence of a receptionist.

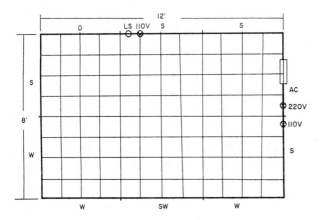

FIGURE 13.8 Prefabricated rooms use a "mix-and-match" approach. The customer selects from five types of 4 ft wide panels: solid (S), with windows (W), with sliding windows (SW), with doors (D), and with a cutout for a window air conditioner (AC). The panels can be pecan wood grain, walnut wood grain, beige, or white on one or both sides. The panels can have a 110 or 220 V outlet or a light switch (LS). The prefabricated rooms can be used also for storage rather than an office. Depending on what is stored, security or fire features can be added, as can blow-out panels in the roof (for flammable or explosive storage).

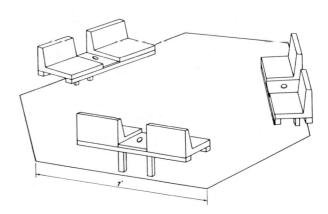

FIGURE 13.9 Lounge seating in an open area has the people face each other at a social distance. The distance is close enough for conversation if desired, but people also can remain private.

DESIGN CHECKLIST: OFFICES

Offices
    Arrangement type (conventional, landscape, open plan)
    Telephone, electric power, computer access
    Visual variety (workstation, floor, walls)
    Noise control (within-office noise, outside-office noise)
    Sightline considerations
    Personal space considerations
    Office reflects user's status
    Visitors can find their way

Meeting/conference room
    Chairs (padded, swivel, tilt, casters)
    Tables (size, shape, support, edges)
    Visual aids (screen type, location, size; equipment storage; marker boards)
    Light control (windows, switch locations, illumination levels during talk)
    Noise control (within-room noise, outside-office noise)
    Amenities

REVIEW QUESTIONS

1. What are the four "items" processed in offices?
2. Give two criteria for office layout.
3. Give the three types of office layout.
4. Who might use a virtual office?
5. What are the two key concepts of office landscape?
6. Give some methods of creating visual variety in the office.
7. Give some methods of creating visual privacy in the office.
8. Give some methods of creating auditory privacy in the office.
9. Who might use a "hot" workstation?
10. Why does the open plan require less ft²/employee than a conventional plan?
11. Why does the open plan require less energy cost than a conventional plan?
12. Briefly describe teleconferencing.
13. Briefly describe EMail.
14. Give five divisions of VDU tasks.
15. Should the backrest of a chair for VDU workstations be high or low?
16. Give two disadvantages of flip charts.
17. What is the difference between a meeting and a class format?
18. Should conference chairs have armrests? If so, why?
19. Briefly discuss shop offices.
20. What is a modem?
21. What is a sociofugal space? What is a sociopedal space? Make a sketch of a two-person office that is sociofugal and make a second sketch of one that is sociopedal.
22. Does a rectangular table encourage confrontation or cooperation?
23. How wide should a projection screen be in a 36 ft long room? For what type of projection should the screen be parallel to the wall?

PROBLEMS AND PROJECTS

13.1 Visit offices in different organizations. How much of the work is (a) processing paper messages, (b) processing computer messages, (c) processing telephone messages, and (d) processing in-person messages?

13.2 Evaluate the design of a VDU workstation. Consider seating, anthropometry of the workstation, and lighting. What does the operator think of your ideas? What does the supervisor think of your ideas? Is there more resistance to change from the operator or the supervisor?

13.3 Design the layout of a real one-person office. What does the user of the office think of your ideas?

13.4 Evaluate the design of an existing conference room. What does the supervisor of that space think of your ideas?

REFERENCES

Bareither, H., and Schillinger, J. *University Space Planning.* Urbana: Univ. of Illinois Press, 1968.

Bennett, C. *Spaces for People*, Englewood Cliffs, N.J.: Prentice-Hall, 1977.

Campbell, D. Interior office design and visitor response. *J. of Applied Psychology*, Vol. 64, 6, 648–653, 1979.

Dean, L., Willis, F., and Hewitt, J. Initial interaction distance among individuals equal and unequal in military rank. *J. of Personality and Social Psychology*, Vol. 32, 2, 294–299, 1975.

Hall, E. The anthropology of space: An organizing model. In *Environmental Psychology.* Proshansky, H., Ittleson, W., and Rivlin, L. Eds. New York: Holt, Rinehart and Winston, 1976.

Hendrick, H. Group decision-making effectiveness as a function of physical arrangement. *Proceedings of 23rd Annual Meeting, Human Factors Society.* Santa Monica, Calif: Human Factors and Ergonomics Society, 111–115, 1979.

Joiner, D. Office territory. *New Society*, Vol. 7, 660–663, 1971.

Konz, S., and Koe, B. The effect of color coding on performance of an alphabetic filing task. *Human Factors*, Vol. 11, 3, 207–212, 1969.

Kryter, K. *The Effects of Noise on Man*, 2nd ed. Orlando, Fla: Academic Press, 1985.

McVey, G. The planning and ergonomic design of technical presentation rooms. *Technical Communication*, 1st Quarter, 23–30, 1988.

McVey, G. Twelve ergonomic guidelines for integrating AV systems into conference room design. *Facilities Planning News*, July 1991.

Osborne, D., and Heath, T. The role of social space requirements in ergonomics. *Applied Ergonomics*, Vol. 10, 2, 99–103, 1979.

Pile, J. *Interiors: 3rd Book of Offices.* New York: Whitney Library of Design, 1978.

Sachs, R. Workstations for the open plan. *Administrative Management*, Vol. 42, 10, 55–61, Oct. 1981.

Sommer, R. *Tight spaces: Hard architecture and how to humanize it.* Englewood Cliffs, N.J.: Prentice-Hall, 1974.

Wilson, P. Information technology. *The Ergonomist*, No. 164, Dec. 1983.

Wrennall, W. Office layout. Chapter 71 in *Handbook of Industrial Engineering*, 2nd ed., G. Salvendy, Ed. New York: Wiley, 1992.

14 SPECIALIZED AREAS

CHAPTER PREVIEW

This chapter describes specialized areas and their special requirements.

CHAPTER CONTENTS

1 Toilets and Locker Rooms
2 Maintenance Shop
3 Battery Charging Area
4 Welding
5 Medical Facilities
6 Food Service
7 Tool Cribs
8 Clean Rooms

KEY CONCEPTS

accessibility requirements
clean spaces
emergency flushing stations
esthetics
line flow (versus scramble
   system)

makeup air
negative air pressure
Occupational Safety & Health
   Administration (OSHA)
utility wall
vending

1 TOILETS AND LOCKER ROOMS

1.1 Toilets

Toilets are relatively permanent facilities and are quite difficult to expand or move. Therefore, plan ahead for a large number of users.

Toilets and locker rooms often are placed on mezzanines to maximize access for employees. (The space under the stairs leading to the mezzanine can serve as a location for vending machines.) Table 14.1 gives the recommended number of toilets for males and females. Table 14.2 gives the **Occupational Safety and Health Administration (OSHA)** requirements. Local building codes may require different numbers. For example, the Uniform Building Code says that "separate facilities shall be provided for each sex when the number of employees exceeds four and both sexes are employed"; Manhattan, Kansas, for example, changed the four to nine.

The toilet should be clean, light, and well ventilated, and the floor should slope to one or more floor drains. Design the toilet entrance, doors, and mirrors so that privacy is maintained. Maintenance is simplified when the toilet walls and floor are tiled. Toilet partitions should be ceiling-hung; urinals and water closets should be wall-mounted. Stall doors give privacy but aid malingering. Equip toilet areas with mirrors; dispensers for paper or cloth towels (air dryers are another choice), soap, and toilet tissue (2 rolls/stall); shelves; coat hooks (both in the room and in water closet enclosures); and cigarette receptacles in the urinal area. A trash can is needed even if air dryers are used in place of paper towels. A faucet may be needed for filling janitorial buckets. Electrical outlets may be needed for shaving (be sure the outlet is near the mirrors) and for

floor polishers. The women's rest room should include a coin-operated sanitary napkin dispenser, napkin disposal receptacles, and a lounge area. The mirrors in a women's rest room should include at least one full-length one, and mirrors should not be placed over washbasins (to minimize congestion and to reduce maintenance problems from loose hair, hairpins, etc., in the drains). Figure 14.1 shows an effective arrangement of rest rooms.

Architectural barriers for the handicapped are illegal. See Figure 14.2. Detailed **accessibility requirements** for the disabled are in the *Federal Register* (29 CFR, part 36), July 26, 1991. The requirements cover not just toilet access but drinking fountains, corridor widths, ramp and stair design, and more. The requirements are quite detailed. For example, the grab bar in a toilet stall must be at least 36 in. long, the stall door opening force must be 5 lb or less, toilet paper dispensers that control delivery or do not permit continuous

TABLE 14.2 Water closets required by OSHA (Subpart J, 1910.141) for employees of each sex. Where toilet rooms will be occupied by no more than one person at a time, can be locked from the inside, and contain at least one water closet, separated toilet facilities for each sex need not be provided.

Number of Employees	Minimum Number of Water Closets
1–15	1
16–35	2
36–55	3
56–80	4
81–110	5
111–150	6
Over 150	1 additional fixture per 40 employees

TABLE 14.1 Minimum number of plumbing fixtures (from Table 7.24.1 of the National Standard Plumbing Code). The Code, however, provides equal numbers for men and women; a better ratio is 2 water closets for women to 1 (water closet + urinals) for men.

Location	People[a]	Water[b] Closets	Lavatories	Drinking Fountains	Showers	Service Sinks
Industrial:	1–10	1	1/2 number	1/75	1/15	1/floor
Heavy, hot or cold conditions	11–25	2	of w.c.	people	people	(where
	26–50	3				applicable)
	51–75	4				
	76–100	5				
	each					
	additional 50	Add 1				
Mercantile:	1–15	1	1/2 number	1/100		1/floor
Office & selling; normal work tem-	16–35	2	of w.c.	people		
peratures	36–55	3				
	56–80	4				
	81–110	5				
	111–150	6				
Service:	1–8	1	1/2 number			
Dressing or locker room used mainly	9–20	2	of w.c.			
at shift change	21–40	3				
	each					
	additional 20	Add 1				

[a]Whenever both sexes are present in approximately equal numbers, multiply the total census by 60% to determine the number of persons for each sex to be provided for.

[b]For men's rest rooms, approximately 33% of the required number of water closets may be urinals.

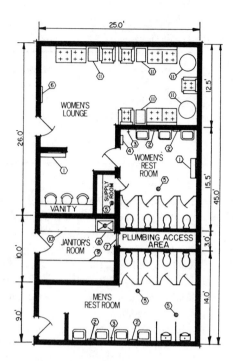

FIGURE 14.1 Recommended restroom arrangement (Rest Rooms, 1971; reprinted with permission of Caterpillar Tractor Co.): (1) mirror and shelf, (2) towel dispenser and waste receptacle, (3) water faucet, (4) sanitary napkin dispenser, (5) floor drain, (6) full-length mirror, (7) pocket door, (8) floor-level mop sink, (9) cabinets, (10) shelves, (11) table with lamp.

paper flow are not permitted, the bottom of mirrors must be no higher than 40 in. above the floor, and so forth.

Service sinks for janitor's closets should have faucets with hose threads and stainless steel rim guards to reduce chipping from pails.

Small "convenience stations" can supplement the major toilet facilities; see Figure 14.3. The justification is the reduced travel time for employees away from their job. (At 3 mph, it takes 1 min to walk 265 ft; at $12/h for labor cost, 1 min costs 20 cents.) Toilets often are in basements

or mezzanines so they are close to people yet out of the product flow.

One drinking fountain (Table 14.1) should be provided for each 75 to 100 people. More might be used to reduce employee travel time. Prohibit common drinking cups. Set the nozzle and pressure at such an angle that the water jet does not splash back down on the nozzle or onto the floor. Guard the nozzle so a mouth or nose can't come in contact with it. The nozzle diameter should be 3/16 to 1/4 in. (Coetzee and Bennett, 1978).

Emergency flushing stations (for eyes and body) are required in some work areas where corrosive materials are used (ANSI Z358.1; OSHA 1910.151; OSHA 1910.94). See Figure 14.4. Eye/face wash units can be freestanding, wall-mounted, or portable (portable units for construction and maintenance) (Weaver, 1983). They also can be combined with emergency showers, which can have one or multiple heads. ANSI 358.1 states "the unit should be located as close to the hazard as possible, and on the same level. The maximum time required to reach the eye wash should be determined by the potential effect of the chemical. For a strong acid or strong caustic, the eyewash should be within 3 m (10 ft) of the hazard." In general, emergency showers, plumbed and self-contained eyewashes, eye/face washes, hand-held drench hoses, and combination units should be "accessible within 10 s and not over 100 ft from hazard." For construction, OSHA 1926.403(a)(b) requires that facilities for quick drenching of the eyes and body shall be provided within 25 ft of the work area for emergency use. Since a worker with chemicals in the eyes can't see, there must be no obstructions, walls, or doorways in the path. Minimum irrigation time for a chemical burn is 15 min;

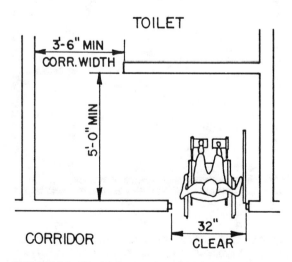

FIGURE 14.2 Wheelchair access to a toilet. It is important to allow enough room for the wheelchair to turn.

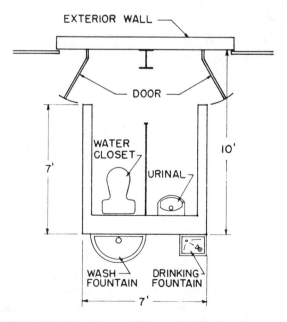

FIGURE 14.3 Convenience station has a urinal and a water closet enclosed by an 8 to 10 ft wall. On the outside is a semicircular washbasin and a drinking fountain.

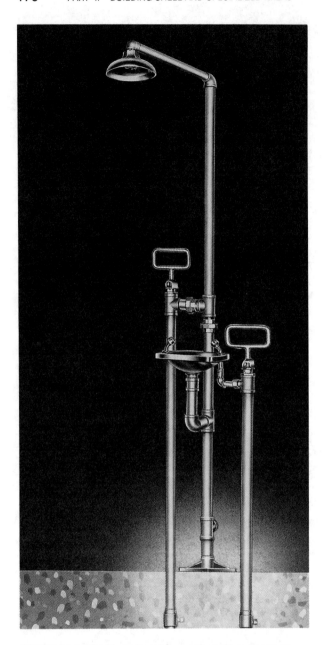

FIGURE 14.4 Flushing stations (here combined with an emergency shower) should not be hand controlled because the hands are needed to force open the eyes. Emergency showers must be designed so water flow remains on without the use of the operator's hands, since often the hands are occupied in removing clothing. (Photo courtesy of Speakman Company.)

remove all clothing with the chemical on it. Note that eyewashes and emergency showers need a drain under them.

1.2 Washrooms and Lockers

Foot-controlled circular wash fountains offer many advantages over conventional individual sinks with hand-activated faucets: They permit group washing, save space, and save water since they cannot be left running after the user leaves the fixture. Sectional control fountains save more water than nonsectional. Compared with multiple sinks, they use just one set

TABLE 14.3 Wash fountain capacity (Courtesy of Bradley Corp.). Allow 48 in. clearance between wash fountains.

Type	Diameter (in.)	Number of Users
Circular	54	8
	36	5
Semicircular	54	4
	36	3
Corner	54	3

of plumbing connections; both full and semicircular wash fountains are available in 36 and 54 in. dia. Individual sinks can save water with metering faucets. Reduced water consumption savings are due primarily to hot water energy savings. See Table 14.3. Allow 24 in. of "people space" around the wash fountain rim or individual sink front. Pop-up drains should not be specified. They are not used and are a maintenance problem.

It should not be necessary to go through the toilet area to get to the locker and shower area; neither should it be

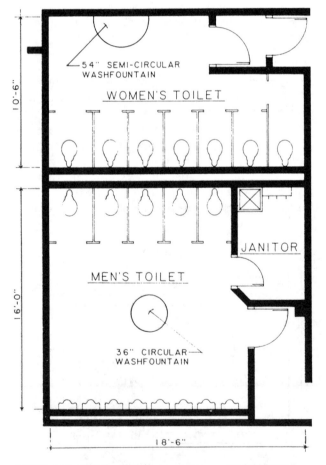

FIGURE 14.5 Toilet area for 50 men and 40 women. The wash fountains eliminate 9 conventional washbasins, the wall the conventional basins are mounted on (such as the urinal wall at the bottom of the figure), and the plumbing connections. (Layout courtesy of Bradley Corp.)

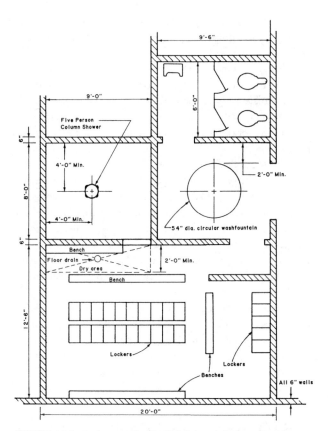

FIGURE 14.6   Locker room for 25 with toilets and showers. (Layout courtesy of Bradley Corp.)

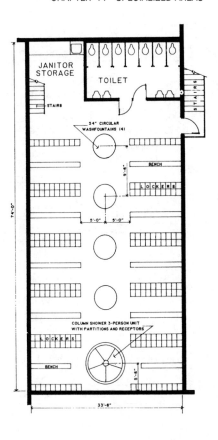

FIGURE 14.8   Locker room for 200 men with toilets and showers. Note the stairs, which imply the use of a mezzanine or basement. For females, the locker room would be unchanged except for elimination of the urinals. (Layout courtesy of Bradley Corp.)

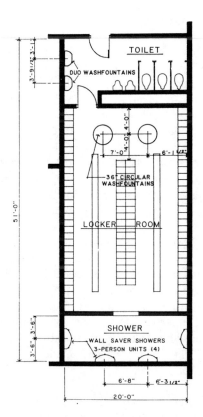

FIGURE 14.7   Locker room for 100 with toilets and showers. (Layout courtesy of Bradley Corp.)

required to go through the locker and shower area to get to the toilet area. See Figures 14.5 through 14.8.

Lockers can be either half or full length; full is recommended. A good size is 12 in. wide × 18 in. deep × 72 in. high. Lockers should be placed on a tile-faced concrete base raised 4 to 6 in. above the floor so that the floor can be washed without rusting the locker or wetting the contents. Use a sloping top to keep items off the top of the lockers. Figure 14.9 shows how to arrange ventilation so that air is forced to go through the lockers, not just the locker room. Table 14.4 gives the amount of ventilation air. Shower change areas should be kept warm—80°F (26.7°C) or more (Rohles and Konz, 1982). Place the lockers so they do not block the ceiling illumination; that is, light the aisles, not the locker tops.

OSHA (subpart J, 1910.141) requires 1 shower/10 employees of each sex required to shower during a shift. When planning individual showers, allow 27 ft² per shower, to provide space for access aisles, toweling area, and the shower stall. Both shower and toweling areas require drains. Have hooks and shelves as well as benches in the toweling area.

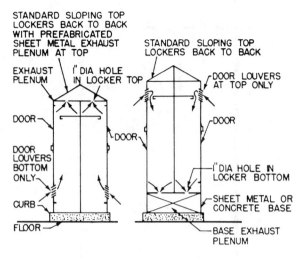

FIGURE 14.9 Locker room ventilation should be designed so air flows through the lockers.

TABLE 14.4 Ventilation for locker rooms, toilets, and shower spaces (ASHRAE, 1991).

Description	Amount
Locker rooms:	
Coat hanging or clean change room for non-laboring shift employees with clean work clothes	1 cfm/ft²
Change room for laboring employees with wet or sweaty clothes	2 cfm/ft²; 7 cfm exhausted from each locker
Change room for heavy laborers or workers assigned to working and cleaning where clothing will be wet or pick up odors	3 cfm/ft²; 10 cfm exhausted from each locker
Toilets	2 cfm/ft²/ at least 25 cfm per toilet facility; 200 cfm minimum
Shower spaces	2 cfm/ft²; at least 50 cfm per shower head; 200 cfm minimum

2 MAINTENANCE SHOP

A key first decision is centralized versus area or local maintenance. Area maintenance has the advantages of faster service and personnel familiar with the machines to be maintained. Centralized maintenance has the advantage of better utilization of people and equipment. A variation is for maintenance people to be stationed centrally but have local, lockable storage facilities. Thus, central people can leave tools and other equipment overnight and report directly to the local job rather than the central area.

In addition to the work area, have a fenced stock area. Provide space for shop tools and equipment; jobs awaiting repair parts; incoming jobs and repair parts; steel and lumber racks; equipment prints and manuals; and oil, paint, and solvent storage. One organization provides lockable wire mesh compartments along the stockroom outside wall. As parts and tools for a work order are accumulated, they are placed in the compartment. Some companies stock fasteners and daily use items in unsecured bins in the work area. Each worker normally should have some tools permanently assigned; thus each should have some permanently assigned lockable storage space.

In addition to specialized tools, most maintenance shops will have a lathe, mill, welding equipment, band saw, cutoff saw, and pedestal grinder. Generally, electrical areas are separated from mechanical areas. Clean areas, such as electronic/instrument/quality control, may require special ventilation and/or air conditioning. Isolate dirty areas such as welding. Design utilities (compressed air, electricity, water) to be widely available.

Material handling, infrequent by production standards, may involve heavy items. Consider an overhead crane that extends from one end of the maintenance area to the other. Compressed air pallets are an alternative. Jib cranes are popular and several may be desired.

3 BATTERY CHARGING AREA

Battery charging rooms have special requirements, including (1) ventilation requirements for the fumes, (2) floors of acid-resistant construction, (3) provisions for flushing and neutralizing spilled electrolyte, (4) facilities within 25 ft to provide quick drenching of eye and body, and (5) fire protection (such as nonsparking materials). Walls should be painted with acid-resistant paint (e.g., epoxy). Fumes in recharging areas are highly explosive (as are fumes in internal combustion refueling areas); therefore, OSHA prohibits smoking in all recharging and refueling areas. See OSHA regulations 1926.403 (battery rooms and battery charging) and 1910.178 (changing and charging storage batteries) and equipment manufacturers for more detail. See Figure 14.10 for an example of a charging-maintenance area.

Batteries are heavy (e.g., 1,200 lb), too heavy to pick up manually. However, rollers under truck batteries tempt some firms to encourage employees to pull them out manually. Even if the rollers on the truck and the rollers on the rack are perfectly aligned horizontally and vertically, manual handling would still be likely to cause back injuries. Two things are necessary: (1) a support platform that can be easily aligned with the truck rollers and (2) mechanical power to pull out the battery. The solution could be a crane (gantry, jib, or bridge) or a cart with a mechanical loader.

Batteries, when removed, are placed at a charging station. Ford (Automatic Washing, 1981) found that washing each battery with water jets was cost-effective. Charging

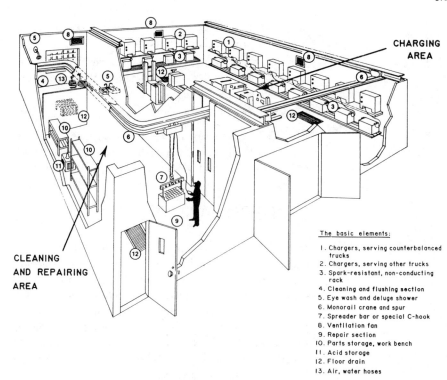

FIGURE 14.10 Charging and maintenance rooms for battery-powered vehicles. Distance between each wall-mounted charger is 10 ft. Each charger should have space for battery storage. This firm used 4 X 8.5 ft to store three batteries/vehicle, with one always in the truck (General Mills, 1983; figure courtesy of Lead Industries Association, Inc.)

CHARGING
AREA

CLEANING
AND REPAIRING
AREA

The basic elements:

1. Chargers, serving counterbalanced trucks
2. Chargers, serving other trucks
3. Spark-resistant, non-conducting rack
4. Cleaning and flushing section
5. Eye wash and deluge shower
6. Monorail crane and spur
7. Spreader bar or special C-hook
8. Ventilation fan
9. Repair section
10. Parts storage, work bench
11. Acid storage
12. Floor drain
13. Air, water hoses

stations probably should be located so no vehicle needs to wait for another to be serviced. For example, provide enough aisle space for a vehicle to pass a vehicle being serviced. Each station probably should have a supply of demineralized water supplied through a hose; automatic cutoffs are popular since they prevent overflow, with its many problems, and expedite filling. Wall-mounted chargers save floor space and reduce the chance of vehicle impact. Battery and charger cables should be reel-mounted to reduce damage from vehicles running over the cables. Local lighting (such as two-lamp fluorescent fixtures) can supplement ceiling lighting to provide adequate light for inspection and maintenance. Using one charging station and set of batteries for each vehicle probably gives the best maintenance control of the batteries. Low-maintenance batteries need a charger that tapers charges initially (decreases current with time) but has a higher rate at the end. If possible, charge during periods of low plant power use; especially avoid peak load times (see Chapter 21).

4 WELDING

Welding areas need considerable ventilation. (Welders have a 40% increase in the relative risk of developing lung cancer as a result of their work experience (NIOSH, 1988).) If roof ventilation is not used, welding areas may have to be on an outside wall. For **makeup air**, use exhaust air from "clean" areas, such as the office, rather than outside air to avoid unnecessary expense of temperature and

humidity control. Welding areas should be under **negative air pressure** so that welding fumes do not spread throughout the factory; thus there should not be open space between the welding area and the remainder of the factory. Keep degreaser fumes away from welding. The arc's ultraviolet radiation can decompose chlorinated hydrocarbons (such as trichloroethylene) into toxic gases (phosgene); death may result from respiratory cardiac arrest (McElroy, 1988, p. 332). Welding and cutting should not be near areas with flammable or combustible vapors, liquids, or dusts. Because of their toxic properties, special exhaust provisions must be made if zinc, bronze, beryllium, chromium, manganese, lead, cadmium, or their compounds are being welded.

Along with the general welding area, the welding operation itself should be enclosed—generally by a booth. The OSHA requirements are given in General Industry: OSHA Safety and Health Standards (29 CFR 1910). Figure 14.11 shows two methods of suspending the movable curtains. Ventilation hoods should suck air at 100 ft/min "in the zone of welding when the hood is at its most remote distance from the point of welding." See Table 14.5. The aspect ratio of the hood (ratio of hood length L to its width W) should be as close to 1 (as square) as possible. The optimal flange width (beyond which velocity increases little) is equal to WL. Flanges help more as the aspect ratio changes from 1. See Chapters 9 and 26.

The floor may be of poured or reinforced concrete but needs to be completely free from moisture. In actual welding areas, concrete may be heat treated or covered with a

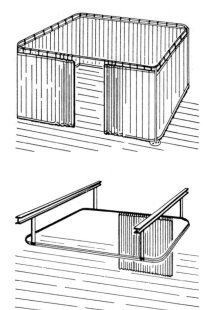

FIGURE 14.11   Welding booths can be suspended from overhead beams (to permit unlimited horizontal access) or can be freestanding (top figure). Curtains and screens should have the bottom 2 ft above the floor; suspension on 2 or 4 wheel trolleys tends to be more satisfactory than on sliding "shower curtain" hooks. The curtains can be made from a variety of special materials. Tinted vinyl filters out ultraviolet welding rays but permits outside light to enter the booth and the welding operation to be seen from the outside (Polentz, 1982).

nonexplosive finish. Areas not exposed to hot metals may be covered with vinyl asbestos for worker comfort. Cover walls and ceilings with a light-colored, nonreflective paint with appreciable percentages of zinc oxide or titanium dioxide. This reduces ultraviolet reflections. The 270-nm wavelength is the most dangerous. For lenses and goggles, polycarbonate lenses not only provide better protection than glass but weigh half as much.

It is better to use a distribution system to bring gases to each welding system than to have tanks at each booth.

Safe applications include the following (for more detail, see OSHA 1910.252):

Keep cylinders away (20 ft) from excessive heat and combustible materials, especially oil and grease.

Separate (20 ft) gas and oxygen cylinders. (A noncombustible barrier at least 5 ft high with a 30 min fire rating can replace the 20 ft.)

TABLE 14.5   Ventilation design for welding booths for 100 ft/min of air flow in the welding zone (Polentz, 1982). A 3 in. wide flange suction opening is assumed.

Distance from Arc or Torch (in.)	Duct Dia. (in.)	Minimum Air Flow (in.³/min)
4 to 6	3	150
6 to 8	3.5	275
8 to 10	4.5	425
10 to 12	5.5	600

Store and use acetylene cylinders only in a vertical, valve-end-up orientation.

Prevent cylinder valves from being knocked off (and thus becoming missiles). In storage, use valve caps; in use, chain tanks to a firm support.

Use valve caps during transport.

The electrical supply system, which naturally must comply with all codes, should be of ample capacity. All welding machines and tables must be properly grounded.

5 MEDICAL FACILITIES

Medical facility requirements depend on the number of employees served and the service provided. American Mutual recommends a nurse when 300 or more people (including office personnel) are employed on one shift. The American Medical Association recommends 1 ft² of medical facilities for each employee on the largest shift (Scope, Objectives, 1971). Other requirements are hot and cold water, a telephone, two rooms (for patient privacy), and bathroom facilities (at least nearby).

Some firms have nurses do pre-employment physicals; others contract these exams to physicians in the community. Nurses may or may not do hearing examinations and dispense safety glasses and shoes. They may or may not do medical insurance paperwork.

Accidents vary drastically between different types of work, especially between office and manufacturing. In addition, only a small percentage (less than 5%) of patients visit the medical facility because of work-related accidents. The rest want an aspirin, have colds, have the flu, cut their finger at home, burned themselves at home, sprained their ankle playing basketball, etc. They also want medical counseling ("I think I'm pregnant, but I'm not married"). Thus, a private area is important. A window between the counseling and treatment areas permits the nurse to monitor the treatment room while counseling.

The following features are part of the medical facility of a specific plant with about 500 employees (100 office, 400 manufacturing): The nurse does pre-employment physicals and hearing checks; insurance paperwork is done in personnel; safety equipment is furnished by local vendors. The facility is run by one nurse. Permanent facilities include two rooms plus a toilet, sink (including eye/face wash unit) outside the toilet, and an audiometer booth. In the two rooms there are nurse's desk and chair, four chairs for patients, two hospital beds, heat lamp, scales, small refrigerator, two-drawer lockable cabinet for medications, three-drawer cabinet for more medications, file cabinet for medical files, and other storage shelves for miscellaneous supplies. In addition, a wheelchair and stretcher are stored in case of emergency.

Boydstun (1979) gives the layout for an occupational health unit for 600 employees. Employers Mutual of

Wausau (Planning, 1970) has plans for 420, 600, and 925 ft². Figure 14.12 shows the 420 ft² plan. Figure 14.13 is a 400 ft² recommendation by American Mutual (First Aid in Industry). For a first aid room with no nurse, Figure 14.14 shows American Mutual's recommendation; Figure 14.15 shows Simonowitz's recommendation.

Waiting patients must have chairs. Place them outside the medical facility if space is limited.

Locate the medical facility

- close to employees (customers)
- close to safety and personnel departments (supervision and paperwork)
- accessible to ambulance

6 FOOD SERVICE

Four types of areas will be discussed: vending areas, snack bar, cafeteria serving, and eating areas. See Box 14.1 for comments on the number of customers.

The food service area needs to be convenient to other services such as toilets, training and conference rooms, and telephones. One alternative to minimize distance is the use of a mezzanine. Esthetics are relatively important. Consider using windows with a view, plants, fountains, and other interesting features.

6.1 Vending Areas

The primary advantage of **vending** machines is the low labor cost. Vending is self-service and the only service labor required is machine stocking and some equipment repair. Low service cost permits service 24 h/day, 7 days/week. A second advantage is the small space requirement, which in combination with low service cost, permits decentralized service. Vending machines can therefore be located at four or five locations within the facility, minimizing user travel and maximizing user convenience. Some firms locate vending machines near supervisors' offices to discourage loitering.

Many types of vending machines are available. I recommended the following mix for serving 100 people:

- one hot drink machine
- two cold drink machines
- two sandwich machines
- one ice cream machine
- one candy/snack machine
- one bill changer (changing $1 and $5 bills)

Other features should include microwave ovens, a sink with faucets, a drinking fountain, a refrigerator for "brown baggers," and condiments (napkins, salt and pepper, ketchup, etc.). Storage (for the condiments) and a serving area will be needed. Signs should be posted giving the name and phone number of the person to contact if a vending machine malfunctions. Trash containers are necessary. Many firms have a separate container for aluminum cans. The can container should have only a can-sized hole in the lid to discourage other types of trash.

See Figure 14.16 for a vending area and eating area for 100 employees. Note that the machines are grouped along a **utility wall** to minimize plumbing and electrical costs.

6.2 Snack Bars

A compromise between a vending area and a cafeteria is a snack bar. Typically, it includes sandwiches (hamburgers, hot dogs, cold sandwiches), french fries, prepackaged salads, desserts (cake, pie, doughnuts) and drinks (cold and hot). Usual requirements are cooking equipment for the sandwiches and fries (grill, microwave, fryer), refrigerator, drink dispensers, counter space, and a cash register.

6.3 Cafeteria Serving

Cafeterias can serve more complicated items than can vending machines, and, of course, they are more personal. The primary problem is the labor cost. In addition, there is the expense of unsold food that may have to be thrown away. Sales must cover these costs. Cafeterias therefore need a considerable number of customers and limited operation time. Generally only one cafeteria per facility can operate, and it is open only a few hours/day—supplemented with vending machines.

One technique to reduce cafeteria labor cost is to subcontract the food service to an outside firm. The subcontractor provides specialized management skills and can pay low food service wages instead of high manufacturing wages.

Cafeterias can be **line flow** or **scramble system**. In line flow, customers pass by all items in a single line. The primary advantage is flow simplicity—the path to take is not confusing. Simplicity is especially important in commercial cafeterias where customers come only occasionally. The primary disadvantage is that the line speed is controlled by the slowest person in the line. People wanting only one or two items can become frustrated.

In the scramble system (see Figure 5.14), the service points (hot food, sandwiches, beverages, desserts, salads) are arranged on islands and customers can "scramble" to whichever service point they desire. The primary advantage is reduced throughput time/customer. Disadvantages are that the system requires more space and that new customers may not be able to understand the flow pattern.

For high-volume cafeterias, separate the price addition and payment functions and have customers pay when they leave the cafeteria. This system reduces bottlenecks in the serving line but requires two people (one for pricing and one for payment) and thus is too expensive for low-volume cafeterias.

6.4 Eating Areas

As mentioned earlier, many firms want a pleasant (esthetic) eating area. **Esthetic** possibilities are windows with a view, carpet on the floor, good noise

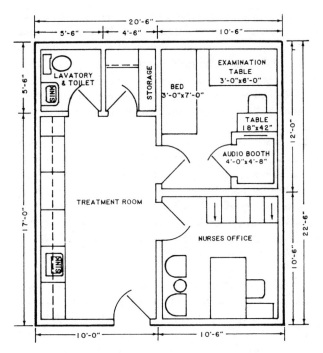

FIGURE 14.12   Health services department for 250 to 500 employees (Planning, 1970).

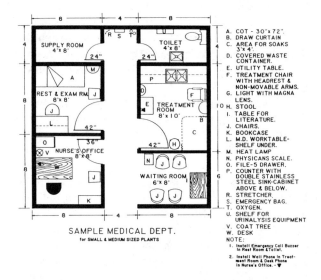

FIGURE 14.13   Health services department for a full-time nurse and part-time physician (First Aid).

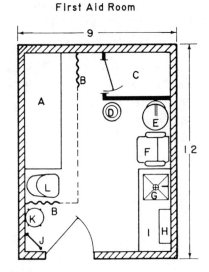

A. Cot
B. Draw Curtain
C. Toilet
D. Waste Receptacle
E. Lamp
F. Treatment Chair

G. Sink
H. First Aid Cabinet
I. Counter
J. Stretcher
K. Stool
L. Chair

FIGURE 14.14   First aid room layout for a small plant (First Aid). Conspicuously post the name, location, and phone number of first aid personnel, hospital, and ambulance service.

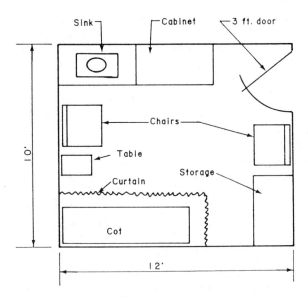

FIGURE 14.15   Nonstaffed first aid room layout for a small plant (J. Simonowitz, personal communication, 1982). It should be near a toilet. Lockable cabinets up to 72" maximize the space utilization.

BOX 14.1 *Number of food service customers*

The size and type of food service facilities to design depend on the expected number of customers. Not all the people working at the facility will use the food service facilities. The percentage (p) using the facility will depend on the following:

- *Pricing policy:* If the food is high priced, the value of p decreases. Firms may subsidize the food prices by not charging for the use of the space or utilities. High prices (or poor quality food) may encourage brown bagging, which would decrease food sales but may not discourage use of the eating areas.
- *Meal break length:* If the meal break is short (30 min), eating outside the facility is less likely and p will increase. If the break is long (60 min), eating outside is more likely. Longer breaks encourage people to do other things (e.g., exercise) and skip the meal.
- *Accessibility:* The time to get to the food service facility and the difficulty in obtaining the food affect facility use. Assuming a person walks about 260 ft/min, if the facility is 500 ft away and the person has to wait for 5 minutes to be served, then $2 + 5 + 2 = 9$ minutes of the break are taken up. The more the time used, the lower will be the value of p. Transport and waiting time is especially critical

for food service use during break times (which typically are only 10–15 min long).

- *Policy on "outside" food:* Will delivery from delicatessens and pizza delivery services be permitted? Vendors may set up mobile facilities immediately outside the plant gates. The fewer the restrictions on outside food, the lower the value of p.
- *Proportion of part-time employees:* Part-time workers will usually decrease the value of p.

Thus, if a facility had 200 people, perhaps only 60% of them would use the food service facilities on a specific day.

An additional consideration is the time distribution of customers within the day. Are the customers on one, two, or three shifts? Are meal times staggered? For example, department A might have lunch from 11:30 to 12:00, department B from 11:45 to 12:15, and department C from 12:00 to 12:30. Reducing the peak load will decrease the facilities required and reduce customer waiting time.

On the other hand, if customers are few, serving cost/customer becomes high. Many cafeterias reduce costs by being open only for a portion of the day with the remaining hours covered with vending machines.

control, artwork on the walls, mural wallpaper, indirect lighting, warm white instead of cool white fluorescent lamps, uneven ceiling fixture spacing (giving more uneven illumination), wood instead of plastic chairs and tables, upholstered chairs, artificial flowers on the tables (changed periodically) and glass plates and tumblers (instead of plastic or paper).

A decision has to be made concerning smoking. Some firms have smoking areas, but many have shifted to a

policy of no smoking anywhere in the facility. If there is a smoking area, it needs extra ventilation.

The eating area should have a mixture of table sizes and shapes. Circular tables accommodate both even- and odd-number groups and encourage conversation. Rectangular tables do not accommodate odd-number groups well but are easier to combine with other tables; that is, you can take three tables of two and make a table for six. Long rectangular tables (seating potential of more than 6) often

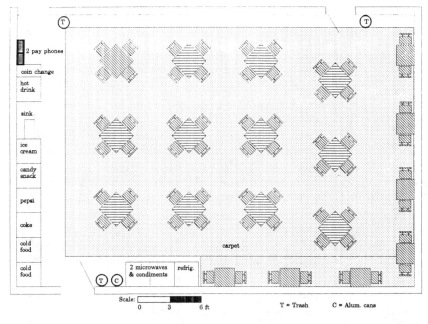

FIGURE 14.16 The eating area for 62 nonsmoking customers has a variety of table sizes and shapes. Note that the drinking fountain and sink are close together to minimize plumbing costs. Trash containers are labeled *T*; aluminum can containers are labeled *C*.

have low utilization because once the table has more than one or two occupants, newcomers feel that they are intruding on the occupants' personal space (see Box 13.3). Tables should be the pedestal type and should be movable.

Trays may be a problem. Many people like to remove their food from the tray while they eat. If there is no convenient location, they will put the tray on another table, thus reducing available table space. One possibility is a tray storage location in a slot under the table. Most industrial cafeterias expect people to bus their own trays when leaving—often requiring them to separate paper, food, and dishes. Trash cans should be plentiful.

Provide one or more locations for condiments, coffee and tea refills, tableware, and tray storage. Also provide a place for coats, hats, and packages, especially if customers come from outside the building. If no place is provided, people will put items on tables and chairs, reducing seating space.

The eating area often is used as a lounge or for business meetings during nonbreak times; more space may therefore be needed. A bulletin board for official notices and one for personal notices should be provided.

Pay telephones should be located near the eating area. If the phones have free local calls, ensure that they do not permit long distance phone calls without a credit card. Phones should be wall mounted and should not be near chairs (to discourage people tying up the phones). A wall-mounted phone book and a roll of paper (toilet paper fashion) for notes are useful.

7 TOOL CRIBS

Tools, dies, and supplies can be stored locally at the machine (no control) or centrally located (under control). In large facilities there may be more than one crib to minimize employee travel time. Multiple cribs increase server cost, since there are more servers, and also may increase inventories, since items are duplicated. Each crib should have two service windows to reduce waiting time, even if one is not always used.

Justifications for tool cribs (versus at-the-machine storage) are

- security for items
- reduced supply item use (less waste with accountability)
- reduced inventory (less duplication: 25 at crib versus 10 at each of 10 machines), saving inventory cost and space
- more accurate inventories due to better records, and thus less chance of stockouts

Space saved by using a central crib probably is not as important as better inventory control (less pilferage and fewer stockouts). The primary expense is the stockroom labor. However, even if storage were done locally at the machine, there still would be some record-keeping expense. Local storage also would require someone to unpackage supplies, organize items, and so forth. Idle stockroom attendant time can be used by assigning various miscellaneous duties to be done during idle time. A common example is tool regrinding. Reduce the idle time of the customer waiting in line by having the customer give the order ahead of time and having the clerk put the items into one of a series of customer lockers built into the crib wall. The locker, which has a lock on the outside, then can be opened by the customer without waiting in line. This technique also permits having a stockroom without full-time staffing. For example, the clerk might be present only in the morning, but customers could get items at any time—even during second or third shift or on weekends.

8 CLEAN ROOMS
(*Clean Spaces*, 1991)

With many manufacturing processes, airborne particles are viewed as a source of contamination where contact between the particulate and the product will cause product failure.

Clean spaces can be Class 100,000, Class 10,000, Class 1,000, Class 100, Class 10, or Class 1. A class 1,000 clean room, for example, has 1,000 particles of size .5 microns and larger per cubic foot. A standard business office is a Class 300,000 environment. In pharmaceutical clean rooms, the emphasis is on control of bacteria and viruses rather than on particles, but most of the same principles apply.

Reducing the number of particles requires controlling external and internal sources: (1) Provide "clean" air to a "sealed" room and (2) minimize the contamination sources within the room.

The clean room is sealed to prevent contamination from the air in the rest of the facility. Thus, entrances have an airlock (a door, a space, and an inner door), windows cannot be opened, pipe openings and electrical outlets are sealed, and so on. Clean rooms generally have considerable window area for psychological reasons, although windows to the outside for daylight awareness should be separated with a corridor from the outside window.

The unidirectional air flow generally is from the ceiling to a floor plenum (vertical flow). This "shower" of air washes workstations; particles find it difficult to move horizontally. Special filters filter the supply air before it enters the room. Makeup air may need to be filtered for chemical contaminants (salt and automobile pollutants), as well as for particles. Positive pressure is maintained in the clean space so that airflow is from the cleanest space to less clean space. Since the departing air is considerably cleaner than normal building air, it usually is filtered and recycled. The air flow volume can be enormous: A Class 100 environment may have 750 air changes/hour. Thus, there is good potential for energy conservation.

Minimizing contamination within the room is a challenge. See Table 14.6. People are the main source of contamination (from skin and hair particles). Some firms use robots rather than people. In the clean space, processes that generate particles (e.g., soldering) should have local exhaust hoods. If the exhaust is hazardous, there should be two fans, both running. An uninterruptable power source should be considered for power. Restrict the entry of maintenance personnel by facing machines into the clean room and having a special access corridor behind the machines for maintenance personnel.

TABLE 14.6 Operation of clean spaces (Clean Spaces, 1991).

Personnel and Garments

- Clean hands and face before entering area.
- Use lotions and lanolin soap to lessen skin particle emission.
- Forbid cosmetics and skin medications.
- Forbid smoking and eating.
- Wear lint-free smocks, coveralls, gloves, and head and shoe covers.

Materials and Equipment

- Clean materials and equipment before entry.
- Do not use pencils or erasers. Use nonshedding paper.
- Handle work parts with gloved hands, finger cots, tweezers, vacuum wands, etc., to reduce transfer of skin oils and particles.
- Shield and exhaust grinding, welding, and soldering operations.
- Use containers to store and transfer materials.

DESIGN CHECKLIST: SPECIALIZED AREAS

Toilets
    Number and location versus people
    Details (electrical, mirrors, shelves, cigarette disposal, towels or dryers, coat hooks)
    Barrier free
Drink fountains
    Number, location
Flushing stations
    Number, location
Lockers and washrooms
    Wash fountains versus multiple sinks
    Clearance
    Water control
    Access
    Showers (individual versus group)
    Lockers (size, type, ventilation)
    Details (drains, sight lines, shelves, benches, hooks)
Maintenance
    Central versus local
    Security
    Storage (materials, tools, supplies)
    Material handling
Battery charging room
    Ventilation
    Emergencies (flushing stations, fire, drains)
    Material handling equipment

Welding
    Booths
    Ventilation
    Floors
    Gas distribution system
    Electrical supply system
Medical
    Types of service provided
    Types of major equipment
    Storage
    Patient comfort and privacy
Food service
    Vending locations
    Vending machine types
    Cafeteria? Kitchen?
    Line flow or scramble?
    Number of seated places
    Table shapes and sizes
    Coat racks and condiments
Tool cribs
    Number of cribs
    Windows/crib
    Lockers for customers
Clean rooms
    Class of cleanroom?
    People contamination minimized?
    Energy conservation

REVIEW QUESTIONS

1. For a machine shop employing 30 males and 10 females, and for an office employing 30 females and 10 males, give the number of water closets, lavatories, drinking fountains, showers, and service sinks.
2. What should be different in a rest room for females, compared with one for males?
3. How can a toilet be made barrier free?
4. For a 400 ft² locker room for employees doing heavy labor, what ventilation volume is recommended?
5. Discuss the advantages and disadvantages of centralized maintenance and the advantages and disadvantages of decentralized maintenance.
6. What are the five special requirements for battery charging rooms?
7. How can ultraviolet radiation in welding areas be reduced?
8. What facilities might be used in a nurse's office in a 500-employee facility?
9. Discuss the advantages of vending machines versus cafeterias.
10. What are the advantages and disadvantages for flow line versus scramble cafeterias?
11. How can you discourage people from putting trash in containers reserved for aluminum cans?
12. Give two advantages of subcontracting the cafeteria food service to an outside firm.
13. Discuss phones near the eating area.
14. Discuss the percentage of people using a food service facility.
15. Discuss air flow in a clean room.

PROBLEMS AND PROJECTS

14.1 What modifications need to be done to make your classroom building and its surroundings more accessible to wheelchair students? What do wheelchair users think of your design? What does the dean think of your design?

14.2 Design a locker room for _____ workers. (The instructor will give the number and sex of the workers).

14.3 Design a welding area. Specify the number of gas, arc, or spot welding stations. Justify your design.

14.4 Design a vending area to serve a department of 50 employees. Justify your design.

REFERENCES

ASHRAE. *HVAC Applications*. Atlanta: American Society of Heating, Refrigerating, and Air Conditioning Engineers, 1991.

Automatic washing adds years to the life of our batteries. *Modern Material Handling*, Vol. 36, 13, 64–65, Sept. 21, 1981.

Boydstun, S. Design of an occupational health unit. *Occupational Health Nursing*, Vol. 27, 1, 7–11, 1979.

Clean spaces. Chapter 16 in *ASHRAE Handbook: HVAC Applications*. Atlanta: American Society of Heating, Refrigerating, and Air Conditioning Engineers, 1991.

Coetzee, K., and Bennett, C. The efficiency of a drinking fountain. *Applied Ergonomics*, Vol. 9, 2, 97–100, 1978.

First Aid in Industry. Wakefield, Mass.: American Mutual Insurance Co.

General Mills' "ideal" charging facilities. *Lead*, Vol. 46, 2, 2–3, 1983.

McElroy, F., Ed. *Accident Prevention Manual—Engineering and Technology*. Chicago: National Safety Council, 1988.

NIOSH. *Criteria for a Recommended Standard: Welding, Brazing and Thermal Cutting*. Publication 88-119, National Technical Information Service, 5285 Port Royal Road, Springfield, VA 22161, 1988.

Planning Employee Health Services. Booklet 04-1079-1 12-70, Wausau, Wisc.: Employers Mutual of Wausau, 1970.

Polentz, L. Welding booths. *Plant Engineering*, Vol. 36, 20, 40–41, Sept. 30, 1982.

Rest rooms (FEG02558-03). *Facilities Planning Guide*. Peoria, Ill.: Caterpillar Tractor Co., 1971.

Rohles, F., and Konz, S. Showering behavior: Implications for water and energy conservation. *ASHRAE Transactions*, Vol. 88, 1, 1982.

Scope, Objectives, and Functions of Occupational Health Programs. Chicago: American Medical Association, 1971.

Weaver, L. Eyewashes and showers: Ensuring effectiveness. *Occupational Health and Safety*, Vol. 52, 8, 13–19, 1983.

15 | WORKSTATIONS

CHAPTER PREVIEW

Just as the human body is an organization of individual organs, a facility is an organization of individual workstations. Principles discussed include the design of the job done at the workstation, the relation of the workstation to other workstations, and the physical design of the workstation.

CHAPTER CONTENTS

1 Design of the Job
2 Workstation Versus Other Workstations
3 Physical Design of the Workstation

KEY CONCEPTS

anthropometric data
automation
availability
cumulative trauma
double tooling
environmental stimulation
    (versus task stimulation)
expensive component, the person
filler jobs and filler people
fixed costs

flexible schedules
get ready, do, put away
idle capacity
job enlargement
job rotation
job sharing
joint movement
maintainability (*MTR*)
material handling cost
mechanization

modularization
modular workstation
normal work area
one on one
permanent part time
pool facilities
recovery time
reliability (*MTBF*)
specialization
working rest

The material in this chapter is a condensed version of Chapters 13, 14, and 16 in *Work Design* (Konz, 1990). Figure 15.1 shows a schematic workstation. This chapter will consider design of the job at the station, the relation of the station to other stations, and the physical design of the station.

1 DESIGN OF THE JOB

Table 15.1 gives eight guidelines concerning design of the job.

1.1 Use Specialization Even Though It Sacrifices Versatility
Specialization has been the key to progress. It tends to decrease cost/unit and to increase quality. The primary problem is to get sufficient sales to justify the specialization since it tends to result in a narrow product line. A successful example of specialization is the fast-food business. The firms offer only a relatively few standardized items. They hope the low price and good quality will attract enough sales to make up for the lack of a large number of items on the menu.

The low price and improved quality of specialization come from special-purpose equipment, material, labor, and organization. Special-purpose equipment tends to give greater speed and capability than general-purpose equipment. The disadvantages are higher capital cost for the equipment and lack of flexibility. Specialized material can be used in either the product or the production equipment; the usual tradeoff is increased capability versus higher cost. Labor productivity and quality generally are better for a specialist. Somewhat surprisingly, in most industries the specialist is paid less than the generalist (e.g., specialist machine operator is paid less than the general machinist), so for the specialist the cost/time is lower as well as time/unit. Although the specialist's job is monotonous, many people want such jobs. From an organization's viewpoint, the challenge is to get sufficient sales from the

TABLE 15.1 Guidelines for job design (Konz, 1990).

1. Use specialization even though it sacrifices versatility.
2. Make several identical items at the same time.
3. Combine operations and functions.
4. Minimize idle capacity.
5. Use filler jobs or filler people.
6. Vary environmental stimulation inversely with task stimulation.
7. Reduce fatigue.
8. Communicate information.

narrow product line. An alternative is to have a broad product line but to organize the manufacturing to get the benefits of specialization. Two examples are group technology and flexible manufacturing systems (see Chapter 4).

1.2 Make Several Identical Items at the Same Time
Tasks can be broken down into three states: **get ready**, **do**, and **put away**. One alternative is to decrease the do time even though the get-ready + put-away time increases. For example, the Post Office forwards mail to people who have moved. When a piece of mail with an incorrect address comes to the carrier, the carrier puts it aside for a special operator. This operator has a diskette for each carrier. The operator enters the first three digits of the last name and the last two numbers of the street address. The computer then prints out the full forwarding address on a peel-off label. The operator peels off the label and puts it on the item. Another alternative is to prorate the get-ready and put-away time over more units. On a macroscopic level, you can reduce get ready and put away by having larger lot sizes.

Get ready and put away also can be reduced at the operator level. In inspection of *n* items for *m* characteristics, for get ready the inspector must do the mental work of fixing in the mind the quality standard for one of the *m* characteristics. The do is the sensing of the object, comparing it against the mental standard, making a decision, and executing the decision. Then, after all items are inspected, the inspector puts away the standard. Inspecting one item for *m* characteristics uses less physical handling than inspecting multiple items for one characteristic but uses more mental agility; inspecting many items for one of the *m* characteristics and then inspecting the items for another of the characteristics, etc., requires more physical handling but less mental agility. If the inspection is easy (the mind can keep track of several characteristics simultaneously), inspect for several characteristics simultaneously. If inspection is difficult, you need to trade off the improved quality of doing one characteristic at a time against its increased inspection time (Su and Konz, 1981).

Get ready and put away of hand tools is a processing example of this principle. Consider a person making hamburgers. If five hamburgers are made at a time, then the pick up and the put down of the spatula can be prorated over five hamburgers instead of one; so can the time to add pickles, onions, etc. The do time increases a little but not as

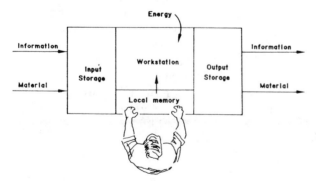

FIGURE 15.1 Schematic workstation shows that both information and material flow into input storage. Then they are transferred to processing where they are transformed, using energy and local memory. Then the information and material go to output storage before being transferred to the next workstation.

much as the get ready and put away time decrease. For more examples, see the comments on multiple items/carrier in Chapter 5.

1.3 Combine Operations and Functions

With planning, several things can be done at the same time. Materials can be multifunctional. One well-known example of multi-function material is a compound that cleans and waxes a surface at the same time—common applications are automobiles and floors. Other compounds contain both fertilizer and weed killers to permit the farmer or gardener to make one application instead of two. An ice cream cone is an example of a container that adds to the product appeal. Many years ago Henry Ford had supplies delivered in special wooden boxes; the boxes became part of the Model T floor. Can you specify the containers that your vendors will ship items to you in so that you can use their container for in-plant handling?

Tools and equipment also can be multifunctional. A claw hammer is a combined tool, combining a nail claw with a hammer. Although tasks are done in sequence, the combined tool eliminates the need to release one tool, find the second tool, grasp it, release it, and regrasp the first tool. The do time is the same whether the tools are combined or not. A pencil and eraser, pliers and wirecutters, and combined drill and countersink are other examples of combined tools which do operations in sequence. Examples of simultaneous processing are lathe form tools that form several surfaces simultaneously, drill presses that drill several holes simultaneously, nut drivers used to turn all the nuts on a race car wheel simultaneously, and a punch press that punches several holes simultaneously.

Much handling is of information rather than of materials. Identification badges can also serve as keys to electronic locks. Carbon copies can reduce repeated copying. Often copying is now done electronically, and so processing and storage can be combined. For example, a name and address when typed on a letter can be stored in the word processor memory for printing on the envelope. Storage for later use also is common on computers and calculators. A widespread application of combining the functions of use and storage is the marriage of the typewriter and the computer into the word processor.

1.4 Minimize Idle Capacity

There are two subdivisions of **idle capacity**: fixed costs and relative costs of different parts of the system.

1.41 Fixed Costs.

Often a major portion of the cost of facilities and personnel is **fixed costs**: It does not vary much as does the usage of the resource. Some examples are capital cost of a factory or warehouse not varying whether the facility is used 8, 16, or 24 h/day; property taxes not varying whether the facility is used 5, 6, or 7 days/week; a cook receiving the same pay whether there are 5 or 100 customers; a secretary receiving the same pay whether 5 or 10 letters are typed/day; an airplane costing the same whether there are 75 or 100 passengers; a professor receiving the same pay whether there are 20 or 40 students/class; a computer costing the same whether used 14 or 20 h/day; a hospital's costs not varying whether there are 70 or 100 patients; a hotel's costs not varying whether there are 300 or 500 guests.

One possibility is to operate the facility more hours/week. This may spread a constant number of customers over more hours, thus requiring fewer facilities. As an example, a machine shop could be run for two shifts instead of one. For the same number of items produced, the two-shift operation would require fewer machines. If a computer is run more hours/day, then a smaller, less expensive computer can be used. Kansas State University runs classes from 7:30 to 5:30 instead of 8 to 12 and 1 to 5 to get better use of the classrooms and thus requires fewer classrooms.

Another possibility is that the longer hours may attract more customers. Service facilities often use this strategy. Examples are resorts open in summer and winter, fast-food restaurants open for breakfast, and discount stores open in the evenings. Note that, even if not all costs are fixed, longer hours may be worthwhile.

A third technique is to **pool facilities** with the concept that peaks and valleys in the different facilities are less likely to coincide. For example, suppose the keying requirements from 10 individuals now go to one person and the keying from 10 others go to a second person. The pool concept is that keying from any of the 20 might go to either of the two. Motor pools are an example for vehicles. With technicians, computers, telephone answering services, and duplicating facilities you can trade off reducing load fluctuations against loss of specialized service.

Another technique, assuming you can determine the schedule, is to revise the time when the output is due. For example, divide the month or year into alphabetic sections. Then bills or shipments for all customers whose names begin with A and B are done first, then those with C and D, etc. Certain intermittent periodic tasks, such as maintenance, can be scheduled in the valleys. Schedule the facility use for nonpeak times. After all, it is not necessary for all surgical operations to be done in the morning, for all shipments from the shipping dock to go on Friday and arrive on Monday, for all shipments to be at the end of the month. Do maintenance and inventory work during slow periods. If customers control when they want to be serviced, you might encourage them to use off-peak times with a price incentive. Common examples are lower prices by barbers on weekdays, lower prices for movies in the afternoon, universities charging less for summer school, lower charges for computer time for runs made during the night, utilities charging less for electricity used during the night or winter, phone companies charging less for long-distance calls during the night. If you have sales or "specials," time them for valleys of demand rather than amplifying the peaks.

1.42 Relative Costs.   You can also minimize idle capacity by considering relations among resources. The general concept is to improve utilization among members of a team.

The term *team* will be defined broadly: Team members include machines, tools, and facilities as well as people. For example, a lift truck driver and the lift truck will be considered to be a team. The concept is that some team members are more expensive than others—thus it is more important to keep some parts of the team busy than other parts.

In the United States, the expensive part of the team generally is human labor. Labor cost (including fringes such as holidays, vacations, Social Security, etc.) ranges from $5/h to $20/h. Although about 1,900 h/yr is representative for most jobs, we will round it out to 2,000 h/yr. Thus, an operator being paid $8/h probably has a cost (assuming 30% fringes) of $10.40/h or $20,800/yr. A minimum wage operator would be about $5/h or $10,000/yr. A machine's costs are for capital, power, and maintenance. Capital cost usually is the most important. A $20,000 machine (such as a lift truck) with a 10-year life would be $2,000/yr or $1/h; a $5,000 machine (such as a jib crane) would be $.25/h; a $1,000 machine (conveyors and a workbench) would be $.05/h. Power costs vary, but most machines are powered by electricity. At 7 cents/ KWH, even a 1 KW (.7 hp) machine costs only $.05/h for power. Maintenance again varies, but few machines require $1,000/yr for maintenance—that is, $.50/work hour. Thus, in the United States, the important thing is to maximize utilization of the **expensive component, the person.**

One possibility is to have duplicate low-cost components to ensure the utilization of the high-cost component. For example, a secretary might make photocopies. If the copier breaks down, have an alternate copier for use until the original is repaired. An alternative is temporarily shifting a person to a different job, such as filing. If a person is not available (e.g., is absent), it may be possible to keep an expensive system (such as an assembly line) going with a substitute worker.

Another possibility is to use one or more inexpensive components, realizing they will be idle some of the time, in order to keep the expensive component busy. In **double tooling** of machine tools, two fixtures are used in order to keep the operator busy. While the part in fixture B is being machined, the operator unloads the part from fixture A and loads a new part. Then, while the part in fixture A is being machined, fixture B is unloaded and loaded. Another example is an executive and a secretary. It is more important to reduce idle time of the executive than of the secretary. In a flow line, a supervisor may keep 30 workers when there is work for only 29 so that illnesses and other absences don't stop the line.

But workers and machines need not be "**one on one.**" They can be fractions: 2/3 worker/machine or 3/4 worker/machine. How can this be done? You can't cut a person into pieces. Part-time workers will be discussed in the next section. But 2/3 worker/machine can be achieved by assigning 2 workers to 3 machines; this results in 2/3 worker/machine or 3/2 machines/worker. The worker simply walks back and forth among the machines (many machines are semiautomatic or automatic and need tending only occasionally). Machines and people also can be shared between functions. For example, a lift truck driver (or vehicle) can work for departments A for 4 h per day, B for 2 h, and C for 2 h. A mechanic or technician can be shared between departments E and F. Although supervisors like to "own" people or equipment (it increases their status as well as making their job easier by giving them complete control), from the organization's viewpoint the system is suboptimal.

1.5 Use Filler Jobs or Filler People

In matching people to work, it is easier to match smaller time segments. For example, if Joe and Pete each work 8 h/day and job A takes 6 h/day, job B takes 5 h/day, job C takes 4 h/day, and job D takes 3 h/day, there is too much work (18 h) for the time available (16 h). If another full-time worker is hired, then there is 24 h available and only 18 h work or 6 h idle. If one of the jobs is not done (say job D), then there is 15 h work for 16 available or 1 h idle. The shorter the jobs, the easier it is to get a balance between work and availability (see Chapter 5 on line balance for more on this topic). To reduce delay time, there are a number of techniques, which fall into two broad categories: (1) adjust the work load to a fixed work force (**filler jobs**) and (2) adjust the work force to a fixed work load (**filler people**).

1.51 Adjust Work Load.   It is easier to adjust jobs when the idle time is in relatively large blocks. For example, 25% idle could occur with a 2 h idle time at the end of 6 h of work; it also could occur as 15 s idle time after 45 s of work in a cycle repeated every minute. Note also that although in theory every worker should be assigned a full day's work (say 8 h), in practice you may wish to give some people "soft" jobs (pregnant women in the last weeks before maternity leave, a person with a broken limb, new employees, very old employees, etc.).

Make a list of short jobs that can be done in the idle time; many routine maintenance and clerical tasks fit this category. Be sure not to have overly precise job descriptions which would restrict people from doing other tasks. Service personnel (receptionists, tool room clerks, etc.) often have considerable time between customers; provide something for them to do while waiting for customers.

1.52 Adjust Work Force.   Scheduling considerations are timing and amount of work.

Set the timing of the start and finish of the job to reduce delay time, especially short delays. If a meeting starts 15 min after the start of work, most people will waste the 15 min. On the other hand, if the meeting is scheduled to finish 15 min before the end of work, it will drag on until quitting time. To shorten meetings, schedule them to overlap lunch or quitting time; to lengthen meetings, start them so that everyone has time available after the meeting.

Staggering starting and stopping time of the employees may help reduce delay time. For example, in a restaurant the waiters should arrive first; the dishwasher may not arrive until 2 h later; the waiters may leave before the dishwasher. Certain days of the week may be busier. Schedule more people for those days. **Flexible schedules** (Joe works 7 A.M. to 3 P.M.; Pete from 9 A.M. to 5 P.M.) can allow the facility to be open for service for 10 h/day instead of 8. There is less coverage only at the start and end of the day but no increase in labor costs. Having workers work 4/40 (4 days of 10 h/day) on a staggered schedule—with some taking their day off on Monday, some on Tuesday, some on Wednesday—is another technique for maximizing service hours while not increasing labor costs. Note that staggered schedules spread the demand on production facilities (workstations) and service facilities (parking lots, toilets, food service facilities, etc.). Flexible schedules require employees to be cross-trained so they can do each other's jobs, but that is desirable in any case. Don't permit vacations during busy periods; schedule them during slack times.

Temporary workers are another possibility. They can be from other departments or even other divisions of the organization. (In Japan, where some workers are on the payroll whether there is work or not, workers sometimes are shifted from manufacturing jobs in a factory to selling jobs in retail stores when business is slack.) Some firms furnish temporary office workers, manual labor, warehouse labor, etc. Hiring a student or teacher for a summer peak is good because the employer is not charged for unemployment compensation when the person leaves at the end of the summer.

Job sharing, that is, two people each working part time, provides benefits to management: (1) It allows for an expanded range of skills and experience; (2) both may work extra hours during peak loads; (3) noncoverage of the job (due to vacations, illnesses, absenteeism) becomes minimal; and (4) if one person leaves, the other provides coverage until a new person is hired.

Permanent part-time workers are another possibility. Part-time jobs are highly sought after. People are hired for less than 8 h/day, 5 days/week. One key advantage is less absenteeism since workers have nonwork time to attend to personal business. Another possible advantage is that people will do tedious or boring work for 4 h/day but not for 8 h. One Kansas firm has a monthly peak load; they hire people who work only the first two weeks of the month. There are many jobs with less than 40 h/week scheduled. For more on alternatives to the traditional 5 day, 40 h schedule, see Nollen (1982).

Many supervisors do not know exactly how long each job will take. Also, people tend to work more slowly when there is a small amount of work. Thus, it is important to assign more work than the time available. If too little is assigned, workers will tend to work more and more slowly, trying to stretch the work to fill the time available. No one wants to come to the boss and say "I'm out of work."

1.6 Vary Environmental Stimulation Inversely with Task Stimulation   People operate best when their **environmental** and **task stimulation** is intermediate—neither too high or too low. Thus, if the task has low stimulation, add stimulation to the task or the environment; if the task has high stimulation, decrease stimulation from the task or the environment.

1.61 Low-Stimulation Tasks. Add stimulation to either the task or the environment. Physical movement is the easiest way to add stimulation to the task. For example, require inspectors to make a physical movement to accept or reject an item; have the night security guard walk around, not just monitor the screen in a warm, dark office. Adding stimulation to the environment is done easily by allowing people to see and talk to each other. See Figure 15.2. Another audio stimulation is the use of background music or, in jobs requiring little conscious attention, even ordinary radio stations. Visual variety can be added by windows (be careful of glare) and by paint. Use a variety of colors, not just battleship gray and industrial green.

1.62 High-Stimulation Tasks. Here the problem is the reverse of the previous one, and the environmental stimulation should be minimized. Most high-stimulation tasks are done in an office environment. What is needed is privacy—primarily audio but also visual. Need for complete privacy generally is intermittent, not continuous. That is, it is needed for an hour or two, once in a while, but not all day every day. For these situations, a private area shared by people in a group on an as-needed basis may be satisfactory. Freedom from distractions (primarily conversation of fellow employees) can be achieved by use of the low-stimulation arrangements of Figure 15.2. Also see Chapter 13 for more on office design.

1.7 Reduce Fatigue   Breaks are given to reduce fatigue. The breaks can be with or without work.

Breaks while working (**working rest**) rest specific parts of the body. Boredom and monotony increase the need for such breaks. Examples of working rest are the following: secretary shifting to running duplicating machine, inspector shifting from mechanical faults to electrical faults, operator shifting from making to maintenance, operator shifting from making to getting or putting away the completed units, bookkeeper also answering the telephone, painter painting walls and then painting trim. In addition to shifting jobs for an individual, shifting jobs between operators

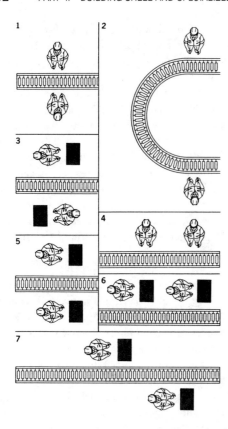

FIGURE 15.2 Workstation arrangements can add or subtract environmental stimulation from fellow employees. Vary stimulation by adjusting orientation in relation to others, by changing distance, and by using barriers such as equipment. The lowest stimulation situation is 7.

(job rotation) is another possibility. As a practical matter it is easier if the workers switched have the same supervisor.

Breaks without working give rest to the whole body, but there is no output while the person rests. Coffee breaks of 10 to 15 min every 4 h period are commonly accepted, although there is a little experimental evidence that they are needed from a physiological viewpoint. Physiologically, the amount of recovery/min declines as the length of the break increases. Thus two 5 min breaks are better than one 10 min break. However, there are practical problems of people stretching breaks beyond their specified limits, and the more breaks they have, the more chances they have to stretch them.

1.8 Communicate Information   Transferring a concept from the mind of an engineer to someone else's mind (operator, skilled trades worker, fellow engineer, technician, boss) is not simple and automatic. There are specific techniques which can improve this communication. See Chapter 29 of *Work Design* (Konz, 1990) and Chapter 7 of this text for some specifics. Be more visual than verbal. Pictures are better than words; demonstrations are generally better than pictures. Pictures and line sketches are useful in that they can show enlarged views and views from special angles; pictures also can be retained by the user for later

reference. Pictures with accompanying text are better than pictures alone. Tables also tend to present information better than text.

Write sentences in active, affirmative ways; avoid negatives and passives. Keep the main topic of the instruction early in the sentence. A problem is that words need to be translated and there are many examples of errors. Grade level required to read writing can be obtained from:

$$GL = .4(A + P) \qquad (1)$$

where

GL = grade level, years

  A = average words/sentence (Treat independent clauses as separate sentences.)

  P = percentage of words with 3 or more syllables (For P, omit (a) capitalized words, except initial word in sentence; (b) easy combinations, such as screwdriver and guestworker; and (c) verbs that reach 3 syllables with the addition of -*es* or -*ed*.)

For example, if sentences average 10 words, $A = 10$. If the percentage of words with 3 or more syllables is 20%, $P = 20$. Then, $GL = .4(10 + 20) = 12$. The writing is at the level of a high-school senior.

Avoid verbal orders. In addition to the previously mentioned translation problem, people forget what they heard (as well as not paying attention in the first place). If verbal orders must be given, have the receiver repeat them back to the giver in the receiver's own words.

2 WORKSTATION VERSUS OTHER WORKSTATIONS

Table 15.2 gives three guidelines concerning relations among stations.

2.1 Decouple Tasks   Chapter 5 treats this topic in detail. Briefly, tasks are decoupled for line balancing and for minimizing the degree to which shocks and disturbances of one task influence another task. Tasks are decoupled with buffers. Chapter 5 gives example design techniques.

2.2 Minimize Material Handling Costs   Material handling adds cost to a product. Reduce the cost by analyzing the components:

TABLE 15.2   Guidelines concerning relations among workstations (Konz, 1990).

1. Decouple tasks.
2. Minimize material handling costs.
3. Optimize system availability.

Material handling cost = capital cost + operating cost

Operating cost = (no. of trips/yr)(cost/trip)

Cost/trip = fixed cost/trip + (variable cost/distance)(distance/trip)

Perhaps the key point of the above equation is that distance/trip is only one part of material handling cost—often a minor part.

Capital cost can be reduced a number of ways. You might buy used equipment instead of new equipment if the application life is short and equipment quality is satisfactory. Perhaps leasing is worth considering. Someone else provides the capital, which is an extra cost, but if capital is tight at the moment it may be worthwhile. Leasing and renting also are worth considering if your time of need is short. If you don't need the equipment full time, perhaps you can share it (and thus the capital cost) with some other user, perhaps even in your own organization. (See Section 1.4 in this chapter on minimizing idle capacity).

Another cost-reduction goal is to reduce the number of trips to zero. Can a trip be replaced by communication (e.g., an in-person sales or maintenance call replaced by a phone call)? Maintenance and repair costs might be reduced by higher quality equipment (a device that breaks down once/30 days instead of once/15). Can the number of trips/period be cut? For example, if product is moved from machine A to machine B once/hour but then waits a day to be processed, it is moved too often.

Fixed cost/trip is (1) information transfer and (2) start and stop. Information transfer costs (paperwork) are being reduced by computerization but still are very large. One possibility is to work by exception—the product is assumed to be moved from A to B unless otherwise noted. No forms need to be completed concerning the move. This is one of the primary advantages of flow line production.

Start and stop (pick up and put down; load and unload; pack and unpack) are costs that do not vary with distance moved. See Figure 6.4. Chapter 8 also discusses distance insensitivity of transport and communication—that is, costs vary little with distance moved. With automated material handling (see Chapter 18 for the discussion of automated guided vehicles and Chapter 17 for the related conveyor systems), distance has little effect on transportation cost. Distance is not completely irrelevant but it is not the only material handling cost factor.

2.3 Optimize System Availability

$$\textbf{Availability} = \frac{\text{uptime}}{\text{total time}} \qquad (2)$$

$$= \frac{\text{uptime}}{\text{uptime} + \text{downtime}}$$

$$= \frac{MTBF}{MTBF + MTR}$$

where

$MTBF$ = mean time between failures = **reliability**

MTR = mean time to repair

= **maintainability**

In calculating optimum availability for a system, use, for each alternative, life cycle costing rather than just initial cost. Life cycle cost = initial capital cost + operation cost over the system life + maintenance cost over the system life + downtime cost over the system life.

There are three strategies to improve availability: Increase reliability (increase uptime), increase maintainability (decrease downtime), and make loss of availability less costly.

2.31 Increase Reliability. Stronger units with less stress last longer. An example of a stronger unit would be a 3 hp motor used on a conveyor when only a 2 hp motor is needed; the additional capacity makes failure less likely (although with additional capital and operating costs). Less stress would be using a 2 hp motor but using special controls to reduce starting problems as well as operating the motor in a less stressful environment (e.g., cooler, cleaner). Another possibility is redundant components in the design—a parallel system instead of a series system. The battery in a gasoline truck may be duplicated so that if one fails, the other works; therefore life should be longer since there is less stress on the two batteries than on one. Redundancy in people generally is too expensive although in special circumstances, such as aircraft pilots, it is used.

2.32 Increase Maintainability. Downtime = fault detection time (time to find out the device doesn't work)

+ fault location time (time to find the problem)

+ logistics time (time to get repair parts)

+ repair time (time to fix unit)

A common strategy is to eliminate fault location, logistics, and repair time from downtime by using a standby unit, which takes over (perhaps with a loss of performance) while the unit is being fixed. This requires planning ahead for spare units. For people, whose downtime is due to excused or unexcused absences, cross-training other people is needed. The assumption is that the standby unit is replaced by the original unit after the original unit is fixed. There is a variation on the standby theme—**modularization.** Instead of repairing the unit, it is discarded and the standby unit replaces it. Small electric motors are a common example—they no longer are repaired but are just replaced. Modules permit relatively quick switching of the defective component and the standby unit, and the switching can be done by less skilled people. In addition, manufacturing techniques and costs may be such that it is neither economic nor even possible for local maintenance people to repair the unit.

2.33 Make Loss of Availability Less Costly. There are two possibilities: Have the downtime occur when you don't need the availability, or have partial function.

Downtime when you don't need the availability really is scheduled downtime (i.e., if you are going to schedule downtime, schedule it for the time that is most convenient). On the other hand, you may have expense for unnecessary maintenance, because the unit may not break down even if you don't do the maintenance (the "don't fix it if it ain't broke" philosophy). Check to see if the maintenance schedule is realistic for your conditions.

Partial function should be part of facility planning. "What if" a machine, process, vendor, or facility is not available because of strike, fire, flood, etc? Thus, two stations, each with an operator, probably will be more reliable than one station with two operators. Splitting an order between two vendors may be a better strategy than giving a larger order to one vendor. In theory, one large unit is more efficient than several smaller units since equipment utilization is better and more specialized labor and equipment can be used. In practice, monopolies (one large unit) are less desirable as the economies of scale are overcome by bureaucracy, lack of competition, and reduced downtime.

3 PHYSICAL DESIGN OF THE WORKSTATION

Table 15.3 gives the 14 guidelines concerning design of a workstation.

3.1 Avoid Static Loads and Fixed Work Postures

People are meant to move. That is, the body works better with a dynamic load than with a static load.

One static-load problem is manual material handling. The key concept is to minimize the lever arm of the object in relation to the centerline of the spine. Thus, carrying a load in a pack high on the back (so its center of gravity is

TABLE 15.3   Guidelines concerning physical design of the workstation (Konz, 1990).

---

1. Avoid static loads and fixed work postures.
2. Reduce cumulative trauma.
3. Set the work height at 50 mm below the elbow.
4. Furnish every employee with an adjustable chair.
5. Use the feet as well as the hands.
6. Use gravity; don't oppose it.
7. Conserve momentum.
8. Use two-handed motions rather than one-hand motions.
9. Use parallel motions for eye control of two-handed motions.
10. Use rowing motions for two-handed motions.
11. Pivot movements about the elbow.
12. Use the preferred hand.
13. Keep arm motions in the normal work area.
14. Let the small woman reach; let the large man fit.

---

close to the spine) is better than carrying it in the hands. However, in general, in the developed countries it is not economical to carry objects manually on a repetitive basis—use lift trucks, conveyors, and the like.

Another problem is the weight of various segments of the body. The head is about 7% of body weight, the trunk about 51%, each hand and arm about 5%, and each leg and foot about 16%. Thus, for a 160 lb person, the head weighs about 11 lb, the trunk about 82 lbs, one hand and arm about 8 lb, and one leg and foot about 26 lb.

For the head, design the workstation so that the head can maintain its normal downward tilt of 20–30°.

For the torso, use chairs instead of having the person stand. Occasional standing, however, is good. If standing is necessary, can a sit-stand chair be used?

For the hand and arm, support not only reduces fatigue but also reduces tremor and thus improves hand accuracy. Arms can rest on the table edge (wrist rests are a common computer keyboard accessory) or chair armrests.

For the legs, most support comes from a chair. If the worksurface is high (and thus the chair seat is high), a footstool or rail may be needed.

3.2 Reduce Cumulative Trauma

From 18 to 64 years of age, more people are disabled from musculoskeletal problems than from any other category of disorder (Putz-Anderson, 1988). **Cumulative trauma** gradually wears away at the body. Symptoms appear sooner for people with weak bodies and later for people with strong bodies, but, with enough trauma, even the strong eventually fail. Cumulative trauma is a function of frequency, recovery time, extreme joint movement, and force.

3.21 Frequency. A job is considered repetitive if basic (fundamental) cycle time is less than 30 s. Four improvement approaches are automation, mechanization, job enlargement, and job rotation.

Automation eliminates the person. Examples include using a robot to load boxes or a palletizer to place cartons on a pallet.

Mechanization retains the person but reduces cycles of a difficult operation. For example, in a given job, a label might be peeled from sticky-back tape. The pinch grip motion might be eliminated if a machine presented the label already separated from the backing tape.

Job enlargement increases the other work a person does, and therefore the elements causing cumulative trauma are done less frequently in a day.

Job rotation spreads the cycles over more people. It is best if the jobs rotated vary (i.e., shift from arm work to mental work or inspection work). For example, in loading a truck, the person unloading the conveyor and the person checking the invoice could switch every 30 min. Another common rotation technique is to use part-time workers. When multiple people use the same workstation, the engineer needs to make the station easily adjustable.

3.22 Recovery Time.

When people work for 8 h/day on a specific task, they have 16 h **recovery time**. There are 2 h recovery per 1 h of work. If they use specific muscles for only 4 h/day (for example if they rotate jobs or work part time), they have 5 h recovery per 1 h of work. If they work 12 h/day, they have 1 h recovery per 1 h of work. Note that a person can recover while still working—if the task done uses different muscles.

3.23 Extreme Joint Movement.

Limit **joint movement** by keeping movements at about the midpoint of the range of motion. Use the following recommendations:

- Don't lift the elbow. The goal is to keep the upper arm vertical.
- Don't reach behind your back. The designer should pay attention to supply and disposal areas at the workstation.
- Don't bend your wrist. The goal is to keep the wrist in a handshake position. Thus, orient jigs and fixtures to minimize wrist bending. Consider tilting tables, boxes, and conveyors.

3.24 Force.

Decrease the amount of force required by using motors and springs. Decrease the time the force is applied; for example, use a trigger to actuate a tool but don't require trigger force to operate the tool. A specific example of this is a gasoline pump with a latch and automatic shutoff (when the tank is full).

3.3 Set the Work Height at 2 in. (50 mm) Below the Elbow

Work height should be based on distance from the elbow, not distance from the floor. Optimum work height is the same for sitting and standing. Since the height of the elbow varies with sitting and standing, and since people vary in size, any workstation that puts the work a fixed distance from the floor—whether the person is sitting or standing, big or little—is bad design.

Note that work height is not table height and that most items (keyboard, items being assembled, etc.) have a thickness of 1 to 5 in. The optimum work height is about 2 in. (50 mm) below the elbow, although for keying work (computers) it is slightly above the elbow (Arndt, 1983).

There are three design approaches: change machine height, change elbow height, and change work height.

1. *Change machine height*: This approach is most useful if only one operator uses the machine for a long time, say one month. It is relatively easy to adjust the height of many benches or tables. See Figure 15.3. Conveyors also can be adjusted in height. See Figure 17.1. A two-level desk (i.e., one level for writing, another for the keyboard) is a common design solution in the office. The same bilevel concept can be used in the shop.

2. *Change elbow height*: Adjustable-height chairs are very common (see the next section). Standing operators can raise their elbow with a platform (often used by machinists) or even with several thicknesses of rugs (reducing fatigue of standing as well as increasing the elbow height).

3. *Change work height*: The height of the work surface can be raised or lowered. Barbers raise children's heads by having children sit on a board across the chair arms. Lowering items may help. Reaching over container sides often causes excess effort. Tip component boxes on their side, put them at a 45° angle, or cut out the side of the box. See Figure 15.4. When loading (as with groceries), lower the bottom of the container by using a low shelf.

3.4 Furnish Every Employee with an Adjustable Chair

The cost of an adjustable industrial chair is very low. Chair cost is entirely a capital cost because operating cost is zero. Assuming a price of $200/chair, a life of 10 years, and 2,000 working hours/year, cost/hour is $200/20,000 = 1 cent/h. If the comparison is between a good chair and a cheap chair, the cost is even less. A typical labor cost (wage + fringes) is $10/h. Thus, if output is improved only .1%, this is worth 1 cent/h. In an 8 h day, .1% is .001 (480 min) = .5 min = 30 s.

Figure 15.5 shows well-designed industrial chairs. The purpose of the chair is to permit the elbows to be adjusted to the work. This means the seat height must be adjustable.

FIGURE 15.3 Modular workstations can be adjusted for varying tasks and workers. Many different accessories can be used. Note the rounded front edge of the work surface and the fluorescent bench-mounted lights. Naturally the work surfaces, shelves, and lights are adjustable in height as well as being variable in type. (Courtesy of Isles Industries.)

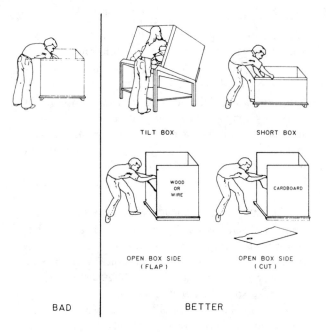

TILT BOX   SHORT BOX

WOOD
OR
WIRE   CARDBOARD

OPEN BOX SIDE   OPEN BOX SIDE
(FLAP)   (CUT)

BAD   BETTER

FIGURE 15.4 Avoid rounded backs. Eliminate the barrier of the box side. Note the back is given support by the arm's being placed on the box side.

One problem can be that the chair is difficult to adjust and so the adjustment is not made. Another problem can be that desks and tables are too thick, preventing the seat from being raised high enough because the top of the thighs hits the underside of the desk or table; therefore specify thin work surface thicknesses, and avoid center drawers and panels.

The second basic adjustment on a chair is the back support. It should adjust up and down as well as in and out. Many designs have an optional spring action which pushes the back support against the back as the user moves on the seat.

The seat itself should be padded, covered with a breathing fabric, be as wide as possible, and not be too deep. Deep chairs (beyond about 16 in.) don't permit some users to simultaneously put their feet on the floor and their back against the backrest. Seats that swivel tend to be more comfortable than fixed seats. Armrests normally are not needed in industrial or office chairs. Casters permit mobility but take away stability. Use 5 supports instead of 4 if the chair has casters.

3.5 Use the Feet as Well as the Hands
The foot can react as quickly as the hands, but, because of the construction of the ankle versus the wrist and the weight of the leg versus that of the arm, it is not as dexterous. Use pedals only when the operator is seated, since use of pedals while standing causes unnecessary strain on the body.

Pedals can be used for power (continuous or discrete) and control (continuous or discrete).

Pedals for continuous power generation no longer have many industrial applications, but discrete power applications are still used. Generally only one foot is used.

There seems to be no particular strength differential between the left and right leg; both legs together give about 110% of one leg. Design so that as few people as possible are excluded because of inadequate strength. Excluding people now has legal consequences due to the Americans with Disabilities Act. Note that populations differ (i.e., women versus men, young versus old, military populations versus civilian populations). Locate pedals so maximal force can be exerted—generally when there is a straight line between the pedal and the back support through the body. If the pedal is used repeatedly, use a wide pedal so the operator can have the option of using either foot.

For continuous or discrete control (such as auto accelerator pedal or machine cycling), bend the ankle by depressing the toe rather than depressing the heel or moving the entire foot and leg. Knee switches use a lateral motion of the knee; the advantage here is that the leg weight isn't moved.

3.6 Use Gravity; Don't Oppose It
Figure 17.1 shows how to use gravity with loading and unloading conveyors. Of course, gravity can move the item along the conveyor itself (chutes, wheel, and roller conveyors; see Chapter 17). Gravity also can be used to move material being worked upon. Examples are paint from a paintbrush, welding beads from a welding rod, solder on a solder joint. Use fixtures that permit the item to rotate and the operator to use a "downhand" position.

3.7 Conserve Momentum
Avoid unnecessary acceleration and deceleration, since it takes time and energy to accelerate/decelerate an arm, leg, or the body. In sports we "follow through" to give maximum velocity and accuracy.

In hand movements, try to make a motion curved rather than have sharp reversals. Sharp objects in the hand's path also cause abrupt decelerations, since the operator must be concerned about injury. Replace sharp edges with curved surfaces.

Momentum also applies to movement of hand carts. One estimate (Strindberg and Petersson, 1972) of maximum push force is .8 (body weight). *Force Limits in Manual Work* (1980) has a series of detailed tables of force limits considering several variables. Push force capability is greater when shoe/floor friction is high, there is enough space to position the body, and the push surface is between the hip and the shoulder (Chaffin et al., 1981). Snook and Ciriello (1991) reported 90% of U. S. male industrial workers can push (at a 95 cm height for 8 h) with an initial force of at least 34 kg and pull with an initial force of at least 32 kg; they can sustain a pushing force of at least 23 kg and pulling force of 24 kg. Since the push force versus pull numbers are not very different and safety on inclines may be a problem, for cart pushing and pulling keep the load lower than the worker at all times. Hand-pull carts should

EXECUTIVE   **MANAGER**   **TASK**   **COUNTER-HEIGHT**   **SIDE**   **SLED**

EOC1 *Executive Chair with arms, swivel tilt control, screw-type seat adjustment. Seat Height: 18"-22 1/2"*

MOC1 *Manager Chair with arms, swivel tilt control, screw-type seat adjustment. Seat Height: 18"-22 1/2"*

TOC1 *Task Chair with arms, swivel, spring back tilt, screw-type seat adjustment. Seat Height: 18"-22 1/2"*

CHC1 *Counter-Height Chair with arms, swivel, spring back tilt, screw-type seat adjustment. Seat Height: 25"-29 1/2"*

SOC1 *Side Chair with arms, memory return swivel, non-adjustable. Seat Height: 18"*

SLC1 *Sled Base Chair with arms, non-adjustable. Seat Height: 17"*

EOC2 *Executive Chair without arms, swivel tilt control, screw-type seat adjustment. Seat Height: 18"-22 1/2"*

MOC2 *Manager Chair without arms, swivel tilt control, screw-type seat adjustment. Seat Height: 18"-22 1/2"*

TOC2 *Task Chair without arms, swivel, spring back tilt, screw-type seat adjustment. Seat Height: 18"-22 1/2"*

CHC2 *Counter-Height Chair without arms, swivel, spring back tilt, screw-type seat adjustment. Seat Height: 25"-29 1/2"*

SOC2 *Side Chair without arms, memory return swivel, non-adjustable. Seat Height: 18"*

EOC3 *Executive Chair with arms, swivel tilt control, pneumatic seat adjustment. Seat Height: 18 1/2"-23 1/2"*

MOC3 *Manager Chair with arms, swivel tilt control, pneumatic seat adjustment. Seat Height: 18 1/2"-23 1/2"*

TOC3 *Task Chair with arms, swivel, spring back tilt, pneumatic seat adjustment. Seat Height: 18 1/2"-23 1/2"*

CHC3 *Counter-Height Chair with arms, swivel, spring back tilt, pneumatic seat adjustment. Seat Height: 25 1/2"-30"*

EOC4 *Executive Chair without arms, swivel tilt control, pneumatic seat adjustment. Seat Height: 18 1/2"-23 1/2"*

MOC4 *Manager Chair without arms, swivel, spring back tilt, pneumatic seat adjustment. Seat Height: 18 1/2"-23 1/2"*

TOC4 *Task Chair without arms, swivel, spring back tilt, pneumatic seat adjustment. Seat Height: 18 1/2"-23 1/2"*

CHC4 *Counter-Height Chair without arms, swivel, spring back tilt, pneumatic seat adjustment. Seat Height: 25 1/2"-30"*

OPTIONS

Order Code: ASC *Articulating Seat Control*

Order Code: GL *Steel Glides replace casters (Reduces seat height by 2 1/2")*

Order Code: RH *Hard Surface Casters*

Order Code: PAL *Polished Aluminum Base*

Order Code: LSU *Lumbar Support added to backrest*

Order Code: KT *Knee Tilt Control*

Order Code: LA *Loop Style, self skinned urethane armrests*

Order Code: SA *Compact, self skinned urethane armrests*

FIGURE 15.5 Chairs have many options. (Figure courtesy of Biofit Seating, Waterville, OH 43566.)

have a vertical plate ahead of the front wheel so a person's heel isn't run over by the wheel when pulling.

See Table 9.2 for floor and shoe coefficients of friction.

3.8 Use Two-Hand Motions Rather than One-Hand Motions

Work output is greater with two hands than with one hand. Physiological cost to the worker is greater with two hands than with one hand, but the increase is not as much as the increase in output. Thus, physiological cost/unit produced is lower when two hands are used.

Speed and accuracy of hand-arm motions were combined into one index, bits/s, by Fitts (1954):

$$I = \log_2 (A/(W/2)) \qquad (3)$$

where

I = information/move, bits

A = amplitude of move

W = width of target in movement direction

The validity of the formula has been demonstrated many times. Typical speed for the preferred (right) hand is 12.9 bits/s, for the left hand 11.7 bits/s, and for both hands working at the same time 21.2 bits/s. Langolf et al. (1976) reported the fingers can process 38 bits/s, the wrists 23, but the arms only 10. The limiting factor in hand-arm movements is not the ability of the brain to command or the ability of the eyes to supervise but the ability of the nerves and muscles to carry out the orders. The spirit is willing but the flesh is weak.

3.9 Use Parallel Motions for Eye Control of Two-Hand Motions

When using two hands simultaneously there are two alternatives:

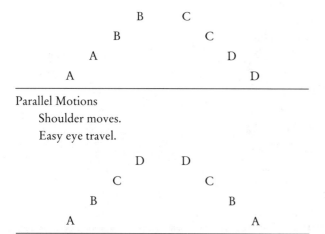

Parallel Motions
  Shoulder moves.
  Easy eye travel.

Symmetrical Motions
  Shoulder steady.
  Difficult eye control.

The general principle is, using both time/unit and physiological cost/unit as criteria, minimize the degree of spread between the hands rather than worry about the motion symmetry—that is, use parallel motions.

3.10 Use Rowing Motions for Two-Hand Motions

Two-hand motions can move in an alternating manner (like bike pedaling with the feet) or in a rowing motion (both hands moving out and back together). Alternation requires more shoulder movement and torso twisting than rowing. Thus rowing is the preferred technique.

3.11 Pivot Movements About the Elbow

Direction of horizontal motion affects speed, accuracy, and physiological cost of the motion. For speed and physiological cost, pivoting about the elbow is best, as is shown in Figure 15.6. Somewhat surprisingly, motions are more accurate when they are cross-body. Since the arm is pivoted about the shoulder, not the nose, put bins ahead of the shoulder, not the nose, for best efficiency.

3.12 Use the Preferred Hand

As was mentioned in Section 3.8, the preferred hand is about 10% faster for reach type motions. The preferred hand also is more accurate in movement and is 5 to 10% stronger. If the hand uses a tool, the advantage in time for the preferred hand is much greater—20 to 30% is a reasonable estimate.

About 10% of the population is left-handed. Thus, workstations should be designed and tools selected considering them also.

In general, have work come into a workstation from the operator's preferred side and leave from the nonpreferred side. The reason is that reach and grasp are more difficult motions than dispose and release. However, if the new item is obtained on the same side as the disposal, a body turn is eliminated.

3.13 Keep Arm Motions in the Normal Work Area

The **normal work area** is not a semicircle but is windshield-wiper shaped, because of the pivoting of the lower arm about the elbow. Figure 15.7 shows the normal work area for males and Figure 15.8 shows the normal work area for females. If motions are outside the normal reach area,

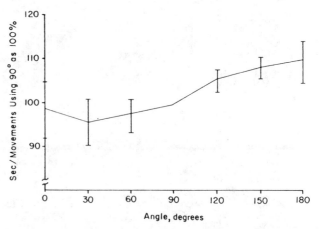

FIGURE 15.6 Movement time increases as the angle changes from 30° (where 0° = 3 o'clock). As a rule of thumb, add 5% more time when the movement is straight ahead and 15% more time when the movement is cross-body (Konz, 1990).

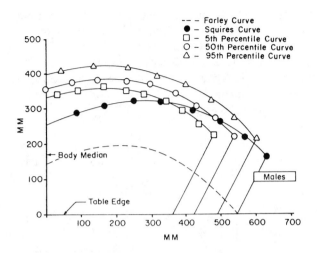

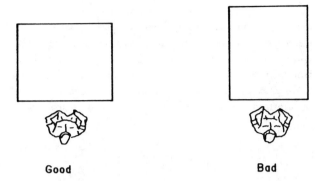

Good **Bad**

FIGURE 15.9   Pallet and bin orientation affects reach and move distances. In addition to initial pallet orientation, consider a turntable under the pallet. After half of the load is completed, rotate the pallet 180°. Long reaches and moves severely stress the back as well as decreasing productivity. See Chapter 20 for the NIOSH lifting formula.

FIGURE 15.7   Normal male work area (right hand) is windshield-wiper shaped.

extra effort and time are required. Figure 15.9 shows how extra reaching can be avoided by proper orientation of pallets and bins.

3.14 Let the Small Woman Reach; Let the Large Man Fit

Because women generally are smaller than men, design so that women can reach objects and controls while at the same time permitting men to fit within the workspace. The design should accommodate most of the user population. The problem is defining *most* and *user population*. *Most* may be 95, 99, or 99.9% of the user population. Figure 15.10 shows that the excluded people can be the upper end of the distribution, split between upper and lower, or lower end. Designing out part of the population makes the designer's job easier. For example, Russian tanks have a squat low profile, which restricts personnel to a maximum height of 66 in. You can design a tote pan that is too heavy for 10% of the population. However, after employees are selected, those who do use the pan carry

larger loads. You can design a machine such that only 1% of the people can't reach the controls or fit in the space.

Thus, the decision is between selection and making the product adjust to the person. In most industrial applications the laws against discrimination and social customs make it more and more difficult to justify a design that excludes anyone. Judges have been making decisions that weak backs are not grounds for not hiring a person—just change the job. Thus, in most situations you must make the jobs adjust (e.g., hoists, adjustable chairs, moving bin locations, etc.). Adjustable machines rather than selection is especially useful when multiple people use the same machine or workstation in a single day. Note also that just because 90% of the work force can lift a specific weight or reach a specific distance, that does not make it necessary to use this maximum weight or distance; less may be better.

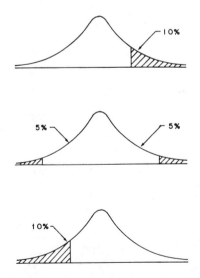

FIGURE 15.10 Excluding 10% can mean the upper percentile, or some from each end, or the lower percentile. For example, with a door, the upper percentile heights are excluded; with education requirements for a job, both low and high education might be excluded; with tote pans, the lowest percentile strengths might be excluded.

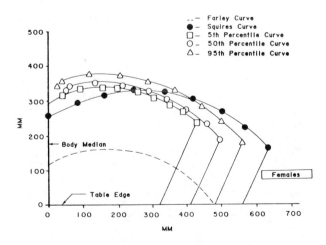

FIGURE 15.8   Normal female work area (right hand) is slightly smaller than for males.

User population also is becoming more broadly defined. The primary change is the changing sex/occupation stereotypes. Engineers no longer are just male; secretaries and nurses no longer are just female; welders, machinists, truck drivers, and others can be of either sex. Thus the designer must design a range from small woman to large man instead of small woman to large woman or small man

to large man. In addition, the immigration of people from Mexico and Asia to the United States has tended to broaden the range of population sizes.

Figure 15.11 and Table 15.4 give values for the U.S. adult population. Jurgens et al. (1990) provide **anthropometric** (human body measurement) **data** for 19 dimensions in 143 regions of the world.

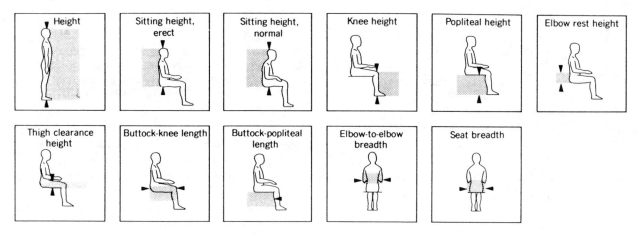

FIGURE 15.11 Key design dimensions are shown above. Table 15.4 gives numerical values for adults.

TABLE 15.4   Body dimensions (cm) of nude adult (age 20–60) civilians in the United States (Kroemer et al., 1986). See Jurgens et al. (1990) for 19 dimensions of populations of 143 countries.

Dimension	5th Percentile		50th Percentile		95th Percentile		Standard Deviation	
	Female	Male	Female	Male	Female	Male	Female	Male
Height: Standing								
Stature	149.5	161.8	160.5	173.6	171.3	184.4	6.6	6.9
Eye	138.3	151.1	148.9	162.4	159.3	172.7	6.4	6.6
Shoulder	121.1	132.3	131.1	148.2	141.9	152.4	6.1	6.1
Elbow	93.6	100.0	101.2	109.9	108.8	119.0	4.6	5.8
Knuckle	64.3	69.8	70.2	75.4	75.9	80.4	3.5	3.2
Height: Sitting								
Height	78.6	84.2	85.0	90.6	90.7	96.7	3.5	3.7
Eye	67.5	72.6	73.3	78.6	78.5	84.4	3.3	3.6
Shoulder	49.2	52.7	55.7	59.4	61.7	65.8	3.8	4.0
Elbow (from seat)	18.1	19.0	23.3	24.3	28.1	29.4	2.9	3.0
Knee	45.2	49.3	49.8	54.3	54.5	59.3	2.7	2.9
Popliteal	35.5	39.2	39.8	44.2	44.3	48.8	2.6	2.8
Thigh clearance	10.6	11.4	13.7	14.4	17.5	17.7	1.8	1.7
Depth/Reach								
Chest depth	21.4	21.4	24.2	24.2	29.7	27.6	2.5	1.9
Elbow-fingertip	35.8	44.1	42.1	47.9	56.0	51.4	2.2	2.2
Forward reach	64.0	76.3	71.0	82.5	79.0	88.3	4.5	5.0
Buttock-knee (sit)	51.8	54.0	56.9	59.4	62.5	64.2	3.1	3.0
Buttock-popliteal (sit)	43.0	44.2	48.1	49.5	53.5	54.8	3.1	3.0
Breadth								
Elbow to elbow	31.5	35.0	38.4	41.7	49.1	50.6	5.4	4.6
Hip breadth (sit)	31.2	30.8	36.4	35.4	43.7	40.6	3.2	2.8

DESIGN CHECKLIST: WORKSTATIONS

Table 15.1
Table 15.2
Table 15.3

REVIEW QUESTIONS

1. List the eight principles of job design.
2. Discuss specialization versus general purpose for manufacture of hamburgers or pizza.
3. Discuss the advantages of making several identical items at the same time for manufacture of hamburgers or pizza.
4. Give three tools that are multifunctional.
5. How is idle capacity reduced by use of pools?
6. Give three examples of how idle facilities can be reduced by scheduling.
7. For a specific person–machine system with which you are familiar, what is the labor cost/hour and what is the machine cost/hour?
8. Give two examples of double tooling.
9. Give three examples of how scheduling work reduces costs.
10. Discuss how temporary workers or permanent part-time workers could be used in a specific organization.
11. Give four advantages to management of job sharing.
12. Give two examples of how stimulation can be added to a task; give two examples how stimulation can be added to an environment.
13. What is the concept of working rest?
14. For a lab report you have previously written, calculate the writing grade level you used.
15. Give the three principles concerning relations among workstations.
16. Why is distance not very important in material handling?
17. Define availability in terms of reliability and maintainability.
18. Give the four components of downtime.
19. For a specific job, describe static loading.
20. List the three design approaches to obtaining optimum work height.
21. What would a good industrial chair cost/hour? What percentage is this of the wage cost of the person using the chair?
22. If a chair has casters, how many supports should it have?
23. How much does each of the following items weigh: your head; your trunk; one hand and arm; one leg and foot?
24. What is the advantage of a knee switch?
25. Sketch the locations of pallet and conveyor for loading a conveyor and for unloading a conveyor.
26. Where should the operator be when moving a cart up a ramp?
27. Calculate the information in bits for a 16 in. move to a 4 in. target.
28. Are parallel or symmetrical motions better?
29. For a reach with the right hand, what angle takes the least time?
30. Why should work enter a workstation from the operator's preferred side?
31. Sketch the shape of the normal work area for the right hand.
32. Should the designer make the product adjust to the person or use personnel selection? Why?

PROBLEMS AND PROJECTS

15.1 For a specific job with which you are familiar, how can idle capacity be minimized? Discuss fixed costs and relative costs. What does the worker think of your ideas? What does the supervisor think of your ideas?

15.2 Determine the writing grade level of a report you have written. Use a sample of about 200 words. Calculate its grade level. Rewrite the section and calculate its new grade level. At the end of each sentence in both passages, indicate the number of words in the sentence. Underline or highlight each word with 3 or more syllables.

15.3 For a specific product in a specific factory, determine the annual material handling cost between two specific operations. Give assumptions. What would the cost be if the distance moved increased by 2? By 20?

15.4 Evaluate the work height of a specific job. What does the worker think of your ideas? What does the supervisor think of your ideas?

15.5 Evaluate the chair of a specific worker. What does the worker think of your ideas? What does the supervisor think of your ideas?

REFERENCES

Arndt, R. Working posture and musculoskeletal problems of video display terminal operator—Review and reappraisal. *Am. Ind. Hygiene Assoc. J.*, Vol. 44, 6, 437–446, 1983.

Chaffin, D., Olson, M., and Garg, A. Volitional postures during maximal push/pull exertions in the sagittal plane. *Proceedings of 25th Annual Meeting of the Human Factors Society*, Rochester, N.Y., 91–95, 1981.

Fitts, P. The information capacity of the human motor system in controlling the tolerance of the movement. *J. of Experimental Psychology*, Vol. 47, 6, 381, 1954.

Force Limits in Manual Work. Guildford, Surrey (England): IPC Science and Technology Press, 1980.

Jurgens, H., Aune, I., and Pieper, U. *International Data on Anthropometry.* Geneva, Switzerland: International Labor Organization, 1990.

Konz, S. *Work Design: Industrial Ergonomics*, 3rd ed. Scottsdale, Ariz.: Publishing Horizons, 1990.

Kroemer, K., Kroemer, H., and Kroemer-Ebert, K. *Engineering Physiology.* Amsterdam: Elsevier, 1986.

Langolf, G., Chaffin, D., and Foulke, J. An investigation of Fitts' law using a wide range of movement amplitudes. *J. of Motor Behavior*, Vol. 8, 2, 118–128, 1976.

Nollen, S. Work schedules. Chapter 11.8 in *Handbook of Industrial Engineering*, G. Salvendy, Ed. New York: Wiley, 1982.

Putz-Anderson, V., Ed. *Cumulative Trauma Disorders.* London: Taylor and Francis, 1988.

Snook, S., and Ciriello, V. The design of manual handling tasks: revised tables of maximum acceptable weights and forces. *Ergonomics*, Vol. 34, 9, 1197–1213, 1991.

Strindberg, L., and Petersson, N. Measurement of force perception in pushing trolleys. *Ergonomics*, Vol. 155, 4, 435–438, 1972.

Su, J., and Konz, S. Evaluation of three methods for inspection of multiple defects/item. *Proceedings of 25th Annual Meeting of the Human Factors Society*, Rochester, N.Y., 627–630, 1981.

MATERIAL HANDLING AMONG WORKSTATIONS

CHAPTER PREVIEW

Unit loads reduce costs. The standard base is the pallet, which comes in many types and sizes. Pallet boards and slipsheets (which save space) are popular for shipping. Loads are stabilized primarily with stretch wrap and strapping. There are many types of unit containers. Bar codes are the most popular automatic identification system.

CHAPTER CONTENTS

1 Unit Load Concept
2 Pallets
3 Pallet Boards and Slipsheets
4 Load Stabilization
5 Unit Containers
6 Bar Codes

KEY CONCEPTS

bar codes

double-face pallets (versus single face)

four-way pallets (versus two way)

intermodal

manufacturer's problem (versus distributor's problem)

pallet board

reusables

scanners (fixed, moving beam)

skid

slipsheet

stabilizing the load

stack only (versus nest only)

strapping

stretch wrap

stringer

tote pan

unit containers

use of the cube

1 UNIT LOAD CONCEPT

Long ago it was discovered that it was more efficient to handle a bunch of things in mass than one by one. The challenge is to implement the concept. We will assume the product is in its primary package (the one the customer sees) and will concentrate on the distribution packaging—the packaging that carries the product from the original producing location to the retail customer. The criteria, expressed as a percentage of the product cost, are transportation cost and protection cost.

There are at least five subgoals:

- minimize handling costs
- use standardized containers
- make efficient use of the cube
- minimize material use
- protect product from damage

Standard containers should be **intermodal**, that is, compatible with factory, shipping medium, warehouse, and retail outlet. Standardization also reduces the number of different sizes, with resulting economies. Efficient **use of the cube** (use of all three dimensions of the space) should be both in storage (at factory, warehouse, and retail) and in transport (within and between locations). Minimum materials use involves frugal use of packaging materials (to reduce cost as well as shipping weight) as well as reuse of pallets and containers. Protecting the product from damage involves not just the weather and leaks, but also accidents and pilferage.

Most unit loads use pallets, although slipsheets are becoming more popular; there are a number of load stabilization techniques. Some applications use unit containers.

2 PALLETS

2.1 Advantages of Pallets
Pallets are used to facilitate handling, movement, and storage of goods in all phases of the distribution cycle. Cost savings come primarily from the use of the cube of storage, reduced handling costs, reduced paperwork costs, and reduced accidents.

2.2 Pallet Design
Pallet design must consider material, types, and other design features.

2.21 Material.
The most common material for pallets is wood, although plywood, plastic, corrugated paper, steel, and wire mesh occasionally are used for special applications. In a typical wooden pallet, the lumber price accounts for 50 to 60% of the finished pallet price. Both hardwoods and softwoods can be used. Hardwoods are used mostly in the eastern part of the United States. In the West, softwood pallets are more common. Wood species

are classified into three groups: Class A, low density; Class B, medium density; and Class C, high density. Woods within a class are considered interchangeable. Pallet users normally don't specify dry lumber; the cost of air- or kiln-dried lumber is high.

2.22 Types.
Figure 16.1 shows four typical pallets. Pallet dimensions are given in inches; the length is given first, then the width, which is the dimension parallel to the deckboards. There are 19 standard ANSI sizes with the 48 × 40 in. accounting for 27% of all pallets (Wooden Pallets, 1978); no other size has over 5%. A 40 × 48 in. can be placed two abreast across the 48 in. dimension in railroad freight cars and two abreast across the 40 in. dimension in most trucks. If possible, use a popular size pallet for all applications.

Table 12.3 gives truck dimensions. Boxcars typically are 102 in. wide, 108 in. high, and 40 or 50 ft. long.

Four-way pallets give better vehicle floor space utilization than **two-way** pallets. For example, with four-way pallets, one row can be loaded with the 48 in. dimension and the next with the 40 in. With two-way pallets, the fork truck (since it can enter only one direction) cannot place the pallets in this manner. Generally, select pallets so the unit loads do not overhang the deckboards; impacts thus are absorbed by the pallet, not the product. If there is an overhang, keep it 2 to 2.5 in. or less.

2.23 Other Design Features.
Use the following methods to obtain better pallet life:

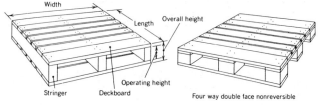

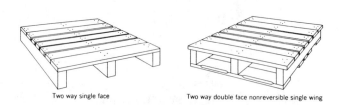

FIGURE 16.1   The most popular wooden pallet is two way, **double face**, and nonreversible. Two decks form the top and bottom surfaces. The bottom deck adds strength and helps distribute tiered loads. For four-way entry (partial), add notches in the **stringers** for lift truck entry on the stringer side. There are a variety of other common designs: **single face**, four-way block, two- and four-way reversible, wing type, and specialty designs. A single-wing pallet has the deck boards extend past three stringers on one side, and a double wing extends past on both sides. A **skid** is a heavy-duty pallet with two stringers but no bottom deck. Typical pallet overall height is 5 in. A common weight is about 50 lb.

1. Use mass, since the ability of an object to absorb stress is inversely proportional to its mass. Heavier deck boards and stringers give longer life. The load often is volume limited, not weight limited. In addition, pallets are designed for mechanical handling systems (not man handling) so weight is no penalty.

2. Butt the first and second end deckboards together to better resist the forks. If some wood on the pallet is of better quality, it should be used on these two boards. Pallets for bagged loads should have a maximum deckboard spacing of 1.5 in. or the bags will sag and be damaged by the fork.

3. Use quality fasteners (bolt and nut, drive screws, annular ring nails)—common nails are not acceptable. Nails and staples driven by electrical or pneumatic equipment are pushed into the wood rather than driven. Thus, they are thinner than hammered nails and split wood less. Coating gun-driven nails with plastic polymers increases their holding power about 1.5 times; threaded nails have about 10 times the holding power of common nails, after the lumber dries. Staples split wood less than nails and can be driven closer to the edge (Hand Nailing, 1981). Nails should be countersunk at least 1/16 in. so they do not snag or rip the load. Use 2 to 4 staggered nails at joints, depending on the deckboard width.

4. Seal the ends of blocks and stringers with bright colored paint to help drivers align their forks and also seal out moisture.

5. Use wood preservatives when the wood is subject to wide variations in moisture or to insect infestation.

6. Design pallets for use conditions. A pallet designed for rack use needs closer dimensional control and has different stresses on it than a floor pallet. A pallet with a 2,000 lb load stacked four high should be designed for 8,000 lb, not 2,000.

7. Chamfer (at least 20 in. long and at 35°) bottom-edge deckboards to reduce damage from fork entry.

8. Brand or stencil each pallet with the supplier's name and purchase date to allow you to relate supplier to pallet performance.

9. Color-code pallets (paint a dot on the end) so the lift truck driver can easily match a pallet with a rack opening size.

The National Wooden Pallet and Container Association (1619 Massachusetts Ave. NW, Washington, DC 20036) has a PC program which evaluates pallet design features and predicts the number of one-way trips before failure. See Table 16.1 for pallet costs. Box 16.1 discusses two plastic pallet applications.

2.3 Pallet Patterns   Figure 16.2 shows items can be loaded onto pallets in different patterns. The simple block pattern has the disadvantage of no cross-stacking to hold the load together. The load can be held together with sheets of paper between layers; by strapping; by shrink, stretch of

TABLE 16.1 Pallet summary (Pallets, 1991).

Type	Cost ($)	Application Note
Corrugated paper	3–10	Expendable, low cost; usually used in shipping
Wood	3–25	Economical, reusable pallet; repair readily at ⅔ cost of new pallet
Plywood	5–50	Continuous uniform load support; relatively light for ease of handling
Plastic	20–100	For product protection, uniform tare weight; steam clean for sanitary applications
Rubber	—	Use in spark-free environments

BOX 16.1   *Plastic pallets save money for Goodyear, Kohler*

Goodyear

A special design pallet costing five times as much as the wood pallet it replaced was economic for Metal Products Division of Goodyear Tire. The pallet operates in a closed-loop system between Akron and Caterpillar Tractor plants in Illinois. The product on the pallet is steel wheels.

The previous method was two wheels, a separator, two more wheels, and then strapping the load. Pallets were stacked two high. Fork trucks moved one pallet at a time (i.e., four wheels). The wooden pallets cost $5.25 apiece and averaged three trips/pallet.

The new plastic pallets cost over $30 apiece. However, they have a molded indentation to hold the wheels so strapping is eliminated. Elimination of strapping time and materials saved $4.72 a trip. There are six wheels/pallet so each fork truck move transports six wheels. The plastic

pallets have proved durable, as 98% were still in service after two years of use.

The $4.72 savings in strapping and materials paid for the special pallet in eight trips. In addition, there was the shipping saving (one third more wheels per truckload) and the fork savings (six/move versus four).

Kohler

Kohler used plastic pallets to ship parts within its Wisconsin facility. The project originally was justified on the $24 plastic pallet lasting over 5 years compared with the $6.35 wooden pallet lasting .75 years. However, some additional benefits were (a) weight counting of inventory became feasible (wooden pallet weight may change from 70 to 40 lbs as it dries out), (b) the pallets only weigh 23 lbs (leading to easier manual handling), (c) there are no nails or splinters to cause injury, and (d) they nest 50%.

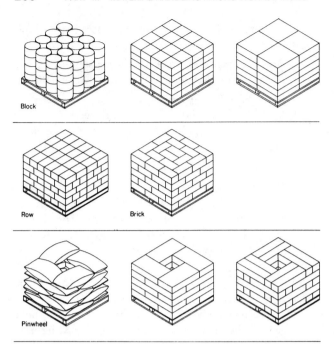

FIGURE 16.2 Pallet patterns are block, row, brick, and pinwheel. Block patterns are for items of equal length and width. The simple block pattern usually has an open center. The row, brick, and pinwheel patterns are for items of unequal length or width. The load is stabilized by alternating the layers 90°. The pinwheel usually has an open center. If the desired palletizing pattern for a specific product is known to the operator, there can be considerable reduction in loading time.

net films; by glue; or by combinations of these methods. Special glues have high adhesion-in-sheer strength to hold the load together, but low adhesion-in-tension to permit the box to be lifted straight up. The row or brick pattern requires a package with a length equal to a multiple of the width; alternating layers 90° creates cross ties for stability. The pinwheel pattern can be used for loads that do not fit a row or brick format; the disadvantage is the open center, which permits the load to shift and doesn't fully use the cube.

The pattern on the pallet also can be determined on the basis of maximizing the number of cartons/pallet. This is a relatively difficult optimization problem because of the many different possible patterns, the influence of carton sizes and shapes in combination with pallet sizes and shapes, and the need for a relatively simple solution that the pallet loader can understand. Downsland (1987) gives an algorithm for the **manufacturer's problem**—loading identical size boxes onto a pallet. Hodgson et al. (1983) and Chen et al. (1991) discuss the **distributor's problem**—loading various size cartons onto a pallet. A key step is to place the carton with the largest area in a corner.

Plastic inserts can be used between the layers of items on the pallet. If these inserts have indentations, they can guide the machine operator so all items are placed in a specific orientation. This standardized orientation may be required if a robot unloads the pallet (it helps humans also).

Pallet loads depend on the pallet size, the height the load is stacked on the pallet, and the load density. A typical load height is 4 ft. Thus, a 42 × 48 in. pallet holds a volume of 3.5 × 4 × 4 = 56 ft³. Load density depends on the density of the individual package and the number of packages in the pallet cube. For example, a vacuum cleaner package might have a volume of 2.2 ft³/package and a weight of 24 lb/package. A 42 × 48 in. pallet would hold about 56/2.2 = 25 units; weight would be about 25 packages × 24 lb/package = 610 lb. If the load did not perfectly fit the cube (most don't), and only 85% of the cube were used, then there would be about 25 × .85 = 21 packages. Package weight would be 21 × 24 = 504 lb. Assuming a typical pallet weight of 50 lb, pallet plus load would be 554 lb. For most products, total loads are under 1,000 lb, although a few go up to 2,500 lb. (For reference, 56 ft³ of water weighs 56 × 62.4 lb/ft³ = 3,500 lb.)

2.4 Manual Loading and Unloading   The following recommendations come from a study by Ray Oakes at Rapistan.

Vertically, the prime working zone is between the knees and the chest. See Figure 17.1. Thus, the operator's feet should be about 8 in. below the bottom of the pallet; this probably will require a platform for the pallet (such as another pallet or a scissors lift). For loading pallets, place the conveyor at hip height; for unloading pallets, place the conveyor at knee height. Variable-height conveyors are good. For maximum productivity, provide foot and leg room beneath the conveyor.

Horizontally, the important thing to emphasize is avoiding reaching more than 24 in. with a load. This means spotting a pallet so the operator reaches across the narrow dimension. See Figure 15.9. A backstop on the pallet can reduce loading problems. The aisle between a pallet and the conveyor should be about 30 in.

2.5 Pallet Maintenance   Preventive maintenance includes equipment on the fork truck. Bumper guards on the heels of the lift truck fork tynes absorb the shock; the load and the top deck do not. A level bubble on fork truck masts helps drivers keep the tynes parallel to the floor and thus reduce broken endboards. For low lift trucks, the pallet jacks sometimes are lifted up when the wheels rest on the top of the bottom board; a stop mounted between the forks will eliminate this problem of ripping off the deckboard.

For repair, three rules of thumb are (1) don't repair a pallet with a fractured stringer, (2) repair cost should not exceed 50% of the cost of a new pallet, and (3) pallet repair specialists seem to do a better job than in-house workers (Burke, 1975). If pallets can't be repaired, burn them for their heat value or haul them to a landfill. If hauling to a landfill, shred them first to reduce their volume and thus the number of trips required.

Pallet storage presents a severe fire hazard. Store pallets outside, away from storage buildings. If they are in buildings, keep the height less than 6 ft. and store them under sprinklers.

The basic problem of pools is that good pallets do not come back; thus, people tend to use a low-quality wooden pallet (that lasts 25 trips) instead of a high-quality one (that lasts 100 trips). A system which favors high-quality pallets, developed in Canada, is gaining favor in the United States (Auguston, 1991). A third party uses a high-quality pallet, which is rented to the manufacturer, who pays a rental fee and a deposit. When a firm receives a shipment, it is billed for the pallet deposit. When the firm returns the pallet to the rental deposit, it receives the deposit refund. The rental company repairs the pallet (if necessary) before returning it to use.

3 PALLET BOARDS AND SLIPSHEETS

Pallets cost money and take up space. **Pallet boards** are pallet-size flat boards 3/4 to 1 in. thick. They cost less and use up 1 in. instead of 5 in. Since they require special handling, they tend to be slave to the system—that is, they are not shipped by truck or railcar. One justification for a pallet board is that it serves as a smooth bearing surface on a roller conveyor. The pallet board is placed on the conveyor and then a pallet or skid rides on the flat surface instead of directly on the rollers. At the takeoff point, special slots are placed on the conveyor. Another use for pallet boards is in rack storage. If the height of a typical slot is 60 in., then 5 in. is needed for the pallet. Cutting this to 1 in. gives a savings of 4/60 = 6.7%. You can either cut the slot height down or stack the material on the board higher. Pallet boards do require a rack with no front cross beam; the pallet board rests on the beams running perpendicular to the aisle (like a drive-in rack). A third use of pallet boards is to store skids or small pallets in a rack system. The pallet board serves as the flat, rigid base.

Pallets cost money and take up space. Slipsheets cost 10 to 30% of a pallet cost and have 1% to 5% of the weight and volume.

The basic **slipsheet** is a thick piece of paper, corrugated fiber, or plastic upon which the load is placed. Originally, sharp forks chiseled under the paper to pick up the load. Now tabs on the sheet are grabbed by special attachments and the sheet and load are pulled onto a platform on the vehicle; the load is removed by a pusher plate. A gripper bar on the truck is used to pull the sheet while simultaneously pushing with the plate (as when putting a load on the floor). See Figure 16.3. The slipsheet paper now comes in different constructions and strength; there also are plastic slipsheets (both one-way and reusable). Tabs on two or three sides permit moving the load from two or three directions.

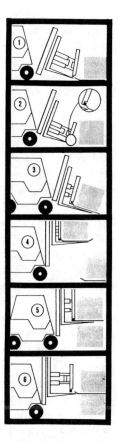

FIGURE 16.3   Slipsheets are used to pull the load onto a lift truck attachment, which then supports the load during lifting and transport. (1–2) Grippers, located at the bottom of the face plate, grasp the slipsheet tab. (3–4) The load then is pulled to the platens. (5–6) To set the load down, the driver opens the gripper and backs up the same rate as the push plate is extended.

Slipsheets are very popular in truck and railcar shipping as they reduce wasted space and weight (many shipments are cube limited rather than weight limited). There also is no return problem.

Slipsheet disadvantages are the slower handling than with pallets and greater load damage within the facility. Also the grabber bar and pusher plate are ahead of the lift truck wheels, and so the rated vehicle load capacity is reduced substantially. Thus, many organizations use pallets for internal handling and slipsheets for shipping. Place a slipsheet on the pallet before loading the pallet in production. Then use the pallet and forks within the plant and the slipsheet and a pusher attachment for shipping.

4 LOAD STABILIZATION

There are four common ways of **stabilizing the load** on a pallet or slipsheets: strapping, shrink wrap, stretch wrap, and adhesives.

4.1 Strapping

There are eight general application areas: closure, unitizing, bundling, pallet loading, skidding, baling, securement, and reinforcement. Straps are steel or plastic. Some advantages of plastic strapping rather than steel are reduced strapping time, lighter weight, safety (fewer breaks and no sharp edges if there is a break), and easier disposal.

Closure with **strapping** allows a container to be reused, gives some reinforcement, and can be used for telescoping containers. Unitizing brings together a number of "bare" items (bricks, lumber, tires) into a uniform shape. There is no rigid base, such as a pallet. Bundling is similar to unitizing but the shape is irregular (e.g., tubing, bar stock, extrusions). Palletizing is putting a load of items—generally corrugated cartons or bags—on a pallet. With steel strapping, tension does not go around sharp corners well, so use corner protectors. However, as pallet loads settle during shipment, the nonshrinking strap tends to become loose and the load unstable. Thus, plastic strapping, which shrinks, has become more popular. Skids are heavy-duty pallets with two stringers but no bottom deck; steel strapping generally holds the heavy loads (paper, aluminum, or steel in coils or plates). Baling holds compressible materials (cotton, fibers, scrap) together in a cube. Securement makes sure a product stays where you want it. One application is around coils to keep them from unwinding. Another application is in railcars to keep poles and machinery secure within the car. Reinforcement is use of strapping to increase a container's structural strength; it also can serve as a handle.

4.2 Shrink Wrap

In shrink wrapping, a film or bag is placed over the object; when heat is applied, it shrinks and thus holds the object. The original machines used considerable energy to generate the heat, but more recent units are more energy efficient. However, most shrink wrap applications are being phased out in favor of stretch wrap.

4.3 Stretch Wrap

Compared with shrink wrapping, **stretch wrapping** has lower material, labor, and energy costs. In stretch wrapping, a film is wound around the object while the film is stretched. Then, after the film end is secured, the film length contracts and the load is held. Stretch wrapping can be done on four, five, or six sides of the load, in single or multiple passes, with nonpermeable film or with stretch netting, which permits passage of air through the holes. There are a large variety of application machines; films come in many different formulations and characteristics.

Both shrink and stretch wrapping allow mixing multiple sizes and shapes of containers into a unit load; it is not necessary to sort boxes into neat stacks for each size, load pallets with the same size box, and square the corners. Thus, a customer may get a mixture of items in one shipment instead of getting multiple homogeneous loads. This substantially reduces labor and shipping costs for small orders.

4.4 Adhesives

Some adhesives have high strength in shear but low strength in tension. Adhesive drops on the four corners of a cardboard carton (or a bead down a bag center) will hold that carton to another carton placed on top of it. However, when a person lifts the top carton, the bond is weak and breaks. There are both hot and cold melt varieties of adhesives.

5 UNIT CONTAINERS

There are a variety of **unit containers**, including welded wire mesh containers, wooden pallet containers, and tote pans. See Table 16.2 for a summary of containers.

5.1 Number of Containers

Reusable containers are used for work in process around the plant, but if this is the only requirement you list, you will have forgotten many other containers:

- containers holding stock in the rework area
- containers holding stock to be scrapped
- containers holding stock or materials that are obsolete
- containers holding supplies
- containers at vendor's or customer's facilities (or in transit)
- containers being repaired
- container spares (work in process of containers)

For flow line production, containers tend to be specialized, and sufficient containers of each type are needed to meet peak production demands (plus the uses listed above). For batch production, general-purpose containers are used; the number of containers for work in process should depend upon average production rather than peak, since not all products will peak at the same time.

5.2 Welded Wire Mesh Containers

The most popular welded wire mesh container is 44 × 54 × 40 in. high, since it maximizes use of truck and railcar cubes. For railcars, two across (when turned lengthwise) fit a boxcar width; stacked three high they fit its height. For trucks, two across (turned widthwise) fit its width; stacked two high they fit its height. Figure 16.4 shows five different types of wire containers. Most units have 2 × 2 in. mesh openings with .26 in. dia steel wire, although smaller or larger mesh openings are available.

5.3 Plastic Shipping Containers

These returnable containers are for shipments from a supplier to a factory, from a factory to a warehouse, or from a warehouse to a retailer. The basic choice is between one-way containers

TABLE 16.2   Container summary (Containers, 1991).

Type	Cost ($)	Design Factors	Application Notes
Shop pans (metal)	10–50	Hopper front and rear, usually with carrying handles; variety of sizes	For handling heavy, small parts; for dipping or draining applications, use expanded or perforated metal designs
Storage bins, hopper front	.5–5	4–20 inch long; 4–10 inch wide; 2–10 inch high	Organize and store small to medium size items; mount on floor, shelves, or racks
Modular containers (plastic)	3–6	Stackable; fit standard racks and carts	Handle small parts and assemblies; good for organizing into families or groupings
Tote boxes	7–14	Often fiberglass; capacities to 3,000 lb	For high total weight (castings, stampings, small heavy parts)
Wood boxes	25–65	Equipped with lids; pallet or skid bases (a 48 × 30 × 30 skid container is $60)	High impact resistance of wood is good for parts handling; readily repaired; commonly used for soft goods and textiles
Wirebound boxes	—	Wooden slats attached to base (wood, plastic, or metal) bound with wire	Handling and shipping of implements assemblies, components; readily built or repaired in the field
Wire: collapsible	–100	High strength/weight ratio; hinged access gates	Hold large, heavy, or irregular shape items; high product visibility; compact storage; maximum stack of 3 or 4
Wire: rigid	140–175	Have corner posts for rigidity (e.g., 3 × 3 in. angle)	Heavy duty for high stacking with order-picking vehicles; loads to 6,000 lbs
Corrugated steel	75–300	Basic unit is a two-piece box welded to platform base or corrugated bottom; variety of solid and perforated lids, stacking legs, and attachments	Heavy-duty handling (castings, stampings, forgings); use with tilting stands and dumping attachments

(throwaways) and reusables. Reusable containers may be corrugated, fiberboard, or plastic. See Table 16.3 for an example calculation. In addition to the economic advantages, there are ecologic advantages to reusable containers. Reusable plastic containers minimize waste but do eventually need replacement. They are usually made of high-density polyethylene (HDPE), however, and can be recycled by returning them to the manufacturer. The reusable containers can have special features such as handles, open or solid bottom and sides, permanent identifiers on the container, etc.

Reusables need to have the container returned. This requires coordination among suppliers, manufacturers, warehouses, and retailers. For example, the shipping manifest needs a two-part column allowing the driver to record the number of containers delivered and the number of containers picked up. Certain color containers or lids may be used to identify companies or products.

A practical reusable container program also requires standardization of containers and reduction to a reasonably small number.

The containers can be stack only, nest only, stack and nest, or collapsible.

Stack only have little or no draft angle. Benefits are maximum use of the container footprint and inside

container cube. The primary disadvantage is the space required for return transportation.

Nest only containers (which generally stay within a facility) are often stored on a supporting surface (static shelving, flow rack, carousel). To be stacked, they need a lid or a bail added. Empty containers can be nested in storage.

Stack and nest containers have a draft angle for nesting; to support the container above they have a bail, a 180° orientation, or a lid. Attached lids save time in handling; separable lids permit easy access to the container interior; recessed lids improve stacking rigidity but limit cube; a high-cube lid improves container capacity. Lids also reduce product damage and theft. Stack and nest containers give good use of the cube for freight; the nesting is good for return freight and for storage of empties.

Collapsibles have no draft angle, and so they maximize use of the container footprint and the inside container cube. For returns, they collapse. A collapse ratio of 3 or 4 to 1 is equivalent to nesting containers.

5.4 Wooden Pallet Containers
These containers use a pallet as a base; the bin has closed sides and no top, the crate has slatted sides and no top, and the box has closed sides and a top (Wooden Pallet, 1980). Sides can be

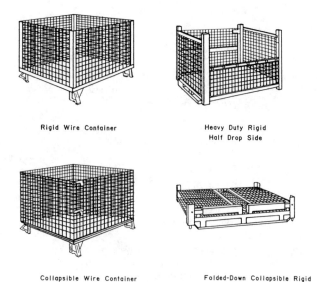

Rigid Wire Container

Heavy Duty Rigid
Half Drop Side

Collapsible Wire Container

Folded-Down Collapsible Rigid

Wrap-Around

FIGURE 16.4 Wire containers come in many varieties (Welded Wire, 1976). Collapsible containers are the most popular since they are the least expensive and fold down to about 30% of their original height. Rigid containers have more strength, durability, and stacking capacity. For even more strength, use heavy-duty rigid models. The version shown has a half drop side to reduce back strain on the person loading and unloading it; gate sides let the complete side open like a door. The collapsible rigid combines strength and collapsibility but at a higher cost. The wrap-around uses any convenient pallet as a base so only the foldable wrap-around needs to be shipped back. (Figure courtesy of Material Handling Engineering.)

collapsible or demountable (sides returned but not base pallet). Figure 16.5 shows a conventional and an interlocking corner. The federal government has done extensive investigation of what works best.

Boxes, Pallet Type, Materials Handling, Wood, Wirebound (MIL-B-43096); Box, Wood and Cleated, Skidded, Load and Bearing Base (MIL-B-26195A); and Pallet,

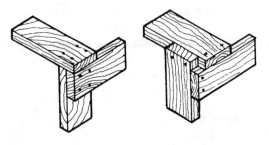

FIGURE 16.5 Interlocking corners (right) are stronger than noninterlocking ones (left) (McElroy, 1980).

TABLE 16.3   Cost analysis of reusable plastic container versus reusable corrugated cardboard for interplant shipment (How To, 1992).

Assumptions
  Project life = 5 years
  Inflation = 5%/year
  Trip = round trip
  260 work days/year
  Ship 1,000 boxes of 4 cubic ft each day
  System has 1,000 boxes at location A, 1,000 in transit, and 1,000 at location B.
  Corrugated life = 30 trips
  Corrugated cost/box = $3.75 (does not include labor cost for setup and breakdown)
  Plastic containers need 10% replacement/yr
  Plastic container cost/container = $17.22

Corrugated
  Annual shipments: 260 days/yr × 1,000 trips/day = 260,000 trips/yr
  Annual container needs: 260,000 trips/30 trips/box = 8,667 boxes
  Annual capital cost: 8,667 boxes × $3.75 = $32,501

Plastic
  Initial capital cost = 3,000 containers × $17.22 = $51,660

Comparison

Year	Reusable Corrugated	Reusable Plastic	Savings Plastic
1	32,501	51,660	(19,099)
2	34,126	5,424	28,702
3	35,832	5,695	30,137
4	37,624	5,980	31,644
5	39,505	6,279	33,226
	179,588	75,038	104,610

Simple payback = $104,610/51,660 = 2.0 years

Materials Handling, Box Type, Wood, Flight Duty (MIL-P-4894) are available from

Superintendent of Documents
U. S. Government Printing Office
Washington, D. C.

For more information on pallets and containers, order Care for Wooden Pallets Can Control Maintenance Cost to You and Pallets and Palletization. Contact:

National Wooden Pallet and Container Association
1619 Massachusetts Ave. NW
Washington, DC 20036

5.5 Tote Pans and Trays   **Tote pans** are relatively small containers used to move material, generally manually. The following is based on studies by Nielsen (1978).

Weight of the tray and contents should not exceed 25 lb; for maximum load, minimize the tray weight. (This level is lower than the NIOSH recommendations in Chapter 20 but the NIOSH weight is based on ideal conditions.)

Tray dimensions should be such that loads are not over 25 lb. Overloading often occurs when trays are used for different applications.

In general, a recommended tray would be 19 in. long and 14 in. wide. Width should not exceed length; maximum length is 20 in. A rectangular shape is best.

Tray depth should not be great because it would interfere with walking and might encourage loads that are too heavy.

Handholds should be provided at both ends of tray length; they should be above the center of gravity. Handles (whether made of the box material, nylon webbing, or steel rod) are better than holes in the side. Holes through the wall "trap" the hand if the lift is above the waist. Holes also permit dust, air, and light to enter and permit contents to escape. The handles should have sufficient bearing surface and no sharp edges.

6 BAR CODES

6.1 Bar Code Alternatives
Table 16.4 briefly compares some alternatives.

6.2 Code Design
The **bar code** is designed to provide a standard machine-readable symbol to track, trace, or count items. Codes are a fast, accurate means of capturing data about a person, place, or thing. The codes can directly give information (part number, item weight, item manufacturer) or indirectly give information through a computer hookup. (Given part number, what is destination?) Codes can be *product* specific, with all items of one product having the same code, or can be item specific, with each individual item having its specific code. In simple codes, a piece of reflective tape is placed on a container or pallet. An optical sensor detects the presence or absence of the tape and a switch is opened or closed. More complex codes contain digits (such as part numbers). These digits then can be displayed for a human to take action or, more commonly, can be used as input to a computer. The computer then can store the information or use it to trigger printers, switches, relays, bells, etc. An example is the Universal Product Code (UPC) which is used in retail grocery stores. The code is translated into a product number. The computer looks up the current price and name; the price and name are printed on a paper tape for the customer. Complicated codes are not limited in length and include letters and other characters as well as symbols. See Box 16.2.

The following borrows extensively from Allais (1982).

The code itself is a rectangular array of bars and spaces in a predetermined pattern. A symbol contains a leading quiet zone (an area with no marking), a start character, one or more data characters (possibly including a check character), a stop character, and a trailing quiet zone. The intercharacter gap is the space between the last element of one character and the first element of the next character within a discrete symbol. In discrete codes, the space and intercharacter gaps are not part of the code and can have wide dimensional tolerances. In continuous codes, both the bar and the space are part of the code structure. Start and stop characters are distinct characters used at the beginning and end of each bar code symbol to provide direction-of-read information to the decoding logic.

There are a variety of codes. The Two of Five Code has been used in warehouse sortation, photofinishing envelope identification, and numbering of airline tickets. It is discrete—all the information is in the bars. Bars are narrow (0 bit) or Wide (1 bit). The character 1 = 10001, 2 = 01001, etc.; all valid characters require two wide and three narrow bars. This makes Two of Five self-checking. If a printing void occurred along a scan line on a narrow bar, there would be only two narrow bars read and it would be a "nonread." If the void occurred on a wide bar, there would be only one wide bar; if there were an ink blob on a narrow bar there would be too few narrow bars. The primary disadvantage of Two of Five is low density (long space for each character).

Interleaved Two of Five is widely used in warehousing, corrugated shipping containers, and the auto industry. It is similar to Two of Five except that both the bars and spaces are coded. Odd-numbered digits are the bars and even-numbered digits are the spaces. Interleaved Two of Five typically has a symbol only 60% as long as in Two of Five. Like Two of Five, it is self-checking. Current specifications for Interleaved Two of Five require an even number of digits; for data with an odd number of digits, a leading zero is added before encoding.

TABLE 16.4 Bar codes versus alternatives (Allais, 1982).

Alternative	Error Rate/ 1,000,000 Characters	Speed, Characters/s	Comment
Bar code and hand-held wand	.3	4[a]	
Keyboard	30,000	2	
Optical character recognition wand	900	Slower than bar code	Can be read by people
Magnetic stripe			Expensive to duplicate code; can't be read from a distance. However, high code density and read/write capability (i.e., change code).

[a] Assume 3 s/wand stroke, 12 characters/code.

Code 3 of 9 (also called CODE 39, a registered trademark of Interface Mechanisms) is the Department of Defense standard symbology and is used for alphanumeric labeling of shipping containers. (Three of the nine characters are narrow.) Code 3 of 9 is an extension of Two of Five to represent the 26 letters, the 10 digits, space, -, ., $, /, +, and %. Code 3 of 9 is discrete and self-checking. Additional characters can be represented by using $, /, +, and % as precedence characters in front of one of the alphabetic characters. For example, +B = b while $B = B. Code 3 of 9 is the most widely used and accepted industrial bar code.

The Universal Product Code (UPC) was established for the supermarket industry to facilitate automatic number scanning with associated price lookup at the point of sale. It is a simplified version of European Article Numbering (EAN). The UPC symbol is divided into a left half and a right half, with six digits in each half. Each individual character consists of 2 bars and 2 spaces in a module (length) of 7; odd characters have 3 of 7 dark or 5 of 7 dark while even characters have 2 of 7 dark or 4 of 7 dark. In addition, each character contains an additional bit of information indicating whether it has an odd or even dark length. There is also a check digit computed from the 11 information digits. In supermarkets, there is additional error security as the read number is compared against product numbers; a leading error number is unlikely to match an actual number. For

---

BOX 16.2   Comparison of 7 selected bar code symbologies (Allais, 1982)

	Interleaved Two of Five	CODE 39	Codabar	CODE 11	UPC/EAN	CODE 128	CODE 93
Date of Inception	1972	1974	1972	1977	1973	1981	1982
Standard Specifications	AIM ANSI	AIM ANSI	CCBBA ANSI		UPCC IAN		
Government Support		DOD					
Corporate Sponsors	C. Identics	Intermec	Welch Allyn	Intermec		C. Identics	Intermec
Most Prominent Application Area	Industry	Industry	Medical	Industry	Retail	New	New
Variable Length	No (1)	Yes	Yes	Yes	No	Yes	Yes
Alphanumeric	No	Yes	No	No	No	Yes	Yes
Discrete	No	Yes	Yes	Yes	No	No	No
Self-Checking	Yes	Yes	Yes	No	Yes	No	No
Constant Char. Width	Yes	Yes	Yes (2)	Yes	Yes	Yes	Yes
Simple Structure (2 element widths)	Yes	Yes	No	No	No	No	No
Number of Data Char. In Set	10	43/128	16	11	10	103/128	47/128
Density (3): Units Per Char.	7–9	13–16	12	8–10	7	11	9
Smallest Nominal Bar-in.	.0075	.0075	.0065	.0075	.0104	.010	.008
Maximum Char. Per Inch	17.8	9.4	10	15	13.7	9.1	13.9
Specified Print Bar Width, in.	.0018	.0017	.0015	.0017	.0014	.0010	.0022
Tolerance at Edge-Edge, in.					.0015	.0014	.0013
Maximum Density Pitch-in.					.0030	.0029	.0013
Does print tolerance leave more than half of the total tolerance for the scanner?	Yes	Yes	No	Yes	No	No	Yes
Data Security (4)	High	High	High	High	Moderate	Moderate?	High

1. Interleaved Two of Five is normally used as a fixed length code. If one or more check digits are verified by the reader, it may be used as a variable length symbol.

2. Using the standard dimensions, Codabar has constant character width. With a variant set of dimensions, width is not constant for all characters.

3. Density calculations for Interleaved Two of Five, CODE 39, and CODE 11 are based on a wide to narrow ratio of 2.25 to 1. Units per character for these symbols is shown as a range corresponding to wide/narrow ratios from 2 to 3. A unit in Codabar is taken to be the average of narrow bars and narrow spaces giving about 12 units per character.

4. High data security corresponds to less than 1 substitution error per several million characters scanned using reasonably good quality printed symbols. Moderate data security corresponds to 1 substitution error per few hundred thousand characters scanned. These values assume no external check digits other than those specified as part of the symbology and no file lookup protection or other system safeguards.

random numbers (such as weights), a second check digit is recommended for the price field. Although the UPC works well in supermarkets, it rarely is recommended or used outside of its retail context.

Reading accuracy can be improved by digit parity checks and a modulo check digit; see Box 16.2. A check digit is calculated by a formula. For example, in a four-digit code, the first digit might be multiplied by the second and divided by the third. The result is the fourth digit. If any of the first three digits are read incorrectly, the comparison of the actual and calculated fourth digit will give a "no read." In some cases, human-readable data are printed next to the machine-readable code. For example, all vendors selling to defense agencies must use the 3 of 9 code and must use Optical Character Recognition, Font A (OCR-A) data above or below the machine code.

6.3 Code-Reading Devices

Codes are read by light pens, fixed beam scanners, or moving beam scanners. The code-reading device may be complemented with a keyboard terminal for manual input of additional information, including preprogrammed messages.

6.31 Light Pens.

Light pens (wands) are moved (stroked) across the code, staying in contact with the code surface. Since they generally are used in nonautomatic systems, a beep tone indicates when the code has been read. Pens with an incandescent (bright) light source are for low contrast and color applications; those with an LED light source are for rough service and long life. Wands can be connected to portable recorders for inventory taking.

6.32 Fixed-Beam Scanners.

Fixed-beam scanners generally are used in automatic systems, although hand-held models are available. The code moves into the reader's field of view. Depth of field is ½ in. to 18 in. When the code is low contrast, such as printed on Kraft carton surfaces, the code and surface are flooded with light (usually fluorescent) to enhance contrast for the read heads. In high-contrast codes, the read head includes a light source (LED or incandescent lamp). With fixed-beam scanners, code location on the item being scanned is critical. In addition, the code is scanned only once.

6.33 Moving-Beam Scanners.

With **moving-beam scanners**, a laser dot is run over the entire object surface in a pattern. Eventually the dot (moving at up to 20,000 in./s) crosses the code—and crosses it several times. Thus, location of the code on the object is not critical. Since the code is read several times, a microprocessor can compare the several readings to improve reading accuracy; this method also permits lower quality printing. Scan widths and depth of field differ among scanner models. Codes also can be read with labels tilted up to 20°; this permits packages to be oriented randomly and be of irregular shape.

6.4 Bar Code Applications

Bar codes are used in conjunction with computers. The codes provide the possibility of relatively fast entry of complicated data with very low error rates; the computer furnishes computational power, memory, and multiple decision options as well as the ability to control various output devices such as conveyor diverters and printers. Generally the computer is a dedicated mini or microcomputer which will communicate with the main computer. The following are some example applications:

Maintenance and repair records: The repair operators stroke a bar code (often on their own ID card) telling the computer who has the item and where it is. Next they stroke a bar code on the item to be repaired. Third they stroke a bar code indicating the item status from a menu (see Figure 16.6).

Shipping inventory records: A ski company put a bar code on each pair of skis. As the shipping box is loaded, the code is stroked for each pair of skis. Then a special code is stroked for the box. The computer then generates a shipping label including the customer's name as well as the serial number of each pair of skis. The computer also can update inventory and produce invoices.

Work-in-process inventory: In a facility making printed circuits, each printed circuit has its own bar code. Operators at 10 of the 25 process steps stroke the code. Thus the organization has exact, up-to-date information on each board. This information will be retained for analysis of failed boards returned from the field.

Component identification in assemblies: Automatic transmissions each are given a bar code. During assembly, the vehicle and transmission codes are stroked. The computer then verifies that there is a proper match.

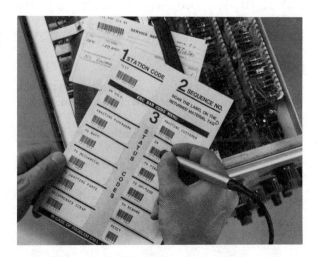

FIGURE 16.6 Menu cards hold a variety of codes next to alternatives such as "on hold," "awaiting paperwork," "to rework." When stroked, the message is entered into the computer quickly and without error. (Photo courtesy of Intermec.)

In addition, if a product recall is necessary, the computer can determine which vehicles received which transmissions.

Inventory verification: Is the number of items actually in stock the same as what the computer thinks is in stock? Periodically a person strokes two codes on each bin; one gives the part number and the other gives bin location. The number of items can be entered manually with a keyboard or by stroking a code on each item.

Item checkout: The item to be checked out has a bar code.

When a person checks the item out, the person's own code is stroked as well as the item code. The computer then can decide whether the item can be issued; if it is issued, the computer will remember who has it.

Variable-path material handling: A code on an item can be read. The computer then actuates various relays and conveyor diverters to send the product to its specified destination. (Be sure to have a diverter path for "no reads.") The diversion can occur from a scanner at the switch point or by allocating a standard amount of time for the item to go from the scanner to the switch.

DESIGN CHECKLIST: UNIT LOADS AND CONTAINERS

Type of unit load to
    Minimize handling costs
    Standardize containers
    Use cube efficiently
    Minimize material use
    Protect product
Pallets
    Material
    Design features
    Pallet loading pattern
    Load/unloading pallet
Pallet board/slipsheets
    Use in place of pallet?
    Type/design

Load stabilization
Unit containers
    Sufficient number
    Standardization
    Type/design
Bar codes
    Which code design
    Code reading device
    Identifying who
    Identifying what
    Identifying where

REVIEW QUESTIONS

1. What is an intermodal container?
2. Lumber accounts for what percentage of a wooden pallet price?
3. What is the most common pallet size in the United States?
4. Which deckboards should be butted together?
5. What is the disadvantage of the simple block pallet pattern?
6. For pallets, what is the difference between the manufacturer's problem and the distributor's problem?
7. Discuss workstation dimensions for manual loading and unloading of pallets.
8. Why are slipsheets popular for shipping but not for within-facility handling?

9. What are the four common ways of stabilizing a load on a pallet?
10. What is the difference between shrink wrap and stretch wrap?
11. Why are shear versus tension characteristics of glue important?
12. Briefly discuss stack, nest, and collapsible containers.
13. How is Code 3 of 9 self-checking?
14. Discuss digit check parity.
15. What are bar code menu cards?
16. Describe (in 20 to 40 words each) three applications for bar codes.

PROBLEMS AND PROJECTS

16.1 For a local factory, determine the number, size, material, and type of every pallet in the facility on a specific day. Any ideas? What does the supervisor think of your ideas?

16.2 Assume a truck is 90 in. wide, 105 in. high, and 44.5 ft long. Assume your product package is 24 × 20 in. and is 12 in. high. (a) If you were to use a pallet, what size would you use? Justify your recommendation with numbers, including pallet pattern and number of packages/truckload. (b) Assume you used a slipsheet. How would you stabilize the load? Give pattern and number of packages/truckload. Justify your recommendation.

16.3 For a specific part on a specific machine, what size and type of tote pan do you recommend? Justify your decision. What does the supervisor think of your ideas?

16.4 Have a debate between two teams on which bar code system (code, method of applying label, reading system, etc.) "our firm" should use. The instructor will give information concerning the firm. There should be a written and oral presentation. Then the class will vote on which system to adopt.

REFERENCES

Allais, D. *Bar Code Symbology*. Lynnwood, Wash.: Intermec, 1982.

Auguston, K. Is the U. S. ready for pallet pooling? *Modern Material Handling*, pp. 76–79, July 4, 1991.

Burke, M. Cost technology: The economics of pallet use. *Material Handling Engineering*, Vol. 30, 2, 67–69, 1975.

Chen, C., Sarin, S., and Ram, B. The pallet packing problem for non-uniform box sizes. *Int. J. of Production Research*, Vol. 29, 10, 1963–1968, 1991.

Containers. *Plant Engineering*, p. 155, July 18, 1991.

Downsland, K. An exact algorithm for the pallet loading problem. *European J. of Operational Research*, Vol. 31, 78–84, 1987.

Hand nailing and stapling of crates (Ref. 41). *Material Handling Engineering*, Vol. 36, 9, 156–159, Sept. 1981.

Hodgson, T., Hughes, D., and Martin-Vega, L. A note on a combined approach to the pallet loading problem. *IIE Transactions*, Vol. 15, 3, 268–271, 1983.

How to Select Shipping Containers. Buckhorn Brochure, Buckhorn, Milford, OH 45150; 1992.

McElroy, R., Ed. *Accident Prevention Manual—Engineering and Technology*. Chicago: National Safety Council, 1980.

Nielsen, W. Some general principles of tray design. Paper at American Industrial Hygiene Association, Los Angeles, 1978.

Pallets. *Plant Engineering*, p. 155, July 18, 1991.

Welded wire mesh containers. *Material Handling Engineering*, Vol. 31, 6, 44–45, June 1976.

Wooden pallet containers (Ref. 36). *Material Handling Engineering*, Vol. 35, 2, 134, 137, Feb. 1980.

Wooden pallets (Ref. 24). *Material Handling Engineering*, Vol. 33, 3, 88–91, 1978.

17 CONVEYORS

IN PART III MATERIAL HANDLING AMONG WORKSTATIONS

CHAPTER PREVIEW

There are many different types of conveyors (fixed-path equipment). This chapter concentrates on only a few, including wheel, roller, pneumatic, belt, and power and free. Conveyors are used not only for transport but also for storage and pacing.

CHAPTER CONTENTS

1 System Considerations
2 Nonpowered (Gravity) Conveyors
3 Powered Conveyors
4 Conveyor Control

KEY CONCEPTS

accumulation pressure
ball transfer section
captive (slave) pallet
cornering wheels
diverging (versus converging)
drive (constant versus variable
 speed)

live rollers
nose over
photocontrols
pitch
plow
power and free conveyor
programmable controller (PC)

slider (versus roller) bed
sortation
transportation (and storage)
 function
trolley (tow) conveyors

1 SYSTEM CONSIDERATIONS

1.1 General   Material handling equipment can be fixed path (or flexible route), intermittent (or continuous), long distance (or short distance), horizontal (or vertical), and so forth.

In general, fixed-path equipment includes conveyors, elevators, and pipes; it usually is permanently installed. Mobile equipment, such as fork trucks (see Chapter 18) can go any distance. Limited-area equipment includes hoists, manipulators, robots, and cranes (see Chapters 19 and 20); it is flexible within a restricted operating area. See Chapter 11 for storage equipment and techniques.

1.2 Conveyors   Conveyors are used for machine-to-machine movement, assembly operations, department-to-department movement, and linkage of production with automated storage and retrieval (AS/R) systems. The American Material Handling Society lists 57 types of conveyors. Only a few of the major ones are described in the following.

Design and selection of conveyors must consider not only their **transportation function** but their **storage function**. The engineer may use a conveyor not only for moving items from point A to B but also, and maybe primarily, as a device to store work in process (buffers, float, banks). These accumulation sections permit each operation to run at its own speed, unhindered by operations behind or ahead of it. (See Chapter 11 for more on storage and Chapter 5 for buffer design.) The interface with machines and computers has become more critical as, increasingly, conveyors are part of an integrated system rather than an independent component. In some applications, speed control is important, such as for carrying product through cleaning, baking, or painting. Conveyors also are used for work pacing since they can present a unit every X seconds. In general it is inefficient to machine-pace a person (see Chapter 5), but nonetheless such systems exist and conveyors are one technique to implement a machine-paced design. Conveyors may also provide a work surface for processing and assembly.

The optimum work height is slightly below the elbow for both sitting and standing work. (Note that fingers often do not work at the bottom of the object; i.e., work height may be above conveyor height.) In general, when manually loading a conveyor, put the conveyor at knee height. When manually unloading a conveyor, put the conveyor at hip height. See Figure 17.1. Provide foot and leg room beneath the conveyor. Don't use lips, since they cause unnecessary lifting. See Figure 17.2.

If the conveyor is 7 ft or more above the floor, provide guards under it to prevent injury from falling objects. Conveyors tend to have many run-in (nip) points; be sure to install all the guards the manufacturer recommends.

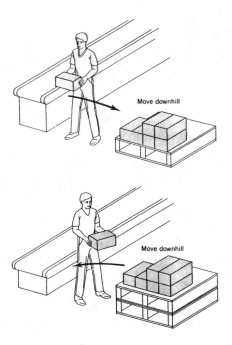

FIGURE 17.1   Move downhill, whether from a conveyor to a pallet or a pallet to a conveyor. When loading a conveyor from the pallet, adjust the conveyor to be low; raise the pallet by having the lift truck driver place it on a platform (such as an empty pallet). When loading a pallet from a conveyor, have the conveyor higher and the pallet on the floor.

2 NONPOWERED (GRAVITY) CONVEYORS

2.1 Chutes   Chutes are cheap to construct, operate, and maintain. Usually they have side walls and sometimes a top. A metal chute, whether straight or spiral, has a lower coefficient of friction than a wooden one but is noisier. Some wooden chutes are lined with metal.

2.2 Wheel Conveyors   See Table 17.1. Gravity wheel conveyors have low capital costs, zero operating costs, and low maintenance costs. They are relatively lightweight and

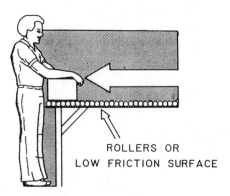

FIGURE 17.2   Slide, don't lift, to reduce back strain. Leave room for the feet. Recommended toe space is 6 in. deep, 6 in. high, and 20 in. wide (DeLaura and Konz, 1990).

TABLE 17.1   Approximate purchase costs in 1992 for wheel, roller, and belt conveyors (Equipment Overview, 1992).

Type	Application Notes	Cost Factors	Typical Pricing
Skate wheel	Convey smooth, lightweight packages, bags, and cartons by gravity (manual push); min. of 6 wheels under each item; rollers set low for guidance; easy to relocate	Width, number of wheels/ft, construction material (steel, alum., galvanized metal), special wheel construction	12 in. wide, 10 wheels/ft = $9/ft, * 90° curve with guard = $120; 24/28 unit = $14/ft, 90° curve with guard = $175
Gravity roller	For applications requiring more uniform conveying surface; transport tote boxes; cartons, baskets, and drums; rollers set low for guidance	Width, roll dia., shaft size, bearings, roll spacing, frame construction	Light to medium duty = $11–25/ft; heavy duty = $35–75/ft**
Powered (live) roller	Moves items horizontally without manual assistance; must be a conveyable container with 3 rollers touching bottom surface; special types used for accumulation; either by zone or along entire line	Width, roller bed construction, drive method (flat or V-belt; continuous chain, roll-to-roll chain, line-shaft belt)	$75–200/ft; belt-driven usually less costly; roll-to-roll chain usually at higher end of cost range
Belt (slider or roller)	Gives complete support under loose materials or in containers; move light/medium-weight loads between operations and departments; can pace work and handle unusual shapes and configurations; belt operates over slider bed or roller bed with roller offering higher capacity	Width, bed construction, ruggedness of terminal machinery, center or end drive; rough top belt used for incline or decline; smooth top belt diverts containers to multiple locations	A common in-plant type (roller bed 23" wide between frames and 9" roll centers) = $45–85/ft; add about $7/ft for slider-bed construction

*Supports are extra; normal duty = $15 apiece, heavy duty = $30.
**More detailed cost estimates are shown below.

Roller		Width Between Conv. Frames, in.	Centerline Roller Spacing, in.	Cost $/ft
Dia, in.	Load Capacity, lb			
1.37	80	15	1.5	25
			4.5	11
2 (1.9)	300	21	3	25
			6	14.50
2.5	600	42	3	75
			8	35

simple and so can be used for temporary as well as permanent installations. Figure 17.3 shows a typical wheel conveyor in which objects flow by gravity or are moved manually.

A 10 ft section of roller conveyor with steel frames, the most common type, would weigh about 90 lb.

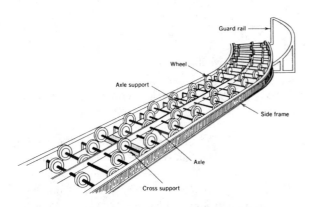

FIGURE 17.3   Wheel conveyors have 5 to 10 ft standard lengths. Standard widths are 12, 15, 18, 24, and 30 in. The wheel pattern is determined by the load. Have three axles and seven wheels under the load at all times. Staggered wheels on axles support the load more evenly. The frame (rather than wheels) determines capacity. A typical 10 ft steel conveyor can hold about 800 lb, while an aluminum one holds about 500 lb. If the wheels are set high so the load clears the side rail, a 6 in. overhang is safe.

Aluminum frames for the 10 ft length would reduce the weight to 50 lb. This light weight and resistance to corrosion must be balanced against steel's greater strength and economy.

Wheels can be steel, aluminum, or plastic. The most common is the general-purpose steel wheel, which has a life approximately 10 times that of an aluminum wheel. Aluminum wheels are used when portability or corrosion resistance is important. Both steel and aluminum wheels can be covered with neoprene or polyvinyl-chloride (PVC). The coating reduces shock, reduces marking or scratching of surfaces, provides more gentle handling of fragile items, reduces noise, and, because of greater contact friction, aids alignment on both curves and straight sections. If the entire wheel is made of polypropylene, it has many of the characteristics of neoprene-coated aluminum with very good corrosion resistance and very light weight.

Wheel selection is a typical tradeoff of initial cost versus performance. For true economy, consider total costs over the life cycle (initial cost plus operating cost plus maintenance cost) rather than just initial cost.

Most flat-bottomed surfaces will convey satisfactorily on wheel conveyors. If the part doesn't have a flat surface, it may ride in a box or on a small pallet. (Since the pallet never leaves the facility, it is called a **captive pallet** or **slave pallet**.) Try to orient pallets so the stringers are at right angles to the wheels; an automatic turntable (which turns the pallets 90°) at the loading and unloading station may be useful. Cylindrical loads can be moved on concave frames.

Pitch, the degree of conveyor decline, is given in in./10 ft section. See Table 17.2. Most applications have a pitch of 3 to 6 (2.5 to 5%). Aluminum wheels, being lighter than steel, require less pitch; heavier containers require less pitch; harder surfaces require less pitch. A level conveyor, with no pitch, is used in some installations. For longer distances, short sections of belt or live roller conveyors are inserted between longer gravity sections.

In addition to standard straight and curved sections, there are some special types. Extendable conveyors are extended gradually into trucks during loading and are retracted gradually during unloading. Spiral conveyors are used for high-density storage (buffers) between machines; a power conveyor raises the units to the top.

2.3 Roller Conveyors

See Figure 17.4. Roller conveyors are used in heavy-duty wheel conveyor applications. Rollers, having more mass than wheels, resist impacts and loads better than wheels. The additional mass requires greater pitch for the same load than do wheel conveyors. Rollers, like wheels, can be covered with plastic or rubber coatings. Roller conveyors tend to be heavier, more rugged, and less portable than wheels; they are not as satisfactory

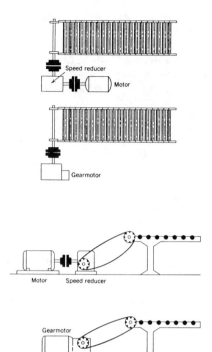

FIGURE 17.4 Roller conveyors can be driven four ways. Have three axles under the load at all times. Rollers generally should be 2 to 3 in. wider than the package.

for very light loads such as empty boxes and do not track loads as well.

2.4 Wheel and Roller Accessories

Special sections complement standard straight sections of wheel and roller conveyors.

Curves are more reliable than right angles for changing directions, since jams and breakdowns are less frequent. For roller conveyor curves, the roller is tapered with the small diameter on the inside radius so that container travel speed is the same on the inside and outside radii. Right angles take up less floor space but often need mechanization (e.g., limit switch and push rod) to get the load to change direction. Note that right-angle transfer generally changes the pallet/container orientation 90°; pallet orientation may be important if it is a two-way pallet instead of a four-way. For manual loading and unloading, orient the pallet so the long axis is parallel to movement and reach distance is minimized. **Ball transfer sections** permit directional transfer by hand to and from machines and conveyors. See Figure 17.5. Since a person can push only about 30 lb easily, maximum load should be about 600 lb if a ball with 5% friction is used (ball frictions range from 3 to 15%). For heavier loads, use turntables; they pivot about a center pivot and the load is supported by the frame. For very heavy loads, use motors, air cylinders, and push rods. Rollovers are circular sections (horizontal axis perpendicular to conveyor direction) which turn a pan or fixture

TABLE 17.2 Guideline for pitch on gravity conveyors (Basic Training, 1982). Avoid loads over 200 lbs.

Item	Approximate Weight (lb)	Conveyor Pitch (in./10 ft)
Barrels	—	5
Baskets	—	5
Boxes, wood	15–25	6.25
	25–50	5
	50–100	3.75
	100–250	3.12
Brick	—	5
Cans (milk)		
Empty		6.25
Full		5
Cartons	3–6	8.75
	6–12	7.5
	12–25	6.25
	25–30	5
Crates	—	5–6.25
Drums	150–300	2.5–3.75
Lumber	—	5
Tote pans	—	2.5–5

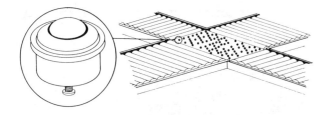

FIGURE 17.5   Ball transfers aid in turning corners, as well as in loading and unloading. (Figure courtesy of Rapidstan Division, Lear Siegler, Inc.)

upside down. Tilters (horizontal axis aligned with conveyor direction) turn the pan or fixture so another face is available.

People and vehicles can cross the conveyors several ways. When the conveyor is recessed into the floor, step plates and floor plates allow safe passage. Gates, either vertically hinged or horizontally pivoted, can be used. If they are hinged, use counterweights or springs. Another option is stairs and a bridge. Never permit people to climb over a conveyor, since the risk of injury is great.

Breaks or stops reduce or stop loads from impacting into each other. A decelerating slope at the discharge end of a gravity conveyor reduces problems.

If the wheels or rollers are set low, the frame sides act as guide rails. If the wheels or rollers are set high, the load can overhang the conveyor width. Check for sufficient clearance and load stability.

Conveyor supports can suspend the conveyor from the ceiling, from the side, or from below. When it is supported from below, portable supports should be on casters if conveyor movement is frequent. Conveyors also can be double- or triple-decked for storage or for use as return lines for empty containers.

3 POWERED CONVEYORS

See Box 17.1 and Table 17.3.

3.1 Pneumatic
Pneumatic conveyors (power chutes) can be divided into dilute-phase and carrier systems.

Dilute-phase systems move a mixture of air and solid. They can be classified as push, pull, and push-pull. Push (positive pressure) systems push material from one entry point to several discharge points. Pull (negative pressure or vacuum) systems move material from several entry points to one discharge point. Push-pull systems are combinations with multiple entry points and multiple discharge points. Traditionally, pneumatic conveyors are used for bulk handling of solid materials (powders, granulars, and pellets). Some more recent applications include moving refuse, scrap, and other varied items.

When bulk products are moved, several types of conveyors are possible. A major advantage of pneumatic conveyors over other conveyors is that the complete enclosure minimizes product contamination and dust. Dust control has been especially important because of the recent emphasis on anti-pollution, environmental regulations, and safety. Pneumatics, in comparison with other conveyors, also have few moving parts and thus minimal maintenance, no loss of material, directional flexibility (turns no problem, vertical as well as horizontal), and a relatively low installation cost. The disadvantages are relatively high capital cost, higher energy operating costs, the fact that material must be flowable, and the fact that they convey only one direction (some other conveyors can be reversed for loading versus unloading). Purchase costs/horsepower drops as blower hp increases. Pneumatic systems are relatively complicated to design. Leave the design up to specialists.

TABLE 17.3   Approximate purchase costs in 1992 for bulk handling conveyors (Equipment Overview, 1992).

Type	Application Notes	Cost Factors	Typical Pricing
Belt	Long-distance, often inclined, transport of bulk solids such as coal, ash, gravel, fertilizer, or sand; fabric or rubber belt typically moving in a troughed configuration over 3 roll idlers	Five basic cost elements: belt, idlers, pulleys and terminals, drive, and structure; additional accessories include belt cleaner, walkways, pulley lagging, gravity takeup, and hood cover	A 15 hp, 24 in. × 100 ft inclined conveyor is common; basic cost, not including accessories, around $15,000
Screw	Pushes loose materials forward in any horizontal direction or along incline using shaft-mounted helix revolving in U-shaped, stationary trough; adapted for dust-tight operation in confined spaces	Bulk material characteristics, drive power, screw size (6, 9, or 12 in. dia), type of construction	10 ft starter section with terminals, end bearings, inlet and outlet screw, and trough ranges from $500–1,800, depending on screw size and construction; cost/foot lower for additional sections
Pneumatic	High-volume transporting of aerated solids in a clean, dust-free operation not confined to straight-line conveying; operation is positive pressure, negative pressure (vacuum), or combination, depending on number of pickup and delivery points	Costs directly related to blower motor hp requirements	Cost/hp drops as blower hp increases; at 3 hp, cost = $1,500–2,000/hp; at 150 to 200 hp, cost = $500–1,000/hp

BOX 17.1   *Conveyor application notes*

The following information comes from Basic Training (1982; 1984) and Considerations (1989).

Package Size

- *Wheel conveyors* (for smooth, flat-bottom loads): Keep at least 10 wheels under a box. Wheels/axle vary, but axles are 3 in. apart.
- *Roller conveyors* (for uneven boxes or those with open bottoms or rims on the bottom): Keep at least 3 rollers under a box or the box will tumble (nose into roller). Box width should be 2 in. less than roller width.
- *Belt conveyors:* Use boxes less than conveyor width. If boxes drag on the conveyor sides, they may be damaged; so boxes less than belt width are best.

Curves

- Allow 2 in. clearance between the box corners and the outside radius. See Figure 17.6.

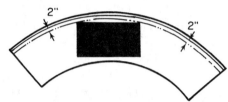

FIGURE 17.6   Clearance of 2 in. must be maintained between the package corners and the outside radius (Basic Training, 1982).

- Boxes will lose orientation on curves using conventional rollers (since the outside travels a greater distance). Orientation is better with tapered rollers, wheel curves, or split rollers (each roller is half the width of the conveyor).
- Long boxes may have an orientation problem on a sharp curve. Make the curve less sharp.
- Mount a guardrail on the outside of curves to prevent boxes from falling off. If a guardrail is used for guidance, mount skate wheels on the rail or the friction may overload the conveyor motor.

Powered Horizontal Conveyors

- Standard speed for a powered conveyor is 65 ft/min and catalog hp usually is based on 65 ft/min.

  HP at desired ft/min = (*HP* @ 65)(desired ft/min)/65

  If a conveyor runs over 200 ft/min, reduce vibration with special conveyor parts and bearings.
- The belt of belt conveyors can rest on a solid steel bed (**slider bed**) or a **roller bed**. Slider beds are rugged and inexpensive but heavier loads cause friction between bed and belt (wearing out the belt and requiring more energy).
- **Live rollers** (i.e., no belt) have low friction and make it easy to move a box on or off the conveyor. Figure 17.7 shows it is difficult to stop a box on a live roller conveyor because the items accumulate (**accumulation pressure**).
- Accumulating conveyors permit small (2–3% of live load) or even zero drive pressure. Zero pressure is

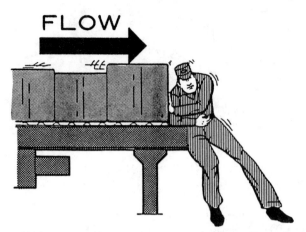

FIGURE 17.7   Pushing by people is not a solution to accumulation pressure (Basic Training, 1982).

achieved by putting sensing rollers at regular intervals (e.g., 24 inches). If the sensing roller is depressed (by a box), the power is turned off in the zone before the sensing roller. See Figure 17.8.

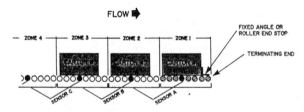

FIGURE 17.8   Remove accumulation pressure by removing the drive pressure. A sensing roller, when depressed, will turn off the rollers in the zone preceding the sensing roller (Basic Training, 1982).

Straight Transfers

- Transfers can be made from a live roller to a belt (or vice versa), from roller to roller, or from belt to belt. For live rollers, power the end rollers with an O ring (Figure 17.9). Minimize box tumbling by ensuring both conveyors are at the same level and by minimizing the gap between the conveyors.

FIGURE 17.9   Power the end rollers on a live roller conveyor with an O ring to produce good transition between conveyors (Basic Training, 1982).

- Have a gravity straight section, equal to package length, before a gravity curve.
- Have a power straight section, equal to package length, before a power curve.
- If the downstream conveyor is faster, it will separate the boxes (increase their gap).

(continued)

BOX 17.1 *Continued.*

- Box gap is:

$$G = C - .5\,BL$$

$$\frac{C_i}{S_i} = \frac{C_f}{S_f}$$

where

G = gap between boxes, in.

C = center distance between boxes, in.

C_i = center distance of boxes on initial conveyor, in.

C_f = center distance of boxes on final conveyor, in.

S_i = speed of initial conveyor, ft/min

S_f = speed of final conveyor, ft/min

BL = box length, in.

For example, if S_i = 100 ft/min, S_f = 200 ft/min, and C_i = 24 in., then C_f = 48 in. If all boxes are 18 in. long, the gap on the final conveyor is 48 – 18 = 30 in.

- Desired gap between boxes depends on the sorting device (see the following). For a plow, make G = plow length; for a 90° push off, make G = 1.5(BL); for a pivoting, flush-mounted diverter, make G = 1(BL).

Diagonal and Parallel Transfers

- For transfer from a live roller to a live roller spur (**diverging**), an included angle of 30° is preferred to 45°. Ordinarily a friction **plow** (deflector) is sufficient (Figure 17.10), but if the box weight is over 50 lb or the angle is 45°, use a powered V-belt plow to reduce box rotation.

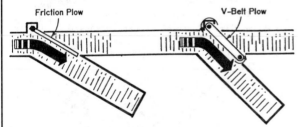

FIGURE 17.10 Plows divert boxes to a spur. Normal design is a 30° spur and a friction plow. Use a powered V-belt plow if the angle is 45° or the box weight is over 50 lb (Basic Training, 1982).

- For transfer from a live roller spur to a live roller main line (**converging**), use a **cornering wheel** (Figure 17.11). Angles of both 30° and 45° are satisfactory. Note that a "traffic cop" is necessary to control for collisions with boxes on the main line.

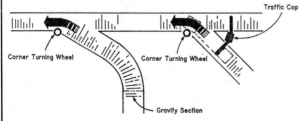

FIGURE 17.11 Converging from a live roller to a live roller requires a cornering wheel; both 30° and 45° are acceptable. Note the straight section (length of the longest box) before the curve. The traffic cop is necessary when there are uncontrolled boxes on the main line (Basic Training, 1982).

- In transfer from a roller bed belt to a live roller spur (diverging), boxes impacting the plow may pull the belt off track. Keep box weight under 15 lb. For heavier boxes, use a special belt with a smooth top but a rough undersurface (to maintain tracking). A 30° angle is preferred over 45°.

- Pop-up wheels or rollers can be used (Figure 17.12) for both diagonal and 90° transfers. See Table 17.4. On command, the powered wheels or rollers pop up and drive the box onto the other conveyor. For heavy items, a rake can pop up also and pull the box off.

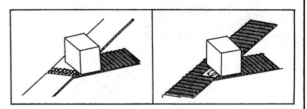

FIGURE 17.12 Pop-up wheels or pop-up rollers rise to direct sorted items to a powered spur (Considerations, 1989).

- Merge two parallel roller conveyors into one with a plow merge (Figure 17.13) for low speed of sorting and a push off for high-speed sorts.

FIGURE 17.13 Plows are a simple way to move two lines onto one (Basic Training, 1982).

TABLE 17.4 Comparison of **sortation** methods (Considerations, 1989).

Category	Category/ Minute	Carton Orientation After Sort[a]	Impact on Carton[b]	Repair/ Maintenance[b]
Class 1 (less than 40 cartons/min)				
Manual	10–15	Perp.	1	NA
Rake puller	10–15	Perp.	3	2
Class 2 (30–80 cartons/min)				
Deflectors	20–40	Perp.	3	1
Diverters	30–40	Perp.	4	2
Pop-up chain and belt	30–45	Para.	3	4
Class 3 (Over 100 cartons/min)				
Pop-up skewed wheels	60–120	Para.	2	3
Pop-up roller	60–150	Para.	2	3
Moving slat	60–150	Para.	1	4
Tilting trays	60–250	Para.	2	3
Tilting slats	60–250	Perp.	5	5

[a] Perp. = perpendicular to flow; Para. = parallel to flow.
[b] 1 = low, 5 = high; NA = not applicable.

(continued)

BOX 17.1 Continued.

90° Transfers

- Transfer onto a belt conveyor should be done only by dropping the box onto the conveyor (Figure 17.14), because transfers at the same level will cause belt tracking problems.

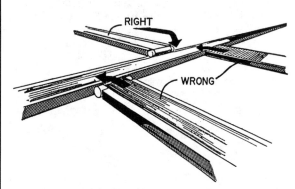

FIGURE 17.14 Transfer onto a belt conveyor from either a belt or roller conveyor causes belt tracking problems. Therefore, have the input conveyor overhang the belt so that the box drops onto the belt (Basic Training, 1982).

- Place a guard opposite the transfer point; consider **cornering wheels** (wheels mounted on the side of the conveyor, e.g., 6 in. above the box base).
- A push-off (a diverter) can be side mounted (Figure 17.15) and push off in one direction, or it can be overhead mounted and push off in either direction. Keep the boxes close to the push-off plate (perhaps with a plow) to reduce impact damage on the box.

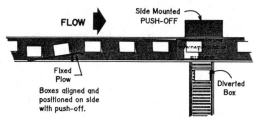

FIGURE 17.15 Push-offs allow transfer to a 90° spur[1]

- Tilt sorting can be done by trays (for fixed-length items) or slats (variable-length items). See Figure 17.16. In the slat system, only the occupied slats tilt. Tilting can be done to either side, and the exit lanes can be close together. Typically the box has a bar code which is read to indicate its sorting location.

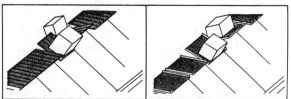

FIGURE 17.16 Tilting slats and tilting trays can sort to spurs on both sides in a short space. Use trays for fixed-length boxes and slats for variable-length boxes (Considerations, 1989).

[1](Basic Training, 1984).

Inclines

- An incline angle of < 25° is normal; 30° is maximum.
- Box tumbling potential can be determined from Figure 17.17.

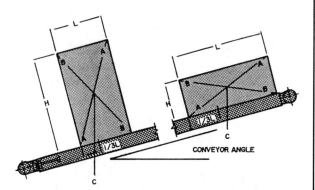

FIGURE 17.17 Determine if a box will tumble by using these steps: (1) Draw the box and conveyor angle; divide the box length into thirds. (2) Find the box center at intersection of lines A and B. (3) Drop a perpendicular from the intersection (line C). If line C falls within the lower 1/3, the box may tumble, depending upon the weight distribution within the box (Basic Training, 1982).

- Place a power feeder belt (as long as the longest package) on the horizontal section just before the incline. See Figure 17.18. The incline belt should have a greater speed than the power feeder belt. If the feeder belt is driven by the incline conveyor, get the speed differential by using a different size sprocket for the chain.

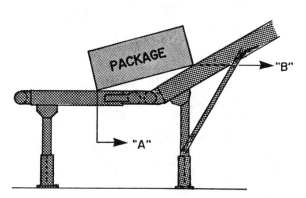

FIGURE 17.18 Power feeders are needed before an incline. The power feeder belt should be smooth (to permit slippage at point A) because point A is moving faster horizontally than point B (Basic Training, 1982).

- At the top of an incline over 10°, where the box transfers to a horizontal conveyor, make the transition smoother between the conveyors by inserting a roller (called a **nose-over**).

Carrier systems use carriers to transport items or paperwork. Banks use them to move money from drive-in stalls to the clerk. Texas Instruments, as an industrial example, transfers electronic parts among nine stations (Smart Pneumatic, 1978). Stations include receiving, warehouse, and various assembly locations. The carriers have dividers and carry an average of four different requisitions per trip. The justification for the system was reduced material handling costs and reduced work-in-process inventories.

3.2 Powered (Live) Roller Conveyors

Live rollers tend to be more rugged and expensive than belt conveyors and more expensive than gravity rollers. Cable or belts drive light and medium-duty live roller conveyors; chains drive heavy-duty styles. A belt or chain can flex only in one plane, so each section (curve, incline, straight, accumulation) needs its own drive unit. Since cables can flex in several planes, one large motor can replace multiple small motors with savings in both capital and operating cost. Live rollers can elevate, lower, or move loads on the level. Often gravity and inclined live roller sections alternate. Diverters can be either the push-off type or pop-up wheels between the rollers.

Most conveyor operations have a nonuniform flow of items. Thus, items accumulate at some point, usually near the discharge end, which often serves as storage. Destructive accumulation pressure can build up on the items. With a live roller system, have individual rollers disengage from the belt when the load is stationary to reduce accumulation pressure. The best solution to accumulation pressure may be a package redesigned to be tougher; it may help outside the plant as well as within. A photocell that shuts off the conveyor drive when items accumulate to a specified point also works. When flow resumes, however, the items move out as a close, dense slug.

3.3 Belt Conveyors

See Table 17.1. With live roller conveyors, the load is carried on rollers driven by a belt; with belt conveyors, the load is carried on the belt supported by the rollers. Belt conveyors have a wide application range. The two primary advantages over gravity conveyors are control of speed and ability to go up and level as well as down. An advantage over other powered conveyors is a continuous surface for carrying the material. Ability to reverse direction is another feature. Another technique is to double-deck a belt conveyor. After an item goes to workstations on the top belt, the empty boxes or completed assemblies return on the lower return belt. Belts, because of their friction, produce little slippage between the belt and package. Thus, they maintain the spacing between successive packages required by open-loop diverter mechanisms. Wheel and roller conveyors have slippage between the item and the wheel or roller, which makes it more difficult to predict the time it will take for a package to reach a diverter after breaking a photocell beam.

Because of their wide usefulness, completely self-contained portable belt conveyors are standard items. See Figure 17.19. Common lengths are 10, 15, 16, 18, 20, 25, and 30 ft, with belt widths of 12 to 24 in. Usually they can run at any angle from −30° to 30°.

Figure 17.20 shows a belt conveyor for bulk loads. When handling bulk loads, the supporting rollers often form a trough; this allows the belt to conform to the rollers and thus carry more material. Material can also be carried on the return run of the belt. Belt conveyor transfer points are dust producers. Enclose these transfer points and use skirts and curtains at openings.

Belts come in a wide range of types, materials, and thicknesses; belt selection should be done by the conveyor manufacturer or replacement distributor.

Browning (1974; 1975) gives detailed design values for belt conveyors which carry pedestrians. Table 17.1 gives some conveyor costs.

3.4 Chain Conveyors

See Table 17.5. With a chain conveyor, an endless chain transmits power from a motor to a carrying surface or unit. The carrying unit can be quite varied. Specific examples of chain conveyors are flight conveyors, apron conveyors, bucket conveyors, and slat conveyors. Flights are "blades" attached perpendicular to the chain. See Figure 17.21. The key advantages of chain conveyors are the positive drive and the many types of possible carrying units. With buckets, for example, the buckets generally are unloaded automatically by being tipped or inverted. In the above chain conveyors the product in or on the unit follows a fixed path, from location A to B to C to D.

FIGURE 17.19 Portable belt conveyors are so popular they come in standard units. (Figure courtesy of Rapidstan Division, Lear Siegler, Inc.)

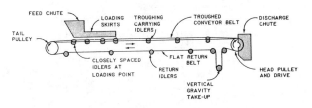

FIGURE 17.20 Bulk belt conveyor schematic (Colijn, 1983).

A special type of chain conveyor, a **trolley** or **tow conveyor**, has a unique feature—a powered trolley on a rail. See Figure 17.22. The trolley is connected to a motor by a chain or cable. (A monorail is an overhead trolley not connected to the motor.) Trolleys on a rail offer several key advantages.

3.41 Advantages. First, the rail can follow almost any path, with both horizontal and vertical curves. In particular, the rail can use the vertical dimensions of a space, an especially valuable feature for storage. Generally, the rail is arranged in a loop and the trolleys go

TABLE 17.5 Approximate purchase costs in 1992 for chain, overhead, in-floor tow, and pallet conveyors (Equipment Overview, 1992).

Type	Application Notes	Cost Factors	Typical Pricing
Chain	Primarily for transporting heavy products using single or multiple chains; can maintain product in fixed position over entire flow path; various chain attachments available for specific conveying jobs	Width and construction based on capacity and load configuration, special features	
Slat	Load-supporting slats attached to chain; handles heavy loads with abrasive surfaces or loads that might damage belt		$500–700/ft for handling loads from 200 to 4,000 lb; heavier duty systems available
Roller flight	Uses rollers instead of slats; for accumulation of heavy loads and rough service applications		Comparable to slat conveyor
Drag line (chain on edge)	Vertical chain pins permit horizontal turns; handles heavy loads moving through subassembly, etc.; multiple drives may be used		$400/ft
Drag line (chain on flat)	Cannot negotiate horizontal turns; frequently imbedded in floor		$300/ft
Overhead	Frees floor space for other uses; products are out of reach and protected from pilferage and damage; suited for sequential processing operations		
Trolley	Series of trolleys supported from overhead track and connected by endless, motor-driven chain or manually pushed, carriers suspended from individual trolleys; for single- or multi-plane conveying; trolleys are programmable	Type of track (enclosed, I beam, patented), cable or chain drive, type of chain, type of drive (caterpillar, sprocket), takeups	Chain installed No. Cost, $/ft 228 40–60 348 90 458 125 678 175
Power and free	Similar to trolley conveyor but load carriers are suspended from 2nd set of trolleys running on an independent or free track; carriers can be disengaged from chain and accumulated or switched onto spurs	Need for components, switches, escapements, steel, controls, types of chain, guards	Chain installed No. Cost, $/ft 228 175 348 225 458 375 678 500–600
Monorail	Trolley rides on single overhead rail; individual load carried on one or more trolleys; can use hoist to provide crane operation; trolley is programmable	Manual or powered operation, type of overhead track, type of hoist if used, number of switches, spurs, carriers	
In-floor tow	Power-driven chain traveling in subfloor trench propels load-carrying wheeled carts along straight runs or combinations of main lines and spurs; various mechanical, proximity, or remote methods or programming available; frees floor space when not in use	Complexity of path, type of programming, new or existing building, speed requirements	$75/ft without spurs; bumpoff (nonpowered) spur costs about $2,500
Pallet or unit load	Similar in design to packaged roller or chain conveyors, but capable of handling heavy (2,000 to 3,000 lb) loads; heavy-duty construction throughout	Load-handling rate and number of directional changes in system	Basic conveyor for 2,000 lb loads = $300/ft; direction change = $3,500; transfer = $3,500; receiving point = $3,500; extra drive = $2,000

around and around—that is, they act like a bus on a route.

Second, the trolley can be connected to a wide variety of carriers. See Figure 17.23. Different trolleys on the same line can use different carriers. Example carts are the common flat wood or steel bed, shelf carts, tilting tray carts, and "moving assembly benches." (If the carriers are different but appear similar, paint each type a different color for ease of operation identification.) Sometimes the rail is overhead and sometimes in the floor. When the carrier is a vehicle such as a cart or wagon, the conveyor is a *tow conveyor*. Tow conveyors are similar to driverless vehicles except that they get their power from the chain rather than having independently powered vehicles.

A third advantage is that the trolley can be disconnected temporarily from the chain. The trolley, with or without load, can be put on a spur track, either powered or nonpowered. In the overhead line version there is the popular **power and free conveyor** (see Figure 17.24), in which the upper line is powered and the lower rail (which carries the trolley) is nonpowered. In another version the rails are side by side. Parking the trolley and loading on a free line or a spur has many advantages. A spur permits one load to pass another load. Parking permits a load to be stored temporarily (such as for buffer storage) and permits a load to be stationary while it is being worked on.

Spurs can be nonpower or power. For nonpower spurs, a level spur depends upon carriers in the powered line pushing themselves into the spur. When the spur is full, the selector button retracts and new carriers go to the next spur, which has a selector button in that same position. If there is no such spur, the carrier will go around the main line loop until the spur has an available position; thus the main line acts as a large buffer. An inclined spur allows the carrier to go to the end of the spur by gravity. Powered spurs can index or run continuously. An indexing spur runs only when a new carrier enters the spur. It moves all carriers forward one position and then shuts off. Continuously

FIGURE 17.22 Trolleys are supported from or within an overhead track or a track in the floor. A basic trolley is shown mounted on an inverted T track.

running spurs pick up a carrier where it enters the spur and move it along the spur until it reaches the end of the line or another carrier (where the tow pin would automatically disconnect).

Temporarily disconnecting the trolley from the chain leads to a potential fourth key advantage (available on some but not all trolley and tow conveyors). Multiple main lines and multiple spurs can allow the trolley to act as a taxi instead of a bus. That is, the carrier can start at any one of many start points and can go to any one of many end points. The route can be appropriate for that specific load instead of following a predetermined loop with a limited number of start and stop points.

3.42 Design Variables. Switching has been mechanized and computerized, so humans need not be present at any of the switch points. For mechanical switching, at each there is a spring-loaded bump at each station. When the

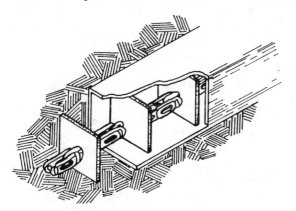

FIGURE 17.21 Flight conveyors have "flights" attached to the chain. These flights pull sludges, slurries, chips, etc. along. If the flight is enclosed in a tube, spillage is eliminated (Fitzpatrick, 1983). An alternative is to substitute a trough for the tube. This reduces friction and capital cost but results in spills, dust, and large radius turns.

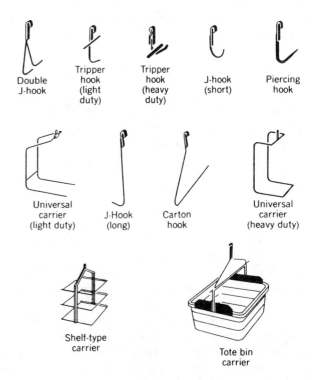

Double J-hook

Tripper hook (light duty)

Tripper hook (heavy duty)

J-hook (short)

Piercing hook

Universal carrier (light duty)

J-Hook (long)

Carton hook

Universal carrier (heavy duty)

Shelf-type carrier

Tote bin carrier

FIGURE 17.23 Many attachments are available with trolley conveyors.

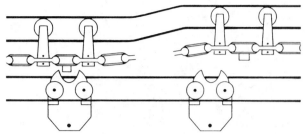

FIGURE 17.24   Powered and free conveyors have a powered track and an unpowered (free) rail. The power track has a series of trolleys, connected and driven by a motor-driven chain. The free track is parallel to the power track—below or to one side. Load carriers are suspended from a set of trolleys on the free rack. Dogs on the power chain mate with extensions on the free track carriers. Typical speeds are 40–60 ft/min.

selector pin hits the bump, the switching control is actuated. Destinations can be set manually by manually dropping selector pins on the front of the carrier. They can also be set automatically by a sensor reading a bar code on the vehicle or package, looking up the code meaning in its memory, finding that it should send that trolley to destination X, and then sending it. Electrical switching using a reed switch is a third option. Many systems permit automatic connecting and disconnecting of the carriers at the start and stop points. For example, for order picking, a carrier could have several picking locations programmed. If a specific location is busy, then the carrier goes on to the next location. Later it will be put back into the main line loop and return to the missed location. For similar self-propelled automatically guided vehicles (AGV), see Chapter 18. Carts with loads also can be stored in automatic storage/retrieval systems. The system then can pick a cart and send it to any destination without any human labor requirement. Thus, automatic handling can be combined with route flexibility—a very powerful combination.

An example trolley system might connect receiving, components areas, assembly, and shipping with spurs for inspection, rework, and spare carriers; in all there may be 100 stations and 500 carriers. Painting, plating, and heat treating can be done either on the main line or on spurs, in either powered or free sections.

Carriers can be either vehicles or suspended carriers. Vehicles need hardened floors under the rail and they vibrate the load. Multiple floor applications need ramps. (To keep a cart level as it goes up and down ramps, use a cart with the rear wheels farther apart than the front wheels. On an up ramp, the rear wheel track is mounted higher than the front wheel track; on a down ramp, the front wheel track is higher than the rear wheel track.) The conveyor generally needs to be a constant height above the floor. When the vehicle is disconnected from the line, storage and ease of movement are better than when using a suspended carrier. Suspended carriers, however, can be stored on spurs in the air—air space is generally the cheapest storage space. Rotating and indexing the carrier may be easier for suspended carriers. (Products might be rotated

continuously while being sprayed, washed, or painted. Indexing attachments turn a load 90° when striking a ripper bar.) Although vehicles can carry a heavier load than can carriers suspended from a single trolley, a rigid bar can be attached to several trolleys and the carrier suspended from the bar so that several trolleys carry the load. One firm uses vehicles that are carried suspended 1 in. above the floor. This system eliminates all vibration problems but permits easy movement of the vehicle when it is disconnected from the line.

Vehicles can be pulled from overhead tracks or tracks in the floor. Most installations use rails in the floor, because of the lower capital cost. Maintenance, however, tends to be easier when the rail is suspended and thus readily accessible.

A suspended self-powered carrier, or monotractor, is now available; it is an aerial version of an automatic guided vehicle. The I beam that carries the chain in the power and free is replaced with electrification for driving the electric motor on each carrier. The monotractors can pick up a carrier from a power and free conveyor, transport it through a monotractor system, and redeposit it into a power and free system.

Since carriers push each other, bumpers are needed. Most simple is a fixed pushing bumper. Next is an energy-absorbing bumper. Safety accumulation bumpers automatically release the cart from the tow pin when the bumper hits an object. They also can apply brakes to prevent roll-back. Accumulation pressure (products on carriers pushing each other on nonpowered spurs and causing damage) can be reduced by a variety of commercially available cart bumpers, interlocks, line disconnects, and the like.

Another option is a self-unloading cart such as a tilt shelf cart or a roller-bed cart interfacing with a roller conveyor.

Trolley conveyors may "surge" and the conveyed products swing instead of moving steadily. Common remedies are conveyor support sway bracing (especially at turns and long, straight runs), repair of failed roller turn rollers or traction wheel turns on the horizontal turns of the conveyor, and use of the correct lubricant (if all else fails, follow the instructions).

3.5 Power

Drives can be **constant speed** or **variable speed**. Most conveyors use constant speed; manufacturing or assembly may require daily or weekly changes, so there a variable-speed motor is used.

Constant-speed motors almost always are AC squirrel cage induction motors because of their low capital and maintenance cost. Generally, the motor and speed reducer are combined into one unit—a gearmotor. If the gearmotor output shaft is parallel to the motor shaft, spur or helical gears are used; efficiency is about 95%. If the two shafts are at right angles, then worm gears are used; efficiency is 55–90%. Right-angle gearmotors are used widely, however, since they are easy to mount to the conveyor frame.

Variable speed generally is accomplished in one of four ways. With an AC motor with two sets of windings and a selector switch, a load can be accelerated or decelerated at the slow speed and operated at the high speed. Second, turning a crank on a sheave or using a variable speed pulley, produces maximum speed change of about 50%. Third, you can electronically control an AC motor by simply turning a dial. Precise speeds can be obtained—important if several conveyors must be synchronized or speeds changed often. Fourth, you can electronically control a DC motor. Ease of operation is the same as with an AC motor. Capital cost is lower but motor maintenance is higher; speed control may not be as good as with an AC.

Most adjustable-speed drives have built-in "soft-start" characteristics (no jerks); this feature not only protects the product but also improves the life of conveyor bearings and transmissions.

On critical conveyors it may be desirable to install two drives adjacent to one another at every drive point. A maintenance operator can transfer the drive chain from a failed drive to the backup drive in about two minutes.

Chain drive sprockets should have at least 15 teeth. Sprockets should have an odd number of teeth, because tooth wear is evenly distributed when each tooth engages the chain once for every two revolutions. (If the sprocket has an even number of teeth, only every other tooth engages the chain.) Lubrication is a must; the proper method—manual, drip, oil bath, or pressure—depends upon the application.

4 CONVEYOR CONTROL

The control system has four basic elements: input interface, logic, memory, and output interface. The input interface takes voltages or currents from switches, relays, and temperature or pressure sensors; it then transforms these power signals into power levels suitable for the control system logic device. The logic takes the input and actuates devices in specific sequences. Simple time-based functions can be controlled by cams, and event-based functions can be controlled by counters. Complicated functions use microcomputers, minicomputers, or programmable controllers. The memory can range from the human operator's brain to computer memories to cams. The output electro-mechanical devices and motors generally need line voltage; the output interface boosts the low-powered logic commands.

The following discussion will cover photocontrols where the information is discrete (usually presence or absence of an item on the conveyor). For situations in which information is more complex, also read the section in Chapter 16 on bar codes.

Photocontrols sense the absence or presence of an object in the light-beam path; they also count and sort.

With timers, they sense motion as well as position (Becker, June, 1981; July 1981; Aug. 1981).

Figure 17.25 shows the six basic elements: power supply, light source, sensor, amplifier, timing and latch/inhibit section, and output.

Most lamps are LED (light-emitting diodes) rather than incandescent, as the LED lamp is considerably more reliable and can be focused more precisely. Often an infrared (800–900 nm) light is used with a sensor which receives only in that band; this makes the system more resistant to signals from ambient light, as well as invisible to the human eye. The LED also can be strobed by pulsing the light at 6,000–7,000 Hz; since the sensor is synchronized to accept light only at that frequency, ambient light rarely causes a signal.

Figure 17.26 shows five light-target-sensor arrangements.

The amplifier can have either a light mode or a dark mode (or a switch permitting either mode). If it is light operated, action takes place upon beam completion; if dark operated, action takes place upon beam interruption.

Figure 17.27 shows how a sensor can be either light operated or dark operated. Figure 17.28 shows a break-beam photocontrol for jam detection on a conveyor carrying cans.

The output can be passive (switch an external circuit) or active (provide a DC pulse for input to an external device such as a computer control system). Some common outputs are to switches, diverters, scales, relays, lights, and push rods. Shielding should be used on the signal conductors; signal conductors and power conductors should not be run in the same conduit.

The preceding discussed photocontrol proximity sensors. Three other types of proximity switches are radio frequency inductive, magnetic bridge inductive, and ultrasonic beam. The sensor also could be contact sensitive (e.g.,

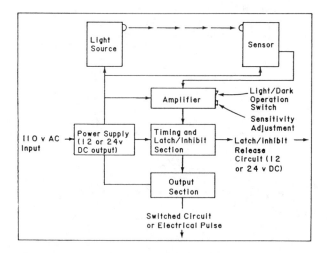

FIGURE 17.25 Typical photoelectric controls have six basic elements—power supply, light source, sensor, amplifier, timing and latch/inhibit, and output (Becker, June 1981).

limit switches). Temperature and pressure sensors can protect components from overloads. In addition, networks of controls often are connected by a **programmable controller (PC)**. The PC basically is a small, rugged, easily programmed computer which can do logic, computation,

and coordination; it tends to have a relatively small memory. The PC, in turn, often is controlled by a higher level computer. Low-level computers tend to control continuously (real time), whereas high-level computers supervise in an intermittent (batch) approach.

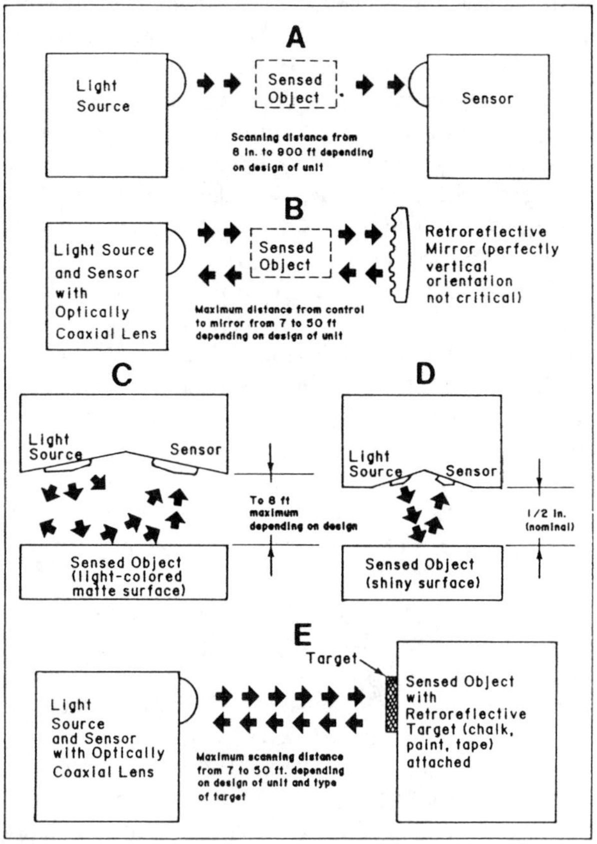

FIGURE 17.26   Five scanning principles are shown (Becker, June 1981): A = beam-break through beam; B = beam-break retroreflective; C = diffuse-reflective proximity; D = specular-reflective proximity; and E = reflective retro-reflective.

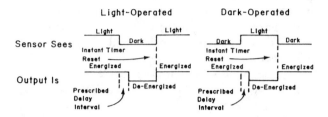

FIGURE 17.27   Light- or dark-operated sensors can be used (Becker, July 1981). "Drop-delay" timing delays output for a prescribed interval after the beam is broken. Drop-delay often is used to postpone action until the object reaches a location, such as moving to a diverter gate or a tilt table on the conveyor.

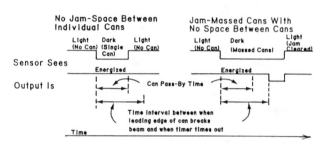

FIGURE 17.28   Jam detection for cans on a conveyor shows no jam (left) and jam (right) (Becker, July 1981).

DESIGN CHECKLIST: CONVEYORS

Conveyor function
    Transportation
    Storage
    Pacing
Load/unload downhill?
Gravity
    Chutes, wheels, rollers
    Portable?
    Load weight, surfaces
    Turns (T or curves)
    Crossing (bridge, gate)

Powered
    Pneumatic
    Roller
    Belts
    Tow conveyors
        Carrier type
        Spurs
        Switching
        Constant versus variable speed
Conveyor control

REVIEW QUESTIONS

1. Give the three types of nonpowered conveyors.
2. Discuss conveyor height and pallet height: (a) when a pallet is loaded from a conveyor and (b) when a conveyor is loaded from a pallet.
3. What is conveyor pitch?
4. When should gravity roller conveyors be used instead of gravity wheel conveyors?
5. For gravity conveyors, when should right angles be used and when curves?
6. If empty cardboard boxes were to be moved, would you recommend a wheel or roller conveyor? Why?
7. How do turntables, rollovers, and tilters differ?
8. For pneumatic conveyors, relate positive and negative pressure systems and the number of entry and discharge points.
9. How can the effects of accumulation pressure be reduced?
10. Sketch a power and free conveyor. Why are they so popular?
11. Discuss "taxi" versus "bus" conveyors.
12. Discuss suspended versus nonsuspended carriers on trolley conveyors.
13. Give the four basic elements of a control system and briefly describe what they do.
14. Give the four common ways of obtaining a variable-speed conveyor.
15. Sketch a control system to detect cans jamming on a conveyor.
16. What does a conveyor plow do?
17. How can you estimate if a box will tumble on an inclined conveyor?
18. What is standard speed for a powered conveyor?
19. Briefly describe the two types of beds for belt conveyors.

PROBLEMS AND PROJECTS

17.1 Visit a local factory or warehouse; observe a specific conveyor. Using a manufacturer's catalog, list all the catalog numbers necessary to purchase the conveyor. What is the approximate price of the conveyor?

17.2 Visit a local factory or warehouse that uses conveyors. How are they powered? How is power transmitted to the conveyor? How is speed controlled? How are items on the conveyor sensed and switched?

17.3 For a specific application, have two people make a presentation why "their" manufacturer's model of a conveyor should be purchased. The class then will vote on whose model will be purchased.

17.4 Read 3 to 5 articles on a topic related to this chapter. Summarize them in a 500-word executive summary. List the references. Don't use articles cited in the text.

REFERENCES

Basic Training Manual No. 1, Application Information for Conveyors. Hytrol Conveyors. Jonesboro, AR 72401; 1982.

Basic Training Manual No. 2, Application Information for Sortation Conveyors. Hytrol Conveyors. Jonesboro, AR 72401; 1984.

Becker, B. Understanding modern photocontrol (1). *Plant Engineering*, Vol. 35, 12, 105–108, June 11, 1981.

Becker, B. Understanding modern photocontrol (2). *Plant Engineering*, Vol. 35, 15, 85–88, July 23, 1981.

Becker, B. Understanding modern photocontrol (3). *Plant Engineering*, Vol. 35, 16, 127–130, August 6, 1981.

Browning, A. City transport of the future: Part 1. *Applied Ergonomics*, Vol. 5, 4, 225–231, 1974.

Browning, A. City transport of the future: Part 2. *Applied Ergonomics*, Vol. 6, 1, 17–22, 1975.

Colijin, H. Designing belt conveyors for bulk materials. *Plant Engineering*, Vol. 37, 25, 80–81, Nov. 23, 1983.

Considerations for Conveyor Sortation Systems. Charlotte, N.C.: Material Handling Alliance, 1989.

DeLaura, D., and Konz, S. Toe space. *Advances in Industrial Ergonomics and Safety II*, B. Das, Ed., London: Taylor and Francis, 1990.

Equipment overview. *Plant Engineering*, Vol. 48, 180–189, July 9, 1992.

Fitzpatrick, D. Selecting scrap metal handling conveyors. *Plant Engineering*, Vol. 37, 21, 76–79, Oct. 13, 1983.

Smart pneumatic tube system. *Material Handling Engineering*, Vol. 32, 2, 58–62, Feb. 1978.

18 WIDE-AREA EQUIPMENT (VEHICLES)

CHAPTER PREVIEW

Vehicles can service an unbounded area. Vehicles are subdivided into operator walks, operator rides, and no operator. Power sources can be human muscle or mechanical (internal combustion and electrical). Vehicle economics is discussed.

CHAPTER CONTENTS

1 Basic Vehicle Types
2 Accessories
3 Power Sources
4 Economics

KEY CONCEPTS

automated guided vehicle (AGV)	hydraulic hand pallet trucks	sleep mode
boxcar special	internal combustion (IC) engine	steer (versus drive) wheels
counterbalanced (versus straddle)	load capacity	two-wheel hand trucks
dual fuel	narrow-width aisles	unit trains
economic service life	order picking	vertical clearance height
floor (platform) trucks	random (versus dedicated) storage	walkie trucks
fork positioners	reach trucks	
free lift	sideshifters	

There are many types of industrial vehicles, so only a brief overview can be given here. In addition, because of technology and the desire of manufacturers to satisfy their customers, variety seems to increase and distinctions blur. For example, when does an order-picking vehicle moving between two high rows of storage racks cease to be a vehicle and become a crane? When does an automated, automatically guided tractor-trailer train cease to be a vehicle and become a conveyor? With this in mind, this material will be divided into basic vehicle types, accessories, power sources, and economics.

1 BASIC VEHICLE TYPES

Basic vehicle types can be distinguished by whether the vehicle moves horizontally only or both horizontally and vertically. Another distinction is whether the operator walks or rides, or whether there is no operator.

1.1 Operator Walks
Within this section, we discuss vehicles of increasing size and power. See Table 18.1.

1.11 Two-Wheel Hand Trucks.
See Figures 18.1 and 18.2. These trucks are used for lightweight, reasonable cube loads infrequently moved short distances horizontally. An industrial application is a wheeled fire extinguisher. Distribution and service applications are for beverage handling, for sales and service workers moving equipment or displays, for baggage transfer, and for appliance moving.

1.12 Floor (Platform) Trucks.
See Figure 18.3. These trucks also are used for lightweight, reasonable cube loads infrequently moved short distances horizontally.

FIGURE 18.1   Two-wheel hand trucks have a wide variety of features. The nose plate slides under the load. Popular accessories are height extension for the back, hooks to hold garments, barrel and pail attachments, stair climbers, and collapsible backs that allow the truck to fit into a car trunk. Since the power is from humans, the emphasis has been on lightweight trucks with hard wheels (for small start-up resistance).

However, since the item generally is lifted manually onto the truck, weights tend to be less than with the two-wheeled truck; the cube, however, can be considerably greater. These are often used for moving materials to and from workstations, for order picking, and for stocking in retail stores.

1.13 Hydraulic Hand Pallet Trucks.
See Figure 18.4. These trucks are used to move lightly loaded pallets short distances (50 ft or less). They lift the pallet only enough to move it, not to stack it. Since they are quite inexpensive, they interface well with powered mobile equipment, freeing the more expensive equipment for long hauls of heavy loads that require stacking and unstacking. For example, they can be kept on board a semitrailer for

TABLE 18.1   Characteristics of transportation vehicles (Mulcahy, 1991).

Vehicle Type	Intersecting Aisle Width (ft)	1991 Cost[1] ($1,000)	Operator's Position	Guidance	Floor	No. of Standard Loads[2]	Power	Travel Speed, mph Loaded	Travel Speed, mph Empty
Truck, manual low lift	10	.5–1	Walk	—	Normal	1	Human	Varies	Varies
Truck, powered walkie	10–12	6–7	Walk	—	Normal	1	Battery	Varies	Varies
Truck, powered walkie rider end control	10–12	8–9	End	—	Normal	1–2	Battery	4.7	6.1
Truck, powered walkie rider mid/end control	10–12	9–10	Middle, end	—	Normal	1–2	Battery	4.7	6.1
Truck, powered driverless pallet	12	45–50	—	Wire optical	Dead level	1–2	Battery	2.3	2.3
Tugger, manual control	10	8–9	Walk, end	—	Normal	5 carts	Battery, gas, LPG, diesel	4.7	6.1
Tugger, driverless	10	45–50	—	Wire optical	Dead level	5 carts	Battery	2.3	2.3
Automated guided vehicle	10–12	45–50	—	Wire optical	Dead level	1–2	Battery	2.1	2.1

[1]Includes battery and charger.
[2]Based on 48 × 40 in. pallet with 2,000 lbs.

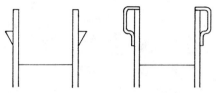

FIGURE 18.2   Protect the hands from contact with other objects. (Note guards on right.)

situations where no docks are available: The pallet truck is used to move pallets to the truck tailgate, where the load is transferred to a lift truck on the ground. There is a wide variety of options, sizes, and models to choose from. One important feature is a pallet entry device, of which several types are available. They are especially useful for empty pallets, which have little weight and thus move away when an operator tries to spear them with the forks. Rather than buy trucks in two or three capacity ranges, most people now just buy the largest capacity (4,500 lb), since the capital cost is low and needs may change over time.

1.14 Powered Walkie Trucks.
See Figure 18.5 and Table 18.1. Travel distances usually are 125 ft or less. Loads carried horizontally usually are less than 6,000 lb; loads lifted vertically are less than 4,000 lb, and the height of lift is less than 16 ft. Because of their relatively light weight and small size (compared with driver-ride vehicles), they are used in crowded, congested areas or areas permitting only light floor loads. Capital cost is about half that of a riding unit of equal capacity, so some firms buy two walkies instead, gaining flexibility and minimizing downtime.

NON-TILT STYLE   FOUR-WHEEL TILT STYLE

A-FRAME STYLE   SHELF STYLE

FIGURE 18.3   Floor (platform) trucks, used for mobile storage, have a variety of features (Floor Trucks, 1977). Push handles, shelves, sides, tubing stakes, etc., can be added or omitted. Welded steel frames and wooden decks are the most popular. The running gear comes in two types: nontilt and balance-tilt. The nontilt design generally has two rigid and two swivel wheels with the wheels at the four corners. Balance-tilt designs have the two fixed wheels centered on the trucks with swivel casters centered on each end. Thus, trucks can turn within their own length and are easy to tilt to clear door sills and other obstructions. Figure courtesy of Material Handling Engineering.

FIGURE 18.4   Hydraulic hand pallet trucks are inexpensive, flexible complements to powered equipment.

Energy cost also is about half that of a comparable rider unit. The operator does not accomplish as much with a walkie, but many firms use a production operator when the need arises rather than using a full-time vehicle operator.

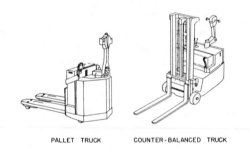

PALLET TRUCK   COUNTER-BALANCED TRUCK

STRADDLE TRUCK

FIGURE 18.5   Powered **walkie trucks** have three basic types: low-lift (pallet), straddle, and counterbalanced. Low-lift (pallet) types are a powered extension of hydraulic hand pallet trucks. The straddle and counterbalanced types solve the balance problem by lifting the load in two different ways. **Straddle trucks** have structural extensions (straddles) to the frame. The load fits inside the straddles. A diagram of the torques shows reaction forces outside the load. The **counterbalanced** type sets up an opposite torque to the load torque by putting weight at the end of a "long" moment arm. The resulting truck tends to be twice the weight and overall length of a straddle model. Thus, straddles can work in smaller spaces and tend to be cheaper; their disadvantage is that the pallet must fit between the straddles. When transporting a load with a counterbalanced type, lower the load to improve stability. Figure courtesy of Material Handling Engineering.

1.2 Operator Rides   These vehicles come in a large variety. Certain types approach conveyors (e.g., order-picking trucks in automatic storage/retrieval units). Stand-up rider trucks, shorter than sit-down models of the same capacity, can work in narrower aisles.

1.21 Trucks.   See Table 18.1. Subdivisions of the "basic" truck category are burden (cargo)/personnel carriers and tractor-trailer units.

Personnel/burden (cargo) carriers. See Figure 18.6. These units usually can carry small amounts of cargo (300 to over 3,000 lb) plus one or more people. Many carriers can tow light loads. The major justification for carrying people is the reduction of walking time through the plant and its resulting fatigue. Typical users would be supervisors, order pickers, expeditors, and maintenance and craft workers. Cargo carriers supplement lift trucks and often are used to move mail, maintenance equipment, and housekeeping equipment. A good concept is to put specialized equipment (such as coolant sludge collection pumps and tanks or welding equipment) on detachable platforms. Then, if either the vehicle or equipment needs repair, the unit can be detached. It also permits one vehicle to carry different equipment at different times. Power sources could be legs (bikes and tricycles), gasoline and propane, or, increasingly popular, electric. Three-wheel vehicles tend to be more maneuverable and four-wheel vehicles more stable.

Tow tractors and trailers.   See Figure 18.7. Lift trucks are expensive. If a load must travel a considerable distance (e.g., 150 ft), consider using a team of material handling units with the loading and unloading done by one unit (a

lift truck) and the transport done by another unit (a tow tractor plus trailers). One tractor can serve many trailers, since tractors are not needed while the trailers are being loaded (unloaded) or are used as temporary storage (order picking, stocking, on-floor storage). Trailers can all be the same or varied. If the tractor is not independently powered but pulled by a chain, it is called a *tow conveyor* (see Chapter 17 for more information). Increasingly the tractor is driverless (see Section 1.3 for more details).

The railroads have been very pleased with **unit trains** for coal. The train and cars are never uncoupled. They are loaded, moved to their destination, unloaded while still coupled (special coupling permits the car to be rotated while coupled), and returned empty. The expense of the return empty is more than recovered by the reduced paperwork cost and better equipment utilization. Some tow tractors are used with trailers (e.g., to move airplanes at airports).

1.22 Lift Trucks.   See Figure 18.8 and Table 18.2. "Use the cube efficiently" is a key storage principle. Other storage principles are given in Chapter 11. To implement these principles, manufacturers have developed many devices.

One technique of using the cube efficiently is to minimize the number of aisles. If racks can hold pallets two deep, then fewer aisles are needed than for one-deep storage. This technique, however, requires being able to put in and take out from the interior position. **Reach trucks**, forks on a scissors arrangement, have been developed to do this. Another technique for efficient cubes is **narrow-width aisles**. Aisle width is set by truck size and maneuverability. Trucks can (and have) been made shorter and narrower. Narrow-aisle vehicles are slower, both horizontally and in lifting speed, than the general-purpose counterbalanced truck. Generally, narrow-aisle trucks are walkies

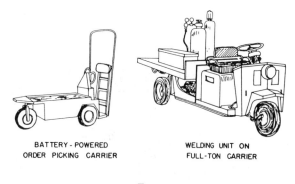

BATTERY - POWERED
ORDER PICKING CARRIER

WELDING UNIT ON
FULL-TON CARRIER

BATTERY - POWERED
THREE - WHEEL CARRIER

FIGURE 18.6   Personnel (burden) carriers generally carry both people and cargo. Shown are a carrier with welding equipment, an order-picking carrier, and a three-wheeled personnel carrier (an industrial version of a golf cart). Figure courtesy of *Material Handling Engineering.*

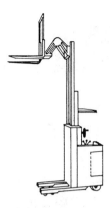

Rider Reach Truck

FIGURE 18.8 Lift trucks move the load vertically and horizontally. Shown is a reach truck, which has the capability to reach into racks without moving the truck itself. See Figure 18.9 for a sit-down counterbalanced truck with different attachments.

or have the operator stand instead of sit. See Chapter 10 for comments on aisles versus counterbalanced, straddle, reach, side-loader, turret, swingmast, and stock-picking trucks.

Retrieval of the load from these racks is known as **order picking**. If less than a unit load is moved, an order picker may ride on a vehicle-mounted platform moving along the racks. Simple platforms move vertically or horizontally in turn; more advanced units can do both simultaneously (i.e., diagonally). Units that move between aisles as well as along the aisle are called *trucks*; units restricted to one aisle are called *cranes*. For more on cranes see Chapter

19; for more on automatic storage/retrieval see Chapter 11; for more on vehicle accessories see Section 2 of this chapter.

1.3 No Driver   In 1992, the cost of operating a lift truck for 7 years (its optimum life) was $198,000. Of that total, 71% was the wage cost of the operator. This economic fact of life has led to efforts to automate the vehicle and thus eliminate the wage cost.

The **automated guided vehicle (AGV)** or "rolling robot" can replace the driver. The AGV can carry the load itself (i.e., as a truck or a unit load) or can be the tractor (towing a train of carts). Unit load AGVs are versatile: They can go forward and backward and deposit or retrieve a load. Their small size minimizes traffic congestion. Tow carts are better over long distances; there may be turning radius problems. They work best when there are just a few pickup and delivery locations.

AGVs are similar to tow conveyors. However, tow conveyors draw power from a cable connected to a central motor, whereas AGVs have a self-contained power source, making them a more flexible alternative.

AGVs are guided magnetically, by wire in the floor, or optically, by paint on the floor detectable under ultraviolet (UV) light. (For human convenience, UV paint can be on a tape laid on the floor, allowing humans to see it also. UV paint can be put on any surface from concrete to carpet.) Beneath the vehicle is a UV light and sensors. One advantage of magnetic guidance is that the wire can serve as a communication link to a computer, although radio is also an alternative.

TABLE 18.2   Characteristics of transportation deposit/retrieval vehicles (Mulcahy, 1991). Also see Table 11.2.

Vehicle Type	Aisle Width[1] (ft)	Transfer Aisle Width (ft)	Stacking Height (ft)	1991 Cost[2] ($1,000)	Operator's Position	Type of Floor	Storage Position	Power	Transactions per Hour[1]	Fork Direction	Height of First Level
Wide aisle											
Stacker, manual push walkie	6–7	6–7	10	5–6	Walk floor	Normal	Floor, rack	Battery	10–12	Front	Floor
Stacker, counterbalanced walkie	11–12	6–8	12	12–13	Walk floor	Normal	Floor, rack	Battery	10–12	Front	Floor
Stacker, straddle walkie	9–10	6–8	12	12–13	Walk floor	Normal	Floor, rack	Battery	10–12	Front	Floor
Lift truck, sitdown counterbalanced	12–13	10–11	18	23–25	Ride, floor level, sit down	Normal	Floor, rack	Battery, gas, diesel, LPG	20–25	Front	Floor
Lift truck, standup counterbalanced	9–11	8–9	19	23–25	Ride, floor level, stand up	Normal	Floor, rack	Battery	20–25	Front	Floor
Narrow aisle											
Straddle	7–8	7.5	20	20–22	Ride, floor level, stand up	Normal	Floor, rack	Battery	18–20	Front	Floor
Straddle reach	8–10	8.5	22	23–25	Ride, floor level, stand up	Normal	Floor, rack	Battery	15–18	Front	Floor
Straddle reach, double deep	9–10	8.5	25	30–34	Ride, floor level, stand up	Normal	Floor, rack	Battery	13–16	Front	Floor

[1]Based on a 48 × 40 in. pallet with 2,000 lbs.
[2]Includes battery and charger.

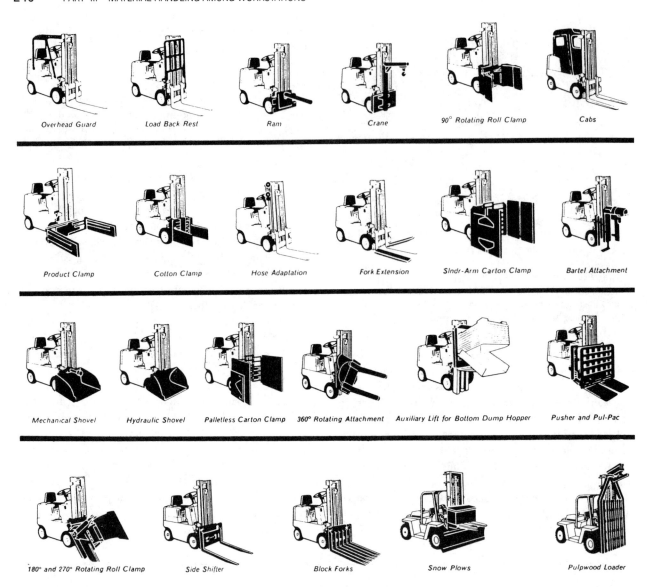

FIGURE 18.9 Attachments for lift trucks include forks, clamps, sideshifters, rotators, rams, push-pull, and fork positioners. A sit-down counterbalanced truck is shown here, but the attachments can be used on other truck types.

A design decision is how "smart" to make the vehicle. The "brain" can be an onboard computer or a central computer that sends directions to the vehicle. If the central computer controls machine tools (central numerical control or CNC) and an automated storage/retrieval system (AS/RS), it could also control the vehicle.

The AGV can be nonprogrammable and follow a fixed route (be like a bus) or can be programmed or directed (be like a taxi).

A mail delivery robot is an example of a nonprogrammable vehicle. It makes all stops on a route, stopping at each location for a specified time period (e.g., 20 s). An employee wanting more unloading time at a stop hits a button on the vehicle; when finished, the person hits a different button to restart the vehicle. Sensors can automatically operate doors and elevators. Safety has not been a problem, because these robots travel only about 1 mile/h and detect obstacles (e.g., pedestrians) with a proximity sensor. (They tend to have personalized names such as "Oscar," "Archie Bumper," "Little Go Beep," and "Harvey Wallbanger.")

The programmable units can be sent to any location, generally over a variety of paths. To minimize congestion, strategically locate by-pass sections. The units avoid collisions at intersections either through an on-board collision-avoidance system or through central computer control. If the AGV is an assembly platform, it may have vertical positioning ability. Moving assembly platforms eliminate the linear assembly line and result in parallel assembly, with great flexibility in scheduling, product mix, and line balancing. Duplicate assembly stations reduce problems of quality, reliability, and downtime.

Battery drain is a potential problem. Two techniques are (1) a **sleep mode** (the vehicle power consumption is reduced whenever the AGV is not moving) and (2) opportunity charging. In opportunity charging, the computer sends the AGV to a recharging station whenever it is not in service. However opportunity charging requires special rapid-charging batteries that need little regular maintenance.

Experience has shown that design of an AGV system benefits from a simulation of the system before installation. The simulation can show traffic bottlenecks, guide-path interferences, inefficient vehicle scheduling, and number of AGVs required.

2 ACCESSORIES

Accessory divisions discussed are attachments, tires, and controls.

2.1 Attachments

See Figure 18.9. Although a vehicle generally uses only one attachment, couplers are available which permit changing attachments in a minute or so—usually without the driver leaving the seat.

The most popular attachment is the fork, which is used with pallets. Clamp attachments eliminate the pallet. Pallets, along with their vertical storage space, can also be eliminated by putting the load on sheets of corrugated cardboard or plastic (slipsheets). A push-pull device grasps the lip of the slipsheet and pulls the load and sheet onto a polished platform on the truck. The push plate pushes the load off onto the floor or on top of another load. See Chapter 16 for more on pallets and slipsheets.

A **sideshifter** is a popular option. It shifts the attachment from side to side so that the operator can engage the attachment to the load without repositioning the truck. **Fork positioners** move the forks closer or farther apart by hydraulics controlled from the operator's seat. See Figure 18.10. They are popular when a variety of pallet sizes are used or when the forks are heavy and thus hard to move by hand. Shaped clamps, often with special facing surfaces, are used for cylindrical loads such as rolls and drums. Flat clamps with adjustable pressure regulators are used to avoid carton crushing. Rotators, teamed with forks or clamps, are useful for dumping. Rams are useful for coils. Because of their mobility, trucks often are the first equipment to reach a small fire. Equip each with a fire extinguisher.

2.2 Tires and Casters

Tires and casters are of two basic types: press-on solid (further divided into standard and cushion) and pneumatic (Torok, 1990). The cushion has a thicker tread, giving improved load-carrying capacity; it provides a softer ride, resulting in longer vehicle life. A lift truck designed for standard tires may not have enough tire clearance for cushion tires.

Pneumatic tires are inflated and are available in tube and tubeless versions. Pneumatics are used mostly for indoor/outdoor operations. If puncturing is a problem, use a pneumatic filled with a soft polyurethane instead of air.

Tires can be made of various compounds. Special compounds resist oil, give extra traction in wet and slippery conditions, or do not mark (used in food applications). Others are static conducting or nonconducting or are designed for high speed. Polyurethane tires combine many of these features (at a premium price) and, with their low rolling resistance, are a good choice for vehicles without power steering. Polyurethane tires are sensitive to heat (very hot environments, high speeds).

Tread patterns for solid tires are either smooth or traction (treads). Pneumatic tires generally have a tread, so they can be used in mud, sand, and snow. Use a smooth tire on a smooth surface and a traction tire on a rough surface.

The tire circumference can be flat (no change in radius with tire width, giving maximum surface contact) or contoured (easy steering). In good conditions, use smooth, contoured tires on the **steer wheels** and smooth, flat tires on the **drive tires**.

Casters (used for hand trucks and carts) are a low capital cost item, so buy the best quality to minimize maintenance and downtime cost. For example, buy casters with sealed bearings. Select casters for the worst load conditions, not just normal conditions. Although steel wheels have the greatest strength and roll easily, they are noisy and hard on floors. Molded rubber and polyurethane are the most popular.

2.3 Controls

For lift trucks, the general trend has been to take away as much discretionary control of the vehicle as possible. Solid state controls, such as a silicon-controlled rectifier, can protect both the drive and the attachment units from operator abuse while improving energy efficiency. A load moment overload device permits a load to be picked up only if it is within truck capabilities. Automatic transmissions are easier to operate than clutches, give smoother load handling, and reduce total truck maintenance costs.

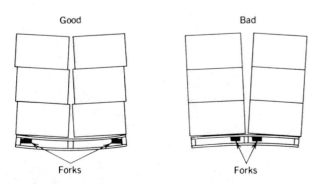

Good   Bad

FIGURE 18.10 Fork positioners aid in load stabilization.

3 POWER SOURCES

The primary types of power are human and mechanical.

3.1 Muscle Powered
Until approximately 1900, most vehicles were powered with muscles, either animal or human. Bicycle pedaling, the most efficient technique of human-generated power, is only 20–25% efficient. That is, for every 100 W metabolized by the person, only 20 W are available from the machine—the 80 W become heat. On a sustained basis a human can maintain a work output of 1/10 hp (75 W) at a cost of 1/2 hp (375 W). Strindberg and Petersson (1972) estimated maximum push force equals .8 (body weight). Snook and Ciriello (1991) reported 90% of U. S. male industrial workers can push (at a 95 cm height for 8 h) with an initial force of at least 34 kg and pull with an initial force of at least 32 kg; they can sustain a pushing force of at least 23 kg and pulling force of 24 kg. As these numbers are not very different and safety on inclines may be a problem, keep cart loads pushed or pulled lower than the worker at all times. Pushing is aided by dry floors and shoes with a good coefficient of friction. For dry floors, rubber cork soles have a coefficient of .83 (averaged over 18 types of floors) and leather soles have .51. See Chapter 9 for floor characteristics.

3.2 Mechanical Powered
The subdivisions of mechanical power are **internal-combustion (IC) engine** (gasoline, liquid petroleum (LP), and diesel) and electric.

To compare power costs on lift trucks, assume 1 gallon of gasoline = 5.35 lb of LP gas = 2 KWH of electricity. The electricity cost is the cost of charging the batteries.

3.21 Internal Combustion.
Gasoline powers the standard IC engine. Diesel offers more fuel efficiency (25 to 30% more miles/gallon and lower cost/gallon) and 20 to 30% longer engine life as well as less downtime and maintenance; the tradeoffs are a higher capital cost engine and more fumes. Diesels are more popular in large engines. LP produces fewer fumes than gasoline, easy conversion from gasoline, and fuel economy (LP gets about 90% of gasoline's mileage and tends to cost about 70% of gasoline); the engine costs more initially. Because of the special equipment for LP fueling, it tends to be used only when 15 or more vehicles use LP. The fueling station should meet National Fire Protection Association Standard 58. Many engines now are **dual-fuel** (gasoline and LP), so the user can switch as fuel costs change. Since fuel costs have risen relative to labor costs, engines have been designed to be more fuel efficient. Fumes have been reduced, and so has required maintenance. As always, however, abusive driving or not doing the required maintenance will degrade performance of even the best engine. Maximum grade for a IC engine lift truck is 15% versus 10% for electric.

3.22 Electric.
As a general rule of thumb, an electric truck, including batteries and charger, will cost twice the purchase price of a comparable IC engine truck. Why do people buy electrics? Their lower energy cost and lower maintenance cost make the total cost over the equipment life considerably lower for electrics. In addition, electrics are quiet and fume free. Electric trucks permit fewer environmental air changes/hour, giving savings in heating and air-conditioning costs. Turner and Estes (1980) report that replacing a propane-powered lift truck with an equivalent electric truck can reduce ventilation by 5,000 ft³/min—a savings in annual heating costs of $750/shift for a plant with typical heating costs.

Battery maintenance over the battery's life costs twice the battery purchase cost. This has increased the use of low-maintenance (less water) batteries and emphasizes the importance of preventing battery overcharge—the main cause of premature battery failure. Formerly, some electric trucks, especially in heavy-duty applications, were not able to complete a shift before requiring recharge but, with improved technology, this is rarely the case today. With fixed-interval charging, if the truck is used just one shift, the battery can be recharged while in the vehicle; if the truck is used multiple shifts, the battery must be removed and a new one inserted at the end of the shift. An alternate, more energy-efficient, procedure is random-interval charging. Batteries are changed and charged whenever they are 80% discharged, as shown by an indicator on the truck control panel. The random-interval procedure is less costly since batteries are used the way they are designed to be used, smaller batteries can be used, and the work load in the charging area is spread. After conversion from an end-of-shift charge schedule to a uniform-depth-of-discharge schedule, AC power for charging was reduced 12 to 18% and peak power 30 to 40% (Knowing Your Batteries, 1979). See Chapter 14 for information on battery charge areas.

4 ECONOMICS

Economic subdivisions are procurement and operation.

4.1 Procurement
This section is divided into specifications and purchasing.

4.11 Specifications.
The first decision to make is the specific type of truck and accessories to purchase. Specify the following:

- lifting capacity (adding a safety factor of 1,000–2,000 lb to the largest anticipated load)
- straddle or counterbalanced
- lifting height
- vehicle power (IC engine or electric)
- accessories

- driver walk, stand or sit
- operational environment (dirt or concrete surface, indoors or outdoors)
- any other requirements

A key specification is the vehicle's load capacity.

Lift trucks operating in trucks and railcars often must be limited in height, length, and width for proper door clearance and tight maneuverability. A **boxcar special** truck has the counterweight placed vertically instead of horizontally; this gives a length from the mast front to the rear of 88 in. (vs. 108 in. for the standard truck) and a height to the top of the roof guard of 79 in. (vs. 81 in.).

Load capacity is given as the pounds of load for a given moment arm. The key equation is the following:

$$TOR = LL(W) + AL(AW) \qquad (1)$$

where

TOR = torque due to load + accessory, in.-lb

LL = load moment arm, in.

= $A + B$

A = distance from center line of front axle to front face of fork

B = Distance from front face of fork to load center of gravity

W = weight of load, lb

AL = distance from center line of front axle to accessory center of gravity, in.

AW = accessory weight, lb

For loads less than 20,000 lb, the rated load assumes that B = 24 in. Thus, if rated load = 4,000 lb, A = 16, and there is no accessory in addition to the fork, permitted TOR = (16 + 24)(4,000) = 160,000 in.-lb.

What is the effect of different load lengths? For lengths of 96 and 40 in. use the following:

Load Length (in.)	B	A	$A + B$	Permissible Load (lb)
96	48	16	64	160,000/64 = 2,500
40	20	16	36	160,000/36 = 4,400

What is the effect of an accessory? Assume AL = 20 in. and AW = 500 lb. Then capacity is reduced by 20(500) = 10,000 in.-lb. In addition, the load is farther out since the accessory took space. If we assume distance from accessory center of gravity and load edge is 1 in., then the results are as follows:

Load Length (in.)	B	A	$A + B$	Permissible Load (lb)
96	48	20 + 1	69	150,000/69 = 2,100
40	20	20 + 1	41	150,000/41 = 3,600

Equation 1 is an approximation when height is less than 13 ft. If it is above 13 ft, decrease permitted TOR by 4,800 in.-lb for every 12 in. of lift. Manufacturers make more precise calculations of TOR considering mast deflections, tire characteristics, etc., but equation 1 will give preliminary truck size determination for estimating purposes.

Free lift is the vertical distance the forks can move from the floor before the mast begins to extend; it is important in low-headroom areas such as trucks, freight cars, or under mezzanines. **Vertical clearance height** can be set by the height of the roof guard or by the collapsed mast height. Wheel base is the distance between the centerlines of the front and rear axles. It is important in aisle width requirements and in clearance at the end of a grade.

4.12 Purchasing.
Although some purchases are to increase the number of vehicles, most purchases are to *replace* a vehicle. When should a vehicle be replaced? Box 18.1 gives an example replacement calculation.

Whyte (1980), who works for a vehicle manufacturer, reports an **economic service life** (the time a vehicle operates near maximum efficiency) of 10,000–12,000 h for an IC engine vehicle and 12,000–15,000 h for an electric. As vehicles operate beyond their economic service life, maintenance and downtime costs increase until replacing the vehicle becomes more economical. In general, a replacement is superior to the unit presently in service, not equal to it. For example, 1990-model IC lift trucks consume about 10% less fuel and require about 10% less maintenance than do 1980 vehicles (Economics, 1991). Whyte reported the mean hours at replacement were 22,800 h for IC vehicles and 26,200 for electrics. He suspected that most vehicles were kept too long.

Tables 18.1 and 18.2 give example purchase costs for new vehicles. The residual value of a six-year-old lift truck averages as 20% of current list price. Used batteries and chargers for electric lift trucks also average about 20%.

About 50% of all lift trucks sold each year are used. "As is" accounts for about 40%, reconditioned about 50%, and rebuilt about 10%. (Reconditioned trucks have malfunctioning parts replaced or repaired; rebuilt trucks are completely disassembled, inspected, repaired, and repainted.)

Of course, leasing also is an option, especially for short terms such as one or two years. Renting is an option for very short periods such as a week or a month.

Table 18.3 gives some operating cost estimates. Fuel costs depend not only on the fuel consumption rate and the fuel cost but also on daily hours of use. For a single shift, a vehicle usually operates 5–7 h, but due to motor idling when the vehicle is not in use, actual working hours may be even lower.

Fueling an IC engine can be done off site or on site. On-site fueling saves money in fuel costs but requires a storage tank (with fire and waste management problems). Fueling on site is recommended if there are over 20 IC vehicles. LP distributors periodically deliver full tanks and pick up empty ones.

The fuel for an electric truck is the battery, so a purchaser effectively buys the fuel with the truck. A typical industrial lead battery used with a 5,000 lb capacity electric truck costs about $4,000. An electric lift truck fleet requires (1) one or more batteries per truck, (2) a battery charger for each truck, (3) battery changing equipment to remove and replace batteries from the trucks, and (4) a charging area. See Chapter 14 for information on charging areas. A charger should be able to recharge a fully discharged battery within 8 h to allow the battery to cool down before returning to service. (Failure to provide an adequate cool-down period warps the battery plates and reduces battery life about one-third.)

TABLE 18.3 Operating costs of lift trucks (Economics, 1991).

	Gasoline	*LPG*	*Diesel*	*Electric*
Fuel cost				
Fuel consumption/shift	7	11	6	31
Fuel cost	1.20	.85	1.10	.09
Fuel cost/shift, $	8.40	9.35	6.60	2.79
Work days/year	260	260	260	260
Annual fuel cost/shift, $	2184	2431	1716	725
Maintenance cost				
Material	1140	1140	1140	700
Labor				
Hours/year	50	50	50	29
Cost/year at $35/h	1750	1750	1750	1015
Total maintenance	2890	2890	2890	1715
Annual operating cost/shift	5074	5321	4606	2440

BOX 18.1 *Equipment replacement justification*

To buy or not to buy—that is the question. It will depend upon the "bottom line"—the return on investment or its inverse, payback. That is, a 25% return on investment is a four-year payback.

Three numbers are needed: (1) capital cost, (2) life of the project, and (3) annual savings. See Fig. 18.11 for a calculation form for the following example. Assume you are trying to decide which, if any, lift truck to buy to replace an existing 3000 lb capacity lift truck. Although the example is for a lift truck, the same procedure can be used for conveyors, cranes, lathes, drill presses, and other equipment.

Capital cost Assume the present truck "Model X" has a present value of $5000 and can be sold one year from now for $4000. Then its capital cost is $1000. Truck Y has an initial cost of $20,000. Accessories add $1800. Freight and shipping add $400, and operator retraining is assumed as $50. Assume no engineering charge and no tax writeoffs. Then Truck Y will cost $22,250. Assume it can be sold for $1250 for scrap at the end of its life.

Project life For the existing equipment, consider the cost to keep it *one* more year. (If you don't replace it, you can repeat the calculation a year from now. It is assumed that operating cost/yr will increase faster than capital costs/yr will decline. In addition, presumably next year will be better for the old item than the years following, so this is the most favorable analysis possible for the existing item.) Thus present truck X will have a life of 1 year. For new items, use the years that you will have the piece of equipment. This could be limited by either the project life or the equipment life. For example, you may be producing this specific item for 5 more years but think the economic life of the vehicle is 10 years, because if you don't make the present product, you will be making something else. In that case, use 10 years.

Capital cost/year Present truck X will cost ($5000 − $4000)/1 = $1000 to keep for one more year. Proposed truck Y will cost ($22,250 − $1250)/10 = $2100/yr. Thus truck Y has an incremental annual capital cost of $2100 − $1000 = $1100/yr over truck X.

Annual costs and benefits Note that to calculate the savings/yr, you can either: (1) calculate the cost of the present and the cost of the proposed and then subtract, or (2) calculate the savings directly. Sometimes one technique is easier and sometimes the other.

What will we get for that $1100/yr difference in capital costs? Fig. 18.11 shows some of the possibilities.

Probably most important will be changes in productivity and quality from the new piece of equipment. After all, manufacturers generally have made some improvements over the years in the equipment and so the new equipment should be more productive than the old. Because labor costs are so high relative to equipment costs, even a small percent change in labor costs becomes very important. Rather than calculate the cost of the present and cost of the old, let's assume that new truck Y is 7% more productive than present truck X. Assuming operator cost of $20,000/yr (wages plus fringes), then .07 (20,000) = $1400/yr. There also may be improvements in costs of other people (assume as $250/yr), product quality (assume as $200/yr), etc. There also may be changes in utility costs (vehicles or other equipment). Assume the present truck X uses $1000/yr of power and the new truck will use an average of $850 (because of more efficient batteries and controls). Maintenance would be higher for the present truck. Assume the present truck X will require $3500 for one more year. Proposed truck Y will require an average of $2800/yr to maintain for the 10 years, low at first and higher later. There are other savings possibilities but they will vary with the type of equipment being analyzed (conveyors, lathes, presses, etc.). For our example, there is a net annual savings of $1400 + 250 + 200 + 150 + 700 = $2700.

Cost/benefit calculations All that remains is to compare the costs and the benefits. The annual benefits are $2700 − 1100 = $1600. The annual capital costs are $1100. Thus return on investment before income taxes is $1600/1100 = 145%; payback = 1/1.45 = .7 year.

The above analysis neglected the effect of interest rates on the capital expenditure and of inflation on the savings; as a rough

(continued)

BOX 18.1 Continued

approximation they can be considered to cancel out. Taxes, both capital and income, are ignored. Typically, the after tax rate of return is 60% of before tax rate of return. However, for large capital expenditures you may wish to do a more refined analysis considering interest costs, inflation, and cash flows; see an engineering economics text such as Jones (1982).

Project __3000 lb lift truck__

Part name_____ Part number_____ Used in Dept. B

Volume:_____ pcs/y _____ pcs/day _____ pcs/h _____ h/pc

Labor cost/h (including fringes) __$10__ Engineer __SK__ Date __Nov 15__

A. One-time (capital) costs	Present	Proposal A	Proposal B
Equipment, $	5000	20,000	_____
Accessories, fixtures, $	_____	1,800	_____
Shipping, installation, $	_____	400	_____
Operator retraining, $	_____	50	_____
Engineering, $	_____	_____	_____
Line 1 Total	5,000	22,250	_____
Line 2 Salvage, scrap, resale value, $	4,000	1,250	_____
Line 3 Application life, years	1	10	_____
B. Capital cost/yr, $/yr			
Line 4 (Line 1 - Line 2)/Line 3	1000	2100	_____
Line 5 Net Savings		1100	_____
C. Annual costs and benefits, $/yr (Cost = -; savings = +)			
Direct labor, $/yr	_____	1400	_____
Indirect labor, $/yr	_____	250	_____
Downtime and maintenance, $/yr	_____	700	_____
Material, direct and indirect, $/yr	_____	_____	_____
Product quality, $/yr	_____	200	_____
Rework and scrap, $/yr	_____	_____	_____
Utilities, $/yr	_____	150	_____
Line 6 Total savings, $/yr	_____	2700	_____
Line 7 D. Cost/benefit calculations			
Annual net savings (Line 6 - Line 5)	_____	1600	_____
Return on investment before taxes, % Line 8 Line 7 x 100/Line 5		145%	_____

FIG. 18.11   Return on investment example for lift trucks.

Maintenance cost is quite difficult to estimate, and many firms do not keep good records. Engine hours are a better estimate of vehicle use than calendar hours, but this method requires an engine-hour meter on the vehicle. A side benefit of an engine-hour meter is the ability to estimate equipment utilization. (For example, if a vehicle had 60 engine hours during May but 21 days × 7.5 h/day of scheduled work time, then utilization was 60/157.5 = 38%.) Vehicle manufacturers point to a strong age effect—an IC engine may require $400 of maintenance in year 1 and $8,000 in year 8.

Maintenance can be contracted out (generally the best alternative if the firm has fewer than 10–15 trucks) or done in-house. If the maintenance is contracted out, vehicle availability may be lower than with in-house maintenance because the vendor may have a longer turnaround time. The firm may need additional vehicles to act as standby units.

Table 18.3 gives some estimated maintenance costs. A rule of thumb is to replace a vehicle before cumulative maintenance costs equal purchase price.

Electric trucks average 20% less downtime than IC engine trucks (Economics, 1991).

4.2　Operation

Efficient storage layout is a key to efficient vehicle operation. For more on storage, see Chapter 11.

A key decision is **random versus fixed (dedicated) storage**. Most installations probably should have *some* items in fixed locations. Random storage, which requires a computer to keep track of where the items are kept, tends to fill space more efficiently. This means the total area can be smaller, thus reducing truck travel distance. In addition, space closest to the end of the aisle can be used well and space at the end of the aisle used rarely—again minimizing travel. Since vehicles can move horizontally much faster than forks move vertically, keep slow-moving items high and fast movers low. (For order picking, it might be better to make two passes down the aisle, one at a high level and one at a low level.)

Long aisles should have occasional cross-aisles to permit trucks to move from aisle to aisle. Wire guidance in aisles not only permits narrower aisles but also allows the vehicle to move automatically between picking points while the operator does other work, such as putting parts into packages. Computers can help by sequencing the picking locations as well as completing paperwork (so the operator keeps the vehicle moving rather than doing paperwork). For small parts, it may be possible to pick for several orders simultaneously. See Chapter 10 for more on aisles.

In any layout, match truck age (in engine hours) inversely with work hours per week. That is, schedule old trucks for few hours per week and new trucks for demanding work. Have some vehicles with couplers so that a variety of attachments can be used.

A lift truck is too expensive for long horizontal moves. It should stay in a zone, and movement between zones should be by tractor-trailer. However, these zones should be short term (changing several times/shift or week) or the issue of ownership may interfere with use. (A supervisor doesn't want to "lend" a truck to another department.) Use over a larger area doesn't permit direct supervision, however, so two-way radio communication may be desirable. Schedule loads within days and within the week. Most loads don't need to be moved immediately. Even out loads on the shipping dock by shipping all week instead of emphasizing Thursday and Friday. Schedule moves to reduce deadheading (returning empty).

More and more the vehicle is computer directed for the best sequence of jobs (system management). Control can be real-time, with communication to the on-board computer by wire or by radio. Batch control also is used when the operator stops at a terminal periodically (say once/hour) to tell the computer what was just accomplished and to receive orders for the next period. For programming simplicity, it may be best to divide the plant into zones.

DESIGN CHECKLIST: WIDE-AREA EQUIPMENT (VEHICLES)

1. What is the load? Weight and volume/load? Container? Packaging? What is accomplished during transport? Storage? Records update? What if container (load, packaging) were changed?
2. Where to? (One source to one destination; one source to many; many to one?) Distance moved? Change in elevation?
3. How is move made? Why not crane? Conveyor? Under what circumstances could a vehicle be replaced by a crane? Conveyor? Why was this specific type of vehicle chosen? Size and model?
4. Why and when is move made? Moves/shift = ? Can move be sooner or later? Effect of more moves/shift? Of less moves/shift? Who decides frequency? Effect of more storage at start and end of move?
5. Who will drive vehicle? One person or multiple? Automated vehicle? Full- or part-time job? Driver's job classification?
6. Maintenance and fuel space provided?

REVIEW QUESTIONS

1. If your organization has fork trucks, why use hydraulic hand pallet trucks also?
2. If your organization has fork trucks, why would it use tow tractors and trailers?
3. How does a tow truck differ from a tow conveyor?
4. List five different lift truck attachments.
5. Electric lift trucks have a capital cost approximately twice that of IC engine trucks. Why do people buy them?
6. Give the formula for availability.

7. How heavy a load can be pushed and pulled by a person?

8. Discuss the advantages and disadvantages of diesel, gasoline, and LP engines.

9. Sketch how a truck with straddles balances the load and how a counterbalanced truck balances the load. Sketch the wheels and the load.

10. How are AGVs guided?

11. Give two techniques to reduce AGV battery drain.

12. How does a "boxcar special" lift truck differ from a standard lift truck?

13. What happens if batteries are not allowed to cool down after charging?

PROBLEMS AND PROJECTS

18.1 For a specific factory or warehouse, make an economic analysis of whether movement of personnel should be by walking, bike, or powered vehicle. What does the supervisor think of your ideas?

18.2 Contact a manufacturer to get data concerning a lift truck with the same capability but different power options (diesel, gasoline, LP, or electric). Make an economic analysis of which power option to use in a specific situation. Justify your recommendation.

18.3 Make an economic analysis of buying versus leasing an automobile.

18.4 For a specific application, have two people make a presentation why "their" manufacturer's model of a vehicle should be purchased. The class then will vote on whose model will be purchased.

REFERENCES

Economics of truck ownership, Educational supplement of Nissan, 425 N. Martingale Road, Schaumburg, IL 60173; 1991.

Floor trucks/platform trucks. *Material Handling Engineering*, Vol. 32, 4, 62–65, (Ref. 18), 1977.

Knowing your batteries better can save your money. *Material Handling Engineering*, Vol. 34, 12, 64–67, Dec. 1979.

Mulcahy, D. Unit handling and storage vehicles. *Plant Engineering*, 56–67, May 2, 1991.

Snook, S., and Ciriello, V. The design of manual handling tasks: revised tables of maximum acceptable weights and forces. *Ergonomics*, Vol. 34, 9, 1197–1213, 1991.

Strindberg, L., and Petersson, N. Measurement of force perception in pushing trolleys. *Ergonomics*, Vol. 15, 4, 435–438, 1972.

Torok, D. Take a hard look at your life truck tires. *Material Handling Engineering*, 64–69, June 1990.

Turner, W., and Estes, C. How best to save energy in materials handling. *Modern Materials Handling*, Vol. 35, 9, 94–99, Sept. 1980.

Whyte, P. Reduce high cost of aging lift trucks. *Industrial Engineering*, Vol. 12, 1, 24–27, Jan. 1980.

19 LIFTING DEVICES

CHAPTER PREVIEW

Lifting devices generally service a limited area. As with conveyors and vehicles, there are many alternatives. Hoists and cranes can be fixed or mobile; they can be powered by muscle, air, or electricity. The load can be attached by supporting, clamping, or surface attaching. Control can range from manual to computer control.

CHAPTER CONTENTS

1 Crane Types
2 Crane Accessories

KEY CONCEPTS

bridge cranes   monorail
hoists   stacker cranes
jib cranes   straddle cranes
lifters and clamps

1 CRANE TYPES

As with vehicles and conveyors, there is a wide variety of overhead lifting equipment. They might be called **hoists** or cranes. A primary division is whether they are fixed (cover a limited area) or are mobile, that is, vehicle mounted. (Balancers, manipulators, and robots are covered in Chapter 20.)

1.1 Fixed (Limited Area)   The small hoist can be powered by hand, by electric motors, or by compressed air. See Figure 19.1 and Table 19.1. Hand-operated hoists, force multipliers, are used primarily in low-frequency-of-lift, low-load situations such as maintenance. The differential chain hoist is the lowest capital cost and least efficient (35% efficiency) hoist. The screw or worm gear hoist is slightly more expensive and efficient (40%). Spur gear hoists are still more expensive and efficient (85%). As hoists are used more frequently and for heavier loads, electric or air motors replace muscle. Air hoists cost more than comparable electric hoists—both in capital and power costs, and are noisier. However, they have variable speed, are spark free, do not overheat on quick repetitive tasks, and simply stall out when overloaded.

Hoist selection depends upon the duty cycle. Table 19.2 gives the five hoist classifications. Class H4 requires close coordination with the hoist supplier if you want the proper hoist. Class H5 is special design, a noncatalog item (McNelis, 1982).

Hoists can be located in one place. An example is a hydraulic hoist used to lift cars at a service station. Similar units are used for loaders, stackers, and positioners in the factory and for dock levelers on the shipping dock. Lift tables are very useful for holding material at the input or

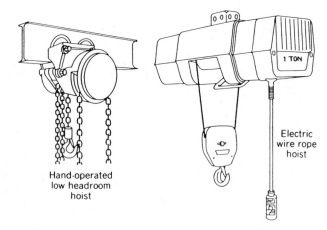

Hand-operated low headroom hoist

Electric wire rope hoist

FIGURE 19.1   Hoists can be manual, electric, or air powered.

output of a machine. In addition to reducing back strain from lifting, they may convert a two-person job into a one-person job. However, since the primary purpose of the lift usually includes moving the object to another location, more commonly hoists will move along a short beam. The result is a **jib crane**. See Figure 19.2 and Table 19.3. A hoist on a long beam is a **monorail** or I-beam trolley. Figure 19.3 shows five different methods of mounting a hoist.

Air-film equipment forces compressed air into a chamber under the load (equipment, pallet, die, etc.). Models with a reservoir can cross gaps and obstructions, such as door sills. A 1 lb force can move a 1,000 lb load. Loads up to 80,000 lb then can be pushed by hand. They are very convenient for assembly lines of heavy equipment since they allow precision location and can be used where there are no cranes. (Lack of cranes can be due to ceiling height, ceiling strength limitations, or crane capital cost.) They also permit easy maneuvering around columns. Entire storage

TABLE 19.1   Hoist characteristics (Hoists, 1991).

Type	Typical Lifting Speed (ft/min)	Load Capacity (tons)	Typical Lifting Height (ft)	Approximate Cost ($)	Application Notes
Hand					
Chain	—	½–50	10–50	300–16,000	Periodic, vertical lifting of loads
Ratchet	—	¼–6	5–10	150–2,000	Hoisting and pulling of loads
Powered					
Electric wire rope	—	—	—	—	Commonly used on large cranes; rope operates over grooved drum
• Light duty	15	½–2	19	3,000–22,000	For less demanding and infrequent or standby operations
• Standard industrial	15	½–20	23	3,000–22,000	Typical lifting applications in manufacturing facilities
Electric chain	15	½–5	10–20	750–5,000	Various lifting applications in smaller handling jobs; chain may trail or store in container
Air chain	Variable	½–2	20	1,300–2,500	Ideal for variable-speed, explosion-proof operations; less overheating
Air wire rope	Variable	2–10	20	5,300–12,000	Similar application features as air chain, usually with greater operating range

TABLE 19.2 Hoist duty service classification (McNelis, 1982). From "Selecting and Applying Overhead Hoists," *Plant Engineering*, 4/1/82.

Hoist Class	Service Classification	Typical Areas of Application
H1	Infrequent or standby	Powerhouses and utilities, infrequent handling. Hoists are used primarily to install and service heavy equipment; loads frequently approach hoist capacity; periods of utilization are infrequent and widely scattered.
H2	Light	Light machine shop, fabrication, and service and maintenance work. Loads and use are distributed, with capacity loads handled infrequently; total running time of equipment does not exceed 10 to 15% of the work period.
H3	Standard	General machine shop, fabrication, assembly, storage, and warehousing. Loads and use are randomly distributed; total running time of equipment does not exceed 15 to 25% of the work period.
H4	Heavy	High-volume handling in steel warehouses, machine shops, fabricating plants, mills, and foundries. Common applications include heat treating and plating operations; total running time of equipment normally approaches 25 to 50% of work period, with loads at or near rated capacity.
H5	Severe	Bulk handling of material in combination with buckets, magnets, or other heavy attachments. Equipment is often cab operated; duty cycles approach continuous operation; user must specify exact details of operation, including weight of attachments.

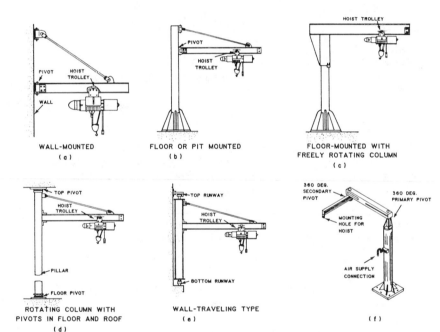

WALL-MOUNTED
(a)

FLOOR OR PIT MOUNTED
(b)

FLOOR-MOUNTED WITH FREELY ROTATING COLUMN
(c)

ROTATING COLUMN WITH PIVOTS IN FLOOR AND ROOF
(d)

WALL-TRAVELING TYPE
(e)

(f)

FIGURE 19.2 Jib cranes come in different types. The wall mounted (*a*) covers a semicircle. Floor mounted on a mast (*b*) also covers 180°, although if the pivots are offset from the column the beam can swing as much as 270°. If the column can rotate (*c, d*), the beam can swing 360°. The wall-traveling type *(e)* covers a rectangle (Kulwiec, 1983). An articulated beam (*f*) can reach around obstructions, such as columns or machines, as well as give smooth lateral motion along a conveyor.

TABLE 19.3 Jib crane characteristics (Cranes, 1991). They are relatively inexpensive, provide 3 degrees of freedom (vertical, radial, and rotary), and are often used for localized activities.

Type	Typical Load Capacity (tons)	Span or Reach (ft)	Approximate Cost ($)	Application Notes
Wall mount	½–10	10–30	500–20,000	Mounted on existing columns; rotates through 180° arc
Pillar mount	½–20	10–20	1,000–30,000	Large area of coverage with 360° rotation; relocated readily
Mast mount	½–5	10–20	800–4,000	Rotates 360°; does not require foundation; must be vertically plumb
Motorized	1–20	10–30	10,000–45,000	Rides on runway mounted to building columns; serves localized operations; keeps center of bay clear for overhead crane and floor traffic

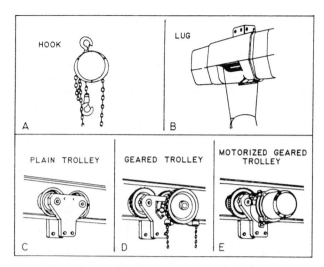

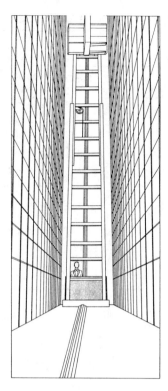

FIGURE 19.3 Hoist-mounting alternatives are shown (McNelis, 1982). View *A* shows hook suspension (good for portability). It can be rigid or swivel. Be careful about swivel hooks as they tend to wrap up the power cord as they rotate. View *B* shows a lug mounting; it can be used for stationary applications or be attached to bolt-type trolleys as in *C*, *D*, and *E*. View *C* is a plain trolley (for less than 3-ton loads). It is moved manually by pushing or pulling the load. It should be kept less than 16 ft above the floor (10–12 ft is better), since the operator at floor level is at the end of a long pendulum and finds it difficult to control the load and hoist. View *D* is a geared trolley. The hand chain, which moves it, permits precise positioning. View *E* is a motorized geared trolley. Use it when there are long distances, repetitive movements, and loads over 3 tons.

FIGURE 19.4 Stacker cranes move on rails in a warehouse.

racks can be air mounted and pushed together to conserve aisle space. Generally, a smooth concrete surface is needed. For one-time moves lay down sheets of plywood or plastic.

Stacker cranes, used in warehousing, move along a column which moves on rails in the floor or ceiling. See Figure 19.4. They are "fork trucks on a rail." Simple models have the operator walk; larger units have the operator ride. The more advanced models are "smarter," with computerized functions. Stacker cranes more and more are developing the same range of accessories as fork trucks—telescoping masts and various attachments (such as grabs and pushers) for lateral motions. A trend has been to increase the vertical capability of the crane as land prices force warehouses to grow vertically rather than horizontally. See automated storage/retrieval systems in Chapter 11 for more information.

Bridge cranes move items within a rectangle. The lifting units move on a "bridge," which moves along parallel rails in the ceiling. See Figures 19.5, 19.6 and Table 19.4. The bridge can cantilever beyond the rails. Bridge cranes minimize aisle space. Split cranes use two lifts per bridge. For example, with two bridge cranes, one works the north third of the bay, one works the south third, and both work the middle. With a split crane, each of the two bridge cranes is split into two; one handles the east half and one the west—giving four cranes in all. They combine only for heavy loads.

Hoists and cranes put stress on the building structure, which typically has a design life of 50 years. Thus, it may

not be possible to install a crane in an existing building, since the original structure was designed to support just the roof and walls. In designing for a crane, consider not just the weight of the hoist plus load on the structure but also the fatigue effects of 50 years' use.

1.2 Mobile (Vehicle Mounted)

Aerial platforms, lift tables, and maintenance lifts are used within and outside the plant. See Figure 19.7. Both towable and self-propelled models are available. The platform can be raised to replace ladders and scaffolds. Check that the collapsed height will pass through your doors.

Mobile hydraulic cranes are used in the yard for general material handling and maintenance as well as occasional construction. See Figure 19.8. The boom generally swivels; it also may telescope. Outriggers (stabilizers) often must be used for stability during the loading and unloading.

Overhead power lines present a safety problem for the person on the ground swinging a load into place (the crane operator normally has an insulated platform). The best alternative is to de-energize the power line. Other alternatives are for the crane operator to use remote control from the ground to get a better position estimate or for a spotter to tell the crane operator where the boom is.

Some cranes have a low silhouette which permits them to be brought indoors to lift machines or for maintenance on the crane. Rough terrain versions are available.

Straddle cranes (rubber tire, self-propelled, mobile gantries) straddle the load to be lifted. See Figure 19.9.

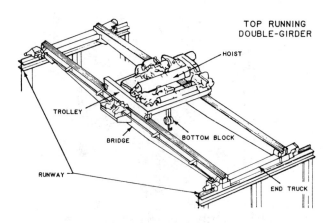

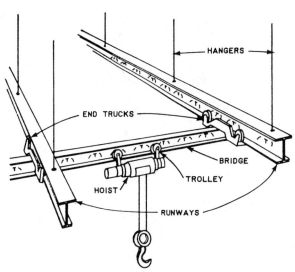

FIGURE 19.5 Top-running, double-girder bridge cranes provide the most headroom because the trolley rides on rails mounted on top of the bridge girders. The double girder gives good load distribution (Kulwiec, 1983). If the trolley rides on the lower flange of the bridge beams, it is an underhung double-girder bridge crane.

FIGURE 19.6 Underhung, single-girder bridge cranes have a maximum span of about 50 ft but can transfer the hoist (and thus load) between tracks (Kulwiec, 1983).

These specialized cranes require minimum aisle space and are very maneuverable.

A helicopter with a winch is a specialized crane.

2 CRANE ACCESSORIES

The first division within accessories is the attachment to the load, the second is the control of the crane.

2.1 Attachment to the Load
There are three ways of attaching to the load: supporting, clamping, and surface attaching.

Supporting attachments are subdivided further into beam and spreader lifters, C-frame lifters, and slide adjusting lifters. See Figure 19.10.

Beam and spreader **lifters** support the load at two or more points, so the load is more stable and the hook-to-load distance is small. Hooks should have latches, which

TABLE 19.4   Overhead bridge crane characteristics (Cranes, 1991). Primary crane for heavy in-plant handling. Provides full coverage of working area of bay.

Type	Typical Load Capacity (tons)	Span or Reach (ft)	Approximate Cost ($1,000)	Application Notes
Top running; single girder	1–25	20–60	6–50	Less costly than double girder, also less headroom; some types used for intermittent service in maintenance areas, machine shops, and pumping stations
Top running; double girder				
Structural girder	1–25	20–60	8–50	Often uses standard hoists; for general in-plant handling
Industrial (welded box girder)	10–150	30–150	25–1,000+	Similar to structural-girder type, but with greater capacities and longer spans
Powerhouse and service	20–250	60–200	50–1,000+	For heavy equipment, intermittent service and slower speeds; typically cab operated
Heavy duty industrial	5–50	20–100	20–500	Custom built; average work cycle is 15–20 lifts/h; dual brakes on hoists
Mill crane	5–350	40–150	200–3,000+	Custom built to rigorous standards for demanding duty cycles in difficult environments
Stacker crane	2–20	30–100	50–1,000	Permits high stacking of structural steel and plate
Underhung; single girder	1–30	10–60	10–50	Somewhat higher capacity than single-girder version; also better headroom
Underhung; double girder	½–25	10–60	6–45	Provides complete area coverage; transfers loads between bays with spurs and interlocks; least available headroom; often supported from roof trusses

FIGURE 19.7   Mobile cranes (platforms) are useful for many maintenance duties.

secure sling loads and also serve as a gauge, showing if the hook has been sprung.

C-frame lifters have a three-sided frame with the load inserted into the open side. They are commonly used to move coils and thus are called *coil lifters* or *hairpin hooks*. When the bottom of the C is two forks, then pallets and skids can be moved. Several C-frames are used for long loads such as extrusions.

Slide adjusting lifters have members that open and close horizontally to place lifting surfaces under the load. Another version is pipe tongs.

Clamping is the second major attaching category. See Figure 19.11. The load weight creates the gripping

FIGURE 19.8   Mobile hydraulic cranes often work with narrow aisles, so maneuverability is critical. Outriggers increase stability during load/unload (Hydraulic Cranes, 1979). Figure courtesy of Material Handling Engineering.

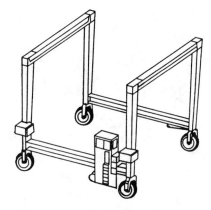

FIGURE 19.9   Straddle cranes minimize aisle space requirements (Straddle Cranes, 1974). Figure courtesy of Material Handling Engineering.

pressures. A mechanical sheet grab picks up steel sheets or structural members. A sling **clamp** (chain dog sling) carries the load on the hook points. An expanding core lifter has a tapered wedge that expands in a hole such as a paper roll core.

Surface-attaching lifters—magnets or vacuum—the third attaching category, speed pickup and release of loads. See Figure 19.12. Since they hold onto top surfaces, they do not interfere with load positioning. Sometimes they are used with other lifters such as beam lifters. Magnets, used with iron and steel, require a reasonably clean, smooth load surface. Derate lifting capacity 50% for corrugated metal (because of lower contact area); loads with holes should be derated also (because of interrupted magnetic path). Magnets have approximately a 4 times greater lifting capability when the magnet surface versus load is horizontal than when vertical (Kulwiec, 1982). Use an emergency battery backup system to prevent load dropping if power fails. Vacuum pads can lift nonferrous, nonporous materials such as aluminum and glass; even concrete and paper can be lifted with additional pumps. Vacuum pads can be rigidly contoured or flexible.

Manipulating lifters use clamping, supporting, or surface-attaching methods but, in addition, turn the load about an axis. One example is a power-driven rotating hook; another is a barrel handler which, after holding a barrel horizontally, tips it to pour out the contents.

2.2 Control of the Crane   The simplest crane, a hand-powered hoist, uses human muscles to furnish power as well as control. However, most cranes are powered by electricity or air and are controlled by push-buttons and servomotors. Usually the push-buttons are suspended in a pendant, a control box hanging at the end of a cable. Another alternative is to put the push-buttons and the operator in a cab with the crane. This arrangement gives the operator good visibility and protection from the environment and can give greater travel speed. Some attachments, however, require someone to "work the floor," that is, attach and detach the load. A relatively recent alternative is radio control by an operator on the floor. The advantage of radio over pendant control is that the operator is free to move about to get the best view and the safest position; there is also an advantage over cab control in that the operator and floor worker may be one person instead of two.

Some cranes are controlled by computers, especially in high-rise rack storage situations. Some computers move the crane from point to point and let the operator make the final positioning; some do the whole job. Many crane minicomputers are tied into production and inventory information stored in larger computers, and so the information can be updated in real time. Sequences of tasks, decision rules (such as the shortest path to the next job), etc. can be programmed into either the minicomputer or the major computer. The crane operator thus becomes a supervisor while the slave does the work.

Electrical power to overhead traveling equipment can be supplied five ways: with coiled cords, cable reels, multi-cable carriers, festooned cables, and enclosed contact conductors. Coiled cords (as with telephones) work for low-voltage, short travel distance applications such as hoists on manually pushed trolleys. For longer travel distances and straight lines, cable reels keep tension on the cable and prevent cable droop. An alternative is to support the cable on rings (similar to curtain rings), so that it is festooned; that is, it hangs in loops. Enclosed contact conductors make power available along their entire length and thus can power multiple units; they also are useful for complex geometry (curves, switches, crossovers, turntables, spurs).

Box 19.1 describes the availability of the Department of Defense standard data system, which includes material handling times.

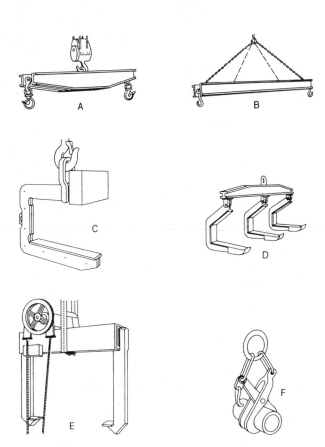

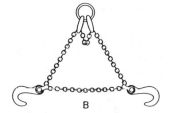

FIGURE 19.11 Clamping attachments hold the load by surface friction or indentation. View A shows a mechanical sheet grab, for sheets, plates and structural members. View B is a sling clamp. The load is on the hook points. Figure courtesy of Material Handling Engineering.

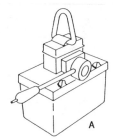

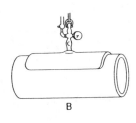

FIGURE 19.12 Surface-attaching lifters use magnets or vacuum. The handle in view A is used to remove the load from the magnet mechanically. View B shows that vacuum pads can be contoured. Figure courtesy of Material Handling Engineering.

FIGURE 19.10 Supporting attachments are divided into beam and spreader lifters, C-frame lifters, and slide adjusting lifters. View A shows a beam lifter and view B a spreader lifter. They add stability to lifted loads and have a minimum hook-to-load distance. Views C and D show C-frame lifters. View E shows a slide adjusting lifter (for coils), and view F shows a pipe tongs. Figure courtesy of Material Handling Engineering.

BOX 19.1   *Time standards for material handling*

The Department of Defense has developed an extensive standard data system. There are nine volumes:

Basic Volume:	*General Guidance* (procedures, allowances, forms, and a master index to operations/elements)
Vol. II:	*Clerical and Sales* (typing, filing, key-punching, etc.)
Vol. III:	*Service Occupations* (janitorial work, building services, etc.)
Vol. IV:	*Farming, Fishing, Forestry* (gardening, groundskeeping)
Vol. V:	*Processing Operations* (plating, pickling, cleaning, heat-treating, etc.)
Vol. VI:	*Machine Trades* (boring, milling, grinding, etc., plus aircraft, vehicle, engine maintenance)
Vol. VII:	*Bench Work* (assembly, forming, molding, sewing, repairing small objects)
Vol. VIII:	*Structural Work* (structural work aligned to building trades)
Vol. IX:	*Miscellaneous Occupations* (transportation services, packaging, warehousing, material handling, etc.)
Vol. X:	*Universal* (get, place, walk, read, write, etc.)

The entire package is available for around $40 (including shipping and handling) from:

Action Office
Defense Civilian Personnel Center
Ste. 302
Falls Church, VA 22041
(703) 756-0314

Example elements/operations from p. 66 of Vol. IX (which has 224 pages) follow:

Operation	Starts	Includes all the Time Necessary to	Ends	Time, $h \times .000\,010$
Sling, remove from part	With reach to loop end	Grasp the loop end, pull loop loose, grasp sling, move sling out of loop, pull sling from under part	With sling free of part and in hand	110
Sling, put around part or object	With move of sling to object	Place a sling around an object, pull end through loop, pull sling tight on object	With sling tight on object	241
Pallet, push on conveyor	With bend to pallet	Bend to and grasp pallet, start pallet in motion, push pallet 4 paces, release and stand up	With stand-up	165
Dock (hydraulic), operate	With a reach to the up button	Operate the hydraulic dock and the machine time to adjust the dock to the truck bed and the time to move the dock away from the truck bed	When the down button is released	2,009

DESIGN CHECKLIST: LIFTING DEVICES

Hoists
    Hand, electric, air
    Duty cycle
    Jib crane?
    Bridge crane
    Building structure strength

Air film
Mobile cranes
Attachment of load
Crane control

REVIEW QUESTIONS

1. What is the mechanical efficiency of a differential chain, screw or worm gear, and spur gear hoist?

2. Give the hoist class for a maintenance hoist used only a few times a year.

3. Why would a hook suspension be used instead of a lug mount?

4. Sketch a jib crane. What is the shape of the area covered for the type you sketched?

5. Sketch a bridge crane. What is the shape of the area covered?

6. Sketch a C-frame lifter attachment.

7. Give the two connecting media for surface-attaching lifters.

8. Give two types of materials for which lifting magnets must be derated.

9. What are the advantages of pendants, radios, and cabs for crane control?

PROBLEMS AND PROJECTS

19.1 Visit a factory. List all the cranes and hoists used. For each, give the type and size as well as accessories used. What is its power source? What is its service classification? How is each attached to the load? How is each controlled?

19.2 Visit a factory and see if you can think of a potential application for air-film equipment. What does the supervisor think of your ideas?

19.3 For a specific application, have two people make a presentation why "their" manufacturer's model of a crane (hoist) should be purchased. The class then will vote on whose model will be purchased.

19.4 Read 3 to 5 articles on a topic related to this chapter. Summarize them in a 500-word executive summary. List the references. Don't use articles cited in the text.

REFERENCES

Cranes. *Plant Engineering*, Vol. 47, 150–151, July 18, 1991.

Hoists, *Plant Engineering*, Vol. 47, 151, July 18, 1991.

Hydraulic cranes. *Material Handling Engineering*, Vol. 34, 9 (Ref. 34), Sept. 1979.

Kulwiec, R. How to select lifting magnets. *Plant Engineering*, Vol. 36, 16, 46–50, Aug. 5, 1982.

Kulwiec, R. Cranes for overhead handling. *Plant Engineering*, Vol. 37, 20, 34–46, Sept. 29, 1983.

McNelis, R. Selecting and applying overhead hoists. *Plant Engineering*, Vol. 36, 7, 65–68, April 1, 1982.

Straddle cranes. *Material Handling Engineering*, Vol. 29, 8 (Ref. 2), August 1974.

20 WORKSTATION MATERIAL HANDLING

CHAPTER PREVIEW

Material handling at a workstation can be manual; guidelines are given. Balancers and manipulators multiply human muscles but an operator is still required. Robots add a "brain," so an operator is not required. Complementing material handling robots (pick and place) are processing robots (welding, spray painting). For high-volume production, use fixed-purpose feeders and positioners.

CHAPTER CONTENTS

1 Manual
2 Balancers and Manipulators
3 Robots
4 Fixed Automation

KEY CONCEPTS

balancers	indexing (dial index) machine	recommended weight limit (RWL)
fixtured pallet	manipulator	remote center compliance (RCC)
flexible (versus fixed) automation	NIOSH *Lifting Guide*	transfer line
gripper (end-effector, end-of-arm tool, EOAT)	processing (versus pick and place) robots	vibratory feeder

1 MANUAL

1.1 NIOSH Guidelines

In 1993 the National Institute of Occupational Safety and Health (NIOSH) revised their *Lifting Guide* (Waters et al., 1993).

The basic concept is to have a load constant of 51 lb (23 kg). This is the maximum that can be lifted or lowered. The load constant is multiplied by various factors (all equal to or less than one) to obtain the **Recommended Weight Limit (RWL)**. If a person lifts less than RWL, NIOSH considers the task acceptable; if the person lifts more than RWL, NIOSH considers the task not acceptable.

1.11 Basic Formula.

The NIOSH formula is:

$$RWL = LC * HM * VM * DM * FM * AM * CM$$

where

RWL = recommended weight limit, lb
 LC = load constant = 51 lbs
 HM = horizontal multiplier, proportion
 VM = vertical multiplier, proportion
 DM = distance multiplier, proportion
 FM = frequency multiplier, proportion
 AM = asymmetry multiplier, proportion
 CM = coupling multiplier, proportion

For metric units, LC = 23 kg. Enter distances in cm.

The NIOSH formula *does not apply* if any of the following occur:

- lifting/lowering with one hand
- lifting/lowering for over 8 h
- lifting/lowering while seated or kneeling
- lifting/lowering in a restricted work space
- lifting/lowering people
- lifting/lowering hot, cold, or contaminated objects
- lifting/lowering while carrying, pushing, or pulling
- lifting/lowering with wheelbarrows or shovels
- lifting is "high speed" (not performed within a 2–4 s time frame)
- unreasonable foot/floor interface (< .4 coefficient of friction between the sole and the floor)
- unfavorable environment (temperature significantly outside 66–79°F (19–26°C) range; relative humidity outside 35–50% range)

The required input information depends upon whether control over the object is needed (1) only at the origin of the movement or (2) at origin and destination.

If control of the object is needed only at the origin, you need:

- initial HORIZONTAL LOCATION of the hands from the ankle midpoint
- initial VERTICAL LOCATION of the hands
- initial ANGLE OF ASYMMETRY of object center
- VERTICAL TRAVEL DISTANCE between the lift origin and destination
- FREQUENCY OF LIFTS per minute
- LIFTING DURATION (h) and RECOVERY TIME (h)
- HAND-CONTAINER COUPLING classification

If control is needed at both origin and destination, you need in addition:

- final HORIZONTAL LOCATION of the hands
- final VERTICAL LOCATION of the hands
- final ANGLE OF ASYMMETRY of object center

1.12 Formulas for Multipliers.

Horizontal multiplier.

$$HM = BIL/H$$

where

HM = horizontal multiplier, proportion
BIL = body interference limit = 10 for inches
= 25 for cm
 H = horizontal distance from the hand's center to the ankle midpoint. The maximum value of H (functional reach limit) = 25 inches (62.5 cm).

The value of HM declines as the hands move farther from the spine. If H <= BIL, then the value of HM = 1.0.

Vertical multiplier.

$$VM = 1 - VC \mid V{-}KH \mid$$

where

VM = vertical multiplier, proportion
VC = vertical constant = .0075 for inches
= .003 for cm
 V = initial vertical height of hand's center. Note that this usually is several inches above the container bottom. The maximum value of V (vertical reach limit) is 70 inches (175 cm).
KH = knuckle height (optimum height) of typical lifter = 30 inches (75 cm). When V <= 30 inches, the lifting is considered as whole body; when V > 30 inches, the lift is considered as upper body.

The value of VM declines from 1.0 for any height departing from optimum knuckle height of 30 inches. The concept is a 22.5% penalty for lifts from the floor or shoulder (60 inch; 150 cm).

Distance multiplier.

DM = .82 + DC/D

where

DM = distance multiplier, proportion

.82 = multiplier at maximum hand height of 70 inches (175 cm)

DC = distance constant   = 1.8 for inches
                          = 4.5 for cm

 D = distance moved vertically, inches or cm

If D <= 10 inches (25 cm), the value of DM = 1.0.

Asymmetry multiplier.

AM = 1 −.0032 A

where

AM = asymmetry multiplier, proportion

 A = angular deviation (degrees) of the midpoint of the two hands (container center) from straight ahead (sagittal plane). A can range from 0 to 135°; ignore direction as clockwise movement is considered equivalent to counterclockwise movement.

The concept is a 30% penalty for a 90° angle.

Frequency multiplier.   See Table 20.1. Lifting frequency can range from less than one in 5 minutes (.2 lifts/min) to 15 lifts/min. The frequency multiplier varies depending on LIFTING DURATION/SESSION and whether the initial vertical location of the hands is above or below typical knuckle height (30 in or 75 cm).

LIFTING DURATION/SESSION in hours has 3 categories:

* <= 1 h with RECOVERY TIME of at least 1.3 (DURATION)

* >   1 h but <= 2 h with RECOVERY TIME of at least .3 (DURATION)

* >   2 h but <= 8 h

During recovery time, the person is resting or has "light" work (such as sitting, standing, walking, monitoring). If a person does not meet the recovery criterion, use the next higher time category.

Coupling multiplier.   See Table 20.2. The coupling multiplier depends upon the height (i.e., whether V is < 30 inches) of the initial and final hand-container coupling and whether the coupling is good, fair, or poor. Table 20.3 defines good, fair, or poor coupling. Table 20.4 defines optimal container design, handle design and hand-hold cut-out design.

1.13 Example Problem with Solution.   Assume a container was moved and control was needed at both origin and destination.

Initial HORIZONTAL LOCATION = 12 in
Initial VERTICAL LOCATION   = 33 in
Final HORIZONTAL LOCATION = 12 in
Final VERTICAL LOCATION   = 22 in
Initial ANGLE OF ASYMMETRY = 5°
Final ANGLE OF ASYMMETRY = 6°
LIFT FREQUENCY   = .5 lifts/min
Lift DURATION/session   = 3 h
RECOVERY time   = 6 h

TABLE 20.1 Frequency multiplier (FM).

| | *Work Duration (Continuous)* | | | | | |
| | =< 8 HRS | | =< 2 HRS | | =< 1 HR | |
Lift/min	V<75cm (30 in.)	V>=75cm (30 in.)	V<75cm (30 in.)	V>=75cm (30 in.)	V<75cm (30 in.)	V>=75cm (30 in.)
=<0.2	.85	.85	.95	.95	1.00	1.00
0.5	.81	.81	.92	.92	.97	.97
1	.75	.75	.88	.88	.94	.94
2	.65	.65	.84	.84	.91	.91
3	.55	.55	.79	.79	.88	.88
4	.45	.45	.72	.72	.84	.84
5	.35	.35	.60	.60	.80	.80
6	.27	.27	.50	.50	.75	.75
7	.22	.22	.42	.42	.70	.70
8	.18	.18	.35	.35	.60	.60
9	.00	.15	.30	.30	.52	.52
10	.00	.13	.26	.26	.45	.45
11	.00	.00	.00	.23	.41	.41
12	.00	.00	.00	.21	.37	.37
13	.00	.00	.00	.00	.00	.34
14	.00	.00	.00	.00	.00	.31
15	.00	.00	.00	.00	.00	.28
>15	.00	.00	.00	.00	.00	.00

TABLE 20.2   Coupling multiplier (CM).

Couplings	V < 75 cm (30 in.)	V >= 75 cm (30 in.)
Good	1.00	1.00
Fair	0.95	1.00
Poor	0.90	0.90

Then   initial   HM =   10/12   = .83

Final   HM =   10/12   = .83

Initial   VM = $1 - .0075 |33-30|$ = .98

Final   VM = $1 - .0075 |22-30|$ = .94

Vertical distance moved = DM = $.82 + 1.8/11$ = .98

Frequency multiplier = FM = .81

Initial asymmetry multiplier   = AM = $1 - .0032$ (5) = .98

Final asymmetry multiplier   = AM = $1 - .0032$ (6) = .98

Initial coupling multiplier = CM = 1.0

Final coupling multiplier   = CM = .95

Thus RWL at lift origin = $51 * .83 * .98 * .98 * .81 *$ $.98 * 1.0 =$

= 32.6 lb

RWL at destination = $51 * .83 * .94 * .98 * .81 *$ $.98 * .95 =$

= 29.7 lb

The 29.7 lb at the destination is limiting as it is lower.

1.2 General Lifting Guidelines

Table 20.5 gives 10 guidelines for occasional lifting. There is one selection guideline, three technique guidelines, and six job design guidelines. Emphasize job design, since it is the most effective. Remember this from the open manhole problem. It is possible to (a) select individuals who can walk around holes, (b) train existing employees to walk around the hole, or (c) design the problem away by putting a cover on the hole. Designing the problem away is best.

See Chapter 15 for more on workstation design.

TABLE 20.3   Hand-to-CONTAINER coupling classification.

1. Good	—	Optimal container design with optimal handles or optimal hand-hold cut-outs
	—	Loose parts or irregular objects with comfortable grip (hand can easily be wrapped around object)
2. Fair	—	Optimal container design with non-optimal handles or non-optimal hand-hold cut-outs
	—	Loose parts with no handles or hand-hold cut-outs
	—	Irregular objects with a grip in which fingers can be flexed about 90 degrees
3. Poor	—	Non-optimal design containers with no handles or hand-hold cut-outs
	—	Loose parts or irregular objects that are bulky or hard to handle

TABLE 20.4   Definitions for Table 20.3.

	Optimal	Non-Optimal
Container design		
Frontal length	=< 40 cm (16 in.)	Non-optimal is
Height	=< 30 cm (12 in.)	failure to meet
Surface	smooth, non-slip	one or more
Edge	non-sharp	optimal condition
Center of mass	symmetric	
Load	stable	
Gloves required	no	
Handle design		
Diameter	1.9–3.8 cm (0.75–1.5 in.)	
Length	>= 11.5 cm (4.5 in.)	
Clearance	>= 5 cm (2 in.)	
Shape	cylindrical	
Hand-hold cut-out design		
Height	>= 3.8 cm (1.5 in.)	
Length	>= 11.5 cm (4.5 in.)	
Clearance	>= 5 cm (2 in.)	
Container thickness	>= 1.1 cm (.43 in.)	
Shape	semi-oval	
Surface	smooth, non-slip	

2 BALANCERS AND MANIPULATORS

Balancers and **manipulators** supplement an operator's muscles for material handling around a workstation. Consider balancers as "armless" manipulators (although they can be trolley mounted). See Figure 20.1. Hand tools often are suspended on balancers.

Manipulators are used to move a product from machine to machine or back and forth from a pallet or conveyor. The gravitational weight of the object is counterbalanced and therefore the operator lifts only a small part (e.g., 1%) of the object's weight. Originally the counterbalancing was with weights; now it tends to be compressed air or a spring. An air-operated balance control system is used when load weights are fairly uniform. Powered pneumatic control systems are used when a wide variety of load

TABLE 20.5   Guidelines for occasional lifting.

Guideline
Select Individual
1. Select strong people based on tests
Teach Technique
2. Bend the knees
3. Don't slip or jerk
4. Don't twist during the move
Design the job
5. Use machines
6. Move small weights often
7. Put a compact load in a convenient container
8. Get a good grip
9. Keep the load close to the body
10. Work at knuckle height

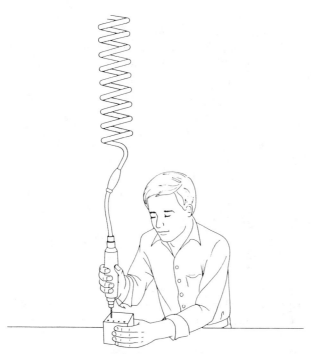

FIGURE 20.1 Balancers (manipulators) permit operators to move loads around a workstation as if the loads had only a small weight.

weights are handled or when all operator lifting effort is eliminated. Triaxial manipulators are used when objects vary (size, configuration, weight) and the pick-up and put-down locations vary. Manipulators (muscle multipliers) fill the gap between hoists and robots; they cost $5,000 to $15,000, whereas robots cost $25,000 and up. Manipulators also are easier to maintain than robots and can handle heavier loads.

In some manipulator (and crane) applications, double hooks can be used. First the operator picks up a raw part from a pallet using one side of the hook. Then the part and hook are moved to the machine. Next the finished part is unloaded from the machine, using the other side of the hook. Then the hook is swung around and the raw part loaded. Finally the finished part is placed in the finished parts tub.

3 ROBOTS

3.1 Types
Robots add a "brain" to the muscle of the manipulator. (Robot comes from the Czech word for workers; Karel Capek's play *R.U.R.* [Rossum's Universal Robots] portrayed workers as robots.) Robots and numerically controlled machine tools are **flexible automation** in contrast with special-purpose transfer devices and machines using cams (**fixed automation**). Robot definitions vary. Japan classifies robots in six categories:

1. manual manipulator—a manipulator worked by an operator

2. fixed-sequence robot—a manipulator that repetitively performs successive steps of a given operation according to a predetermined sequence, condition, and position; the set information cannot be changed easily

3. variable-sequence robot—a manipulator similar to that in category 2, but whose set information can be changed easily

4. playback robot—a manipulator that can produce, from memory, operations originally executed under human control (A human operator initially operates the robot to input instructions. Information relating to sequence, conditions, and positions is put into the memory.)

5. NC (numerical control) robot—a manipulator that can perform a given task according to the sequence, conditions, and positions commanded via numerical data, using punched tapes, cards, or digital switches

6. intelligent robot—one that uses sensory perception (visual and/or tactile) to detect changes by itself in the work environment and, using its decision-making capability, can proceed with its operation accordingly.

In the United States, the Robotics Institute of America defines a robot as a reprogrammable, multifunctional manipulator that moves parts, tools, or specialized devices through variable programmed motions for the performance of a variety of tasks.

Although robots in *Star Wars* moved on wheels or legs, most present-day robots have a fixed base. Figure 20.2 shows four design alternatives. Polar robots have a spherical work envelope in which the arm pivots about two fixed points and telescopes. Cylindrical robots have a cylindrical work envelope in which the arm rotates about a fixed point, moves up and down vertically, and telescopes horizontally. Cartesian robots have a rectangular work envelope in which the arm moves horizontally to the side, up and down vertically, and telescopes horizontally. In practice, the rectangle can be quite long, say 100 ft. Revolute (articulated) robots have a cylindrical work envelope with a domed end and the joints mimic a human arm. Polar, cylindrical, and revolute robots also can be mounted on a movable platform to cover more space.

At the end of the robot "arm" is the "wrist." Wrist axes include roll (also called *swivel*—rotation in a plane perpendicular to the end of the arm), pitch (also called *bend*—rotation in a vertical plane through the arm), and *yaw* (rotation in a horizontal plane through the arm). Wrist axes don't change the work envelope much but permit tool orientation to a specific point.

Attached to the wrist is the "hand," also called **gripper, end-effector**, or **end-of-arm-tool (EOAT)**. The payload of the EOAT includes the weight of the EOAT plus the weight of the object; most payload capacities are given for the weight placed right at the mounting plate. There are two types of EOATs: (1) grippers and (2) tools—

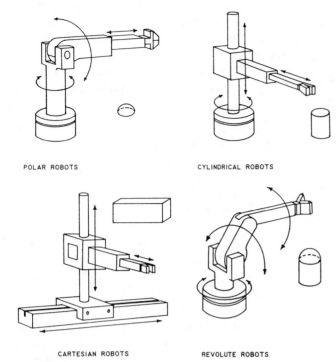

POLAR ROBOTS   CYLINDRICAL ROBOTS

CARTESIAN ROBOTS   REVOLUTE ROBOTS

FIGURE 20.2   Work envelope shapes for robot designs (Industrial Robots, 1983). Each has three axes of motion. The envelope for a polar robot is a hemisphere; for cylindrical, a cylinder; for cartesian, a box; and for revolute, a cylinder with a dome. Three additional movements can be gained by the wrist or gripper; pitch (rotation in the vertical plane through the arm axis), yaw (rotation in the horizontal plane through the arm axis), and roll (rotation about the arm axis).

spray guns, routers, sanders, grinders, drills, screwdrivers, heating torches, impact wrenches. Many robots can change their EOAT automatically. Whether the change is automatic or manual, there needs to be a specific mechanism or technique so that the EOAT is in exactly the same position and orientation before and after the change. Otherwise the program will have to be modified after each change. As with any complex device, downtime may be a problem. Maintainability will be increased by having a spare EOAT so that repairs can be done while the robot continues operating.

There are four possibilities for multiple grippers: (1) a gripper that can pick up more than one item at a time; (2) a robot arm with two grippers on one end; (3) a robot arm pivoted in the middle, with grippers on each end; and (4) two robot arms operated by one controller.

An example of the first is a gripper that can pick up six bottles at a time to load into a carton. An example of the second is a robot with two grippers on one end of the arm. One removes the finished part from a machine and then rotates out of the way while a second gripper places a part into the machine to be processed. Another example of the second type is a spray painting robot with two guns on one arm so that two parts can be painted at one time; in addition the servo could give 420° of wrist rotation, versus only about 210° for humans. An example of the third possibility

is a press-to-press transfer in which the arm must make a 180° swing. A gripper on each end of the arm may reduce cycle time by 2 to 3 s. An example of the fourth is two standard three-axis robots loading and unloading a forging press. There are two advantages of the "two arms with one head": (1) cycle time is reduced over that of one robot and (2) eliminating one control unit and one power unit saves about $30,000 in capital cost.

Dual grippers are most useful when you want to (1) minimize machine downtime, (2) have high production rates, or (3) have a large variation in part characteristics. Category (3) has two subcategories: (a) part variation due to before machining versus after machining and (b) parts variation due to different types of parts.

Robots can be mounted on tracks as well as on the floor or overhead. That is, "picker to the part" as well as "parts to the picker" can be used.

Robots are powered pneumatically, hydraulically, and electrically. Pneumatic tends to be cheapest, hydraulic is for heavier loads, and electric is for precise positioning. Positioning ability can be divided into two categories: accuracy (constant error) and repeatability (random error). For example, assume you wanted to place a part at position $X = 1.500$. If the unit placed it, on the average, at 1.501, then you would have an accuracy of .001. If the range of locations varied from 1.495 to 1.505, then you would have repeatability of −.005 to +.005.

3.2 Applications

Robots tend to be put into structured, repetitive jobs; the job need not be continuous. Robot applications can be put into two general categories: **processing** and **pick and place**.

Processing is subdivided into welding and painting, and pick and place includes assembly, machine loading, and foundry.

Processing applications generally are spray painting and welding (spot and arc). The robot is a "workhead" and the part is transported by other devices. There has been much publicity about these "steel-collar" workers. There is little difference conceptually between numerical control machines in machining operations and robots for processing. Perhaps the biggest difference is the NC machines are programmed digitally whereas robots are programmed by analog (leading the arm through the pattern).

Pick and place applications use the majority of robots: loading and unloading of pallets and containers, transferring parts from one production machine to another, and loading and unloading conveyors. The robot transports the part, and other machines do the processing. Generally, the robots are programmed by moving the arm through the motion sequence desired for each type of item (walking it through). This path then is stored in memory. Then when that specific motion is required, the robot repeats it.

Space requirements for robots tend to be considerably less than for human operators due to elimination of space required for human comfort and safety and to better use of vertical space (humans can't reach 10 ft high). Thus, in justifying the cost, consider space savings and energy savings (less climate control) in addition to labor savings.

Figure 20.3 shows a robot picking and placing in a machining cell. As with humans, a U shape is a good design of a cell. See Chapter 4 for more comments on cells.

Robots used for loading/unloading and for welding tend to be used for two- or three-shift operations. Spray-painting robots tend to be used on one-shift operations. Justifications are improved quality, operator relief from a disagreeable task, and paint savings of 10 to 30%.

In contrast to the well-established technologies of "hard" automation, the installation, debugging, and systems work with robots tends to be very expensive. The cost of the robot itself often is less than 50% of the cost of the installed operating system. Use backup capability if possible (i.e., leave manual spray painting gun in area where robot is installed) since, although robots have relatively long times between failure, when they do fail it takes considerable time to fix them.

At present the primary problems are in the "hand" (vacuum, magnetic, or mechanical) and the lack of sensory input. The initial location and orientation of the item and final location and orientation of the container need to be more exact for the robot than they are in most industrial situations. Robots generally cannot pick a randomly oriented part out of a bin or off a conveyor. Thus, parts positioners and locators are much more important for a robot than for a human. For positioning, use vibratory feeders (see Figure 20.7) and self-locating features. One possibility is a **remote center compliance (RCC)** device. The RCC, which can be mounted on the robot between the tool mounting plate and the gripper or can be mounted under the item to be gripped, floats mechanically, and so precise alignment is not as necessary. Cones, chamfers, and countersinks are additional examples of making positioning requirements less strict.

At present, most robots have no vision capability. Of the few that do see, almost all see in two dimensions—that is, they consider all objects as plates. Binary visual systems translate each point of the image into black or white (object appears as silhouette). Gray scale systems allow different shades of gray to be assigned to the image points. Infrared light sources minimize problems due to ambient light. Unfortunately, most real problems are three dimensional, so there is much work to be done before robot vision systems become widely applied.

Use a predetermined time system (designed for humans) to describe the reaches, grasps, moves, positions, and disengages that the robot needs to do. Nof and Lechtman (1982) describe a predetermined time system for robots. Work elements are divided into motion elements, sensing elements, gripper elements, and delay elements. These predetermined time analyses will give you a better idea of exactly what you want the robot to do.

4 FIXED AUTOMATION

4.1 Transport
For high-volume production, transport between workstations can be dedicated and special purpose. When the workstations are relatively compact (a "workhead"), it is practical to arrange them along the perimeter of a circle. See Figure 20.4. Then transport

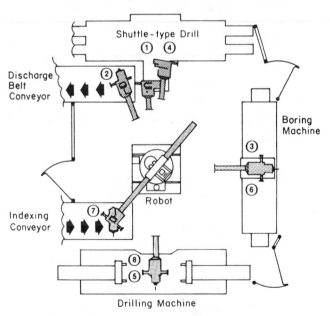

FIGURE 20.3 Robot in a machining cell (Robot Loading, 1982). Two 20 lb differential-housing castings are machined at one time on each machine. V block locators are used. The numbers in the figure show the robot's movement sequence. The entire cell is supervised by one operator who is responsible for quality, tool changes, and correcting malfunctions as well as for loading parts into a fixtured pallet on the indexing conveyor. Note the barriers and gates to restrict human access to the robot's movement area. Light curtains and pressure-sensitive floor mats also could be used.

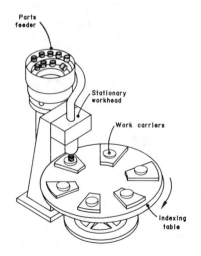

FIGURE 20.4 Rotary indexing table with one stationary workhead (Boothroyd et al., 1982). In practice, there may be several workheads; but with more heads, downtime becomes a problem since a defective component at any head stops the entire machine. (Boothroyd et al. examine this problem in detail, considering, among other things, component quality and cost of rejected assemblies.)

between stations can be done by moving a rotating table (an **indexing machine** or **dial index machine**). Note that each piece moves from each station at the same time (synchronous transfer) and that it is difficult to put buffers between stations.

When workstations are larger, the stations are arranged in a **transfer line**. This line need not be straight; it can be L or U shaped. See Figure 20.5. Transport can be by (1) walking beam; (2) chain drive, with fixtures attached to the chain; and (3) powered roller conveyors, with fixtures moving on rollers (Groover, 1980). With the walking beam and chain drive, all transfers are synchronous and buffers are difficult to implement. Powered rollers permit buffers.

4.2 Location and Orientation

Generally, location and orientation of the unit being transferred are a challenge. The general strategy is that, once you locate the unit at the first station, maintain that location during the transfer to the next station. In order for this to be done, the fixture that the unit is in is moved with the part still in it to the next station in a special pallet—a **fixtured pallet** (generally made of machined steel). After the pallet reaches the next station, it is located by stops against its machined surfaces. Then the next machine can process the unit (1) while it remains in the fixture or (2) after the unit has been transferred temporarily to the machine by some dedicated device.

Location and orientation of the unit at the first station (and components to be added at subsequent stations) also are a problem. One solution is to have a human take randomly jumbled parts from a pallet, conveyor, or bin and load a fixture. Another common alternative is to have the human load a magazine, which in turn loads the machine or fixture. (Ideally this magazine would be loaded as the last operation where the component is made.) The magazine can be loaded intermittently (e.g., once/10 min) instead of synchronously with the machine (e.g., once/30 s). Figure 20.6 shows some feeding mechanisms. Orientation of the parts also can be done manually, which is common for

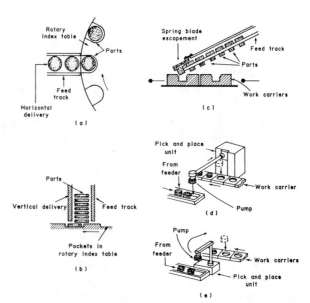

FIGURE 20.6   Feeding mechanisms for volume production (Gay, 1973). Rotary index tables can be fed (*a*) horizontally (spring actuated) or (*b*) vertically (gravity). The rotary table is the escapement (a mechanical device that separates, on demand, only one or a predetermined number of parts from the feeding system). Parts release also can be (*c*) due to the movement of the work carrier. The item also can be picked up and released magnetically or with vacuum. See (*d*) for a linear transfer and (*e*) for a rotary transfer. It also is possible to transfer while changing part orientation (e.g., 90° or 180°).

low-volume applications. For high-volume applications, automatic orientation is desirable. Tool engineers have a variety of strategies and devices. Figure 20.7 shows one common approach—a **vibratory feeder**. Figure 20.8 shows an example vibratory feeder track. Boothroyd et al. (1982) give a detailed analysis of vibratory feeders as well as other types of feeders. Box 20.1 comments on using two feeders in series.

Boothroyd et al. (1982), in Chapter 7, give an excellent analysis of when fixed or flexible mechanization is economic. Design of the product components is especially important.

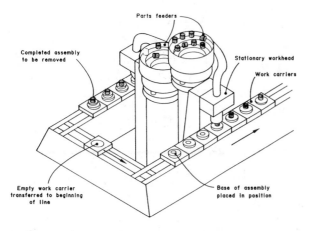

FIGURE 20.5   In-line transfer (horizontal plane) with shunting work carriers (Boothroyd et al., 1982). In this design, the carriers need to be precisely positioned at each workhead but there can be buffers between heads.

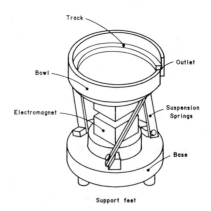

FIGURE 20.7   Vibratory feeders take a random pile of items and orient them automatically (Boothroyd et al., 1982). The items are dumped into the bowl and then vibrate up the track to the outlet. A feed track normally connects the outlet and the machine head. The feed track can be considered a magazine.

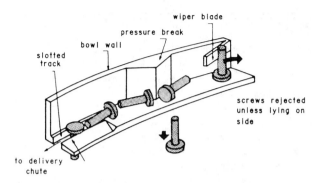

(a) Orientation of screws in vibratory bowl.

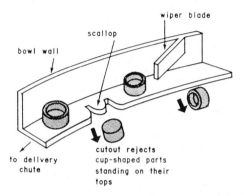

(b) Orientation of cup-shaped parts.

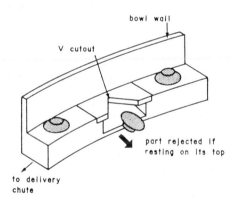

(c) Orientation of truncated cones.

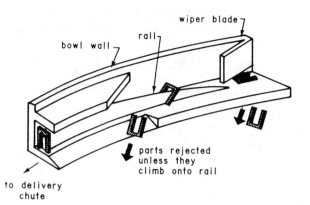

(d)

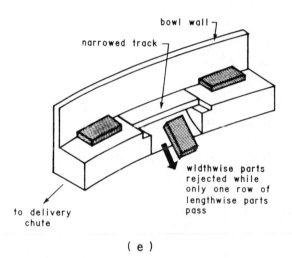

(e)

FIGURE 20.8   Orientation in vibratory feeders can be done several ways (Boothroyd et al., 1982). The general strategy is to reject parts back into the central pile unless they are in the desired orientation. Example (a) shows orientation of a screw. The wiper blade rejects all screws not lying flat. The pressure break has two functions: (1) It passes only screws in single file with head or shank leading; and (2) it returns excess parts to the bowl if the delivery chute becomes full. The slot and slope pass screws with head up. Example (b) shows orientation of cup-shaped parts. Example (c) shows orientation of truncated cones. Example (d) shows orientation of U-shaped parts. Example (e) shows orientation of rectangular blocks.

BOX 20.1   *Parts feeders as inspectors (Feeders Solve . . ., 1981)*

Parts feeders are for production rates in the thousands and machine cycles in seconds. However, they assume that the parts fed have no defects. In practice, you will find defects. For example, for screws you will find screws without slots, without heads, without threads, with partial shanks, with deformed heads, screws of the wrong size, etc. These defects, if passed by the feeder to the unit, cause defective units. If not passed by the feeder, they jam the feeder or cause endless recycling of rejects.

One useful technique is to use two feeders in sequence. The first feeder "qualifies" the parts for the second feeder,

which feeds the machine. Thus the second feeder rarely jams so machine downtime is reduced.

The first (inspection) feeder usually is remote from the second feeder. A good location is just after the machine that makes the parts, so feedback is immediate. The first feeder should have reject chutes so rejected parts are separated rather than falling back into the central supply. This feeder inspection is not a substitute for precision parts gauging. However, if a close tolerance gauge is automatically fed, use a preliminary sorting feeder to minimize downtime from jamming caused by gross oversize and deformed parts.

DESIGN CHECKLIST: WORKSTATION MATERIAL HANDLING

Manual material handling
    NIOSH guidelines
        Does lift meet assumptions?
        RWL
        Reduction of lifting strain
    Table 20.5 guidelines
Can a balancer/manipulator help?
Robot
    Pick and place or processing
    Type

    Shape of area covered (Figure 20.2)
    EOAT design
    Power type
Fixed automation
    Synchronous versus nonsynchronous transfer
    Buffer design (see Chapter 5)
    Location and orientation
        Previously oriented
        Vibratory feeders

REVIEW QUESTIONS

1. For lifting, does NIOSH consider selection or job design more important?
2. What are the assumptions of the NIOSH lifting guidelines?
3. What are the six modifying factors of the ideal weight?
4. What is the recommended solution to the "open manhole" problem?
5. What are the differences between a balancer, a manipulator, and a robot?

6. List the six Japanese categories of manipulators and robots.
7. What is the difference between a cylindrical and revolute work envelope?
8. What is an EOAT?
9. Some robots are transporters and some are workheads. Give examples of each.
10. What does a remote center compliance device do?

PROBLEMS AND PROJECTS

20.1 Using the NIOSH guidelines, describe in numbers a lifting situation that realistically could occur in a job. Make some realistic modifications to the job and show how they affect the RWL.
20.2 Evaluate a manual material handling job using Table 20.5.
20.3 Make a time analysis (using a predetermined time system such as MTM) of loading/unloading a machine using a single hook and a double hook. Assuming the double hook costs $100 more than a single hook and will last 5 years,

how much time will have to be saved on each cycle to make a double hook a worthwhile purchase?
20.4 Make an economic analysis of a job that might use a robot. Using a manufacturer's catalog, which model and type robot would you recommend? What is the capital cost and operating cost of doing the job without the robot? What is the capital cost and operating cost of doing the job with the robot? Justify your recommendation.

REFERENCES

Boothroyd, G., Poli, C., and Murch, L. *Automatic Assembly*. New York: Marcel Dekker, 1982.

Feeders solve a lot of problems. *Production Engineering*, Vol. 28, 3, 66–70, March 1981.

Gay, D. Ways to place and transport parts. *Automation*, 80–84, June 1973.

Groover, M. *Automation, Production Systems, and Computer Aided Manufacturing*. Englewood Cliffs, N.J.: Prentice-Hall, 1980.

Industrial robots. *Modern Material Handling,* Vol. 38, 4, 45–53, March 7, 1983.

Nof, S., and Lechtman, H. Robot time and motion systems provided means of evaluating alternate robot work methods. *Industrial Engineering*, Vol. 14, 4, 38–48, April 1982.

Robot loading/unloading pays off in machining cell. *Robotics Today*, Vol. 4, 1, 69–71, Feb. 1982.

Waters, T. *Users Guide to Lifting Guideline* (NIOSH), Washington, D. C.: Supt. of Documents, 1993.

Waters, T., Putz-Anderson, V., Garg, A., and Fine, L. Revised NIOSH equation for the design and evaluation of manual lifting tasks. *Ergonomics*, Vol. 36, 7, 749–776, 1993.

IV SERVICES AND ENVIRONMENT

CHAPTER PREVIEW

Utilities are the veins, arteries, and nerves of the plant body. Electricity and compressed air are the power sources. Water is used for processing as well as for liquid waste disposal. Communication is primarily through the phone network, although specialized computer networks are growing in importance.

CHAPTER CONTENTS

1 Electricity
2 Compressed Air
3 Water
4 Communication Systems

KEY CONCEPTS

aftercooler and separator
air inlet
air receiver
beepers (pagers)
biological oxygen demand (BOD)
brownouts
busway (versus wire in conduit)
capacitors
chemical oxygen demand (COD)
cogenerator
compressor
dedicated lines
demand charge
demand rachet clause

dirty power
dryers
electronic mail (EMail)
electrostatic discharge (ESD)
fax
filter-regulator-lubricator (FRL)
fuses and circuit breakers
high-efficiency motors
in-plant generation
interval splitting
load deferring
local area network (LAN)
PA system
power factor

process water
sanitary (versus storm) sewer
standard cubic feet (SCF) of air
static electricity
substation
teleconferencing
torque (of motors)
uninterruptible power supply
 (UPS)
videophone
WATS line
wet wall

1 ELECTRICITY

1.1 Power Users

Electricity is of primary importance. Motors furnish the power from machines; artificial lighting is the dominant source of illumination; heating, ventilating, and air conditioning (HVAC) equipment (fans, blowers, heaters, refrigeration, etc.) determines the thermal environment; and electricity permits communication through telephones and computers.

Motors consume over 50% of electricity generated in the United States (about 65–70% of industrial electricity); lighting consumes about 25%–20% direct, plus another 5% in cooling equipment to compensate for unwanted heat from lights (Fickett et al., 1990).

Table 21.1 gives electrical power requirements of some common machines. See Box 21.1 for selection of motors. See Chapter 24 for techniques to minimize the use of energy.

1.2 Costs of Power

The electrical power charge per KWH depends on the amount of power used, the power factor, the demand charge, and (in some cases) time-of-day or season charges.

1.21 Amount of Power.

Big power users pay less per KWH. For example, the energy charge might be 8 cents/KWH for the first 500 KWH per kilovolt – ampere (KVA) of billing capacity, 7 cents/KWH for the next 2,000 KWH, 6.5 cents/KWH for the next 2,500 KWH, and so on.

TABLE 21.1 Energy requirements of some machines (1 hp = 746 W).

Watts	Machine
	Office
10	Table radio (Sony ICF-9740W)
15	Fluorescent desk lamp
60	Oscillating fan
400	Slide projector (Kodak)
400	Personal computer (IBM PC-XT)
800	Transparency projector (3M)
800	Laser printer (HP Laserjet III)
	Shop
370	Bench belt grinder (Portable-Cable Type CN2)
880	Arc welder (Lincoln Model R3S-250)
1,200	Engine lathe (Hendly)
1,300	Mill (Milwaukee #2)
2,600	Mill (Brown and Sharpe #2)
2,900	Band saw (Do All 26)
3,700	Belt grinder (Portable-Cable Type G8)
3,700	Air compressor (20 ft³/min)
4,100	Welder (Miller CP 200)
5,600	Lathe (Monarch 1220K; 15" swing)
11,800	Machining center (Maho 600-S)
20,000	Machining center (Mori Seiki SL-15M)

1.22 Power Factor.

Power factor is explained in Figure 21.1. Some utilities bill for KWH and a power factor charge and others just for KVA (which has the same result). Power factor loss is caused by AC devices which operate on the principle of electromagnetic induction (induction motors, polyphase motors, welding machines, electric

---

POWER FACTOR VECTOR RELATIONSHIP

KVA, or kilovolt-amperes, is the resultant of KVAR (kilovolt = amperes = reactive) and KW, and it determines the heating burden that will be imposed on all elements of the power system. All power system components (conductors, transformers, switch gear, etc.) must be sized to withstand the KVA load to which they will be subjected.

KVAR is the "phantom power" component of KVA. It provides only the magnetic *force* needed for operation of the device, and performs no work.

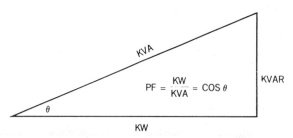

$$PF = \frac{KW}{KVA} = \cos\theta$$

KW is the work-performing component of KVA and is the only component that varies directly with the work performed and the energy consumed.

θ is the phase angle between KW and KVA; power factor, expressed as a decimal, is the cosine of the phase angle.

FIGURE 21.1 One KVAR of capacitors will cancel the effects of 1 KVAR of the inductive KVAR, or phantom power, component that causes low power factor. If all of the phantom power component is eliminated, the phase angle diminishes to zero, and power factor is 1.00, or 100 percent. At that point the system need carry only the current needed to supply the energy requirements of the load. KW varies directly with the true power requirements of the load, but the KVAR input to an electromagnetic device remains virtually constant from no load to full load. Thus, lightly loaded motors operate at extremely low power factor. It is seldom practical to improve power factor to more than about 95 to 97%, because as θ approaches zero, diminishing returns are realized, with a given block of capacitors having less effect on diminishing the phase angle.

BOX 21.1 *Motor selection*

The standard industrial motor is the AC three-phase squirrel cage induction motor. (On early models, the rotor assembly viewed from the end resembled a squirrel cage.) It is essentially a constant speed motor but can drive variable speed devices or do speed matching, usually with an adjustable frequency drive. For most industrial applications, 1 hp (and larger) motors use three-phase power. Where three phase is not available, single phase is used for motors of less than 10 hp. Table 21.2 shows it is worthwhile to use **high-efficiency motors**. For example, a 1 hp output industrial efficiency motor would use 19,840 KW in 20,000 h. A high-efficiency motor costing $27 more saves 2,080 KW at a capital cost of $.014/KWH. Thus, if electricity costs more than 1.4 cents/KWH, buy the high-efficiency motor.

When specifying a motor, give frame number, locked rotor code letter, and National Electrical Manufacturers Association (NEMA) design letter (Motors, 1991).

Frame number (usually accompanied by a letter suffix) defines physical dimensions of the motor.

Locked rotor code letter defines motor starting current in KVA/hp for use in motor wiring and control and protective equipment.

NEMA design letter governs torque and slip. Standard motors are designed to operate in ambient temperatures up to 40°C (104°F). Intermittent-duty motors operate less than 60 min/24 h; continuous-duty motors operate 60 min or more. Design B is the standard, accounting for 80% of integral hp induction motors sold to industrial plants. Enclosures can be drip-proof, totally enclosed fan-cooled (TEFC) or totally enclosed nonventilated (TENV). Typical applications are fans, blowers, pumps, and machine tools. Design A has higher maximum torque and starting current. Design C is for hard-to-start applications, typically conveyors, compressors, and plunger pumps. Design D is for hard-to-start applications with short-time duty: cranes, hoists, elevators, and punch presses and shears (high-inertial devices).

For a typical 50 hp, 460 V, drip-proof, 1,800 rpm, NEMA Design B motor with locked rotor code F, Figures 21.2, 21.3, and 21.4 show the effect of speed, load, and voltage (Motors, 1991).

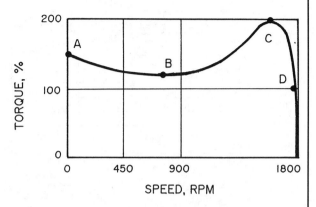

FIGURE 21.2 Starting **torque** (*A*) is available to break the load free from rest. Pull-up torque (*B*) is the minimum torque during motor acceleration. Breakdown torque (*C*) is the maximum torque. Full-load torque (*D*) is the torque at full load speed. (Adapted from Motors, 1991.)

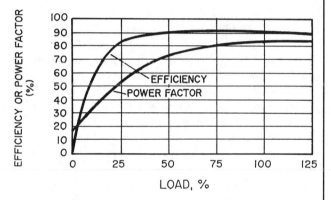

FIGURE 21.3 Load affects both torque and power factor. Efficiency is relatively unaffected by light loads (until the load becomes less than 40%). Power factor, however, drops rapidly below 75% load. (Adapted from Motors, 1991.)

TABLE 21.2 High-efficiency motors generally have a capital cost/kwh saved less than the cost of electricity. It is assumed the motor operates 20,000 h in its life. The data are for 3-phase open drip-proof motors (Grainger Catalog, 1991). The efficiency is average efficiency, not NEMA nominal efficiency. The benefits of high-efficiency motors tend to be greater at higher hp.

| HP Output | Efficiency | | | Ind. Duty KWH in 20,000 h | KW Saved with High Eff. | Price ($) | | | | $/KWH saved |
	Ind. Duty	High Eff.	Delta Eff.			Ind. Duty	High Eff.	Delta	%	
.33	66.1	78	11.9	7,740	1,430	196	228	32	16	.022
.5	75.9	81	5.1	9,830	620	222	258	36	16	.158
.75	75.3	82	6.7	14,860	1,210	248	275	27	11	.022
1	75.2	84	6.8	19,840	2,080	254	283	29	11	.014
1.5	77	86	9	29,060	3,040	282	310	28	10	.009
2	81.4	87	5.6	36,660	2,360	310	350	40	13	.017
3	82.5	86.5	4	54,250	2,500	248	285	37	15	.015
5	85.5	86.5	1	87,250	1,010	273	315	42	15	.042
7.5	87.5	90.2	2.7	128,000	4,000	389	447	58	15	.015
10.0	87.5	90.2	2.7	170,000	5,100	478	550	72	15	.014
								Average	14	

(continued)

BOX 21.1   Continued

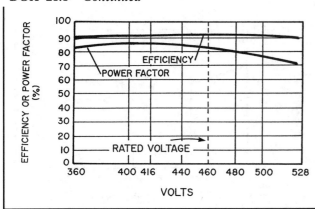

FIGURE 21.4   Voltage changes affect power factor more than efficiency. A 10% change in 460 V would give 414 to 506 V. (Adapted from Motors, 1991.)

brakes and clutches, arc and induction furnaces); power factors range from .5 to .9. Fluorescent lamps with built-in capacitors have power factors of about .96, and resistive loads (e.g., heaters) have power factors of 1. Synchronous motors have leading power factors (i.e., over 1). With lagging power factor, the utility must have a larger capacity upstream for transformers, power sources, etc., so they make the customer pay for it.

Increasing the power factor, typically by **capacitors**, reduces the electric bill and increases electrical capacity downstream of the capacitors. If a reduced power bill is your only goal, put the capacitors at the plant input substation. (KVAR output of capacitors varies with the voltage squared, so cost/KVAR is lower at higher voltages.) To increase plant electrical capacity, put capacitors farther downstream.

Capacitors can be bulk (static) or automatic power factor controllers (dynamic). An automatic power factor controller might have six 50-KVAR capacitors and one 25-KVAR capacitor. The unit then would insert anywhere from 25 to 325 KVARs, in steps of 25 KVARs. Commonly, the automatic system (which is more expensive/KVAR than static capacitors) would be installed with a base of static capacitors. For example, a base of 1,000 KVARs might be topped with the 325-KVAR automatic system.

For low load conditions, bulk capacitors may overcompensate and make the power factor leading instead of lagging. Not only do power companies penalize firms for leading power factors, some electronic equipment and adjustable frequency drives cannot operate with leading power factors. Therefore, permit switching off of the total block of static capacitors.

1.23 Demand Charge.   The utility has fixed costs that must be paid regardless of the amount of energy supplied to the customer. For example, they must have capacity to furnish power at 3 P.M. in the summer (peak load); this capacity is much greater than that needed at 3 A.M. See Box 21.2 for an example demand charge calculation. Note that the demand charge is greater than the energy charge.

The utility demand meter continuously integrates power consumption over a prescribed demand interval—usually 15 min, but occasionally 7.5 min. The **demand charge** is set by the maximum rate during the billing period. Thus, demand charge for July may be set by the power usage from 3:00 to 3:15 on July 12th. Some utilities also have a **demand ratchet clause** that makes the peak for any one month carry forward for a year. For example, the demand for any month shall not be less than 75% of the maximum demand in the previous year. Thus you might pay in January for a hot day the previous summer. Clearly, it is very worthwhile to reduce peaks of power usage.

There are three approaches to reducing peaks: load deferring, interval splitting, and in-plant generation.

In **load deferring** and **interval splitting**, it is necessary to know the timing of both load peaks and demand charge periods. In load deferring, electric furnaces, battery chargers, air compressors, fans, and pumps (i.e., loads that do not affect production immediately) are turned on and off at the proper time. Spethmann (1981) gives some sophisticated strategies for load shedding and reinstatement. In interval splitting, a load is turned on halfway through the demand period so only half of it counts.

In-plant generation is use of internal equipment to generate power. The Public Utility Regulatory Policies Act of 1978 requires utilities to buy surplus power from **cogenerators**. Thus, if your firm uses a boiler to generate heat or steam for a process or to burn waste, it can sell any electricity generated beyond its immediate needs to the utility. This has improved the economics of in-plant generation of electricity and encouraged more firms to install it. It is best to cogenerate during times of peak demand, thereby reducing the demand from the utility. Another possibility is to run emergency power-generation equipment (e.g., diesel generators) during peak demand times. The equipment needs to be tested periodically in any case; testing can be done when it gives the most benefit. If you agree ahead of time to run your own generators during peak demand periods (power sharing), the utility often will reduce the annual energy charge rate.

BOX 21.2 *Example power calculations*

Usage

Small plant with one-shift operation.
Monthly energy use: 200,000 KWH
Power factor (PF): .65
Monthly peak demand: 1,500 KW
Peak reactive power: 1,500 @ .65 = 1,753 KVAR

Rate Schedule A

Energy charge: .05 $/KWH
Demand charge: 15 $/KW of adjusted demand
    Adjusted demand = .8 + .6 (KVARH/KWH)

Rate Schedule B

Energy charge: .07 $/KWH
Demand charge: 15 $/KW
Power factor clause: Total bill is

• increased .1% for .01 PF deviations below .85
• decreased .1% for .01 PF deviations above .85.

Determining Charges

Schedule A

Energy charge: 200,000 KWH × .05 $/KWH = $10,000
Demand charge:
    PF = .65 (θ = arc cos .65 = 49.46°)
    KVARH = 200,000 (tan 49.46) = 233,826

Multiplier = .8 + .6(233,826/200,000) = 1.502
Demand charge = $15/KW (1,500 KW)(1.502) = $33,795
Total charge = 10,000 + 33,795 = $43,795

Schedule B

Energy charge: 200,000 KWH × .07 $/KWH = $14,000
Demand charge:
    $15/KW × 1,500 = $22,500
    Multiplier = 1 + .001(.85–.65)/(.01) = 1.02
Total charge = (14,000 + 22,500)(1.02) = $37,230

Capacitor Need

How many capacitors are needed to raise PF from .65 to .95?

1. Peak load of 1,500 KW:

    Present @ .65 = 1,753 KVAR
    Proposed @ .95 =   493 KVAR
                      1,260 KVAR at peak

2. Base load: The base load could be estimated (assuming negligible power consumption during nonworking hours and working 22 days of 8 h/month) as 200,000 KWH/176 h = 1,136 KW. At .65 PF, this is 1,328 KVAR; at .95 PF, this is 373 KVAR. Thus, 955 KVAR of capacitors would be needed to maintain a .95 PF on an average day.

1.24 Time-of-Day and Seasonal. For most firms, the demand charge is due to the first shift, so electricity use on other shifts only incurs the energy charge. In addition, in an effort to level demand load, some utilities give discounts for electrical demand at off-peak use. This may encourage a firm with heavy electrical use to process items on the second shift.

1.3 Electrical Distribution
Electricity is stepped down at substations and then distributed locally.

1.31 Substations. Key to electrical distribution is that power in a conductor varies with voltage squared. Thus, changing voltage from 120 V to 480 V increases power by 16, not 4. Utilities usually transmit power at high voltages (69 to 345 KV) and then step it down to 4.16 or 13.8 KV for in-plant distribution.

At secondary substations within the plant, voltage is stepped down further to 480Y/277 V or 208Y/120 V. (Y is a way of making electrical connection.) Three-phase 480 or 208 VAC is used for motors, single-phase 277 or 120 is used for lighting, and single-phase 120 is used for receptacle outlets (Palko, 1990).

A **substation** consists of three elements: (1) switching and protection on the primary side, (2) the transformer, and (3) switching and protection on the secondary side. Unit substations have all three elements together; free-standing substations have a transformer standing alone but connected by cabling or busway to the other substation elements. Free-standing substations (with the transformer located outside on a concrete pad) have been growing in popularity due to fire protection and liquid containment considerations.

In some locations, the main substation may not be owned by the utility but by the user (although maintained by the utility) at substantial savings (Keithly, 1989).

1.32 Local Distribution. **Wire in conduit** is the alternative for low capital cost, low flexibility, and low density of use. **Busway** is better when installation labor costs are high, when equipment needs may change, and when equipment power requirements are high. The prefabricated busway sections (typically 10 ft long with 5 tap positions on each side) require less installation time than wire in conduit. Where no taps are needed, use feeder busway, which has a lower capital cost than plug-in busway. A plug-in busway section can be substituted later if needs change.

Near the point of use, there may be additional electrical control equipment such as a motor control center and circuit breaker panels. Note that circuit breaker standards of North America (ANSI) are not the same as those of other countries (Bridger, 1992).

In offices and conference rooms, plentiful electrical outlets facilitate furniture rearrangement. Flush-mounted floor outlets minimize risk of tripping over electrical cords connected to a wall outlet. Equipment needs to be connected with power wiring and electronic wiring. Some alternatives are under-floor raceway and flat conductor cable (FCC) wiring between the floor and the carpet. Where power and electronic wiring cross, the two cables must be separated by a grounded layer of metal shielding; this requirement is met inherently if the electronic cable is above the power cable because the power cable is required to have a top shield.

In corridors, closets, and toilets, provide outlets for maintenance activities.

1.4 Clean, Stable Power

1.41 Problem. Factory automation has increased the need for clean, stable power. Electronic equipment is especially vulnerable, but other equipment also can be affected. For example a 10% reduction in voltage reduces induction heater output 19%, possibly increasing quality problems.

Dirty power can come from the utility or from internal sources. Dirty power can be (1) voltage fluctuation, (2) frequency deviation, (3) transient surges, (4) electrical noise, or (5) power outages.

Utilities set the nominal voltage at their line midpoint. Depending on its location in relation to the midpoint, your facility may chronically get a lower voltage (if beyond the midpoint) or higher voltage (if closer to the power source). The same problem can occur within a plant with long cable runs.

Low voltage from the utility (**brownouts**) can occur when the utility's demand exceeds its ability to supply power. It may cut voltage up to 5% for several hours or even days. Transient voltage dips (sags) also can be caused by routine circuit switching at utility substations. Typically, sags occur twice as often as surges. Sags affect computers and, especially, electronic controllers (e.g., those used for variable-speed motors). Transient surges can occur from outside (such as lightning) or inside (motor start or turn off, power-factor capacitor switching, or pulsed loads such as X-rays, lasers, welders). Transients cause three types of problems to electronic equipment: (1) component failures, (2) upsets (corruption or loss of data), and (3) latent failures, in which the component is weakened and will fail later (Hutchins and Clark, 1991). **Electrostatic discharge (ESD)** damage during packing, maintenance, and other activities may not be noticed because some semiconductors can be damaged by an ESD of 1,000 V but a fingertip cannot feel an ESD less than 3,500 V (Hutchins and Clark, 1991). Electrical noise is subdivided into radio frequency interference (radar, microwave, and lightning) and electromagnetic interference (switches, relays, and arcing connectors). Power outages can be anywhere in the network but most commonly are caused by interrupted power lines (from lightning, ice).

1.42 Solutions. Naturally, solutions depend upon the problems. Consistent low or high voltage can be adjusted by changing the taps on the transformer. For fluctuating voltages, a voltage regulator can help. Electromechanical types work for most equipment but static types probably will be needed for electronic equipment.

Dedicated lines reduce the effect of transients from equipment such as motors. Transients also can be reduced by a transient voltage surge suppressor (TVSS). Suppressors at the service entrance guard against external transients (e.g., lightning), and point-of-use suppressors protect against local transients. Lightning (negative charge) zags downward in discrete steps of about 150 ft. As the stepped leader nears the ground, positive point streamer currents are attracted to it and strain upward to it from pointed objects. When the leader is about one step's distance from the ground, a positive point-discharge current rushes to meet it. The goal of lightning rods is that the positive current should come from the grounded lightning rod rather than from the building itself. Maximum spacing is 20 ft on high points of the roof and all projections; a terminal must be located within 2 ft of a corner, roof edge, or gable end. The ground should have at least two ½ in. dia electrodes at opposite corners of the building; additional electrodes are recommended every 100 ft along the perimeter (Frydenlund, 1990). For more information, see the National Fire Protection Code 78 or contact the Lightning Protection Institute, P.O. Box 1039, Woodstock, IL 60098.

Permanently wired suppressors in receptacles give about 6 times the protection of portable plug-in types. To prevent inducing power system disturbances into data lines, don't locate power and data lines in the same cable tray or raceway.

Static electricity is a voltage transient generated when materials of high electrical resistance are moved against each other (such as the human body and a synthetic fiber rug). Problems range from electrical circuit damage to computer program interference, to annoyance. Some solutions are as follows:

- Raise the humidity (above 40% rh in the air) to reduce static electricity two ways. First, the moisture forms a film on surfaces, which allows the voltage to flow to ground. Second, the moisture is a lubricator, reducing friction between moving bodies.
- Make carpets conductive. Total electronic capability carpets have carbon fibers that conduct electricity. Topical treatments can be added to existing carpets. Treatments last only about 2 months and will increase carpet soiling.
- Place an antistatic floor mat under the operator's chair; use a cord to connect the mat to ground. A typical price is $150–200. Alternatively, use an antistatic desk mat (typical price of $30) under a computer terminal and keyboard; ground with a cord. The operator also can be connected to ground with a static-bleed wrist strap; this technique often is used for electronic assembly.
- Discourage the use of nylons, slips, and polyester clothing. Cotton T-shirts and jeans or chino slacks discourage static electricity. Shoes also can have static-dissipating soles.

Fuses and **circuit breakers** open the circuit when it is overloaded. A circuit breaker can reset after the overload is over, whereas a fuse cannot. A thermal circuit breaker depends upon a bimetallic element to heat up and so requires a finite time to operate; both thermal-magnetic and hydraulic-magnetic circuit breakers have no appreciable time delay for large overloads.

An **uninterruptible power supply (UPS)** is useful—especially for voltage dips. In a forward-transfer plan, the load normally draws from the utility. The UPS (in standby mode) is transferred on line in 25–200 ms. In the on-line (reverse-transfer) plan, the UPS is on line at all times and the utility power is switched to direct use only if the UPS fails. The primary advantage of the reverse plan is that the UPS includes an inverter (and thus furnishes clean, stable power); the disadvantages are a small power loss from the UPS and that the UPS is under load (and thus wearing out). UPS batteries usually are lead acid, although nickel-cadmium is used for self-contained UPS systems (Migliaro, 1991).

1.5 Electrical Safety

Table 21.3 gives the effect of various levels of current and Table 21.4 gives some electrical safety tips.

Ohm's law says:

$$I = E/R$$

where

I = current, amps

E = voltage, volts

R = resistance, ohms

For a given voltage, the key variable is resistance. Without protective clothing, the key human resistance is of the skin. Dry, clean, unbroken skin varies from 100,000 to 600,000 ohms, depending on its thickness; wet or broken it may be 500 ohms. Contact of 120 volts with a dry finger with 400,000 ohms would give 120/400,000 = .3 milliamps; contact with a wet finger with 15,000 ohms would give 120/15,000 = 8 milliamps. Thus, for safety, keep the skin dry and dirty (dirt can double the skin ohms). Don't

increase conductivity with metal objects such as watches, rings, buckles, lighters, keys, or coins in pockets.

Although the principles for electrical safety are well known, there still are many deaths and injuries.

2 COMPRESSED AIR

2.1 Reasons for Using Compressed Air

Compressed air is used primarily as a power source (hand tools, air motors, engine starters) as well as for breathing air (in protective garments), for instrumentation, and as a carrier (as in spray painting and pneumatic material handling). Air motors and valves are cheap to operate. For example, a cylinder with a 2 in. bore and 4 in. stroke operating at 70 psi will use .08 SCF/cycle. (**SCF, standard cubic feet**, according to the American Society of Mechanical Engineers (ASME), is air at 14.7 psia and 68°F and 36% rh; SCF is used rather than actual cubic feet in all compressed air calculations.)

Table 21.5 shows compressed air requirements for some common tools. Note that tools usually do not require air all the time—a typical duty cycle is 25%. Thus, when sizing the compressed air system, use the average use. An air storage tank (air receiver) buffers the compressor from fluctuation demands.

TABLE 21.4 Electrical safety tips (Hammer, 1989).

Deenergize the circuit (don't forget to discharge capacitance stored charge).

Use Ground Fault Circuit Interrupters (GFCI). A GFCI monitors the circuit. If it senses imbalance in the current, it breaks the circuit. It does not work for line-to-line contact, only line-to-ground.

Insulate with distance (isolate). Put distance between people and current. Barriers can replace physical distance.

Insulate equipment—knobs, controls, handles, tools, shafts.

Insulate the person—insulating material to stand/sit on, nonconductive shoes, rubber gloves.

Warn people: active warning = lights, sounds; passive = signs, colored backgrounds.

TABLE 21.3 Effects of 60 cycle AC current (Hammer, 1989). DC currents, for the same effects, are 3 to 5 times the AC value. Frequencies of 20–100 Hz are especially dangerous as they cause ventricular fibrillation.

Milliamperes	Effect
1	Perceptible shock.
5–25	Lose muscle control. For 60 Hz, "let go" current (the current at which people still can let go) depends on weight. Typical values are 6 for women and 9 for men.
25–75	Very painful and injurious. Death if paralysis lasts over 3 min.
75–300	Death if for over $\frac{1}{4}$ s (ventricular fibrillation).
2,500	Clamps (stops) heart. Burns to skin and internal organs. Immediately applied resuscitation may succeed.

TABLE 21.5 Compressed air requirements (Air Compressor, 1990).

Type of Tool	Common Location	Cubic Ft/Minute (cfm) Required per Tool When Operating	Time Used (%)	Total (cfm)
Grinders (8 in.)	Cleaning	50	50	25
Chippers	Cleaning	30	50	15
Hoists	Shipping	40	20	8
Nutsetter (large)	Assembly	30	25	7.5
Screwdriver	Shipping	35	20	7
Blowguns, chucks	Machine shop	25	25	6.25
Woodborer	Shipping	30	20	6
Hoists	Cleaning	35	50	3.5
Screwdrivers (small)	Assembly	12	25	3

Annual operating cost typically exceeds capital cost by a factor of 3 to 4; thus, system efficiency should receive a higher priority than capital cost (Air Compressor, 1990). Compressed air costs (capital + operating + maintenance) about 25–30 cents per 1,000 cu ft of free air ingested by the compressor.

2.2 Compressed Air Systems

2.21 Central Versus Local. The first decision to make is whether to have one centralized location or multiple locations. Although feasible, piping air to remote locations (different buildings or within a large building) results in piping losses and condensation in the cooled air. Local compressors also can be designed for local air needs (e.g., cleaner air, higher or lower pressures). For example, normal plant air is 80–100 psi. If air pressure in an area only needs to be 30 psi, it may be more cost effective to produce 30 psi with a local compressor than to reduce 80 psi air to 30 psi. Multiple remote compressors also can back each other up (assuming the system is interconnected). See Figure 21.5. However, multiple locations require multiple foundations and utility connections.

2.22 Number of Compressors. Because failure of the compressed air system may shut the plant down, one approach is to have one compressor with full load capacity (e.g., 200 hp, delivering 1,000 standard cubic ft./minute SCFM) with a second compressor of the same size as standby. Another approach is two 100 hp compressors of 500 SCFM running and one compressor of 500 SCFM as standby. If the air requirements are lower during second and third shift, there might be five 50 hp compressors of 250 SCFM with four running and one standby during the first shift; during the second and third shift perhaps only one needs to run.

Note that multiple compressors can be spread throughout the facility (with appropriate interconnections) or located at one central site.

2.23 System Design. The system consists of a number of components: (1) inlet air filtration, (2) compressor, (3) aftercooler and separator, (4) dryers, (5) air storage, (6) piping, (7) filter-regulator-lubricator unit, and (8) using elements (valves, cylinders, motors).

The **air inlet** should be outside of the building because (1) cool inlet air increases compression efficiency (see Table 21.6) and (2) taking air from inside increases air-conditioning load. Locate the intake 8 to 10 ft above the ground. Avoid intakes over hot pavements or on the sunny side of the building. The intake should be pointed toward the ground (or shielded) to prevent entry of rain or snow. Also, keep the intake air as clean as possible. Keep away from compressor engine exhaust, vehicle fumes, smoke, paint exhausts, and moisture sources. Contaminants also enter

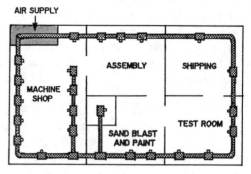

Centralized Air Compressor System

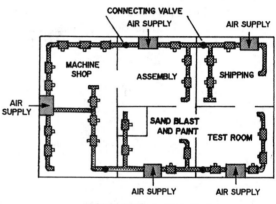

Multiple Air Compressor System

FIGURE 21.5 Compressed air systems can be centralized or at multiple locations. If multiple, use connecting valves so the systems can back up each other. Note the compressor is located on the outside wall so outside air can be used for the air supply.

through leaks (even though pressure inside the system is higher than outside), and the aspirator (jet pump) effect, molecular flow effect, and the shortened diffusion path effects.

Centrifugal **compressors** are used for large air volume requirements (4,000 CFM); reciprocating and rotary positive displacement compressors are for smaller requirements. Because of the noise from a compressor, it should be

TABLE 21.6 Cool air intake for compressors improves efficiency; hot air penalizes efficiency (Baughn, 1992).

Air Intake Temp. (F)	*Required Intake Vol. (cfm)*	*HP Change (%)*
30	925	−7.5
40	943	−5.7
50	962	−3.8
60	981	−1.9
70	1,000	Reference
80	1,020	2.0
90	1,040	4.0
100	1,060	6.0
110	1,080	8.0

located in a sound-controlled location. For energy efficiency, use a high-efficiency motor and synthetic lubricants. A variable-speed drive on a screw-type compressor gives good efficiency for variable demand requirements. Air compressors deliver about 4–5 CFM/hp at 100 psig discharge pressure.

The next component is the **aftercooler and separator**. Compression heats the air (180–350°F). When it is cooled down to about 100°F, moisture saturates the cooler air. The heat removed from the air can be recycled through air or liquid heat exchangers. There are three uses for the recovered heat: space heating, water heating, and both space and water heating. Example applications of space heating are supplemental or complete heating of the building, accelerated drying of cleaned or painted items, preheat air for furnaces (increases efficiency), and warm air curtains. In the summer, excess heat may be dumped outside. Example applications of water heating are to supplement building heating, for cleaning and laundry processes, and for chemical and pharmaceutical processes. After recovering heat, there will still be too much moisture in the line, so you will need a dryer.

In general, refrigerated **dryers** have the lowest capital cost and the lowest operating costs. Use whenever the pressure dew point is above 33°F (Murray, 1991). Pressure dew point refers to the temperature at which moisture will condense in the pipe while under pressure. At 100 psig, it is much higher (40–50°F) than atmospheric dew point. Thus, you need to follow all your lines, paying close attention to lines going through air-conditioned areas, outside areas, and unheated areas and to lines in front of open doors, to find the lowest temperature the lines are exposed to. If you have a pressure dew point below 33°F, use a desiccant dryer. A third type of dryer, a deliquescent dryer, has the lowest capital cost of all but, since the air is dried by passing through salt, downstream equipment tends to become corroded. The purpose of a dryer is to remove *liquid* water, not water vapor. If 60°F air is necessary, air dried to 35°F is not better (although more expensive) than 50°F air.

An air storage tank (**air receiver**) reduces pulsations in the discharge line and permits compressor size to be based on average air demand instead of peak demand. A large storage tank also permits turning the air compressor off during peak electrical demand periods to avoid peak demand charges. Receivers should have about 1 gallon capacity for each cfm of compressor capacity.

Air lines (and electrical and water lines) should be overhead and exposed, rather than underground, for simplicity of maintenance, service, and change.

A **filter-regulator-lubricator (FRL)** should be installed upstream of each pneumatic hookup. See Figure 21.6. The filter protects the pneumatic equipment against water, oil, and dirt. For equipment requiring finer filtrations, use two filters in series (such as a 5 micron prefilter and a .03 micron secondary filter). The regulator saves energy by allowing supply pressure to be reduced to optimum equipment pressure (it may not be needed if primary pressure is within 10 psig of equipment pressure). The lubricator—used only on devices not requiring extra filtration—delivers an air mist to the tank. The mist not only reduces maintenance but also can reduce noise several dbA (Oviatt, 1981).

Air from a normal compressor is not sufficiently pure for breathing. The OSHA respirator standard is as follows: atmospheric oxygen content, maximum of 5 mg of condensed hydrocarbons per m^3 of gas, maximum of 5,000 ppm of CO_2, and 20 ppm of CO. If combustion engines drive the compressor, be sure the compressor intake is remote (and upwind) from the engine exhaust. Additional processing is necessary to remove moisture and contaminants. Using personal protective garments, this is generally done locally just before use rather than purifying the entire system's air.

Air-using devices usually have a rated operating pressure of 90 psig. For every 1 psi pressure decrease, device effectiveness drops 1.4%. Performance increases with pressure over 90 psig, but at the cost of greater tool wear (Foss, 1981). Hose length should not exceed 20 ft or there will be excessive line friction loss. OSHA Standard 1910.242(b) states, "Compressed air shall not be used for cleaning purposes except where reduced to less than 30 psi and then only with effective chip guarding and personal protective equipment."

Noise from rotary tools comes primarily from the exhaust. Since the noise in rotary tool exhausts is proportional to velocity cubed, a 50% reduction in flow velocity reduces noise 9 dbA. To reduce exhaust noise, the exhaust can be piped away (very effective, but most useful for nonportable equipment), expanded through a muffler, or released through a diffuse and tortuous path. A quick

FIGURE 21.6   Filter-regulator-lubricator (FRL) units protect the equipment.

retrofit on noisy old tools is perforated metal (preferably with hole diameter equal to metal thickness) over the exhaust port. It can reduce noise 10 dbA (Oviatt, 1981).

3 WATER

Figure 21.7 shows an overview of water supply and disposal in a facility. Plumbing costs about 10% of a building cost.

3.1 Water Supply

See Table 21.7. In most circumstances, the supply is from a public main and is not a problem. However, for fire fighting, an emergency supply may be needed. Two alternatives are tanks and open reservoirs (ponds). Protect water from freezing by using a heater for tanks and an agitator or a bubbler near pond intakes.

Quality of incoming water may not be satisfactory; various treatments (such as sedimentation, water softening, or complete demineralization) may be necessary for some uses.

If some of the supply water is recycled from within the facility, suspended contaminants may have to be removed with filters. A typical arrangement is a coarse screen, a fine screen, and periodic backflushing to clean the filters. Note that most spray painting systems use a water wash to separate overspray from the air. Painting booths will therefore need a water supply and a drain. The collected paint sludge is a hazardous waste.

3.2 Piping Systems

There should be two separate systems leading to a sanitary sewer and a **storm sewer**. Toilets, washrooms and showers, laundries, and kitchens connect to the **sanitary sewer**. Drinking fountains, eyewash units, and safety showers should be located for worker convenience and can be connected to either the sanitary or the storm sewer. If there is no kitchen or laundry waste, a rough estimate is 20 gallons/person-shift of waste water to the sanitary sewer (Powers, 1973). See Toilets and Locker Rooms in Chapter 14 for fixture information.

Typical parts of a water supply system are the building main (runs horizontally, usually below the floor), riser (runs vertically), and horizontal fixture branches (runs horizontally in the wall). Note that flow rate is a function of pipe area, so a 1 in. dia pipe will have 25% of the flow rate of a 2 in. dia pipe. A **wet wall** has plumbing pipes. Minimize the amount of wet walls (e.g., by placing male and female toilet areas back to back).

TABLE 21.7   Water supply and distribution checklist. (Adapted from Facilities Requirement Checklist, 1991.)

Source and Capacity
    Municipal mains
    Impoundage or natural source

Water Demands (gallons per day, gpd and/or gallons per minute, gpm)
    Domestic water (potable)
        • Drinking fountains
        • Toilet areas
        • Food areas
    Process water
        • Manufacturing processes
        • Boiler feedwater
    Fire protection water
        • Hydrants and hose pipes
        • Sprinkler systems
    Air conditioning cooling water
    Lawn sprinkling

Water System Equipment and Capacities
    Meters and controls
    Pumps
        • Domestic water
        • Process water
        • Fire and sprinkler pumps
        • Air conditioning water pumps
    Water treatment systems
        • Softeners
        • Distilled water
    Water heaters
        • Domestic hot water
        • Boiler feedwater
        • Special processes
    Water coolers
        • Process
        • Drinking water

Storage and Distribution
    Storage
        • Reservoir or artificial lake
        • Tank
        • Working pressures
    Storage capacity
        • Domestic
        • Reserve for fire
    Distribution

Water Supply

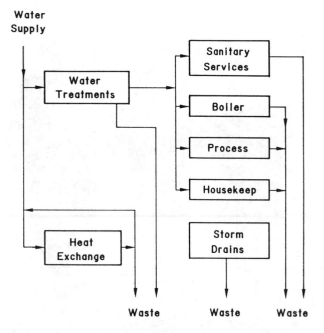

FIGURE 21.7   Overview of water supply and use (Powers, 1973).

Hot water in pipes cools rapidly, so minimize the distance between the water heater and the use areas. Water for hand washing should be 110°F. If kitchens or other areas require higher temperature water, boost it locally rather than using the higher temperature water throughout the building. Water for water heaters can be preheated by process heat, heat from air conditioning condensers, or solar panels. If hot water is used for washing floors and equipment (as in a food industry plant), the daily hot water usage may be thousands of gallons.

3.3 Water Discharge

See Table 21.8. Establish which process wastewater streams (1) can be discharged to the storm sewer without treatment, (2) can be discharged to the sanitary sewer without treatment, and (3) must be treated before discharge. Monitoring typically includes **chemical oxygen demand (COD)** and **biological oxygen demand (BOD)**. Microbial growths (algae, bacteria, and fungi) may

TABLE 21.8 Sanitary and storm drainage system checklist (Adapted from Facilities Requirement Checklist, 1991.)

SANITARY SYSTEM
Domestic Sewage
Locations
 • Toilets
 • Drinking fountains
 • Vending machines
 • Food areas
Quantity (gpd)
Collection system
 • Piping and manholes
 • Pumps
Treatment and disposal
 • Type of treatment plant
 • Effluent disposal
 • Raw sewage to municipal system
Industrial Wastes
Types and quantities (gpd)
 • Acids
 • Oils and fuels
 • Others
Collection system
 • Piping
 • Pumps
Treatment and/or disposal
STORM DRAINAGE SYSTEM
Quantity of Runoff
Paved areas
Roofs
Ground areas
Collection System
Inlets, catch basins, manholes
Piping
Drainage ditches
Subdrainage
Outfall and Disposal
Headwalls
Outfall
Disposal
 • Stream
 • Municipal system

have to be controlled with a microbicide treatment. The waste trapped by filters may have to be treated as hazardous waste.

Municipalities generally base their sewerage charge on water consumption. See Table 21.9. Meter process water separately so that the firm has a lower sewerage charge.

Water for heat exchange is the greatest industrial use of water. The "once-through" process is widely used. It requires an adequate supply of water, and the cooling water must be discharged to storm sewers, which requires a discharge permit from the U.S. Environmental Protection Agency (EPA). When heat exchangers are used to cool a product, provide water sampling points downstream of the exchanger, because a small leak can cause a great deal of contamination. The rising cost of water and expanding environmental controls have increased the interest in recycling water (i.e., a closed-loop system). Generally, recycling requires some monitoring and perhaps filtering of the water before reuse.

Boilers generally do not require demineralized water, but various compounds are added to the boiler water. Boiler maintenance water (from boiler blow-down, ash sluice water, boiler cleaning) is quite contaminated and special precautions need to be taken for disposal.

Process water is water that comes in contact with the raw materials or product. It may contain just dirt but often is fairly contaminated.

Water going into floor drains often becomes contaminated from product dust, spills, and so forth. Roof and yard water also may become contaminated since rain or snow picks up product dust and spills, especially around tank cars and drums. Monitor water for contaminants and install the necessary controls. The EPA began regulating storm-water runoff (including snow-melt runoff) in 1990. A typical permit application requires 29 h to complete (Johnson, 1992).

TABLE 21.9 Monthly sewer and water charges in Manhattan, Kansas, in April 1992.

Sewer
 Basic fee: $4.78 for any amount up to 200 cu ft
 Increment: .72 for each additional 100 cu ft
 Pollution: $S = SV(.006\,255)(A(BOD - 300) + B(SS - 350))$
where

S	=	surcharge, $
.006 255	=	.00075 hundred cu ft in a million gallons × 8.34 pounds/gallon of water
A	=	unit charge for BOD ($/lb) = $.039
BOD	=	biological oxygen demand, parts per million by weight
300	=	allowable BOD, parts per million by weight
B	=	unit charge for suspended solids ($/lb)
SS	=	suspended solids index, parts per million by weight
350	=	allowable suspended solids, parts per million by weight

Water

$6.10	=	basic fee for any amount up to 200 cu ft
+ next 1,800 cu ft	=	$1.78 per 100 cu ft
+ next 38,000 cu ft	=	$1.20 per 100 cu ft
+ all over 40,000 cu ft	=	.97 per 100 cu ft

4 COMMUNICATION SYSTEMS

See Table 21.10 for communication and signal systems.

4.1 Public-Address (PA) System

A wired **PA system** can be used for paging, announcements, and background music. Communication is one way rather than two way. A multizone system allows paging only specific areas. Speakers should provide sound 15 db over the ambient noise. Horn speakers should be used except for background music. The less powerful cone speakers improve music tones.

Telephones with speaker phones can be used as a paging system in areas with phones.

4.2 Radio

A wireless (i.e., FM radio) system also can be used. The communication is two way. Newer units are small enough to be helmet mounted and so could be used for people at a construction site. Radio is popular for vehicle dispatching and control (taxis and fork trucks).

For fork truck dispatching, radios often are connected to an onboard portable computer terminal. The terminals can accept and store or transmit data via their own keyboard, scanner, etc. (Systems Integration, 1991). The pallet (incoming or outgoing) should have a bar code with the product, the quantity, and the purchase order number. Some benefits are

TABLE 21.10 Communications and signal system checklist (Adapted from Facilities Requirement Checklist, 1991.)

Public Address System
Central speaker
Type and distribution
Background music system

Radio
Vehicle dispatching
Beepers
Cellular phones

TV Systems
Security
Manufacturing process
Education and news

Alarm Systems
Security and plant protection
Fire and storm system

Telephone System
Central switching
Plant telephone system
  • Distribution
  • Outside lines
  • Long-distance direct lines
  • Internal lines
  • Recording systems
Public telephones

Computer Systems
Local area networks
Outside plant

- minimum human data entry (reducing errors)
- immediate inventory updates
- computer-generated information to driver on putaway location (dock or warehouse)

People also can receive messages (but not transmit) using **beepers** (also called **pagers**). The least expensive models simply beep and then the person goes to a telephone and calls a designated number and asks for the message. More expensive models have a small screen and can give a short message (e.g., "call home" or "call 532-5606"). The original models had a limited range, but they can connect with a nationwide network.

In an interesting application, beeper transmitters can be connected to contacts on a door (e.g., on a vehicle), and when the door is opened, a message is sent. Other versions for cars track the vehicle if it is stolen by transmitting a signal for the police to locate it.

An advanced beeper is a cellular phone. It is much like a conventional phone, but it does not require a wire. Thus, it can be vehicle mounted (automobile, boat, etc.). Air Fone permits calls to commercial aircraft passengers. Cellular phones have become very popular in countries that have a poor telephone infrastructure. Fax machines can be connected to cellular phones as well as to hard-wired phones.

In 1992, cellular phones were marketed with data transmission capability in addition to voice capability, allowing transmission of computer data files.

4.3 TV Systems

TV systems are primarily used for security purposes. A camera with low-light capability monitors a specific area. Normally they are fixed focus and stationary, but available models can oscillate over an area and can even zoom on command. Often, the signal is run through a VCR so that a record is available. The tape can be reused, if desired.

TV cameras also can monitor a production process if it is noisy, dangerous, is difficult to view, and so forth.

Videotapes are often played through VCRs for education and training, but TV signals also can be used. A firm may wish to be connected to a network for news and entertainment. Typically, the network connection is in the breakroom or a conference room. Increasingly, firms are setting up video-conferencing facilities to reduce travel costs.

4.4 Alarm Systems

Fire alarm systems are discussed in Chapter 9. Security is discussed in Chapter 10.

4.5 Telephone System

The telephone system has become more economical over the years, not only in relative cost but also in absolute cost. See Table 21.11. Telephone communications are of three types: incoming, outgoing, or within plant.

4.51 Incoming Calls. The receiver of an incoming call has a number of potentially useful options: transferring, forwarding automatically, holding, conference calls,

TABLE 21.11   Relative cost of mail and long-distance phone calls since 1919.

	First-Class Letter Within the United States (cents)	3-Minute Phone Call, New York City to San Francisco, During Day (cents)	Ratio
1885	2		
1917	3		
1919	2	2,070	1,035
1932	3		
1945	3	400	135
1958	4	400	100
1963	5		
1965	5	200	40
1968	6		
1970	6	140*	23
1971	8		
1974	10		
1975	13		
1978	15		
1981 (March)	18	158*	9
1981 (November)	20	158*	8
1985	22	158*	7
1987	25	107*	5
1990	25	75*	3
1992	29	75*	2.6

*Customer dialed.

speaker phones, music or advertising for the waiting listener, automatic switching to an extension if calling party pushes a designated button, and so forth.

One of the most useful options is to have the receiver pay for the long-distance call; these inward wide area telephone service (**WATS**) **line** numbers are given 800 prefixes. The primary advantage of inward WATS lines is that the customer will make a free call but is reluctant to make a pay call.

If the person called is not present, a secretary can take a message. This function is now automated with answering machines in many organizations. Answering machines also can route calls to specific extensions (if the caller has a touch tone phone and follows the instructions).

4.52 Outgoing Calls.   Outward WATS lines are also popular. The charge/min is substantially less than the non-WATS rate; the rate usually is the same, independent of distance, within the United States and Canada. Callers normally are charged a minimum time of 1 minute per call, although very heavy users can get call charges in portions of a minute. If there is heavy traffic between a few points, such as plants and national headquarters, dedicated lines (tie-lines) are useful.

Starting in the mid-1980s, **fax** service became widely available. Compared with a letter, a fax offers speed and low cost. For local faxes (within a city), there is no telephone cost. To determine the cost of long-distance faxes, you need the transmission rate and the cost per minute. Fax

machines take 20–30 s/page and if a firm's WATS rate is 20 cents/min, then a one-page fax costs 20 cents and a two-page fax also costs 20 cents. Both are less than a 29-cent stamp. Compared with a phone call, a fax has the advantages of leaving a permanent record and of not requiring the receiver to be present. Automated fax machines can send the same message to multiple destinations and redial if the receiver is busy.

Employees may use the organization's telephones for personal business; the problem occurs for both long-distance and local calls. The cost is twofold: telephone charges and wage costs. A wide variety of control equipment can be used, ranging from installing telephone locks to restricting the regions that phones can call (e.g., within plant only, local, regional WATS) to using computerized call analysis (lists of numbers called, time used on call, etc.). Many firms have pay phones installed near breakrooms.

Teleconferencing (talk among three or more phones) is a low-cost alternative to travel. The most common approach is audio only. A teleconference is set up with the teleconferencing operator. At the prearranged time, each participant calls one number and the conference begins. The cost is low and any phone can be used. See Box 21.3.

Another alternative is **videophones**. Key to the development is signal compression. After a video camera shoots a scene, special digital signal processing (DSP) chips convert up to 15 frames/s of millions of individual picture elements (pixels). Other chips select the key pixels (e.g., the speaker's face) while ignoring static pixels (e.g., the wall). Complex images (e.g., plaid shirts) cause shakier pictures. The picture (2–10 frames/s) then is sent over ordinary telephone lines and reassembled at the receiver. Because the equipment costs about $30,000, only one conference room is set up at each facility and people must go to the conference room to make or receive a call. Coast-to-coast calls are about $30/h.

Telephone networks also are used extensively to transmit data between computers. The sending computer connects to the line through a modem, which converts the digital signal to an analog signal transmitted on the phone line. The signal is reconverted at the receiving end. Advanced telephone data circuits are digital, eliminating the use of the modem.

Electronic mail (EMail) is a network of mainframe computers around the world connected through the telephone system. A person composes a message on a local terminal (usually a PC connected to a mainframe) and sends the message through the local mainframe, which then sends it to the receiving terminal. Messages usually are free to the sending organization because it has already paid for the computer and the telephone charges are low (messages are sent in fractions of a second).

4.53 Within Facility.   A special type of telephone is a sound-powered phone—it uses no electricity and thus can be used in explosive environments. Two or more

---

BOX 21.3   *Teleconferencing recommendations (Jenkins, 1982)*

Teleconferencing is an intensified face-to-face meeting. It requires an organized, clear agenda; a strong leader; freedom from noise and distraction; and a sense of sharing a common purpose.

Visual aids need to be planned beforehand. For audio conferences, send them by mail ahead of time. For electronic blackboard and facsimile, plan for their transmission time. Even with full motion video, rehearsals need to be made to consider lighting, props, and other things. (People tend to judge the conference against a professional TV program instead of a conventional meeting, so they are very critical of minor flaws that they would ignore in a face-to-face meeting.)

As with any meeting, each person should have a copy of the agenda.

For audio conferences, each participant should have photos and background sheets on all the participants. Start speaking with your name (unless your voice is known to all).

---

phones are connected with a two-wire circuit. Another application is two portable phones (connected by 12–100 ft cords) for people who need to communicate but have difficulty due to obstructions or noise.

4.6 Computer Networks

Most information (a letter, a form) used to be processed on a typewriter. The paper (letter, form) then went to another location. In the late 1970s and 1980s, the word processor replaced the typewriter (in most organizations). The word processor, however, still is a "stand alone" device, producing paper that must go to another location. The next step has been to send messages electronically rather than on paper.

One possibility (a fax, short for facsimile) is to produce the message on paper and send it electronically over telephone lines.

Another possibility is not to use paper (or a disk) at all but to send the message electronically from the word processor/computer to another word processor/computer. There are two means of transmission: wire and radio (which is fairly rare but useful for portable computers or for use in inadequately wired buildings). Wire transmission can be through the telephone system (a common carrier) or a dedicated (nonutility-owned) system. Since much communication takes place within a building or local complex, many organizations have connected their word processors/computers in a **local area network (LAN)**. A typical distance would be 200–500 m. (In contrast, wide area networks (WAN) connect computers at different sites within a city or in various cities.) It is possible to connect a local network with other local networks. Repeaters restore voltage levels and sharpen signals, permitting a LAN to extend over a longer distance. Bridges connect LANs of the same type; gateways connect dissimilar LANs or connect a LAN with a non-LAN service.

Four aspects of LANs will be discussed: modulation method, transmission medium, access method, and arrangement (also called *architecture* or *topology*).

The modulation method can be baseband (one signal transmitted), which is the common method, or broadband (multiple signals, e.g., voice, data, and video, over the same transmission medium).

The transmission medium can be twisted pair, coaxial cable, or optical fiber. Twisted pair (telephone wire) is low cost but is susceptible to electrical interference and has low data transmission rates (1 megabit/s). In the 1990s, improvements in LAN protocols made increased use of twisted pair feasible. Coaxial cable (used in over 90% of LANs) has reasonable costs and good freedom from noise. Optical fiber is the high-cost, high-performance alternative.

The three common methods of accessing the network are contention, time slot, and token passing. Contention devices listen to the circuit and send a message only when no other message is being transmitted. If a "collision" occurs between messages, the messages will be garbled and the messages will be retransmitted. In the time slot method, each station is permitted to transmit only during a designated time slot. In the token method, a special message (a free token) is sent to determine if the network is available. If it is, then a busy token is issued and the busy token and message are sent together. When the message is received, the network is ready to receive a new free token.

Figure 21.8 shows three arrangements—bus, ring, and star. A key characteristic of the bus arrangement is that failure of a node does not cause the system to fail; thus, bus systems are usually reliable. A bus is similar to broadcasting in that any station's transmission is received by every other station. A ring arrangement connects the ends of the bus. When a station transmits, only the following station on the network receives the message. Because the message may have to be retransmitted a number of times to reach its destination and because node failure means some stations are not accessible, reliability is not as good as with a bus. In the star arrangement, each node communicates only with the central node. The central node has to be very reliable. The central node can be passive (broadcasting to every other node) or active (transmitting only to the designated destination).

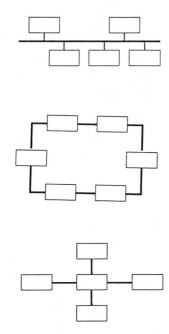

FIGURE 21.8 LAN architecture commonly is (1) bus, (2) ring, or (3) star. For small offices, consider the low-tech "sneaker net," in which people simply carry information around on pieces of paper or floppy disks.

DESIGN CHECKLIST: UTILITY NETWORKS

Electricity
    Energy-efficient motors
    Power management
    Distribution
    Clean, stable power
    Safety
Compressed air
    Cost
    System design
Water
    Where used
    Recycle

Communication system
    PA system
    Two-way radio
    TV systems
    Telephone system
        Incoming calls
        Outgoing calls
        Within facility
        Teleconferencing
        Computer networks
        Local area network

REVIEW QUESTIONS

1. Make a sketch showing the relationship of KW and KVA and power factor.
2. Briefly describe how power factor could be controlled with a combination of static and dynamic capacitors.
3. What is a demand ratchet charge?
4. Give the three approaches to reducing power peaks.
5. Why is it good to do work with heavy electrical demand on the second shift?
6. What are the three elements of an electrical substation?
7. When power and electronic wiring cross, which should be on top? Why?
8. How does voltage depend upon distance from the generator?
9. How do lightning rods work?
10. Give four solutions to static electricity.
11. Approximately what percent difference in electrical efficiency can be expected from a high-efficiency motor? Does the percent saving increase or decrease with motor size?
12. If a person with skin resistance of 10,000 ohms touches a 120 volt line, will this exceed "let go" current?

13. What is a ground fault circuit interrupter?
14. What are the advantages of multiple air compressors?
15. What is the difference between a forward-transfer UPS and a reverse-transfer UPS?
16. If an air-powered screwdriver used 20 ft³/min and electrical power cost was 5 cents/KWH, what is its operating cost/hour? Assume no capital cost for the compressor.
17. What is an air receiver used for?
18. What is a "wet wall"?
19. What is process water?
20. What do you estimate as the telephone cost of sending a two-page fax? Show calculations.
21. Should a compressor intake have hot air or cool air?
22. What is a local area network (LAN)?
23. Briefly discuss the bus, ring, and star networks.
24. Is the typical motor design A, design B, design C, or design D?
25. Is air from a normal compressor suitable for respirator use?
26. What are inward WATS, outward WATS, and tie-lines?

PROBLEMS AND PROJECTS

21.1 Assume you are about to select one of two motors. It is 5 HP but comes in an energy-efficient and standard version. The motor will run 3,800 h/yr at 90% of full load. For the utility rates in your town, what is the annual difference in power cost?

21.2 Visit a factory. How many gallons of water are used/month? What is the monthly cost? Where is it used? Is there any potential for recycling? How are contamination problems minimized?

21.3 What is the cost of a long-distance phone call to a destination 5, 50, 500, and 5,000 miles away (i.e., international)? What is the cost for various distances if you have a WATS line?

21.4 Updating the data of Table 21.11, when do you predict postage and a 3-minute phone call from New York to San Francisco will have equal costs? Use WATS rates for the phone call.

REFERENCES

Air compressor systems: Centralized or multiple. *Plant Engineering*, Vol. 46, 42–54, Nov. 22, 1990.

Baughn, E. Cutting the cost of compressed air. *Plant Engineering*, Vol. 48, 85–87, Jan. 16, 1992.

Bridger, B. Comparing ANSI and IEC power circuit breaker current-rating standards. *Plant Engineering*, Vol. 48, 81–84, March 5, 1992.

Facilities requirement checklist. *Plant Engineering*, Vol. 47, 178–181, July 18, 1991.

Fickett, A., Gellings, C., and Lovins, A. Efficient use of electricity. *Scientific American*, Vol. 266, 65–74, Sept. 1990.

Foss, R. Fundamentals of compressed air systems. *Plant Engineering*, Vol. 35, 10, 73–77, May 14, 1981.

Frydenlund, M. Understanding lightning protection. *Plant Engineering*, Vol. 46, 54–58, Dec. 13, 1990.

Grainger Catalog, 333 Knightsbridge, North Suburban, IL 60251 (1-800-323-0620), 1991.

Hammer, W. *Occupational Safety Management and Engineering*, 4th ed. Englewood Cliffs, N.J.: Prentice-Hall, 1989.

Hutchins, D., and Clark, O. Protecting electronic equipment from transient voltages. *Plant Engineering*, Vol. 47, 81–85, Nov. 7, 1991.

Jenkins, T. What it takes to teleconference successfully. *Administrative Management*, Vol. 43, 10, 28–29, 42, 44, Oct. 1982.

Johnson, D. Storm. *Plant Engineering*, Vol. 48, 74–79, March 5, 1992.

Keithly, D. Cutting costs through substation ownership. *Plant Engineering*, Vol. 45, 64–66, June 22, 1989.

Migliaro, M. Specifying batteries for UPS systems. *Plant Engineering*. Vol. 47, 100–102, March 21, 1991.

Motors. *Plant Engineering*, Vol. 47, 25–31, July 18, 1991.

Murray, B. Choosing compressed air drying equipment. *Plant Engineering*, Vol. 47, 112–114, March 21, 1991.

Oviatt, M. Energy conservation and noise control in pneumatic devices and systems. *Plant Engineering*, Vol. 35, 275–278, 1981.

Palko, E. Secondary substation transformer options. *Plant Engineering*, Vol. 46, 52–58, Feb. 8, 1990.

Powers, T. Control of industrial water emissions. Chapter 44 in *The Industrial Environment*. Washington, D.C.: Supt. of Documents, 1973.

Spethmann, D. Electrical energy management. *ASHRAE Journal*, Vol. 23, 7, 31–35, July 1981.

Systems integration in warehousing: Controlling your lift truck operation. *Modern Material Handling*, Vol. 46, 72–74, Feb. 1991.

IN PART IV SERVICES AND ENVIRONMENT

CHAPTER PREVIEW

After a consideration of basic illumination principles and illumination sources, specific techniques are given for general lighting, task lighting, and special lighting.

CHAPTER CONTENTS

1 Lighting Basics
2 Sources
3 General Lighting
4 Task Lighting
5 Special Lighting

KEY CONCEPTS

coefficient of utilization (CU)	luminaire dirt depreciation (LDD)	reflectance
contrast	luminaires (fixtures)	security lighting
emergency lighting	luminance	task factors
illuminance	people factors	video display terminal (VDT)
inspection (shape versus surface)	polarization	wavelength
lamp lumen depreciation (LLD)	quality of light	zonal cavity method
Life Safety Code	quantity of light	

1 LIGHTING BASICS

1.1 Relative Cost

The recent emphasis on energy conservation in lighting should not obscure a fundamental fact: *Light is cheap; labor is expensive.* Cost of light depends upon the type of light used (fluorescent, high-pressure sodium, etc.), lighting geometry (primarily distance from the task to lamp), whether local or general area lighting is used, etc. Box 22.1 shows cost calculations for high-pressure sodium (HPS) lighting in a high-bay manufacturing area. Annual cost is about .38 $/ft².

Labor cost/ft² depends upon the labor cost and the worker density. Assume a wage rate of $8/h and fringe costs of 35%, for a labor cost of 8 × 1.35 = $10.80/h. If the worker works 240 days of 8 h/day, annual cost is $20,736/yr. Assume 1 worker/300 ft². Then annual labor cost/ft² is 69.12/.38 = 180 times as expensive as lighting costs. In an office, fluorescent lighting, which is slightly more expensive than high-pressure sodium lighting, is the most common. However, 1 worker/100 ft² is typical, so in the office the ratio of labor cost to lighting cost usually is over 300. The point is, don't try to save a penny by cutting down the lighting if it costs a dollar on labor productivity.

1.2 IES Illuminance Recommendations

Table 22.1 defines various illumination units. Visibility of a task is affected by **quantity of light (illuminance)**, by **quality of light** (glare, direction, color), by **task factors** (size and contrast of the object, time available, speed and accuracy needed), and by **people factors** (how good are the eyes?).

Traditionally, higher illuminances have been used to overcome light quality, task, and people problems. This brute force approach still works but, as the cost of energy has increased, more elegant solutions concerning light quality, task, and people factors have become more popular.

The Illuminating Engineering Society (IES) formerly recommended a specified level of illuminance for various tasks. The present recommendation procedure is more subtle. Activities are broken down into nine categories

(IES, 1987). See Table 22.2. At the low illuminance end, there are relatively easy tasks done occasionally. At the high end, difficult tasks are done repeatedly. Table 22.3 further explains some of the nine categories. However, in addition to the general description of the task, the speed and accuracy required also are specified. More light is needed when speed and accuracy are more important. Vision generally declines after age 40, so more light is needed for people over 40. (Give your grandmother a larger light bulb for Christmas!) Environmental luminance also is considered—with more light needed when environmental reflectance is low. Table 22.4 shows how the factors affect the illuminance levels in Table 22.3. Although the new procedure does not give a "fine-tuned" illumination level, at least this first draft is more appropriate than the former procedure of ignoring quality of light, task factors, and people factors.

1.3 Quality of Light and Task Factors

Figure 22.1 shows how performance is affected by contrast in the task. Once contrast is above .3, improved contrast makes little difference. However, below .3 even a little improvement in contrast helps greatly. Contrast can be improved by modifying the lighting or the task.

Changing the lighting polarization or incidence angle to improve contrast is difficult, since the best lighting design may depend upon the task. Incorporate versatility into the lighting by use of an adjustable desk lamp. The operator would get the best visibility by trial-and-error movement of the lamp.

Changing task contrast is the second option. For reading materials, use new printer ribbons, clear photo-duplications, pens not pencils, black pens not blue, felt tip pens instead of ballpoints, white paper for computer print-outs instead of green, white paper instead of brown or manila paper, and so on.

Environmental contrast, on the other hand, should be minimized to reduce the adjustments of the eye. Table 22.5 gives recommendations relative to environmental contrast.

BOX 22.1   Example lighting costs for a high-bay manufacturing area. The example uses 400-W high-pressure sodium (445 W with ballast) and assumes a fixture life of 15 years (adapted from Turner, 1983). From Modern Materials Handling, Copyright 1983, by Cahners Publishing Company, Division of Reed Holdings, Inc.

Energy
(56 lamps) × (445 W/lamp) (1 KW/1000 W) (4250 H/yr)
    = 105,910 KWH/yr
(105,910 KWH/yr) ($.04/KWH) = $4236 $/yr
Demand KW (lights are on during peak hours)
(56 lamps) (1 KW/1000 W) (445 W/lamp) = 24.9 KW/
    month
(24.9 KW/month) ($5.91/KW-month) (12 months) =
    $1766
Fixtures
(56 fixtures) (300 $/fixture) (1/15) = $1120

Lamp Replacement
($46 $/bulb) (4250 H/yr/24,000 H/bulb) (56 bulbs) =
    $456

Total yearly cost = $4326 + 1767 + 1120 + 456 =
    $7668

Using an area lighted of 20,000 ft², the annual cost =
    $7668/20,000 = .38 $/ft². Note that energy cost is
    $4236 + 1766 = $6002 of the $7668 total cost or
    about 80%.

TABLE 22.1 Units and definitions of illumination (Konz, 1990).

Quality	Unit	Definition and Comments
Luminous intensity, l	candela	Light intensity within a very small angle, in a specified direction (lumen/steradian) Candela = 4π lumens
Luminous flux, φ	lumen	Light flux, irrespective of direction; generally used to 1. express total light output of source 2. express amount incident on a surface
Illuminance, dφ/dA	lux	1 lumen/m² = 1 lux = .093 footcandle 1 lumen/ft² = 1 footcandle = 10.8 lux
Luminance (brightness)	nit	Luminance is independent of the distance of observation as candelas from the object and area of the object perceived by the eye decrease at the same rate with distance. 1 candela/m² = 1 nit = .29 footLambert 1 candela/ft²π = 1 footLambert = 3.43 nits
Reflectance	unitless	Percentage of light reflected from a surface

Typical Reflectance		Recommended Reflectances	
Object	*%*	*Object*	*%*
Mirrored glass	80–90	Ceilings	80–90
White matte paint	75–90	Walls	40–60
Porcelain enamel	60–90	Furniture and	
Aluminum paint	60–70	equipment	25–45
Newsprint, concrete	55	Floor	20–40
Dull brass, dull copper	35		
Cast and		*Munsell*	*Reflectance*
galvanized iron	25	*Value*	*%*
Good quality		10 =	100
printer's ink	15	9 =	100
Black paint	3–5	8 =	78
		7 =	58
		6 =	40
		5 =	24
		4 =	19
		3 =	6

Quality	Unit	Definition and Comments
Contrast	unitless	$$C = \frac{\text{luminance of brighter} - \text{luminance of darker}}{\text{luminance of brighter}}$$
Wavelength	nanometers	The distance between successive waves (a "side view" of light). Wavelength determines the color hue. Saturation is the concentration of the dominant wavelength (the degree to which the dominant wavelength predominates in a stimulus). Of the 60 octaves of the electromagnetic radiation, the human eye detects radiation in the octave from 380 to 760 nanometers.
Polarization	degrees	Transverse vibrations of the wave (an "end view" of the light). Most light is a mixture; horizontally polarized light reflected from a surface causes glare.

Note that these are maximum values—a smaller ratio is better. Office desktops, typewriters, and **video display terminals (VDTs)** should be relatively light in color to minimize the contrast with white paper. For machinery, paint the stationary and moving parts different colors; the background should be darker than the task. Use high-reflectance matte (not gloss) surfaces; avoid polished metal or glossy surfaces. Of course, walls and machinery need not be painted in only battleship gray or industrial green. Light colors on the ceilings, walls, and floors not only reuse the light but also reduce brightness contrast (see Table 9.1). Areas around windows (such as sashes and dividers) should be light colors to reduce contrast.

1.4 People Factors
Many people do not have 20-20 corrected vision. Jacobsen (1952) reported 32 of 132 inspectors needed either glasses or a prescription change. Ungar (1971) studied industrial workers. For those 15 to 30

years old, 20% had faulty eyesight; for 31 to 40 it was 25%; for 41 to 60 it was 50%; and for 61 to 70 it was 70%. Ferguson et al. (1974) reported that 37% of workers wearing glasses needed a new prescription and 69% of those without glasses needed glasses.

Thus, employers should require periodic visual examinations if vision is a critical part of the job. A plant nurse can do a preliminary screening.

2 SOURCES

The ideal illumination source (which does not exist) would be free, give the desired amount of light upon demand, and have high quality (color, glare, highlighting, contrast). The first basic choice is between artificial illumination and sunlight. Although sunlight can act as a supplement for

TABLE 22.2   Recommended target maintained illuminance (lux) for interior industrial lighting (IES, 1987). See Table 22.3 for examples; see Table 22.4 for factor calculations.

Reference Work Plane	Type of Activity	Illuminance Category	Illuminance (lux)		
			Total Factor −3 or −2	Total Factor −1, 0, +1	Total Factor +2 or +3
General lighting throughout spaces	Public spaces with dark surroundings	A	20	30	50
	Simple orientation for short temporary visits	B	50	75	100
	Working spaces where visual tasks are only occasionally performed	C	100	150	200
Illuminance on task	Performance of visual tasks of high contrast or large size	D	200	300	500
	Performance of visual tasks of medium contrast or small size	E	500	750	1000
	Performance of visual tasks of low contrast or very small size	F	1000	1500	2000
Illuminance on task obtained by a combination of general and local (supplementary) lighting	Performance of visual tasks of low contrast and very small size over a prolonged period	G	2000	3000	5000
	Performance of very prolonged and exacting visual tasks	H	5000	7500	10,000
	Performance of very special visual tasks of extremely low contrast and small size	I	10,000	15,000	20,000

artificial light in some office situations, there are many problems with sunlight as an illumination source. See Chapter 9 for comments concerning windows and atriums.

2.1 Lamps

Table 22.6 summarizes typical lamps. For offices, fluorescent lamps are the dominant lamp. Although incandescent lamps are used for special effects, they have only about 25% of the lumens/watt of fluorescents as well as having quite short lives. High-pressure sodium (HPS) is beginning to be used, but it presents color problems; special selection of room colors is needed. Lin and Bennett (1983) conclude that sodium sources shouldn't be used where faces are viewed substantially.

For industrial and warehouse applications, high-pressure sodium and multivapor are gradually replacing fluorescent. Mercury and incandescent are only for special situations. Lumens/watt is not the only driving force for HPS lighting; its long life and high percentage of lumen maintenance (lamps don't dim much with age) also help. The primary problem is their color and their restrike time in case of power failure. Some firms mix multivapor and HPS to get better color. Restrike time can be overcome by using a second set of lamps (fluorescent or quartz) for emergency lighting. Fluorescent lamps have a larger surface area and thus are less bright, giving less direct and indirect glare (reflections). They also have good color and immediate restrike. Mercury now is used primarily in low-wattage situations where low mounting heights and low amount of illumination are important.

2.2 Luminaires

Luminaires (fixtures) distribute light (up versus down) in six ways, but direct (all down) and semidirect (mostly down) are the most used. Semidirect tends to be best since (1) the light on the ceiling reduces brightness contrast and (2) semidirect luminaires tend to stay cleaner (i.e., lose less light) as air can move upward through the fixture. "Beam spread" is a further description of the downward component. See Table 22.7. Wider beam spreads give more overlapping (better illumination on vertical surfaces and less dependence on a single lamp). High mounting heights tend to use narrower beams. For example, if the bottom of the fixture were 33 ft above the floor and work height were 3 ft, then effective mounting height would be 30 ft. If a concentrating beam were used, then spacing between fixtures should be less than (.30)(.5) = 15 ft. On the other hand, the fixture should not achieve a broad spread by poor shielding. The shielding angle—the angle between a horizontal line and the line of sight at which the source first becomes visible—should be greater than 25°, preferably approaching 45°.

Since it is difficult to exactly foresee the future, fixtures should be selected that are relatively easy to relocate within the area. A variety of plug-in designs is available.

3   GENERAL LIGHTING

3.1 Amount of Light

One possibility is to light an entire area uniformly. This gives maximum flexibility in

TABLE 22.3 Selected task examples of illuminance categories for use with Table 22.2 (IES, 1987). The IES also has examples for sports and recreation, outdoor facilities, and transportation vehicles.

Illuminance	Task/Activity
B	Dance halls and discotheques
	Restaurant: dining
	Corridors in nursing areas at night
	Library: inactive stacks
	Offices: micro-fiche reader, video display terminal
	Farm shed: machine storage
C	Lobbies: bank
	Drafting: light table
	Elevators
	Corridors in nursing areas in daytime; hotel corridors and stairs
	Locker rooms, toilets
	Offices; lounges, lobbies, reception areas
D	Conference rooms
	Restaurant: cashier
	Nursing station
	Hotel: lobby, reading area
	Library: active stacks, circulation desk, book repair
	Office: AV area; duplicating area; reading ink writing, newsprint, typed originals, 8–10-point type
	Assembly: simple
	Inspection: simple
	Machine shop: rough bench or machine work
	Material handling: wrapping, labeling, stock picking
	Painting: dipping, simple spraying
	Welding: orientation
	Woodworking: most areas
E	Barber shops and beauty parlors
	Laboratories: science
	Restaurant: kitchen
	Nursing desk
	Hotel: front desk
	Library: card files
	Offices: mail sorting, reading 6-point type or phonebooks
	Assembly: moderately difficult
	Inspection: moderately difficult
	Machine shop: medium bench or machine work
	Painting: fine hand painting and finishing
	Sheet metal work: most areas
	Woodworking: fine bench and machine work, fine finishing
F	Operating room: general
	Classroom: demonstration
	Graphic design: charting, mapping, layout, artwork
	Assembly: difficult
	Inspection: difficult
	Sheet metal: scribing
G	Autopsy table
	Cloth products: sewing, cutting
	Inspection: very difficult
	Machine shop: fine bench or machine work, fine polishing
	Painting: extra fine hand painting and finishing
H	Dental suite: oral cavity
	Surgical task lighting
	Inspection: exacting
	Machine shops: extra fine work
	Welding: precision manual arc-welding

TABLE 22.4 Factors for Table 22.2 (IES, 1987). If the task was in illuminance category D, the worker age was "Under 40," speed/accuracy were "Important," and task background reflectance was "30 to 70," then the components would be –1, 0, and 0; the total factor would be their sum, –1, so the middle column in Table 22.2 would be used and the recommended illuminance would be 300 lux.

	Factor		
Variable	–1	0	+1
For illuminance categories A, B, C (general lighting throughout spaces)			
Occupant ages, year	Under 40	40–55	Over 55
Average weighted room surface reflectance (%)	Over 70	30–70	Under 30
For illuminance categories D, E, F, G, H, I (illuminance on task)			
Worker ages, years	Under 40	40–55	Over 55
Speed and/or accuracy	Not important	Important	Critical
Task background reflectance (%)	Over 70	30–70	Under 30

arranging the workstations and machines in the area; it eliminates the need to move fixtures if the area is rearranged; it also allows use of large lamps, which have higher lumens/watt than small lamps. However, it costs more for lamps, luminaires, and power and is uninteresting esthetically.

The basic equation is

$$I(A) = (N_1)(N_2)(L) \qquad (1)$$

where

I = illuminance in area, lux

A = area illuminated, m²

N_1 = number of luminaires

N_2 = number of lamps/luminaire

L = lumens/lamp

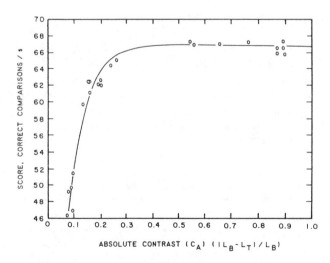

FIGURE 22.1 Contrast of the task above .3 makes little difference in performance, but when the contrast is poor, even a little improvement helps considerably (Rea, 1981). L_T = luminance of the task target, and L_B = luminance of the background (i.e., white pages). Reprinted from the April 1981 issue of *Journal of IES* with permission of the Illuminating Engineering Society of North America.

TABLE 22.5 Recommended maximum luminance ratios (IES, 1987).

Between		Environment[b]		
		A	B	C
Tasks and	Adjacent darker surroundings	3 to 1	3 to 1	5 to 1
	Adjacent lighter surroundings	1 to 3	1 to 3	1 to 5
Tasks and	More remote darker surroundings	10 to 1	20 to 1	[a]
	More remote lighter surroundings	1 to 10	1 to 20	[a]
Luminaires, and windows, skylights, etc.	Surfaces adjacent to them	20 to 1	[a]	[a]
Anywhere with normal field of view		40 to 1		

[a]Luminance control not practical.

[b] A = Interior areas. Reflectance of entire space can be controlled in line with recommendations for optimum seeing conditions.

  B = Areas. Reflectances of immediate work can be controlled, but control of remote surround is limited.

  C = Indoor and outside areas. It is completely impractical to control reflectances and difficult to alter environmental conditions.

For example, if you wished 750 lux in a 2,500 m² area and were going to use luminaires with 1 lamp/luminaire and each lamp were rated as 22,500 lumens, then $N_1 = 83$ luminaires required.

However, equation 1 needs to be modified for three types of losses. The first loss is the absorption of the light by the room surfaces. The **coefficient of utilization (CU)** adjusts for (1) the shape and reflectance of the area lighted (the room cavity), (2) the shape and reflectance of the area above the lamps (the ceiling cavity), (3) the shape and reflectance of the area below the surface lighted (the floor cavity), and (4) the light pattern of a specific lamp in a specific luminaire. The second loss is **lamp lumen depreciation (LLD)**, the loss of light output of a lamp with age. The third loss is **luminaire dirt depreciation (LDD)**, the loss of light output due to dirt on the luminaire. The resulting equation is

$$I(A) = CU(LLD)(LDD)(N_1)(N_2)(L) \qquad (2)$$

To solve the above equation (the **zonal cavity method**), consult the *IES Handbook* (IES, 1987) for detailed values of the variables. Lamp and fixture manufacturers also can furnish values, generally along with a computer program to solve the equation for any design you suggest.

For preliminary calculations, assume that $(CU)(LLD)(LDD) = .5$. Another rough calculation technique uses Table 22.8. For example, if 150 W HPS lamps are used with 15 × 15 spacing, room width = 150 ft, and mounting height = 25 ft (i.e., width = 6 [MH]), then on the work surface illuminance will be about 490 lux. If the spacing is doubled in one direction, the lux is cut to 50%; if the spacing is doubled in both directions, the lux is cut to 25%.

Once the number of lamps and luminaires is determined, the spacing decision remains. For uniform lighting, the checkerboard pattern on the "same color squares" is efficient. However, this tends to overlight the center and underlight the perimeter, so the spacing can be widened in the center and reduced on the perimeter. End-to-end luminaires are an inefficient pattern. Lighting next to windows should be controlled separately (automatically or manually).

In many areas (such as private offices, warehouses, and manufacturing areas with relatively fixed equipment), the locations of the specific work areas to be lighted are known. In these situations, specifically locate the luminaires. For example, in warehouses put the lights over the aisles, not over the racks. In offices put the luminaires where the light does not reflect into the user's eyes (check with a mirror on the work surface). See Figure 22.2.

3.2 Energy Conservation   See Chapter 24.

TABLE 22.6. Lamp characteristics of typical industrial lamps. (Courtesy of General Electric Lighting Business Group.)

Type of Lamp	Watts	Lumens/Watt Initial	Mean	Lumen Maintenance (%)	Rated Life (h)	Restrike (min)	Relative Cost
High-pressure sodium	35–1,000	64–140	58–126	90–92	24,000	1–2	Low
Metal halide	175–1,000	80–115	57–92	71–83	10,000–20,000	10–15	Medium
Fluorescent	28–215	74–100	49–92	66–92	12,000–20,000+	Immediate	Medium
Mercury	50–1,000	32–63	24–43	57–84	16,000–24,000+	3–6	High
Incandescent	.100–1,500	17–24	15–23	90–95	750–2,000	Immediate	High

TABLE 22.7 Luminaire beam spreads versus fixture spacing (S)/mounting height (MH) (General Electric, 1983). Table courtesy of General Electric Lighting Business Group.

Luminance Classification	*S/MH* *above Work Plane*
Highly concentrating	Up to 0.5
Concentrating	0.5 to 0.7
Medium spread	0.7 to 1.0
Spread	1.0 to 1.5
Widespread	Over 1.5

4 TASK LIGHTING

Rather than furnish all the light from the ceiling, often it is better to furnish some light locally—task lighting. The goal can be reduction of energy costs for local control of light characteristics such as color and direction, made possible because the light can be perfectly positioned to the task. Energy savings come from the lower overall wattage used in an area (saving also the air-conditioning load) and the individual control of each light, permitting individual lights to be turned off.

If a local source is used, glare control is essential. Direct glare usually is not a problem, but indirect glare (reflection) needs to be carefully controlled. For example, put low-reflectance materials such as cloth on the insides of inspection booths and in VDT work areas. If possible, the local lamp should be adjustable in position. Figure 22.3 shows some alternatives.

4.1 Inspection   For color contrast, the **inspection** can be for the object **shape** or the object **surface** characteristics.

To detect shape, maximize the contrast of the task and the background. For example, if buttons are inspected for holes, the table color could contrast with the button color. Another technique is a mask with the specified item shape and transillumination which allows a thin border of light to show for an in-tolerance part. When you are looking for shape, object orientation is critical. To demonstrate this, try to recognize a print of a face turned upside down. For printed material, letters and background should have maximum contrast. White on black is a slightly higher contrast than black on white, so reverse slides are better for viewing, but for ordinary printing use black on white. Avoid "arty" low-contrast letters (black on red, red on brown, etc.). They are not only less legible but less attractive (Konz et al., 1972).

To detect surface characteristics such as color and texture, minimize the contrast of the task against the background. Pearls sold at retail are displayed on black velvet; the maximum contrast makes it difficult to detect color differences between pearls. When pearl merchants buy from each other, they display the pearls on white cloth to maximize color differences. Thus, if you are sorting green beans for color, sort on a table painted the green of a good bean.

The color of the light also is important. Color differences in red material are emphasized by sources strong in blue light, and in blue material by sources strong in red (Misra and Bennett, 1981). When inspecting for a color, specify whether the lamp used is incandescent, cool white fluorescent, HPS, etc., since the colors appear quite different under different lamps.

4.2 Lighting for VDT Areas   The following material is from Konz, 1990.

TABLE 22.8 Approximate lux (maintained) on work surface for various lamps and spacing. Room width (*W*) = 6 (Mounting Height). If *W* = 3 *MH*, reduce lux 15% and increase wattage 15%; if *W* = *MH*, reduce lux 50% and increase wattage 50% (General Electric, 1983). Table courtesy of General Electric Lighting Business Group.

Lamp	Watts	Spacing (ft)				
		10 × 10	*15 × 15*	*20 × 20*	*25 × 25*	*30 × 30*
High-pressure	70	350	150	100	–	–
sodium	100	550	250	150	100	–
	150	950	490	270	160	110
	250	1900	860	490	320	220
	400	3200	1500	810	540	380
	1000	–	–	2300	1500	1000
Multivapor	175	920	380	220	160	110
	400	2200	970	540	380	270
	1000	–	3200	1800	1130	810

Continuous Rows of 2-Lamp Fluorescent (Cool White) Lamps in Fixtures

Lamp	W	Spacings[a]				
		6 ft	*8 ft*	*10 ft*	*12 ft*	*15 ft*
Rapid start	40	1300	970	760	650	540
Slimline	75	1300	970	760	650	540
High output	110	2000	1500	1200	970	810
Power grove[R]	215	3200	2400	1900	1600	1300

[a]Spacings assumed within maximums established by fixture manufacturers.

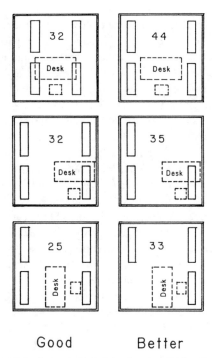

Good   Better

FIGURE 22.2   Reduced reflected glare improves lighting (Goodbar, 1982). The numerical index is equivalent sphere illumination (ESI), a lighting index which combines the luminance level and reflected glare into a single index. The upper pair shows the effect of moving the lamps outside the desk projection to reduce glare. The middle and lower pairs show how visibility can be improved with three fixtures in the room instead of four—the reduction in amount of light is more than compensated for by less glare. Reprinted from the Nov. 1982 issue of *Lighting Design and Application* with permission of the Illuminating Engineering Society of North America.

A video display terminal (VDT) differs from conventional paper-reading tasks. The screen is vertical instead of horizontal but, more important, increased light *reduces* the readability of the display. Thus, if paper and screen both must be used (and that is the norm), there is a potential lighting conflict between enhancing visibility of the paper and enhancing visibility of the screen.

Of the two tasks (paper and screen), the screen task is the more difficult because the electronic letters are less sharp and the letter/background contrast is worse, and because there is some flicker. Reading rates are often 25% slower with screens than with paper.

Light below 100 lux enhances screen visibility; above 500 lux it enhances print legibility. If task lighting is not used, a compromise general lighting is 200–500 lux.

Reducing glare from general lighting or windows is essential. Use the mirror test to determine glare. Put a mirror on the screen and look at it from the operator's eye position. Any bright spots in the mirror are glare sources to be eliminated.

Example glare sources and solutions are ceiling luminaires (turn off the specific light or shield the bulb with high cutoff reflectors), windows (reorient screen or cover window with opaque blinds or curtains), white paper mounted on a room divider (move posted material or

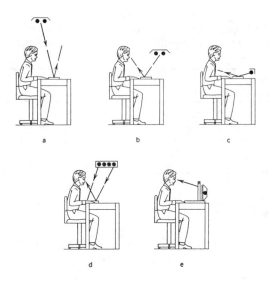

FIGURE 22.3   Placement of supplementary luminaires can use five techniques: (*a*) luminaire located to prevent reflected glare—reflected light does not coincide with angle of view (check with mirror on work surface); (*b*) reflected light coinciding with angle of view; (*c*) low-angle lighting to emphasize surface irregularities; (*d*) large-area surface source and pattern reflected toward the eye; and (*e*) transillumination from diverse sources.

change divider angle), walls over 50% reflectance (replace with low-reflectance paint, perhaps even wall carpet), white shirt of operator (wear darker shirts), or light-colored phones or objects (move them). The keyboard should be on a matte surface; keys should be matte also.

Figure 22.4 shows a workstation with a number of adjustment features. Adjusting screen height and angle can reduce glare. Also, various antiglare treatments can be applied to the screen itself.

A common solution to lighting conflicts is to use task lighting, with relatively low (350 lux) general lighting supplemented by higher illumination (e.g., 600 lux) on documents. Illuminate the non-VDT areas at a higher level (e.g., 800 lux) so that the office will not be overly dark.

VDT work restricts workers' posture, causing neck and shoulder problems. The problem is most severe for older people and people wearing bifocal or trifocal glasses. The best solution is special glasses made for the viewing distance used at the screen. Optometrists use 22 in. (56 cm) for this distance but can make it closer for special applications (such as a laptop computer). The range at which letters are in focus or readable will vary with the individual (e.g., a focal length of 22 in. and a visibility range of 13–25 in.). Note that these glasses are useless for work outside the workstation, so the person will need additional general-purpose glasses.

For more on VDTs, see Chapter 13.

5 SPECIAL LIGHTING

5.1 Warehouse Aisle Lighting

As was discussed in Chapter 11, items generally are stored in racks. The

FIGURE 22.4 Adjustability is the key to a well-designed VDT workstation. The keyboard and the display are on separate tables. Each can be adjusted in height. In addition, depending upon which is purchased, there is either a display table tilt option (to reduce glare on the screen), or a swivel option (when the display is shared between workers). The keyboard table can be moved in and out as well as up and down. Note the rounded edges and corners on the furniture. The source documents are on an adjustable holder. (Photograph of IBM Synergetix [tm] Furniture courtesy of International Business Machines Corporation.)

challenge is to light not the top of the racks but the vertical surfaces—generally in long, narrow aisles. If stacking height < aisle width, then treat the area as an open bay.

The use of high racks makes high-intensity discharge (HID) lamps preferred over the former fluorescent lighting, since HID luminaires have better directional characteristics. Different directional patterns are available, such as medium symmetrical, wide symmetrical, and directional asymmetrical (4 lobes—2 up and down the aisle and 2 into racks). For luminaire mounting heights below 16 ft, type of HID luminaire makes relatively little difference. Above 16 ft, the asymmetrical generally is best. A high-reflectance aisle floor helps. Frier and Frier (1980a) recommend mounting the luminaires above the aisle instead of above the racks. Use a maximum distance above the rack top of .5 (aisle width); otherwise too much light is wasted on the rack tops. A large number of low-watt sources will give a more even distribution than a small number of high-watt sources.

5.2 Security Lighting

The purpose of **security lighting** is prevention of crime. (For more on security, see Chapter 10). The light should discourage intruders (reduce "offense") and improve detectability when entry is effected (improve "defense"). Design the lighting security system as if the guard or TV camera were the audience at a play and the intruder were the actor. Give the audience a good view of the actor; conversely the actor should not be able to see the audience. (Physiologically, the intruder, with dark-adapted eyes, will be faced with a light barrier; tension will increase pupil dilation, which increases the effect of the light.) Design the "stage lighting" to minimize glare for the guards, cameras, or neighbors in the plant vicinity.

Table 22.9 gives recommended security lighting illuminances. High-pressure sodium lighting is the recommended source since it gives high lumens/watt and is compatible with most TV cameras. Generally, low-light cameras are used except in well-lighted areas, such as interiors, doors, and gates, where daylight cameras are satisfactory. If the power fails, the lights go out; high-intensity discharge lights will stay off for 10 to 15 min after power comes back on (see Table 22.6). Thus some source of emergency power is desirable; see Chapter 21 on standby power.

In general, the defender (the guard) should not be visible to the intruder. Thus, a gatehouse should be lit so as to conceal the interior. That is, it should be relatively dim within the gatehouse, relatively bright outside.

If there is no on-site defender (i.e., you depend upon police observers), the lighting system can be turned around and the building exterior lit directly.

Garages and parking areas need about 10 lux. For garages, place the fixtures to illuminate the between-car walk spaces rather than the area over the car stalls.

5.3 Emergency Lighting

Emergency lighting is designed to permit people to leave the building if the normal power supply fails. Although local codes also may apply, the **Life Safety Code** specifies 10 lux on the floor along the route people might take leaving the building (*Life Safety Code*, 1988). Frier and Frier (1980b) recommend 30 lux for congested or critical areas such as corridor intersections, tops of stairs, dangerous machinery, etc. The vertical surface around the exit door should be 20 to 50 lux. See Table 22.10.

TABLE 22.9 Recommended area illuminances (lux) for security lighting, measured on the horizontal plane (Baker and Lyons, 1978).

	District Brightness		
Risk	High (Adjacent Main Road Lighting; Lighted Adjacent Land; Floodlighting)	Medium (Adjacent Secondary Road Lighting)	Low (No Adjacent Lighting on Adjoining Property or Nearby Roads)
Extreme	20–30	10–20	5–15
High	10–20	5–15	2–10
Moderate	5–15	2–10	1–5

TABLE 22.10 Minimum illuminance (lux) levels for safety are the absolute minimum at any time and at any location on any plane where safety is related to seeing conditions (IES, 1987).

Hazards Requiring Visual Detection	Normal Activity Level	
	Low	High
Slight	5	11
High	22	54

The emergency lighting can be either completely separate lights with batteries or the normal lights with a backup power source. Separate lights tend to be projector and reflector (PAR) lamps, a battery, and a device to turn the light on when the power fails. Use PAR with wide horizontal patterns for open areas and with long narrow beams for corridors and stairs. The normal light can have a local battery (e.g., units in AC fluorescent fixtures) or a central battery. There also can be a motor-generator. See uninterruptible power in Chapter 21. Note that HID lamps need 1 to 15 min restrike time (see Table 22.6) and another 3 to 5 min before coming to full intensity; thus, you may need some fluorescent lamps in HID areas. Another option is a quartz lamp (about the size of a lipstick case) in some HID fixtures. After a momentary power interruption, the quartz lamp automatically lights until the HID lamp cools down, restrikes, and regains 60% of its full light output. Exit signs can be self-powered but, in much of the country, codes permit the exit sign to be illuminated externally by the emergency lighting—this should be less expensive than having both emergency lights and internally powered exit signs.

DESIGN CHECKLIST: ILLUMINATION

Cost of lighting versus labor
Recommended lighting quantity
Quality of light/task factors
People factors
Illumination source
General versus task lighting
Inspection

Lighting for VDT areas
  Mirror test
  Adjust workstation
  Adjust ambient lighting
Security lighting
Emergency lighting

REVIEW QUESTIONS

1. What is the relative cost of labor versus lighting?
2. What illuminance would you recommend for illuminance category D if (a) worker age was 30, speed and accuracy were important, and task background reflectance was 50%? (b) worker age was 50?
3. What illuminance would you recommend for a mail sorting area if worker age was 35, speed and accuracy were critical, and task background reflectance was 25%?
4. Give five different techniques of improving contrast for office tasks.
5. Why should window sashes be light colors?
6. Why isn't high-pressure sodium lighting used more in offices?
7. What is the restrike time of fluorescent, multivapor, and high-pressure sodium lamps?
8. If a concentrating beam is located 25 ft above the work surface, how close should the lamps be to each other?

9. What is coefficient of utilization, lamp lumen depreciation, and lamp dirt depreciation?

10. Discuss task/background contrast when inspecting for shape and when inspecting for surface characteristics.

11. What color light should be used if you wish to inspect for color differences in red material?

12. How is the mirror test used for VDTs?

13. What is the maximum distance recommended between the top of a rack and the lamp? Assume an 8 ft aisle.

14. How much illuminance should there be for security lighting in a high-risk area where there is adjacent secondary road lighting?

15. Do exit signs need to be self-illuminated?

PROBLEMS AND PROJECTS

22.1 Using Table 22.3, determine the recommended lighting levels for the IE office of your university. Use a light meter to determine actual light levels. Any ideas? What does the department head think of your ideas?

22.2 Design a ceiling lighting system to light a classroom uniformly. (Is this a good goal?) Assume $CU \times LLD \times LDD = .5$. Justify your design.

22.3 Visit a local factory and observe an inspection operation (it might be combined with making the object). Any ideas? What does the supervisor think of your ideas? What does the operator think of your ideas?

22.4 Evaluate the lighting at a VDT workstation. Consider both lighting of the source documents and the screen and glare. Does the operator agree with your ideas? Does the supervisor?

REFERENCES

Baker, J., and Lyons, S. Lighting for the security of premises. *Lighting Research and Technology*, Vol. 10, 1, 10–18, 1978.

Ferguson, D., Major, G., and Keldoulis, T. Vision at work. *Applied Ergonomics*, Vol. 5, 2, 84–93, 1974.

Frier, J., and Frier, M. Design techniques for long, narrow areas. Chapter 7 in *Industrial Lighting Systems*. New York: McGraw-Hill, 1980a.

Frier, J., and Frier, M. Emergency lighting systems. Chapter 14 in *Industrial Lighting Systems*. New York: McGraw-Hill, 1980b.

General Electric. *Industrial Lighting* (Booklet 201-31309). Nela Park, Ohio, 1983.

Goodbar, I. The application of the ESI system to office lighting. *Lighting Design and Application*, Vol. 12, 11, 19–27, 1982.

IES. *IES Lighting Handbook: Application Volume.* New York: IES, 1987.

Jacobsen, H. A study of inspector accuracy. *Industrial Quality Control*, Vol. 9, 2, 16–25, 1952.

Konz, S. The eye and illumination. Chapter 19 in *Work Design*. Scottsdale, Ariz.: Publishing Horizons, 1990.

Konz, S., Chawla, S., Sathaye, S., and Shah, P. Attractiveness and legibility of various colors when printed on cardboard. *Ergonomics*, Vol. 15, 2, 189–194, 1972.

Life Safety Code, An American National Standard. Boston: National Fire Protection Association, 1978.

Lin, A., and Bennett, C. Lamps for lighting people. *Lighting Design and Application*, Vol. 13, 2, 42–44, Feb. 1983.

Misra, S., and Bennett, C. Lighting for a visual inspection task. *Proceedings of the Human Factors Society*, Rochester, N.Y., 631–633, 1981.

Rea, M. Visual performance with realistic methods of changing contrast. *Journal of IES*, Vol. 10, 3, 164–177, April 1981.

Turner, W. Application of high pressure sodium lighting. *Modern Materials Handling.* Vol. 38, 8, 25, May 20, 1983.

Ungar, P. Sight at work. *Work Study*, 46–48, March 1971.

CHAPTER PREVIEW

Very loud noise affects performance, but in most industrial environments the problems are annoyance, speech interference, and, ultimately, loss of hearing. Noise reduction falls into the categories of plan ahead, modify the noise source, modify the sound wave, and personal protection.

CHAPTER CONTENTS

1 Noise: Definition and Standards
2 Effects of Noise
3 Noise Reduction

KEY CONCEPTS

acoustic shadow
damping
decibels
dosimeters
driving force
free field

frequencies
masking noise
octave band
speech interference level (SIL)
transmission loss (TL)

1 NOISE: DEFINITION AND STANDARDS

Noise is expressed in **decibels** (dB):

$$SPL = 20 \log_{10} \frac{P}{P_0} \qquad (1)$$

where

SPL = sound pressure level, dB

P = sound pressure level of noise, N/m^2

P_0 = sound pressure level reference, N/m^2

    = 0 dB

    = .000 020 N/m^2

A 10 dB increase represents a tenfold multiplication of sound. Thus 50 dB is 10 times greater than 40 dB, 100 dB is 1,000,000 times greater than 40 dB. In addition, the difference between 80 dB and 90 dB is much greater than the difference between 60 and 70 dB.

Although using the log of a ratio permits a very wide range of the ratio, P/P_0, to be compressed to a small range for SPL, it makes it difficult to understand how noises combine. Rather than equations, most people use a quick graphical technique. (See Figure 23.1.) For example, two machines, each with 90 dB, produce a combined total of 90 + 3 = 93. A 90 + 96 gives 96 + 1 = 97. Or, looking at it another way, if an area has two identical machines giving a combined 93 dB, then completely silencing one drops the noise to only 90 dB. For the second example of a 90 and 96 dB machine giving a 97 total, then silencing the 90 drops the noise to 96 dBA (dBA, decibels on the A scale). As a rule of thumb, a sound will be twice as loud (subjectively) when noise increases by 6 to 10 dB; an increase of 3 dB is just noticeable (subjectively), even though the pressure is doubled.

Sound **frequencies** are divided into **octave bands**. In sound meters that record in octave bands, bands are 22–44 Hz, 44–88, 88–177, 177–355, 355–710, 710–1,420, 1,420–2,480, 2,480–5,860, and 5,860–11,360. An octave band analyzer divides a noise into nine component bands. For most applications, however, people don't want to be given nine numbers to describe a noise. They want one number. There are various ways to combine the nine numbers into one number. The most popular, the dBA scale, corresponds to the 40 phon equal loudness contour.

Most noise is not steady state. Meter response speed can be set as "slow" or "fast"; the slow setting knocks the tops off the spikes from transient noises (lasting a fraction of a second) and is a good representation for most situations. (Use impulse meters if you wish to measure the transient noises.) The U.S. noise standards (see Table 23.1) give permitted dBA levels for various time periods. All continuous, intermittent, and impulsive sound levels from 80 to 130 dB shall be integrated into the computation. Noise **dosimeters** record noise in proportion to the noise standard and automatically compute the dose as the sum of the fractional exposure at the noise level. That is, if the noise is 90 dB, it records at 100%; if it is at 95, it records at 200%, etc. Many devices have an indicator if any exposure is over the 115 dBA limit. The dosimeter microphone should be placed on or near the collar, at the center of the shoulder and directly under the ear. The measurement is not valid if the microphone is placed on the shirt pocket or the waist.

2 EFFECTS OF NOISE

2.1 Annoyance
Just as many decisions are made to air-condition a space using comfort as a criterion rather than performance, many decisions are made to reduce noise in a space based on annoyance rather than performance. Table 23.2 gives tolerable limits for some environments.

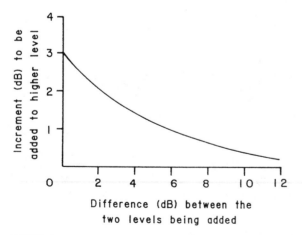

FIGURE 23.1 Addition of decibels in a free field requires three steps. First, take the difference between the two dB levels. Second, enter difference on horizontal axis, go up to curve, and left to increment to add. Third, add increment to higher of two dB levels. For example: 80 + 80 = 0 difference; 0 difference yields a 3 increment; 80 + 3 = 83. Also 86 + 80 = 6; 6 yields 1; 86 + 1 = 87.

TABLE 23.1 Maximum daily noise level exposures permitted in the United States by the Occupational Health and Safety Administration. The tradeoff of exposure time versus noise is 5 dBA for each doubling (halving) of exposure time, noise beyond 115 is not permitted; noise below 85 is considered to have zero effect.

Noise (dBA)	Maximum Exposure/Day (hours)
85	16
90	8
95	4
100	2
105	1
110	0.5
115	0.25

TABLE 23.2 Tolerable limits (dBA) in various rooms for noise (continuously present from 7 A.M. to 10 P.M. (Kryter, 1985).

dBA	Type of Space
28	Broadcast studio, concert hall.
33	Theaters for drama (500 seats, no amplification).
35	Music rooms, school rooms (no amplification). Very quiet office (telephone use satisfactory), executive offices, and conference rooms for 50 people.
40	Homes, motion picture theaters, hospitals, churches, courtrooms, libraries.
45	Drafting, meeting rooms (sound amplification).
47	Retail stores.
48	Satisfactory for conferences at a 2 to 2.5 m table; telephone use satisfactory; normal voice 2 to 6 m; medium-sized offices and industrial offices.
50	Secretarial offices (mostly typing).
55	Satisfactory for conferences at a 1 to 1.5 m table; telephone use occasionally slightly difficult; normal voice 1 to 2 m; raised voice 2 to 4 m; large engineering and drafting rooms, restaurants.
63	Unsatisfactory for conferences of more than two or three people; telephone use slightly difficult; normal voice .3 to .6 m; raised voice 1 to 2 m. Secretarial areas (typing); accounting areas (business machines); blueprint rooms.
65	"Very noisy"; office environment unsatisfactory; telephone use difficult.

Table 23.3 gives tradeoffs for estimating community reaction to noise.

2.2 Performance

There is no evidence that productivity is lower when work is done in a high noise level (say 100 dBA)—unless the person is working at maximum mental performance or speech communication is part of the job.

TABLE 23.3 Adjustments to noise levels when you are attempting to predict community annoyance to noise (Goodfriend, 1973).

Situation	Adjustment in dBA Level (dBA)
Very quiet suburban	+ 5
Suburban	0
Residential urban	− 5
Urban near some industry	− 10
Daytime only	− 5
Nighttime	0
Continuous spectrum	0
Pure tone(s) present	+ 5
Smoothy temporal character	0
Impulsive	− 5
Prior similar exposure	0
Some prior exposure	− 5
Signal present	
20% of the time	− 5
5% of the time	− 10
2% of the time	− 15

Figure 23.2 shows how to estimate how noise levels will interfere with speech. The specific index, **speech interference level** or *SIL*, is the arithmetic mean of the dB readings in the octave bands centered at 500, 1,000, and 2,000 Hz. In general, $SIL = dBA - 7$. From Figure 23.2, at 1 m, when maximum vocal effort is used, noise level is about $93 - 7 = 86$ dBA; if you shout to be understood, noise level is about $75 - 7 = 68$ dBA.

2.3 Health

There are many difficulties in setting a noise level that will not impair health. Table 23.1 gives the present U.S. regulations. If the noise *TLV* (threshold limit value) were set on the same basis as *TLV*s for chemical substances, it would need to be about 65 dBA (Stekelenburg, 1982). Thus, the 90 dBA for 8 h is set more by politics and economics than by medical facts. The law in the United States requires a hearing conservation program for all employees exposed to a time-weighted average of 85 dBA.

3 NOISE REDUCTION

To make conversations more private, you may wish to *increase* the environmental noise—that is, use **masking noise.** Examples are fountains, background music, fans, air conditioners, and fluorescent ballast hum. Firms even sell special white-noise generators.

However, in most situations noise is to be reduced. Use the following sequence: (1) plan ahead, (2) modify the existing noise source, (3) modify the sound wave, and (4) use personal protection.

3.1 Plan Ahead

Prevention subcategories are to substitute less noisy processes, purchase less noisy equipment, use quieter materials and construction, and separate people from noise.

Examples of substituting less noisy processes are (a) reducing the use of impact tools such as riveting

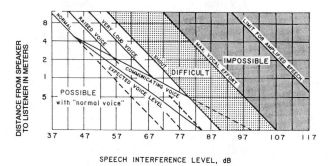

FIGURE 23.2 Speech interference levels (*SIL*) = dBA − 7. The figure shows the amount of vocal effort required as a function of *SIL* and distance. The figure's "vocal effort" is based on males with average voice strengths, facing the listener, with no reflecting surfaces nearby, and the spoken material not being familiar to the listener. For U.S. telephones, telephone use is satisfactory when SIL is less than 65 dB, difficult from 65 to 80, and impossible above 80. Subtract 5 dB for calls outside a single exchange (Peterson and Gross, 1972).

(by using welding) and chipping (by using grinding), (b) replacing internal-combustion engines with electric motors (lift trucks, lawnmowers), and (c) replacing gear transmissions with belt transmissions.

Purchasing less noisy equipment requires specifying the noise units, levels, and conditions. For example, *SPL* for machine with auxiliary equipment shall not exceed 85 dBA (slow response) at the operator location when installed as specified.

Using quieter construction is shown in Figures 23.3, 23.4, and 23.5. A general goal is to minimize turbulence and vibration. Figure 23.6 shows isolation of machine foundations. Isolation not only reduces noise transmission but improves accuracy on adjacent machines.

Separating people from noise is done by distance and by barriers. In a **free field** (no reflecting surfaces and initial point 1 to 3 ft from the source), noise decreases by 6 dB for every doubling of distance. Thus, increasing the distance helps. A barrier *between* the noise and the ear helps. Figure 23.7 shows how a double-wall construction is better than a single wall of the same mass. Avoid acoustic leaks in the walls due to doors, windows, air ducts, electrical connections, etc. A barrier *behind* the noise tends to focus it. See Figure 23.8. Thus, the best place for a loudspeaker (and worst for a noisy machine) is in the corner.

3.2 Modify the Existing Noise Source

The pound of cure is more expensive than the ounce of prevention. As was shown in Figure 23.1, start with the loudest noise first, since quieter noises contribute little to the total. The four subcategories are (1) to reduce the driving force, (2) to reduce response of vibrating surfaces, (3) to change direction of the noise, and (4) to minimize velocity and turbulence of air.

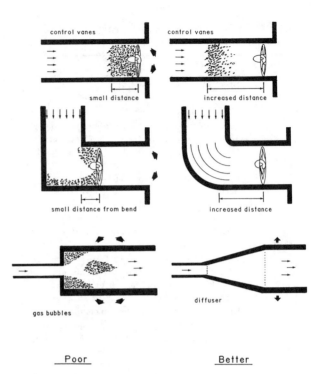

FIGURE 23.4 Smooth flow makes less noise than turbulent flow. Smooth flow requires (a) no abrupt directional changes, (b) no abrupt volume changes, and (c) distance for turbulence to die down.

There are many examples of reducing the **driving force**. Perhaps easiest to do is maintenance. Tighten loose screws, lubricate bearings, rebalance equipment, replace leaky hoses. This will improve equipment life and accuracy as well as reducing noise. If the force can be exerted over a longer time period, noise will be reduced. Figure 23.9

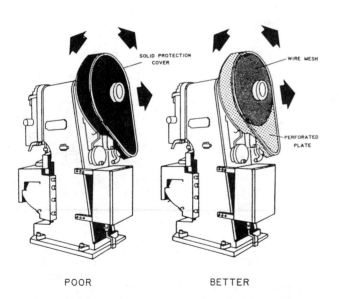

FIGURE 23.3 Small vibrating surfaces give less noise than large surfaces (little versus big drums). Thus, a perforated cover gives less noise than a solid one.

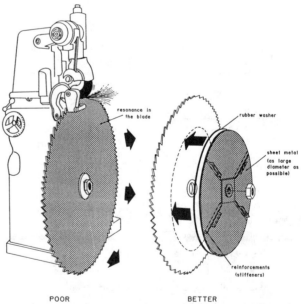

FIGURE 23.5 Damped surfaces make less noise. **Damping** can be due to shape only (see the stiffeners), materials only (such as plastic versus metal bins), or combined shape and materials (see "sandwich" construction at right: blade, rubber washer, sheet metal).

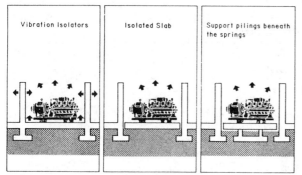

FIGURE 23.6 Machine foundations transmit vibrations and thus noise. Lighter machines can rest on various vibration isolators (see Chapter 9). Heavier machines may need cuts in the concrete floor so that the slab itself is isolated. If the ground is clay, it may be necessary to have the slab rest on pilings.

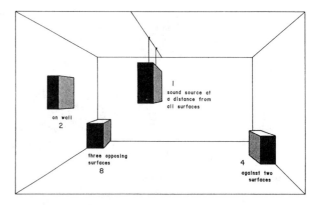

FIGURE 23.8 Barriers behind noise sources concentrate noise. A noise source suspended in the room center radiates through a full sphere but in the corner only through one-eighth of a sphere.

demonstrates this concept for punches, gear teeth (helical and bevel gears are quieter than spur), and shears. Note that presses and shears can be smaller if the force is distributed over time.

Examples of reducing response of vibrating surfaces generally involve increasing the stiffness or changing the material of the "drumhead." Add stiffeners to chutes and vibrating metal sheets. Substitute wood and plastic for metal (e.g., in tote pans, gears). Another possibility is isolation of the vibration path. Figure 23.10 shows isolation of pipes.

Changing direction of the noise is especially useful with high-pitched (over 1,000 Hz) noise. A benefit of up to 5 dB can be obtained by turning the machine or exhaust. Consider turning an exhaust vertically upward instead of horizontally toward the operator. Often the operator can turn so that the noise comes from the rear (second best is noise from the front, worst is from the side).

Minimizing velocity and turbulence of air can be done a number of ways. If the air volume is important, deliver the same volume through a larger opening, thus lowering velocity. Turbulence is caused by air blowing over sharp edges. Use fillets and rounded edges on machinery in an air stream.

3.3 Modify the Sound Wave

Modifying the sound involves "catching" the sound after it leaves the source and before it reaches the ear. Confining and absorbing, unfortunately, tend to be expensive and the sound reduction small.

To reduce noise by 10 dB, intensity must drop to 10%, by 20 dB to 1%, by 30 dB to .1%. Thus, enclosures must have minimum openings; this is difficult to implement in practice.

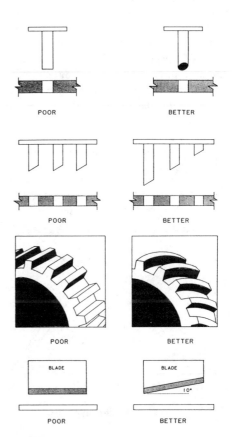

FIGURE 23.9 Exert force over a longer time period. Examples are angles on punches, staggering multiple punches, helical and bevel gears versus spur gears, standard and progressive shears.

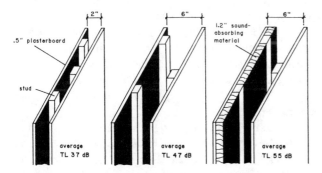

FIGURE 23.7 Double-wall construction is more effective than single-wall construction. Air gaps give good transmission losses. Note the penalty when the stud contacts both walls.

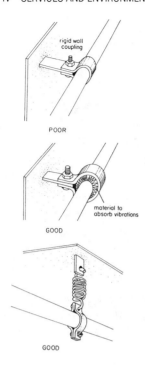

FIGURE 23.10 Isolate vibrations in pipes.

TABLE 23.4 Sound absorption coefficients (α) of typical surfaces; α is the ratio of sound energy absorbed by the surface to sound energy incident. Note the effects of frequency. (Adapted from Hill, 1973.)

Material	Frequency (Hz)			
	125	500	1000	4000
Brick: glazed	.01	.01	.01	.02
Brick: unglazed	.03	.03	.01	.07
Concrete block: coarse	.36	.31	.29	.25
Concrete block: painted	.10	.06	.07	.08
Floor: carpet, heavy with 40 oz pad	.10	.14	.37	.65
Floor: linoleum, rubber, or cork tile on concrete	.02	.03	.03	.02
Floor: wood	.15	.10	.07	.07
Glass fiber: mounted with impervious backing	.14	.67	.97	.85
Glass fiber: mounted with impervious backing, 3 lbs/ft^3, 3 in. thick	.43	.99	.98	.93
Glass: window	.35	.18	.12	.04
Plaster on brick or tile	.01	.02	.03	.05
Plaster on lath	.14	.06	.04	.03
Plywood paneling, 3/8 in.	.28	.17	.09	.11
Steel	.02	.02	.02	.02

Transmission loss (TL) of an enclosure wall depends upon the wall design. Single walls are not as effective as double walls. See Figure 23.7. A single wall also has a resonance frequency (at which TL is sharply reduced). See Figure 23.11. TL improves with wall weight but it is nice to absorb the noise as well as merely reflecting it. Table 23.4 gives sound absorption coefficients of different materials. (For good enclosure design it is necessary to record noise by octave band so that the proper material can be selected.)

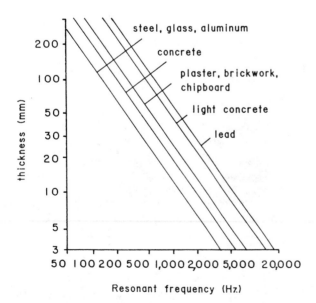

FIGURE 23.11 Resonant frequency depends upon both wall thickness and material. A 50 mm (2 in.) concrete wall would have a resonant frequency of about 500 Hz.

Simple changes can be important. Unglazed brick absorbs 3 times as much as glazed brick; unpainted concrete block absorbs 3 to 5 times as much as painted concrete block; a wood floor absorbs 3 to 15 times as much as a concrete floor (a heavy carpet can absorb up to 30 times as much as concrete at certain frequencies).

If the noise is over 1,000 Hz, it is more directional and it may be possible for the worker to be in the **acoustic shadow**. See Figure 23.12.

3.4 Use Personal Protection
Protection concerns the ear itself. The first possibility is to remove the ear from the noise. Table 23.1 gives the tradeoffs. For each halving of exposure time, the noise level can be increased 5 dBA.

The second possibility is to stay in the noise but cover the eardrum. Ear plugs plug the ear canal; muffs cover the entire ear and tend to give better protection, especially at frequencies above 400 Hz. Muffs also are visible from a distance so the supervisor knows who is protected. However, they are hot, heavy, and muss the hair; glasses reduce their seal; and the head is in a spring. Ear plugs are light, small (they get lost, so use the kind with the plugs connected by a cord), and do not affect appearance. A disadvantage is that supervisors can't tell who is wearing protection.

If audiograms detect a standard threshold shift of 10 dB (average, for either ear, of readings at 2,000, 3,000, and 4,000 Hz), employees must be fitted or refitted with adequate hearing protection, shown how to use them, and required to wear them.

The goal is "Ears alive at 65!"

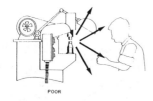

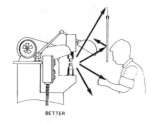

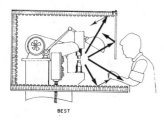

FIGURE 23.12 High-frequency noise is directional and the worker can be in the acoustic shadow. Reflection and absorption is better than just reflection.

DESIGN CHECKLIST: NOISE

Criterion
    Annoyance
    Performance
    Health

Noise reduction
    Plan ahead
    Modify existing noise source
    Modify sound wave
    Personal protection

REVIEW QUESTIONS

1. What is the total dB from 80 dB + 80 dB? From 80 + 90?
2. Why should you eliminate the loudest noise first? Discuss using an example of a 96 dB and a 90 dB noise.
3. What is the noise standard in the United States for 8 h exposure? What is the tradeoff for each doubling or halving of time? What is the maximum permitted noise level?
4. What is the tolerable noise level for secretarial offices?
5. What is the relation between speech interference level and dBA?
6. When would you want to increase noise?
7. Give three examples of how to reduce turbulent flow.
8. Are acoustic shadows more useful for noises above 1,000 Hz or below 1,000 Hz?
9. Where in a room is the best place for a loudspeaker (and worst for a noisy machine)?
10. Give four examples of spreading energy over a longer time in order to reduce noise.
11. Is it best for noise to strike a person from the front, side, or rear? Which is worst?
12. Does a painted wall absorb more or less noise than an unpainted wall?
13. Do ear plugs or earmuffs tend to reduce noise more?

PROBLEMS AND PROJECTS

23.1 Using a sound level meter, measure the sound level (dBA) of 10 different situations. Include a small sketch with each situation.

23.2 Using a sound level meter inside a building, measure the sound level (dBA) at different distances from a noise source. Plot dBA versus distance. What do you conclude?

23.3 Investigate a noisy situation. What could be done to reduce the noise or the noise effect? What does the supervisor think of your ideas?

23.4 Using a sound level meter inside a building, measure the sound level (dBA) coming from a single noise source as you move into the acoustic shadow (e.g., by going behind another machine or around a corner). What do you conclude?

REFERENCES

Goodfriend, L. Control of community noises from industrial sources. Chapter 46 in *The Industrial Environment—Its Evaluation and Control*. Washington, D.C.: Supt. of Documents, 1973.

Hill, V. Control of noise exposure. Chapter 37 in *The Industrial Environment—Its Evaluation and Control*. Washington, D.C.: Supt. of Documents, 1973.

Kryter, K. *The Effects of Noise on Man*. Orlando, Fla.: Academic Press, 1985.

Peterson, A., and Gross, E. *Handbook of Noise Measurement*. Concord, Mass.: GenRad, Inc., 1972.

Stekelenburg, M. Noise at work—Tolerable limits and medical control. *Am. Ind. Hygiene Association J*. Vol. 43, 6, 403–410, 1982.

CHAPTER PREVIEW

Traditionally we substituted energy for labor. Because of energy costs and environmental concerns, more attention now is focused on energy minimization. The three major targets of energy conservation are lighting, HVAC, and motors.

CHAPTER CONTENTS

1 Overview
2 Lighting
3 HVAC
4 Motors

KEY CONCEPTS

belts (V versus synchronous)
coefficient of performance
conditioned air
efficacy
energy-efficient motors
gearmotor

illuminance
indirect light
internal metabolism
life-cycle costs
luminaire dirt depreciation
    (LDD)

occupancy sensors
skin (of building)
spot (versus group) replacement
task lighting
transmission (of energy)
zone (of building)

1 OVERVIEW

1.1 Why?
Saving money is not the only reason for conserving energy. National policy may encourage less use of oil or preserving the environment. For example, replacing a 75 W incandescent lamp with an 18 W fluorescent lamp will not change the amount of light but will save 57 W. In the 10,000 h life of the fluorescent lamp, this is 57 KWH. To produce the 57 KWH, a coal-fired power plant would burn 770 lbs of coal (releasing 1,600 lbs of CO_2 and 18 lbs of sulphur to the atmosphere); an oil-fired plant would burn 62 gallons of oil (Fickett et al., 1990).

1.2 How?
Getting individuals to conserve energy voluntarily is difficult. Many employees see no personal benefit from conservation activities. What is needed is a management policy to conserve energy. As with any other policy, results depend on the intensity of the management's emphasis on the policy. One individual must be responsible for the energy conservation program, because if it is everyone's job, it is no one's job.

Maintenance is a key consideration in energy efficiency. Another consideration is to include utility costs in purchasing decisions. That is, purchases should be based on **life-cycle costs**, not just initial capital cost.

Small firms may lack the capital for some energy saving projects, but most utilities now provide capital for energy conservation. In addition, energy supply services will provide free consulting and will furnish capital for equipment in return for 50–70% of the savings until payback is achieved.

Table 24.1 gives the energy utilization index by building type.

Three major targets of energy conservation are lighting; heating, ventilation, and air conditioning (HVAC); and motors.

2 LIGHTING

Lighting ergonomics is discussed in Chapter 22. This section will discuss minimizing energy use for lighting. In the United States, lighting consumes about 25% of electricity—about 20% directly, plus 5% for cooling of unwanted heat from lamps (Fickett et al., 1990). Designers should be careful not to degrade the necessary lighting. The goal is to cut fat, not muscle and bone.

Figure 24.1 shows that energy consumption is a function of power and time. Section 2.1 will discuss reducing power; Section 2.2 will discuss reducing time (Fisher et al., 1990).

2.1 Reducing Lighting Power
The luminous environment, the physical environment, equipment selection, and design and maintenance procedures are important factors in reducing lighting power.

TABLE 24.1   Average energy utilization index (EUI) by building type (U.S. Dept. of Energy, 1978).

Annual BTU/ sq ft	Building Type
55,000	Assembly
90,000	Education
210,000	Food sales
200,000	Food service
220,000	Health care
110,000	Lodging
80,000	Retail
106,000	Office
125,000	Public safety
55,000	Warehouse

2.11 Luminous Environment.
The luminous environment includes visual task requirements, visual comfort, and color.

Requirements. Visual tasks require illumination (see Chapter 22). Safety requirements for illumination should not be compromised for energy conservation. Consider grouping tasks by illumination level; that is, place

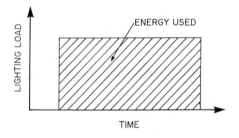

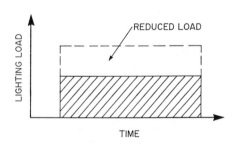

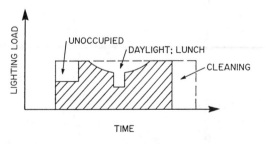

FIGURE 24.1   Reduce lighting energy by (1) reducing the connected load or (2) reducing the time the load is on (Fisher et al., 1990).

high-illuminance tasks in one area and low-illuminance tasks in another. Obtain high **illuminance** with task lighting (on separate controls) and ambient lighting. Note that the high illuminances (category D and above) given in Chapter 22 are based on lower general lighting supplemented by task lights. It generally is more cost-effective to reduce task difficulty (e.g., increase object size, contrast, viewing time) than to increase illuminance. See Chapter 22.

Comfort. Lighting comfort is primarily a glare problem. See Chapter 22 for some solutions.

Color. Color of an object depends on the human eye, the surface of the object, and the color of the light. There may be a temptation to use lamps with high lumens/watt (such as sodium lamps) without considering their poor color rendition. For example, is detection of an object enough or is identification of the object's color also necessary?

2.12 Physical Environment.
Physical environment considerations are geometry, reflectances, daylight, and voltages.

Geometry. Geometry influences both direct light and indirect light. A distant light is dim; therefore, for high-ceiling rooms, suspend ceiling luminaires below the ceiling and use task lights. Light is used more efficiently for low cavity ratios (where the cavity ratio is a ratio of wall area/floor area). Although opaque interior walls and partitions increase visual and auditory privacy, these obstructions give inefficient use of flux and cause disturbing shadows. Consider nonopaque (transparent and translucent) walls and partitions—especially above the line of sight or where normal walls would restrict passage of sunlight into the building interior. Transparent surfaces pass more light than translucent surfaces, but they also give less visual privacy. Partial-height partitions, instead of full-height walls, will require more attention to good acoustic design (see Chapter 23).

Reflectances. **Indirect light** is reflected light. Light that does not reflect is lost. Room surfaces (ceilings, walls, floors, furnishings) should be a light color. In general, light and bright surfaces are considered more esthetic and pleasant than dark (and possibly dreary) surfaces. Often forgotten is the reflectance of the luminaire. Light that does not emerge from the luminaire is wasted. Luminaires vary in the amount of light they trap. In addition, luminaire surfaces get dirty and should be cleaned (see Box 24.1). Filtering the air supply thus provides better illumination as well as cleaner air for breathing.

Daylight. With careful design, replacing some artificial illumination with daylight may be possible. See Chapter 9 for a discussion of windows. Take care to avoid glare, heat gain (during the summer), and heat loss (during the winter). Generally, automatic controls are needed to dim artificial lights near windows when there is sufficient daylight.

Voltage. High-voltage distribution systems (such as 480/277) have low line losses, so use them when codes permit. Lamp and ballast life are degraded by voltage fluctuations. Ballasts should have power factors of .9 or above.

2.13 Equipment Selection.
Equipment includes lamps, ballasts, and luminaires. The goal is system **efficacy** (lumen output from the luminaire versus energy used by the system).

Lamps. Lamps should be selected from multiple criteria of optical characteristics (line versus point source), life, color, and so forth—not just lumens/watt. See Table 22.6.

Ballasts. Standard ballasts use about 17% of fluorescent system power. They can no longer be sold (Hagar, 1991), but, since they have a 10–15 year life, they are in many facilities. Improved electromagnetic ballasts use a better grade of steel core and copper windings, and so they run cooler, resulting in less wasted energy (also less heat for air conditioning to remove); since ballast life is doubled for every 10°C reduction of ballast temperature, the improved ballasts last 20–25 years. Another option is a ballast which disconnects the cathode-heater circuit after the lamp arc is struck (saving about 2 W/lamp). Another option is an electronic ballast, which excites the lamp's phosphors at 25,000 Hz instead of 60 Hz, so the lamp needs less power to operate (in addition to using less power for the ballast).

Luminaires. As pointed out in Chapter 22, light distribution of luminaires is in six classes: direct (90–100% down), semi-direct (60–90% down), general diffuse (40–60% down), direct–indirect (40–60% down), semi-indirect (10–40% down) and indirect (0–10% down). Indirect lighting uses more energy because it reflects light off absorbing surfaces. The direct versus indirect decision should be decided primarily with lighting ergonomics (glare, shadows, modeling, task versus general lighting, etc.) rather than with energy economics.

After deciding which of the six classes the luminaire will come from, select the specific model. More efficient luminaires have higher initial internal reflectance, better geometry to let the light escape from the luminaire, and lower **luminaire dirt depreciation** (LDD) values—that is, the initial reflectance degrades more slowly.

2.14 Design and Maintenance Procedures.
Design and maintenance of lighting systems affect energy use.

Design. Perhaps the most important decision concerns the room's brightness pattern. Key decisions are general versus task lighting and location of ceiling luminaires. If **task lighting** is used, general room illuminance should be one-third of general illuminance. Grouping the high-illuminance tasks in a small area permits lower general illuminance in the remaining area.

Light utilization improves with fewer opaque surfaces (partitions, interior walls, tall pieces of furniture) and more reflectances (luminaire itself, ceiling, walls, floors, furnishings).

BOX 24.1   *Lighting maintenance*

To reduce the cost of lighting, first attack the giant of energy costs. Energy accounts for approximately 90% of the total cost of illumination. The remaining costs include the lighting fixture, the lamp and ballast, and the maintenance of replacing lamps and cleaning fixtures. Several techniques can reduce maintenance costs.

Lamp and Ballast Replacement

There are two general strategies: **spot replacement** and **group replacement**. In spot replacement, each lamp or ballast is replaced when it fails. In group replacement, all the lamps are replaced at one time, even though most still have some life left. The tradeoff is between discarding a lamp which has some life left versus the labor cost of replacing the lamp.

Lamps tend to have relatively low failure rates until mean lamp life. See Figure 24.2.

For spot relamping, it takes .25 to .5 h to bring a ladder, remove the old bulb, replace the new bulb, replace the ladder, and dispose of the bulb. This can be reduced somewhat if a bulb is not replaced immediately upon failure but rather every few months when the area is checked for burnt out lamps and ballasts.

Spot relamping is done in between group relamping. Typically, calculations show group relamping should be done at 60–70% of lamp life. Assume that lamps burn for 3,200 h/year in a single shift operation. For fluorescent lamps with 20,000 h life, 60% would be 12,000 h. This would occur at 12,000/3,200 = 3.7 years. As a practical matter, group replacement should take place at even annual intervals, so use a 4-year replacement cycle.

Group replacement does require a considerable workload at one time. Therefore, many firms subcontract this work. To minimize disruption, the work can be done during vacation shutdowns or during the evening.

When installing lamps, date the lamp. Then if it fails prematurely, it can be returned for a warranty claim. During group relamping, the newest used bulbs can be kept for spot relamping.

Consider replacing ballasts when group relamping. The old, inefficient standard ballasts have a 15-year life, but the new, efficient ballasts have a 20–25-year life.

Cleaning Fixtures

Over time, the light output from a fixture declines as the surfaces become dirty and thus less reflective. Luminaire dirt depreciation (LDD) depends on the type of fixture and the "degree of dirt" of the environment. The Illuminating Engineering Society has developed dirt depreciation curves

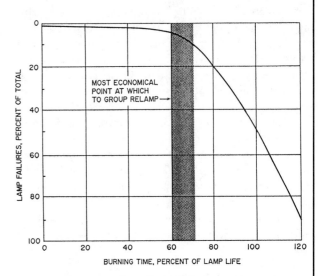

FIGURE 24.2   Lamp survival declines drastically after about 75% of rated life (Williams, 1989). At mean life, about 5% of fluorescents have failed; at 2 times mean life, about 50% have failed; and at 2.5 times mean life, almost 100% have failed. The IES Energy Management Committee (Fisher et al., 1990) recommends replacement at 70–80% of life versus the 60–70% recommendation by Williams (who works for a lamp manufacturer).

for five levels of dirt (very clean, clean, medium, dirty, very dirty) for six general types of luminaires. These curves are given in the *IES Lighting Handbook* (IES, 1984) and also can be obtained from lighting manufacturers. Typical values of LDD are .65 to .90 (Williams, 1990); that is, a typical luminaire loses 10–35% of its capability.

Architects consider LDD in their plans and typically overlight an area so that even a dirty fixture will deliver enough light to meet the specifications. If the fixtures are clean, there can be fewer of them.

Several strategies can reduce cleaning and energy costs. Use fixtures which do not get as dirty (e.g., sealed fixtures). Reduce the amount of dirt in the luminaire environment (e.g., with air filters). Clean the luminaires regularly. Fixtures usually are cleaned when bulbs are replaced, but fixtures could be cleaned more often. For example, fixtures might be cleaned every two years and lamps replaced every four. Use automatic dimming circuits. When the lamps are clean, the circuit dims them; as they get dirty their output gradually increases.

Selective switching permits lighting portions of the general area; the more lamps on a single switch, the larger the potential for energy savings.

The lighting design should be evaluated on life-cycle costs/benefits of the building system. Life-cycle costs include capital, operating, and maintenance costs. Costs could be minimized by not installing the system or by not turning it on. Clearly, what is needed is not minimizing cost but minimizing cost to achieve the desired benefits.

The building system is emphasized (rather than the lighting system) because lighting affects the performance of

the building occupants; lighting also affects the building heating and cooling. The penalty of heat from lighting in the summer generally outweighs the benefit of heat in the winter.

Maintenance.   The organization should have a policy on luminaire cleaning and lamp replacement. See Box 24.1.

2.2 Reducing Lighting Time

The previous section dealt with reducing the amount of electrical load; this section deals with reducing the operating time of the lighting system. Factors discussed are occupancy, cleaning, and daylight.

2.21 Occupancy.

The last person who leaves should turn off the lights. Enforcing this policy is difficult. Thus, consider a number of automated strategies.

The key consideration is turning the lights off. **Occupancy sensors** (infrared, ultrasonic, microwave or sound activated) turn lights on and off; they are common in areas such as library stacks, warehouses, offices, conference rooms, storage rooms, and lavatories. Most devices have adjustments for range (area) of sensing, time delay, and sensitivity.

Another possibility is a "manual on, automatic off" system. A timer turns the lights off, either after an interval of being on (e.g., 10 minutes in the lavatory) or at a specific time (say 5:05 P.M. in the office). For HID (high intensity discharge) lamps (which have a delay before restrike), the light levels fluctuate from normal to low rather than normal to off. There should be an over-ride capability. A warning can be included (e.g., turning off half the lights and then the remainder after a 60 s delay). Mechanical timers should have a spring reserve to keep the clock running if there is a power failure. Astronomical clocks compensate for changing times of sunrise and sunset; they are useful for outdoor and lobby lighting.

2.22 Cleaning.

If the area can be cleaned during normal operations, then the lights need not be on for additional time for the cleaning crew. (It also is easier to recruit cleaning personnel to work during the day than the evening or night.) Illuminance required for cleaning is relatively low (100–200 lux), so lights used for cleaning can be set at a low level. Two light switches would be necessary—one for cleaning and one for normal operations. The cleaning crew should be able to turn lights on in small areas (i.e., lighting specific offices, not entire areas or floors). The lights should be turned off when the regular work crew leaves and turned on manually by the cleaning crew.

2.23 Daylight.

Some areas (offices, lobbies) have artificial lighting supplemented by daylight. When daylight illuminance is high, artificial illumination can be reduced. A photocell can detect the total light on the task, and a control circuit can then dim or increase the artificial illuminance. The number and placement of the sensors and the lights controlled are critical, since daylight varies greatly with distance from the windows and with interior shielding (bookcases, etc.). Fluorescent lamps need a special dimming ballast.

3 HVAC

HVAC is a major consumer of energy. This section discusses minimizing HVAC energy. See Chapter 26 for the ergonomics of toxicology, ventilation, and interior climate.

3.1 Building Design

Important elements are building shape and orientation, building construction, and building equipment.

3.11 Building Shape and Orientation.

For additional detail, see Chapter 9. The walls and roof are a building's **skin**. The larger the ratio of skin to building volume, the greater the energy exchange with the exterior environment. Square buildings have the minimum surface area.

For rectangular buildings, orient the long axis and the side with the most windows to the south (if in the northern hemisphere). Compared with east or west orientation, southern orientation produces higher solar gain during the winter and lower solar gain during the summer. West orientation and west-facing windows have a negative impact on peak air conditioning loads during the summer. Orientation is especially important in a sunny climate.

3.12 Building Construction.

An energy-efficient building has minimum energy exchange with the exterior environment. Consider the walls, roof, windows, and doors. See Chapter 9 for more comments on energy-efficient walls and roofs. Note that walls can be shaded by deciduous trees or vines to reduce summer heat gain. Shading works best on low-mass buildings in sunny climates. See Chapter 9 for comments on windows and Chapter 10 for comments on doors.

3.13 Building Equipment.

The building will have equipment to add heat in the winter (furnaces), subtract heat in the summer (air conditioning), and move air between the interior and exterior climates. In addition, air is moved to redistribute internal heat and to exhaust contaminants.

Furnaces can be central or local (space heaters), as can air conditioners. Central units are usually more efficient but less efficient in distribution. The efficiency of electric heating and cooling equipment is specified by the **coefficient of performance** (ratio at a specified single temperature of output energy to source energy supplied). Heat pumps, which move heat from a lower temperature source to a higher temperature area, have a COP of 2–4.

In high-ceiling areas, bring the conditioned air in low to condition the occupied space. Adding heat near the ceiling is akin to throwing ping-pong balls into water: The heat doesn't penetrate very far down. Local heating and cooling is especially efficient when there is a low occupant density per cubic foot of space (e.g., warehouses, docks, hangars). Heat typically is distributed with hot air, although hot water is an option. Cooling is distributed with cool air. In many cases, the air is also purified and its humidity is raised or lowered, producing **conditioned air**.

When air or water is exchanged from exterior to interior environments, or vice versa, heat exchangers ("economizers") can recycle the energy—adding heat and moisture to cold outside air and cooling and dehumidifying warm summer air. In some cases, this energy can be stored (as hot water or ice) for later use. Thermal storage usually increases the overall system energy use (because of storage losses) but reduces energy costs during peak periods and allows downsizing of fans, pumps, and ducts.

Air also can be exchanged within a building, for example, from a warm ceiling area to the floor or from the building interior to the building exterior. Most buildings have an **internal metabolism** produced by people, machines, and lights; if this heat is moved to the perimeter, the furnace will have very little load.

3.2 Building Operation

A key to HVAC energy efficiency is to tailor the HVAC output to building occupancy. That is, if the ventilation is to remove welding fumes but no welding fumes are generated, then the fan ventilating the welding area should be turned off. If the office is heated to keep people warm but no people are there because it is Thanksgiving Day, then the furnace should be turned down. Rooms used occasionally (lunchroom, conference rooms, utility closets, etc.) should have ventilation tied in with the light switch; the switch should be manual on, automatic off (with a time delay).

In general, turn on heating and cooling equipment about 1 h before people arrive and turn it off about .5 h before they leave. In some circumstances, it may be desirable to turn the air conditioning on during the middle of the night to cool the building when the external air is coolest and the power rates are lowest.

Buildings should be divided into **zones** so that heating and cooling can be appropriate for the occupancy, local heat generation, and local heat loss.

Efficient management of HVAC equipment requires computer control because people do not make all the necessary adjustments at the proper times. (A computerized energy management system typically controls not only HVAC but also lighting and motors.)

Set hot water temperatures for faucets at 120°F and use flow restrictors; a dishwasher may require a booster to 150°F. Set the air thermostat at 78°F in the summer and 70°F in the winter.

Gas pilot lights should be replaced by electronic ignitions or should at least be turned off during the summer.

When there are multiple furnaces, chillers, and boilers, operate one unit at 100% capacity before bringing the next unit on line.

4 MOTORS

Motor selection is discussed in Box 21.1. For energy-efficient motor use, use efficient motors, use efficient energy transmission, and operate the motor efficiently.

4.1 Use Efficient Motors

Efficient motors must be the right size and type and must be energy efficient.

4.11 Use the Right Size and Type. Assume a mechanical load is 4.8 hp. Using the standard 15% safety factor, the motor requirement then becomes 4.8 × 1.15 = 5.5 hp. But since standard-size motors are used, a 7.5 hp motor might be purchased—so load is 4.8/7.5 = 64% of capacity. Thus, using safety factors plus only standard sizes can lead to oversized motors. Oversized motors present power factor problems (see Figure 21.3), as well as high starting and breakdown currents (see Figure 21.2). The additional capital cost for nonstandard-size motors may be repaid very quickly by the lower operating costs of using them close to full load.

4.12 Use Energy-Efficient Motors. Table 21.2 shows typical full-load efficiencies of standard and energy-efficient AC motors. The average difference is about 5%. **Energy-efficient motors** run cooler than standard motors (motor heating varies as the square of motor current), and so they have longer insulation life and more time between rewinds. Motor efficiency usually is lower after a rewind since the heat in the burnout ovens (used to remove the windings) generally damages the insulation in the stator core; the motor efficiency after rewind will be better than before rewind but not as good as that of a new motor. High-efficiency motors also have higher power factors, which is good. The basic tradeoff is the higher capital cost (about 15% more) versus the lower operating cost. Thus, high-efficiency motors generally are recommended for large motors operating more hours/year. Use standard motors for small motors and intermittent operation.

Annual power cost is

$$PCOST = .746 \, (HP)(PFL)(H)(CKWH)/EFF$$

where

$PCOST$	=	power cost, $/yr
.746	=	KWH/HP
HP	=	rated motor horsepower
PFL	=	percentage of full load
H	=	hours operated/year
$CKWH$	=	cost of KWH of electricity, $/KWH$
EFF	=	efficiency of motor, proportion

For example, assume a 50 hp motor runs at 90% of full load for 4,000 h/year at a power cost of $.07/KWH. Assume a standard motor has 90.4% efficiency and a high-efficiency motor has 95.0%. Then annual power cost would be .746(50)(.9)(4,000)(.07)/.904 = $10,398 and $9,894—an annual savings of $504 for the high-efficiency motor. A detailed analysis would consider time value of money, differences in motor lives, changing power costs over the years, and other factors. Generally, the calculations will show it is worthwhile to purchase high-efficiency motors when the existing motor is replaced but it is not worthwhile to replace existing motors before they need replacement.

4.2 Use Efficient Energy Transmission

A motor drives a mechanical device, such as a fan or pump. Although the mechanical device sometimes is connected directly to the motor shaft, generally there is an intermediate device—a **transmission**.

One very inefficient technique is to run the motor at constant speed but throttle the output of the mechanical unit with a partially closed valve or damper; this is akin to driving a car with one foot on the gas and the other on the brake.

The motor may be connected to the device with **V-belts** or **synchronous belts**. V-belts slip—especially during start-up. Synchronous belts have teeth on the inside of the belt and engage with mating grooves on a pulley or sprocket; an example is an automobile timing belt. Balmer (1981–1982) reported that for 30 hp motors used for air handlers, V-belts had a 7 s ampere surge to 222 amps from a normal of 20 amps while synchronous belts had a 1 s surge of 190 amps; V-belts had 5% less cubic feet/min fan output than did synchronous belts; V-belts also had 4.5 h/year maintenance cost versus 1.5 for synchronous; and replacing the V-belts with synchronous belts had a 19-month payback.

Synchronous belts have efficiencies of about 98%, which does not change with time. V-belts have initial efficiencies of 95–98% but slip during starts and stretch during their life. Unless maintained and retensioned, efficiency may drop by 5–10% (Wallin, 1988). V-belts also may have higher tension, resulting in lower shaft bearing life. Synchronous belts and sprockets for a 75 hp motor cost $1,000 versus $340 for V-belts and sheaves. Assuming 5% slip for V-belts and electricity at $.08/KWH, converting to synchronous belts gave a 6-month payback for a 75 hp motor operating 168 h/week.

The motor and speed reducer can be combined into one unit—a **gearmotor**. If the gearmotor output shaft is parallel to the motor shaft, spur or helical gears are used, and efficiency is about 95%. If the two shafts are at right angles, then worm gears (efficiency 55–90%) are used. Right-angle gearmotors, however, are easier to mount in some applications.

An adjustable-speed drive (typical efficiency of 95%) can minimize many of these transmission losses, especially if the device rpm is greater than the motor rpm (i.e., the alternate design uses speed-increasing gears). Typical applications are fans and pumps; example savings are 20–30% of the energy. An AC invertor converts AC voltage to DC voltage, which then is smoothed by a filter network. Then it is converted back to AC at the desired voltage to produce the desired rpm. One disadvantage of invertors at low speeds is inadequate fan cooling; possible solutions are high-quality motor insulations, high-efficiency motors (which run cooler), or a supplemental cooling system.

4.3 Operate the Motor Efficiently

Turn the motor off when it is not needed. For example, turn off conveyors during coffee breaks and lunch, and turn off ventilating fans for unoccupied buildings (nights, weekends). Generally, turning equipment off automatically through a computer is best. Turning on should be staggered to minimize peak loads.

DESIGN CHECKLIST: ENERGY MINIMIZATION

Is the goal saving money or ecological?
Who is in charge?

Lighting: Power
    Task modification
    Glare
    Color

    Lighting geometry
    Reflectances
    Daylight
    Voltages

    Ballasts
    Luminaries

    Brightness pattern
    Lamp cleaning and replacement

Lighting: Time
    Turning off lights
    Lighting while cleaning
    Dimmers

HVAC
    Building shape and orientation
    Energy efficient windows
    Energy efficient doors
    HVAC equipment
    Operating policy of HVAC

Motors
    Proper size and type motor
    High efficiency motor?
    Transmissions
    Turning motor off

REVIEW QUESTIONS

1. In the United States, lighting consumes about what percentage of electrical use?
2. Is light used more effectively in high or low cavity ratio rooms?
3. Should ballasts have low power factors or high power factors?
4. Discuss the problem of turning off lights.
5. Discuss lighting for cleaning crews.
6. What are HVAC economizers?
7. What is the average difference in efficiency in a standard and an energy-efficient motor?
8. What is the difference between a V-belt and a synchronous belt?
9. Why are gearmotors more efficient if the output shaft is parallel?
10. Why should lamps be dated when installed?
11. If a lighting fixture has a luminaire dirt depreciation of .8, what percentage of the light is trapped in the fixture?
12. How could automatic dimming circuits compensate for LDD?
13. Sketch reducing lighting energy by (1) reducing the connected load and (2) reducing the time the load is on.

PROBLEMS AND PROJECTS

24.1 What recommendations do you have for reducing the energy consumption of the IE office of your university? Break your analysis into lighting, HVAC, and motors. What does the department head think of your ideas?

24.2 Do project 24.1 for a retail store. What does the store owner think of your ideas?

24.3 Do project 24.1 for a factory. (Unless you have a team, you may want to concentrate on only lighting, HVAC, or motors and not all three.) What does the plant manager think of your ideas?

24.4 Make an analysis of the time that lights are on in the IE department of your university. What do you recommend to get the lights turned out when no one is there? What does the department head think of your ideas?

24.5 Make an economic analysis of the present lamp cleaning and replacement policy in your university's IE department. What do you recommend? What does the department head think of your ideas?

24.6 Select the largest motor in a building and do an economic analysis on whether it should be replaced with an energy-efficient motor. (The largest motor may be in the HVAC system.) What does the building owner think of your analysis?

REFERENCES

Balmer, T. V-belt drives and energy. *Energy Engineering*. 37–41, Dec.–Jan. 1981–1982.

Fickett, A., Gellings, C., and Lovins, A. Efficient use of electricity. *Scientific American*, 65–79, Sept. 1990.

Fisher, W., et al. (IES Energy Management Committee). IES design considerations for effective building lighting energy utilization. *Journal of the Illuminating Engineering Society*, 165–185, Winter 1990.

Hagar, J. Sorting out fluorescent ballast options. *Plant Engineering*, Vol. 48, 45–47, Jan. 10, 1991.

IES. *IES Lighting Handbook* (Reference Volume). New York: Illuminating Engineering Society, 1984.

U.S. Department of Energy, *Instructions for Energy Auditors*. Washington D.C., 1978.

Wallin, A. Synchronous belt drives save energy. *Plant Engineering*, Vol. 45, 50–52, April 28, 1988.

Williams, H. Evaluating the pros and cons of group relamping. *Plant Engineering*, Vol. 46, 36–39, Sept. 28, 1989.

Williams, H. Planned maintenance reduces lighting installation and operating costs. *Plant Engineering*, Vol. 47, 64–67, Jan. 11, 1990.

25 WASTE MANAGEMENT

CHAPTER PREVIEW

Waste management is driven by both economic and ecological forces. Hazardous waste should receive special attention. The first step is to minimize the amount of waste in the product, supplies, and packaging. The second step is to minimize the problems of waste collection, separation, treatment, and storage. The third step is to dispose of the waste.

CHAPTER CONTENTS

1 Overview
2 Waste Minimization
3 Collection/Separation, Treatment, and Storage
4 Disposal

KEY CONCEPTS

chemical (versus physical, biological) process
concentrate (of liquids)
corrosive
countercurrent rinsing
cradle-to-grave policy
degreasing
detoxification
dewatering
drag in (and out)
earthhappy packaging
emergency plans

exempt small-quantity generator
flash point
hazardous waste
ignitable
incineration
keep it pure guideline
landfill
mass balances
material exchange service
material safety data sheet
minimization (of waste)
neutralization

pickling
product life
reactive
saving trees
stills (for solvents)
toxic
transfer efficiency (of painting)
volatile organic compounds (VOCs)
volume reduction

1 OVERVIEW

1.1 Waste in General

Minimizing the cost of raw materials and supplies used to make a product makes good economic sense for the firm. From the viewpoint of society, less use of materials means that less has to be produced, which has less impact on the environment and conserves more resources for the future.

Another dimension of waste is disposal. The firm benefits from reduced waste because disposal costs are thereby reduced. Society benefits because there is less impact on the environment and scarce resources (landfill space and clean air, soil, and water) are conserved.

Not all waste is equal. More attention should be paid to radioactive, infectious, and hazardous waste. Note that minimization of hazardous waste may also improve health and safety of the employees.

1.2 Hazardous Waste

The Environmental Protection Agency (EPA) legal definition of **hazardous waste** does not include some things (radioactive materials, asbestos, infectious waste, PCBs) that are covered by other laws. Hazardous waste is defined by the EPA on the basis of four characteristics.

1. **Ignitable.** The material poses a fire hazard (flash point of less than 140°F) during routine management. The **flash point** is the lowest temperature at which a liquid fuel will give off enough vapor to form a momentarily ignitable mixture with air. Vapor pressure at this temperature is inadequate to sustain continuous burning after ignition. The lowest temperature at which continuous burning occurs is the fire point.

2. **Corrosive.** The material has the ability to corrode standard containers (steel at $\geq .25$ in./year at 130°F). This is understood as a pH ≤ 2 or ≥ 12.5 (pH neutral $= 7$; $< 7 =$ acid; $> 7 =$ basic).

3. **Reactive.** The material is normally unstable. It is capable of explosion, reacts violently with water, or is capable of generating toxic gases.

4. **Toxic.** The material contains certain toxic constituents at levels greater than those specified in the EPA regulations. The EPA has made some lists of hazardous materials:

   - *Toxicity character leaching procedure (TCLP) list:* 7 heavy metals (e.g., 5.0 mg/L of lead), 7 pesticides (e.g., .03 mg/L of chlordane), and 25 organic solvents (e.g., .5 mg/L of benzene)
   - *K list:* source-specific compounds, for example, a by-product of a specific pesticide
   - *F list:* generic compounds that occur in many processes, for example, solvents, distilled bottoms
   - *P list:* acutely hazardous commercial chemicals, for example, arsenic, cyanide
   - *U list:* less active hazardous commercial chemicals

The EPA considers a firm with less than 100 kg/month of hazardous waste as a conditionally **exempt small-quantity generator.** A firm with 100–1,000 kg/month is a small-quantity generator, and one with more than 1,000 kg/month is a large-quantity generator. However, producing 1 kg/month from the *P* list makes the firm a large-quantity generator. States may have more stringent requirements.

For hazardous waste, the EPA follows a **"cradle-to-grave"** policy. Once hazardous waste is associated with a firm, the firm is financially responsible for the waste as long as the waste exists—even if the waste has been sold to another firm (say for disposal). The legal, liability, and disposal costs of hazardous waste have led (as intended) to considerable effort to minimize the generation of hazardous waste. In short, be wise: minimize.

1.3 General Approaches

The first step in waste management is to minimize the amount of waste (in the product, the supplies, and the packaging). The second step is to minimize the problems of waste collection, separation, treatment, and storage. The third step is to dispose of the waste.

2 WASTE MINIMIZATION

Waste **minimization** will be discussed in three areas: the product, the supplies, and the packaging. Raymond (1990) provides detailed waste minimization checklists for construction, drycleaning/laundries, educational/vocational facilities, electroplating, fabricated metal manufacturing, laboratories, metal finishing, pesticide users, photography, printed circuit board manufacturing, printing, and vehicle maintenance.

2.1 Product

Product waste can be minimized by extending product life, changing product materials, and reducing manufacturing waste.

2.11 Extend Product Life. Disposal of a product at the end of its life can be a major problem. (At present this is the responsibility of society as a whole rather than the manufacturer. However, in 1992, Germany passed a law requiring automobile manufacturers to dispose of their cars at the end of the cars' life.)

Products can be industrial goods (a computer) or consumer goods. Consumer goods can be hard goods (TV sets, toys), soft goods (clothing), and food (orange juice). Industrial goods and hard goods are often made of metal, plastic, and wood and do not biodegrade easily. Soft goods do degrade, but over long periods (decades). Waste (kitchen and human) degrades rapidly and often is disposed of through sanitary sewers.

The primary way to minimize product waste is to increase **product life.** For example, a spoon could be used

5,000 times instead of once; a car could last 10 years instead of 7; a shirt could last for 100 wearings instead of 50; food could spoil in 10 days instead of 6; rechargeable batteries could be used instead of disposable batteries, and so on. Increasing product life minimizes both the materials used in making the product and, eventually, the product disposal problems. Other problems exist, however. People dispose of products that have not degraded physically. For example, a child may outgrow shoes or toys, a person may no longer like the style of clothing, a computer may become obsolete, etc. In addition, for many years the emphasis has been on the reduction of human labor (ads promise that products are "quick," "easy," "convenient"), but minimization of *material* waste often increases human labor (causing "waste" of time). Thus, reducing waste by extending product life is not a technical problem only.

2.12 Change Product Materials. Consider reducing **chemical** waste by changing to a **physical** or **biological** **process.** For example, instead of degreasing ferrous metal parts with a hot caustic-solution bath, consider a bake-out oven. Consider reducing use of chemicals for cleaning through mechanical agitation (air or water jets, brushes, ultrasonics). One utility now uses carp instead of chemicals to do the weeding in canals. Insecticides can be replaced with electrical insect killers (bug zappers) or predators (birds). A textile plant uses ultraviolet light instead of a biocide in its cooling towers.

If the waste is hazardous, it may be possible to change the material so the waste is nonhazardous. For example, some inks used on cardboard have cadmium; they can be replaced with inks without cadmium. Wastes from inks that have a petroleum distillate base are hazardous waste. Thus, try to switch to inks, dyes, and paint that are water or soybean oil based and that do not contain heavy metal pigments. Some compounds are safe in the product itself but make recycling of the product at the end of its life difficult. For example, GE has reformulated some plastics to eliminate some bromine compounds (used for flame retardation) and cadmium (gives yellow color and high-temperature resistance). They are safe in the plastic but cause problems if the plastic is melted or placed in acid.

Even if the waste is nonhazardous, a change may be desirable. For example, vacuum forming of plastics (instead of hand layup of fiberglass) reduces use of acetone for a solvent, as well as producing a significant labor savings.

2.13 Reduce Manufacturing Waste. Figure 25.1 shows four examples of punching an item from strip stock. In the new examples of stock layout, less material is wasted. A special width of sheet (or diameter of lathe stock or width of paper) may have a higher initial cost than the standard size, but, considering extra operations and scrap disposal costs, the special may be better.

In the office, four examples of **saving trees** are to photocopy on both sides of a sheet of paper; to replace letters (and envelopes) with electronic communication, such as EMail and faxes; to file correspondence electronically rather than with photocopies; and to reuse envelopes for internal correspondence.

Determine the weight of the finished product as a percentage of the raw material. This tried-and-true technique (determine **mass balances**) was originally developed for cost reductions but can also be used for waste minimization. For example, varnish is used on electric motor coils. Weigh the coils with and without varnish. Then multiply the varnish weight/motor by the number of motors to determine what amount of varnish is used. Then compare this with the actual amount of varnish purchased. A plant found that they were throwing away several pounds of varnish per drum because they were not allowing the drum to drain long enough. If liquids are received in railroad tankcars, make sure the track is level if the pumping point is in the car center. An alternative is to put the tankcar on a grade and pump from the low end.

Consider redesigning the process. Waste should not be considered an unavoidable result of the manufacturing process but as a measure of its efficiency. When 3M redesigned a process producing 4,200 lb of waste sludge/ year, they saved $150,000/year. Monsanto slightly changed its method of making an adhesive; the former waste became part of the product and several hundred thousand dollars/year of disposal costs were eliminated. Monsanto cut air and water emissions of PDCB, a carcinogenic chemical used in making mothballs, by 90% (1,000,000 lbs) by cooling the plant's waste vapor and capturing the

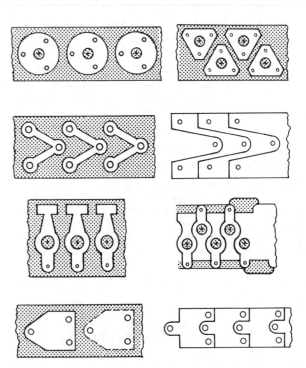

FIGURE 25.1 Layout of items within a strip can reduce waste. Changing nonfunctional perimeter shape and reversing the item are two good strategies.

crystallized chemical for reuse before it was emitted. When Union Carbide made polyethylene without 5 lb/ton of waste, they reduced disposal costs and also saved energy costs. Although reducing the scrap rate usually is done for cost reduction reasons, a lower scrap rate also reduces waste disposal problems.

Waste can be caused by purchasing too much of a material or supply. For example, assume your engineering staff want to try out a new solvent. They need 10 gallons. If purchasing buys a 55-gallon drum, then 45 gallons of hazardous waste may need to be disposed of—at a price of $500/drum or more. (The first disposal alternative, however, is to see if the supplier will take it back.)

Waste can be caused by spills and leaks. The cost of the material itself is relatively insignificant compared to the administrative costs of disposal, especially if the material is hazardous. See Box 25.1. Hazardous waste dust that settles on paved areas (roads, aprons, parking ramps, etc.) can become hazardous runoff and is very hard to clean up once it enters the sewer system. One oil firm found the simple solution was to sweep up the dust (with a power sweeper) rather than hosing down the area or letting the rain wash it down.

2.2 Supplies
For supply waste minimization, important considerations are (1) solvents, (2) coolants and lubricants, (3) metal finishing, and (4) painting and coating.

2.21 Solvents.
Solvent cleaning and **degreasing** is the process of using an organic solvent to remove unwanted grease, oils, and other organic films from surfaces (often in preparation for painting). The pollutants generated include (1) the liquid waste solvent and degreasing compounds containing unwanted film materials and (2) the air emissions containing volatile solvents. Three processes are cold cleaning (brushing on the solvent or dipping in a dip tank), vapor degreasing (heated solvent in a tank produces vapors; the item is placed in the vapor; the condensed solvent then drips back into the liquid and takes with it the removed dirt and grease) and precision cleaning (typically with Freon on instruments) (Higgins, 1989).

The first step is to eliminate the solvent. For example, if the compound being removed is a die lubricant, is the longer die life worth the cleaning cost? Perhaps the punches and dies can be redesigned to require less lubricant. If electrostatic painting is used instead of conventional spray painting, less solvent is needed for equipment cleanup. If the paint is water based instead of solvent based, solvents are not needed for thinning or equipment cleaning; the waste water also is significantly less toxic.

The solvent can be replaced with a less hazardous solvent, protecting the employee's health and the environment.

For example, AT&T traditionally cleaned circuit boards with chlorofluorocarbons (CFCs). CFCs, however, deplete the earth's ozone. AT&T has switched to a detergent (similar to household detergents) and cut cleaning costs from 25 cents/sq foot of board to 15 cents. The Torrington Company formerly used 1,1,1-trichloroethane to remove stamping and quenching oils from bearings. A hot water alkaline cleaner was installed and payback was 1 year (Higgins, 1989). Higgins et al. (1991), in a study of a Colorado manufacturer, suggested replacing vapor degreasing using TCA with an aqueous cleaner; the payback was 1 year. In another application, carriers contaminated with photoresist were cleaned with acetone baths and an IPA bath. The photoresist could be removed, in a dishwasher-type unit, with high-pressure, heated deionized water and a mild detergent; the payback was 25 months. In another application, Freon TWD-602 and TF were used for cleaning fingerprints, dust, and solder residue. The cleaner was replaced with a mixture of biodegradable organic solvents that were nonhazardous; the payback was 1 month.

BOX 25.1 *Emergency planning for hazardous materials (Griffin, 1988)*

Hazardous materials may be released in an on-site spill or accident; **emergency plans** must be developed in advance. The plans have to consider regulations from a multitude of agencies (local fire departments, city codes, state codes, federal codes). The National Fire Protection Association also has guidelines.

Business plans

- List hospitals and other medical resources that would be used.
- List actions that will mitigate, prevent, or abate hazards.
- Have written procedures to notify employees and the surrounding public of a release of a hazardous material.
- Implement a hazardous waste training plan (initial and refresher) for employees.

State and local plans

- Plan to notify and coordinate with relevant agencies (fire,

police, medical, public health, etc.).

- Plan to notify the public (radio, TV, telephones, amateur radio, etc.).
- Delineate responsibilities for site security, fire suppression, evacuation, etc.
- Train emergency response employees in the following areas:
    emergency procedures
    health and safety procedures
    equipment use and maintenance
    obtaining aid from other organizations
    medical resources
    evacuation
    monitoring and decontamination
    first aid
    making information available to the public
    reducing psychological stress

The point is that traditional processes should be examined periodically. The preceding processes should have been changed just for cost reduction reasons. Change should not come only when an environmental problem is discovered. The philosophy of "if it isn't broke, don't fix it" protects many obsolete and inefficient techniques.

If a hazardous solvent cannot be replaced, perhaps it can be reduced. For example, if the item is cleaned before the substance has time to harden or oxidize, cleaning is easier. (Just think of dishes being washed in a dishwasher.) A rinse cycle before the wash cycle may also help. Another possibility is to use partially spent solvent in a prerinse tank before the item enters the clean solvent tank.

In vapor degreasers, the solvent can be saved by a number of equipment modifications. See Box 25.2. Solvent recycling is discussed in Section 4.1.

2.22 Coolants and Lubricants.
Metalworking operations (boring, drilling, forging, grinding, heat treating, reaming, sawing, shaping, stamping, and turning) often involve high-pressure, metal-on-metal contact between tools and workpieces. The resulting heat and friction can cause excessive wear on tools and undesirable metallurgical transformations in the workpieces. A variety of cutting oils and coolants have been developed to reduce friction (cooling the tool and workpiece) and remove chips from the workpiece.

Fluids can degrade for a number of reasons: (1) accumulation of metal particles or shavings (swarf), grease, or dirt; (2) evaporation of water (changing the concentration); (3) breakdown of additives and microbial contamination; and (4) contamination from tramp oil (hydraulic or other oils) (Higgins, 1989).

The primary approach is to extend the life of the fluids. Two strategies are to reduce contamination/deterioration and to remove contaminants and reuse the fluid.

Reduce contamination/deterioration. Have the supplier deliver fluids just in time to reduce deterioration while in storage. When diluting, use demineralized water because water with a high mineral content can degrade the ability of the fluid to emulsify oils (which may lead to corrosion and bacteria growth). Hydraulic oils (usually soluble in water) greatly enhance bacterial growth, so minimize hydraulic fluid leakage into the coolant. Table 25.1 gives some maintenance tips.

Remove contaminants and reuse. A key is to minimize the variety of fluids to be cleaned. Make a computerized list of the fluids used for each machine. Because there may be dermatitis and health problems associated with fluids (Glass, 1989), a **material safety data sheet** (MSDS) should be obtained for each fluid and biocide (used to control microbial growth) on the list. The fluids may contain hazardous ingredients (e.g., chlorine, sulfur and phosphorous compounds, phenols, etc.), which make the cutting fluid a hazardous waste. Thus, select products which minimize the use of hazardous compounds. From a recycling viewpoint, costs are much lower when the variety of liquids is minimized. Coolant management is simplified if a central reservoir can serve multiple machines rather than having an individual reservoir on each machine. A key question is whether fluid cleaning will be done on-site or by an outside vendor.

BOX 25.2   Guidelines for new solvent vapor degreasers (Holtz, 1980)

Control Equipment

Minimize openings—easily opened and closed covers on open-top units; downtime covers and silhouetted openings on conveyorized units.

Safety switches—condenser flow switch and thermostat (shuts off sump heat if coolant stops circulating or temperature is too high); spray safety switch (shuts off spray pump if vapor level drops excessively); vapor level control thermostat (shuts off sump heat if vapor level rises too high).

Major control devices—refrigerated chiller or carbon adsorption system or freeboard ratio $\geq .75$ (open-top units only) or system demonstrated to be equal to or better than those listed.

Drying tunnel or tumbling basket—to minimize drag-out (conveyorized units only).

Labels—permanent, conspicuous label summarizing operating procedures.

Operating Requirements

Keep cover closed except when processing loads through degreaser (open-top units).

Minimize solvent drag-out with these measures:
   Rack parts for drainage.
   Move parts no faster than 11 fpm.
   Leave parts in vapor zone at least 30 s or until condensation ceases (open-top units).
   Tip pools of solvent out of degreased parts (open-top units).
   Allow parts to dry in degreaser for 15 s or until visually dry (open-top units).

Do not degrease porous substances such as cloth, leather, wood, or rope.

Do not allow work load to occupy more than half of the degreaser's open-top area (open-top units).

Do not spray above the vapor level.

Repair solvent leaks immediately or shut down the degreaser.

Store waste in closed containers. Do not dispose of or transfer to another party in any way such that more than 20% evaporates.

Maintain exhaust ventilation below 65 cfm per ft^2 of degreaser open area unless necessary to meet OSHA requirements.
   Use no ventilation equipment near the degreaser opening.

Do not allow water to be visible in solvent exiting the water separator.

TABLE 25.1 Maintenance of coolants and lubricants. (Adapted from Higgins, 1989.)

- Minimize leaks into the reservoir. Hydraulic, lubricating, and gear oil seals need to be checked and maintained.

- Minimize solids in the reservoir. Solids can be product debris (chips, dirt) or personal debris (gum, paper). Screens and filters (rough and fine) reduce input of product solids. Dirt can come from the product or from dust settling from the air. Maintain good housekeeping.

- Control the pH. Ruane (1989) says that since normal skin is slightly acidic (pH 6.8), ideal pH would be close to this level. Unfortunately, in order to provide good corrosion protection (particularly on ferrous metals) and bacterial stability, most operating (diluted) fluids have a pH between 8.5 and 9.5. Higher pHs tend to cause excessive defatting of the skin, and mists of these high-pH coolants can cause severe nose and throat irritations.

- Control the concentration. A trained maintenance technician with specialized measuring equipment will do a better job than unskilled, unequipped machine operators. Use high-quality makeup water. Control addition of the concentrate. Keep a control chart on the concentration and pH.

Some of the steps are physical separation (settling, flotation, straining, filtration, centrifugation, and ultrafiltration) and thermal processes (pasteurization and distillation) (Higgins, 1989). Simple settling tanks or oil/water separators can separate swarf and unemulsified oils from coolant; basket strainers can remove coarse solids from a recirculating fluid; magnetic separators can remove ferrous metal fines (cyclone-type can remove nonferrous); and cartridge filters can remove finer solids and metal fines. Centrifuges work faster than settling tanks. The heavy particles move to the outside and lighter oils remain on the inside, and the clean cutting fluid is between these two dirty layers. Ultrafiltration usually requires pretreatment by other processes to keep the membranes from becoming fouled.

For straight oil wastes, use pasteurization (to 140–250°F) to kill bacteria. For oil/water mixtures, distillation will remove water as well as kill bacteria.

2.23 Metal Finishing. Metal finishing involves applying a functional, protective, or decorative coating to a metal part; examples are metal plating (chrome, nickel, cadmium) and conversion coating (such as chromating). Metal is removed in etching and chemical milling. In printed circuit board manufacture, metal (copper, gold, silver) is added or removed from plastic.

Check to see that the amount of compound used is compatible with that specified. For example, at a tank where parts were dipped in brightener in one electric motor plant, the supervisor told the operator that the parts were not bright enough. The operator therefore used more brightener. In this case, more was 5 times the required amount. The overuse was discovered several years later when someone investigated why brightener expense was so high (purchase cost was $500/drum).

Cleaning processes are used to prepare parts for coating and to improve coating adhesion; examples are cleaning, degreasing, and **pickling** (acidic removal of surface oxides). After processing, the products are rinsed in a dip tank. The waste streams are discarded process solutions, aerosols and vapors, spills, and rinse water. Rinse water is the primary waste stream (Higgins, 1989).

Reduce waste by substituting less hazardous materials, reducing or eliminating tank dumping, minimizing drag out, modifying rinsing, recovering metals from rinse waters, and improving housekeeping (Higgins, 1989).

Substituting less hazardous materials. Traditionally, zinc, cadmium, brass, and precious metals were plated using alkaline cyanide baths. The high costs of cyanide waste disposal and protection of the operator have led to alternative noncyanide techniques. Typically, the increased operating cost of these techniques is more than balanced by the lower waste disposal costs.

Reducing/eliminating tank dumping. Minimize impurities entering the plating solution (**drag in**) by ensuring adequate drip time of items before they enter the plating tank. A warm water rinse before plating reduces drag in of cleaning compounds. The plating bath life can be extended if impurities are removed from the plating bath. Suspended solids can be removed with cartridge filtration. Carbonates, the principal impurities in cyanide baths, can be removed by chemical precipitation.

Minimizing drag-out. Plating solutions can be **dragged out** of the tank by the part or rack. There are four general approaches: (1) increase drip time, (2) consider design of parts and racks, (3) decrease solution viscosity, and (4) decrease solution surface tension.

Increasing drip time gives more time for liquid to drain from the part and rack. In a batch process, allow a reasonable time for the parts to drip back into the plating tank. Use drip troughs between tanks. Figure 25.2 shows two possibilities for when the conveyor is moving continuously. Higgins (1989) described a small plating shop in which the plater kept a plastic bucket of distilled water next to each plating bath. After plating, he dipped each rack of parts in the bucket before rinsing in a flowing rinse tank. Each morning, before starting work, he would empty the bucket from the previous day into the plating tank (recovering the chemical) and refill the bucket with distilled water.

Design racks and parts to minimize pockets, cavities, and depressions which can "bail" from the tank. Minimize the surface area of the rack which goes into the solution. Place parts on the racks so they do not bail solution from the tank. Arrange parts with blind holes and concave surfaces face down or rotate them on the rack. An air jet (air knife) can blast solution off parts or racks. If plating tank evaporation is sufficient, the part can be sprayed with rinse water.

Decrease solution viscosity by increasing solution temperature or by reducing chemical concentration of the

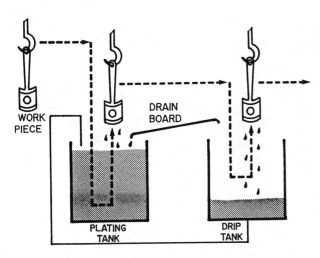

FIGURE 25.2 Minimize drag out losses by (1) using a drain trough under the conveyor or (2) using a drip tank following the plating tank. Note that the conveyor makes a vertical movement over the drip tank (to increase the effective holding time) (Higgins, 1989). If the solution volume in the drip tank is larger than normal evaporation loss in the plating tank, increase plating tank evaporation by running it at an elevated temperature during nonworking hours. Another alternative is an atmospheric evaporator.

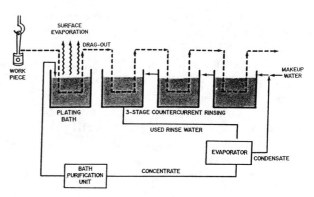

FIGURE 25.3 Rinsing with a dead rinse and **countercurrent rinsing** (Higgins, 1989). Solution from the still (dead) rinse tank is returned to the plating solution (evaporation permitting). Concentration in each flowing rinse tank can decrease by a factor of 10. Use of three countercurrent tanks in series typically reduces water use by 95% over a single rinse tank. Thus, if the plating bath contained 40,000 mg/L of dissolved solids, the three rinse tanks could be 4,000, 400, and 40 mg/L. For a drag-out rate of 1 gal/h, a single rinse tank would need 1,000 gal/h, whereas the three counterflow rinse tanks would need 10 gal/h. Counterflow rinsing does require extra capital costs and space for the additional tanks. Because of the low flow rates, use aeration to improve solution mixing.

bath. Since a lower chemical concentration in the solution generally reduces the viscosity, operating the solution at the lower end of the specification range will drag out less solution and the solution that is removed will be less concentrated.

Decrease surface tension by adding nonionic wetting agents or by increasing solution temperature. If velocity of withdrawal from the solution is reduced, the thickness of the drag-out layer will be reduced due to surface tension effects.

Modifying rinsing. The objective is to reduce the amount of water so that the waste water is concentrated, thus reducing treatment costs (in general). See Figure 25.3.

Recovering metals from rinse waters. Recovered metals may be reused in the plating solution. However, reuse assumes that there is no contamination due to reprocessing (e.g., the solution may attack pumps, pipes, and gaskets made of non–corrosion-resistant materials and thus add a contaminant to the solution). The cleaned rinse water may or may not be able to be recycled to the process—depending on whether it has picked up contaminants, is the proper pH, and so on.

Three possibilities are ion exchange, reverse osmosis, and electrodialysis.

Improving housekeeping. An advantage of improved housekeeping is minimal capital costs. Examples of improved housekeeping include the following:

- Repair and maintain leaking tanks, pumps, and valves.
- Prevent accidental bath dumps by installing high-level alarms on all plating and rinse tanks.
- Minimize volume of water used in cleanup
- Remove anodes from tanks when plating is not being performed. Dissolution of anodes during nonuse can

result in a high concentration; then a portion of the bath must be discarded to reduce the concentration.

2.24 Painting and Coating. Rather than being applied by a brush or roller, most industrial painting is applied as a liquid spray in a spray booth. See Figure 25.4. Painting generates two sources of hazardous waste: paint sludges and waste solvents (liquid and vapors).

For conventional spray painting (35–80 psi), 40% of the paint is deposited on the surface; the remainder (overspray) is sprayed into the air and bounces off the surface. Thus, the transfer efficiency is 40%. The air from a paint booth typically is exhausted through a water scrubber, which separates the paint from the air. The paint sludge is accumulated and put into drums for hazardous waste disposal.

The spray-painting operator should be videotaped and analyzed in slow motion to determine an efficient spraying pattern.

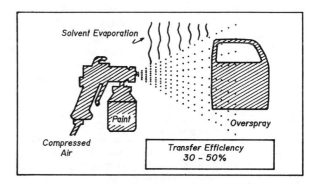

FIGURE 25.4 Conventional spray painting has a low transfer efficiency. Solvents evaporate from the painted surface and from cleanup operations.

The solvent (from the painted surface and from the overspray) evaporates into the air. These **volatile organic compounds (VOCs)** are limited by the EPA to 420 g/L for paints that cure below 90°C and 360 g/L for paints curing above 90°C (Higgins, 1989). The interaction of VOCs, nitrogen dioxide, and sunlight produces ground level ozone. To reduce VOCs, California, Texas, and New York have passed regulations (as of November 1991) requiring at least a 65% transfer efficiency; 26 other states have comparable legislation pending.

Another source of solvent waste is the solvent used to clean the painting equipment and booth.

There are a number of possible painting alternatives. The **transfer efficiency** can be increased (overspray decreased). The solvents can be reduced. The paint and the solvents can be less toxic. Five alternatives are (1) powder coating, (2) waterborne coating, (3) high solids, (4) electrostatic, and (5) airless and pressure atomized air-assisted (Higgins, 1989). Table 25.2 gives their transfer efficiencies.

Powder coating deposits heat-fusible powders on metallic substates, which then are cured (e.g., 350°F for 30 min). See Figure 25.5. In addition to the high transfer efficiency, there are almost no VOCs from the paint or the cleanup. Three variations are electrostatic dry powder painting, fluidized bed method, and plasma spraying. A disadvantage is the curing temperatures. Aluminum alloys, for example, will lose strength.

Waterborne coatings replace the organic solvent with water. The VOC emissions from the paint are eliminated and those from the cleanup are reduced. Two disadvantages of waterborne coatings are (1) that the surface must be completely free of oil-type films or the paint will not adhere well and (2) that the drying time is fairly long and may require ovens.

High-solids (25–50%) painting uses solvents but has more solids in the spray. Less paint used means fewer VOCs to evaporate, and less overspray means fewer VOCs used in cleanup. A disadvantage is that surface preparation for removal of oil is critical.

Electrostatic painting (see Figure 25.6) uses some solvent but attracts paint to the surface electrostatically. The transfer efficiency is high, and thus cleanup is minimized. A disadvantage is that the surface must be electrically conductive (metallic). A variation of electrostatic painting is electrocoating (dipping electrically conductive items into a

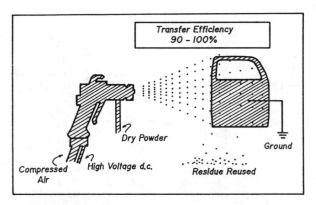

FIGURE 25.5 Powder coating has a high transfer efficiency. There are no solvents from the painted surface and little from cleanup operations.

tank of ionized paint). A disadvantage of electrocoating is that the system can apply only one coat (the electrocoated surface prevents further electrodeposition).

Airless painting has higher transfer efficiency than conventional spray painting, so there is less overspray and therefore fewer VOC emissions. Disadvantages are that the paint application needs more control and that the paint finish is coarser.

Another option is high-volume, low pressure (HVLP), which uses warm, dry, low-pressure (10 psi) air. HVLP has transfer efficiency over 65%.

2.3 Packaging

The primary purpose of packaging is to protect something during transport and storage. Additional purposes include theft reduction, improved sales, appearance of the product, and ease of use. Consistent with these purposes, the waste minimization hierarchy of preferences is

- reduction
- reuse
- recycling/disposal

In general, packaging is not hazardous.

2.31 Reduction.

The concept of reduction is to minimize the volume (weight) of the package in relation to the volume (weight) of the product.

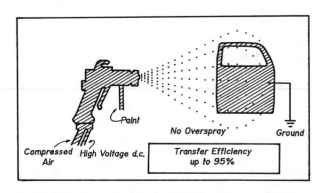

FIGURE 25.6 Electrostatic spray painting has a high transfer efficiency. With less overspray, cleanup solvents are reduced.

TABLE 25.2   Transfer efficiency of painting processes (Higgins, 1989).

Transfer Efficiency (%)	Painting Method
30–60	Air atomized, conventional
65–70	Pressure atomized, conventional
65–85	Air atomized, electrostatic
85–95	Centrifugally atomized, electrostatic
90–99	Powder coating

One possibility is to ship larger packages. For example, instead of packages containing 5 lb of product, use 10-lb packages; instead of boxes containing 10 units, use boxes containing 20 units. Scott Tissue now furnishes a jumbo roll of toilet paper for hospitals, airports, and office buildings; it replaces 11 rolled cardboard cores and reduces weight by 35%. However, large packages may cause handling and storage problems for the customer. Pepsi, for example, tried 3 L bottles of Pepsi, but people didn't buy them because they didn't fit in refrigerators.

Another possibility is to shrink the product. For example, ship concentrated liquid products and let the customer add the water. The disadvantage is that the customer no longer has a ready-to-use product. Converting **concentrate** to a ready-to-use product requires time for the customer and presents potential quality problems in reconstituting the liquid. Thus, the customer must receive some benefit from the concentrate (such as a lower cost). The concentration strategy may find its greatest acceptance within firms (rather than with retail consumers). For example, McDonalds now ships orange juice concentrate instead of ready-to-drink juice.

Steelcase (Foam, 1992) eliminated cardboard cartons when shipping office chairs. They use a molded polyurethane foam cushion over the chair back, a base platform, and plastic film wrapping. In addition to reducing 1,200 tons of waste/year, they were able to ship 58% more chairs in the same size truck and reduce the packaging cost (materials and labor). Retailers send the foam cushions back to Steelcase for reuse.

Consider placing more cartons on a pallet (to minimize pallet waste). Use an additional layer on an existing pallet, or use a larger pallet.

2.32 Reuse.
Returnable pallets are an example of reuse. See Chapter 16.

Returning containers is greatly facilitated if they can nest or collapse. For example, Hoover Corp. formerly transported parts between its plants in cardboard containers but has switched to collapsible (3 to 1 ratio) plastic containers.

Ford Motor formerly received ashtray assemblies from suppliers packaged in a partitioned corrugated paper box. They now come in a long-life partitioned plastic box. Ashtray painting was improved due to elimination of paper dust. The plastic box is returned upside down to minimize dirt and moisture accumulation. Note that reuse of containers may require changing standard procedures. See Chapter 16 for more on plastic versus corrugated returnables.

2.33 Recycling/Disposal.
Some packaging can create demand, encouraging recycling of other products. For example, Burger King has replaced bleached paper food bags with light grey bags made from recycled newsprint. It is called "**earthhappy packaging**."

Design packaging to be recycled. For example, minimize corrugated paper stapled to wood or glued to foam. Eliminate inks with hazardous materials. Molded paper pulp can replace plastic foam.

3 COLLECTION/SEPARATION, TREATMENT, AND STORAGE

The waste remaining after minimization must be collected/separated, treated, and stored before disposal.

3.1 Collection/Separation
For recycling and disposal, the ideal is a pure (minimum impurity), nonhazardous material. Thus, when collecting a material, the first guideline is **keep it pure**—don't mix waste streams. Note that the law states that mixing *any* quantity of a hazardous waste with a nonhazardous waste renders the entire mixture hazardous. Don't mix chemical wastes without testing compatibility. Clearly label containers.

For example, recycling aluminum cans is easier if they are not mixed with other trash. Put a separate container for aluminum cans next to other trash containers so that people will sort aluminum cans from other trash. You may have to use a lid with a can-sized hole to keep out other trash.

In the factory the same keep it pure principle applies. Keep chips and scrap of specific materials (low-carbon steel, alloy steel, etc.) separate. If it is not economical to store by exact type of material, try to keep a class of material together (e.g., all steel, all aluminum, all brass). This strategy requires considerable planning and employee training, as well as separate storage containers and enough space.

Keep it pure also applies to liquids. As previously discussed, it is easier to keep the mixture pure if you have fewer types to mix. For example, Roux (1983) analyzed his plant's cutting oils and reduced 30 types to 1; coolants were also cut to 1. The reduced variety permitted bulk purchasing and use of bulk tanks with individual compartments, which not only reduced packaging waste but also produced a lower cost/gallon, reduced the size of the drum storage area, and reduced the amount of drum deposits. The chips are collected in color-coded bins—one color for each type of oil. The chips drain before being placed in a wringer.

Waste oil collection must be planned. In one factory, workers dumped waste oil down the sanitary sewer drain because they didn't want to bother walking to the designated disposal point. The company had to have covers welded over the drains. The point here is that people don't always follow instructions—especially if it requires extra work.

If oil is mixed with water (as from cleaning a floor), there should be a sump in the room center with the floor sloped 2° from all four corners. Oil and water can be separated by storing for 2 or 3 days in a vertical tank. Water can be pumped from the bottom and oil from the top. The intermediate "cuff" (a mixture of oil, water, and dirt) is disposed of.

For large amounts of oil (e.g., frequent draining of a machine or engine), drain the oil into an oil drain cart. The oil in the cart then is emptied into floor-level receptacles. See Figure 25.7. The receptacles have 4 in. ID underground piping leading to a buried double-walled accumulation tank outside the building. The pipe should not be plastic, since solvents may attack it. The pipe should have a 2% grade, no 90° bends, and cleanouts at critical points to simplify maintenance (*Facilities Planning Guide*, 1977).

Drain systems become chemical storage systems. Be sure you know what is going to mix in your drains. The chemicals may react in the pipes or even destroy the pipe.

Used packaging materials are a large source of waste. Unfortunately, corrugated paper may be mixed with staples, wood, wire, plastic foams, adhesives, and product. Cold glue or gummed labels don't reduce the corrugated value much. Staples and wire can be separated at the pulping

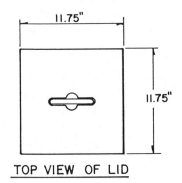

TOP VIEW OF LID

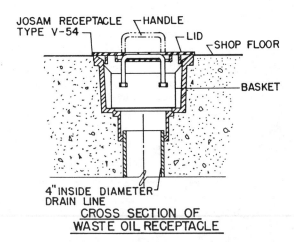

CROSS SECTION OF WASTE OIL RECEPTACLE

FIGURE 25.7 Waste oil receptacles should be approximately 12 in. square and 6 in. deep with an internal strainer and a removable mesh-type basket (*Facilities Planning Guide*, 1977). Reprinted with permission from Caterpillar Tractor Co. A solid cover will keep out more dirt than a mesh grate. Keep receptacles at least 8 ft away from columns so the drain cart can be maneuvered. The receptacle top should be flush with the floor to minimize spills and simplify handling.

mill by weight or magnetically, although at a cost. Reinforced tape, latex and hot melt glues, and plastics are very difficult to remove. Traces of product (e.g., from the inside of a bag) make the mixture useless.

One firm in Kansas subcontracts paper sorting to a facility employing the mentally retarded. It thus serves a social need of providing jobs and eliminates a monotonous, unchallenging job for its own workers.

You must decide whether to use central or local waste collection. For storage of hazardous waste, central usually is better. Consider material handling, equipment, storage space, and job design.

If the waste can be transferred safely on an automated device (conveyor, towveyor, automatically guided vehicle), then distance is not too important. (If hazardous material is moved in a pipe, the pipe should be colored yellow with the legend in black letters.) If the waste is expensive or inconvenient to transport, local collection may be best.

Do you want to store the waste at one central location or multiple local locations? If the waste is hazardous, government paperwork requirements encourage one location, discouraging the use of multiple part-time people keeping records at multiple locations. Safety and fire codes and opportunity for bulking also encourage centralization.

Space for hazardous waste should allow room to separate different types of waste and incompatible wastes. Storage of flammable material should be at least 50 ft from the property line (for large quantities). Signs indicating a hazardous waste accumulation area are required. If the hazardous substances are liquids, consider secondary containment (auxiliary tanks or pits to catch materials and floors divided by concrete barriers) in case of spills. These special requirements encourage use of one central area instead of multiple areas.

If the same collection method is used, the amount of labor required at the storage areas will be the same whether there is one disposal site or many; however, a central location may justify mechanization. For example, chip handling may be mechanized, which may lead to additional changes. For example, adding a side loop to clean and recycle coolant may be worthwhile.

Personnel factors can also be involved. Waste handling is often a low-status job which is difficult to fill. Do you want to have 5 people at local areas each spend 60 min/shift or 1 person at a central location spend 300 minutes/shift? Another consideration is that hazardous waste handlers need specialized training and personal protective equipment.

3.2 Treatment   Treatment can be volume reduction, detoxification/sterilization, neutralization, and biological.

3.21 Volume Reduction.   A simple **volume-reduction** treatment is **dewatering**. For example, paint sludge can be dewatered so more fits in a 55 gallon drum.

Since disposal costs usually are per drum, it is very expensive to pay hundreds of dollars per drum for disposal when much of the drum contents is just water. Note also that disposal of liquid wastes in hazardous waste landfills is banned. The liquid may have to be absorbed (at additional cost) by various materials (fly ash, lime, clay, vermiculite, and zeolite).

For industrial wastewater sludge, dewatering can be done by vacuum filters (not very common), belt presses (economical for large plants), and plate-and-frame filter presses (for small volumes of sludge) (Higgins, 1989).

Another example of concentration is filtering wastewater to concentrate the waste into a sludge. The idea is to dispose of 100 gallons of contaminated liquid instead of 10,000 gallons. A pesticide company had an annual shipment of 29,000 lbs of pesticide-contaminated waste water. They installed four 55-gallon drums of activated charcoal and pumped the wastewater through the charcoal, decontaminating the liquid. The contaminated charcoal is disposed of once a year.

Solids also can have volume reduced. Consider crushing, grinding, and shredding. For example, shredding is a common procedure for pallets, scrap wood, old tires, tree limbs, lumber waste, and plastic and glass containers. If you shred to 25% of the original volume, then 25% of the truck trips are needed to haul the waste away.

Incineration is another volume-reduction technique. See Section 4 for more comments on incineration.

3.22 Detoxification/Sterilization. The goal in **detoxification** and sterilization is to change a waste from hazardous to nonhazardous. One possibility is to tie up the hazardous compound in a nonleachable solid matrix (stabilization or solidification) so that it can pass the EPA leaching test. Another possibility is to destroy the hazardous constituent chemically or thermally (Higgins, 1989).

Sterilization (e.g., of hospital waste) can be done by steam or incineration.

3.23 Neutralization. One **neutralization** goal is to change the pH so that it is within the 2.5 (acid) to 12.5 (basic) range that makes it nonhazardous waste instead of hazardous waste. The pH also might be adjusted to a value at which other processes are more efficient. For example, in removing copper from a solution with a chelating ion exchange resin, a pH of 4 to 5 is best. In another example, when Tinker Air Force Base waste stream operated at a pH of 7.5 to 8.5 instead of 2.5 to 3, chrome reduction was equally as effective, but sludge volume was reduced by 2/3—saving $1,000 per day (Higgins, 1989).

Precipitation may be used with neutralization. For example, soluble copper in an acid solution can be precipitated effectively by adjusting the pH to slightly above 7 (neutral).

Acids (to neutralize bases) and bases (to neutralize acids) often are obtained from other firms' waste streams.

For example, a cement plant may get rid of its fly ash (which is basic) and help some other firm neutralize its acid waste. There are 12 **material exchange services** in North America, located in Edmonton, Sacramento, Mississauga (Ontario), Grand Rapids, Indianapolis, Springfield (Illinois), Newark, Winnipeg, Helena, Syracuse, Charlotte, and Tallahassee. For information, contact Industrial Material Exchange Service, 2200 Churchill Road, P.O. Box 19276, Springfield, IL 62794-9274.

3.24 Biological Treatment. Organic compounds can be destroyed by biological treatment. An example is the landfarming of oil sludges from petroleum refining. Another example is biodegradable material such as yard waste (grass clippings).

The slow action of nature can be speeded by composting, in which air and chemical constituents are arranged to get the best design. Stipp (1991) describes how a Maine composting consultant firm (their motto is "Compost happens") composted dead chickens. The chicken flesh, high in nitrogen, had to be cut with carbon to biodegrade well. The final "recipe" was to add, for every dead chicken, a gallon of sawdust (from a local sawmill) and a gallon of chicken manure. Let sit for two months, stirring daily. Serve to garden plants. They have even developed a way to compost explosive sludge contaminated with TNT. Another firm (McCoy, 1991) reports horse manure is excellent for processing water-based ink and paint residues.

3.3 Storage
Emergency plans are described in Box 25.1. Although at present no consensus exists, the regulatory trend is to isolate storage space for hazardous materials from habitable space and to limit the quantity stored in a single location. An example of the quantity limitation is 10 drums (Van Valkenburgh, 1991).

A popular technique is a portable building. A building of less than 120 sq ft is exempt from many building permit processes and can be used indoors or outside. Outside storage has many advantages.

Most regulations require facilities to have a secondary containment method for leaks and spills. Usually this is accomplished with a sump space beneath a grated floor.

For storing flammables, use a double-skinned, noncombustible wall with 1.5 in. air space. Fire resistance can be up to 4 h. Explosion relief panels should be in the roof rather than side walls because side walls can become missiles (Van Valkenburgh, 1991).

The fire hazard of volatile liquids can be controlled by using underground storage tanks. However, corrosion leaks are a serious problem. Some common approaches are (1) impervious pavement with a curb around the tank to control spills, (2) corrosion-resistant materials (such as fiberglass or steel with cathodic protection), and (3) double-wall tanks with monitoring of the intervening space (Young, 1988).

4 DISPOSAL

The previous sections dealt with minimizing waste and with collecting, processing, and storing the remainder. The final step is disposal. The four possibilities are (1) recycling/reuse, (2) incineration (waste to energy), (3) landfill (disposal into land), and (4) disposal into sewers (water).

4.1 Recycling (Reuse)

The simplest recycling is to reuse the material at the same site, but recycling also is done by others.

4.11 Reuse at Same Site.

A simple example of recycling is using a mulching lawn mower instead of bagging yard waste and hauling it to the landfill. Another example is returning cutoff casting sprues (in a foundry) to be remelted. Typically there would be some cleaning or processing before the item can be reused.

The word *still* has connotations of moonshiners and prohibition, but **stills** are also used in recycling solvents. Used solvent is poured or pumped into an insulated and sealed boiling chamber. The chamber then is heated electrically to bring the solvent to its boiling point. The solvent vapors then drift out into the condenser. Solvent droplets run off water-cooled tubes into a reclaimed-solvent container. The contaminated particles in the solvent concentrate in the boiling chamber as "still bottoms."

You cannot mix solvents in a conventional still, so keep waste solvents segregated and set up a schedule for distillation for each type of solvent. In addition, special provisions must be taken for solvents with boiling points over 200°F.

The initial investment for a 15-gallon still is about $5,000. On-site distillation of 100 gallons of 1,1,1-trichloroethane saved $250/month and had a payback of less than 2 years (*Kansas Generator of Hazardous Wastes*, 1990).

4.12 Reuse by Others.

Unfortunately, the recycling concept has been oversold as a solution to pollution. The idea is often presented as if someone was going to pay you for your trash. A better perspective is to picture yourself as the recipient of the material. Why would you want it and how much are you going to charge to take it?

However, there are some success stories of one firm's trash being another firm's treasure. The Industrial Materials Exchange was mentioned previously as a source of materials to neutralize or absorb waste. It also could dispose of your wastes. For example, you may be able to send acidic waste to another firm (which then would use it to neutralize basic waste).

Broken or shredded pallets can be a fuel source. Silver from photographic and X-ray films has a ready market. Grease from kitchens can be sold to soap companies. Vehicle oil can be processed and reused—either as oil or as a fuel. Alfalfa dust can be pelletized with steam (with the possible addition of other nutrients such as animal fat) to make animal feed.

An important problem in recycling is to identify the material composition of a specific item. Table 25.3 lists the identification numbers of various plastics. Recycling complex mixtures of materials (e.g., steering wheels, electronic equipment) is so expensive that it rarely is done.

4.2 Incineration (Waste to Energy)

There are four advantages of **incineration**: heat (energy) generation, volume reduction, minimum need for sorting/processing, and detoxification.

Heat generation depends on the material's composition and moisture content. See Table 25.4. For trash, use 8,500 BTU/lb; for garbage (kitchen waste), use 2,500 BTU/lb; for animal solids, use 1,000 BTU/lb.

Generated heat may be used for steam, heating, and electricity generation. If the heat is used, then the income can be used to reduce the disposal cost. (Engineers should challenge incineration without heat recovery because it is equivalent to burning money.)

Incineration savings can be large. For example, in 1981 Boeing announced that by burning plant waste, it would provide 60% of the heating energy for a 6,000,000 ft² factory; previously it had spent $330,000/year to have the trash hauled off. The new boiler/trash burning system had a 35% return on investment.

Incineration also reduces volume considerably. Typically, incombustible solids are less than 10% (often less than 5%) of the original volume. It is important to minimize air pollution (typically particulates and acid), which may require neutralization of the vapors and use of scrubbers.

A third advantage of incineration is that only minimum sorting/processing is required. Naturally, some sorting is beneficial, but incineration is relatively "robust."

TABLE 25.3 Codes for plastics.

Number	Material	Example Use
1	PET (Polyethylene-tere)	Mouthwashes, soft drink boxes, cooking oils
2	HDPE (High-density polyethylene)	Milk jugs, 2 L soft drinks
3	Vinyl/PVC (Polyvinyl chloride)	Meat packaging
4	LDPE (Low-density polyethylene)	Grocery bags, bread bags
5	PP (Polypropylene)	Yogurt, shampoo, straws
6	PS (Polystyrene)	Styrofoam, egg containers
7	Other plastics	

TABLE 25.4 Heating values of some industrial wastes (Air Pollution Control, 1991).

BTU/lb	Waste
17,000	Asphalt
11,300	Coated milk cartons
10,000	Coffee grounds
7,000	Corrugated boxes
8,600	Formaldehyde (melamine)
15,000	Formaldehyde (phenol)
7,600	Formaldehyde (urea)
10,000	Latex
5,200	Magazines
17,300	Naphthalene
8,000	Newspapers
11,600	Nylon
18,000	Paraffin
15,100	Pitch
24,100	Polyethylene
21,100	Polypropylene
18,400	Polystyrene
8,000	Polyurethane
13,000	Polyurethane (foamed)
9,800	Polyvinyl acetate
8,600	Polyvinyl chloride
5,000	Saran
7,300	Sawdust (pressed)
8,500	Starch
6,700	Sugar (dextrose, glucose)
7,100	Sugar (sucrose)
16,900	Vegetable oils
7,600	Wood (beech, birch)
7,200	Wood (oak)
8,000	Wood (pine)

A fourth advantage, not always recognized, is that, under proper combustion conditions, hazardous or infectious waste can be made nonhazardous through the heat of combustion. In prolytic incineration, the combined organic and inorganic material is heated, without oxygen, to 800–1,600°F. Then the decontaminated inorganics (carbons, ash, and inert salts) are removed for recycling or landfill and the toxic organic fumes enter a second chamber where they are burned at 1,500–3,500°F to detoxify the compounds.

4.3 Landfill

Some industrial hazardous wastes are treated or disposed of in on-site **landfills** rather than commercial landfills. Landfill design is complex and not for amateurs. Some things to consider are record-keeping, waste stabilization, landfill design, land-water monitoring, leachate collection, runoff collection, ground-water monitoring, and landfill closure. Also, consult a lawyer to minimize litigation.

Commercial landfills consider all these problems for you—at an increasingly steep price as the laws become more and more strict and neighbors lodge more and more protests over landfills (NIMBY, "not in my back yard").

Disposal into a landfill includes transportation to the landfill (especially expensive for hazardous materials), a tippage fee at the landfill, and unknown future legal expenses. If the landfill owner goes bankrupt, then the responsibility for the waste reverts to the original owner (the cradle-to-grave policy), so items sold and buried 20 years ago can come back to haunt you.

Hazardous waste is not the only waste that can come back to cause lawsuits. One maker of hydraulic hose cuts all defective hose into short lengths to make it nonusable before putting it into the landfill. The firm worries that people will dig up the rejected hose, use it, be injured, and sue the company.

4.4 Disposal into Sewers (Water)

Disposal into streams and lakes generally is unacceptable and often is illegal. The pollutant usually must be discharged into a municipal sewer. Municipal sewers often prohibit some materials (e.g., toxic or infectious waste). Other materials may be accepted, with some limits, for example, pH of 6–9, suspended solids over 350 mg/L, biological oxygen demand (BOD) over 300. If the material is not in compliance with the limits, it may be accepted with an additional fee or it may not be accepted. For example, the Clean Water Act sets the upper discharge limit as a pH of 9. See Table 21.9 for the formula used in Manhattan, Kansas, for BOD and suspended solids. Preprocessing waste before it enters the sanitary sewer may therefore be either necessary or desirable.

Neutralization to 6–9 pH is normally the first step. Be sure that overdosing (e.g., adding too much lime to acidic waste) does not cause more problems than it cures. Neutralization treatments also can cause precipitation of various compounds into a sludge (which must be processed and disposed of). Suspended solids can be reduced by using screens (coarse, then fine), followed by sedimentation-flotation. Biological oxygen demand is reduced by bacterial action; there are a wide variety of anaerobic and aerobic treatments.

For pollution, the concise statement is "Be wise: minimize."

DESIGN CHECKLIST: WASTE MANAGEMENT

Is the waste infectious, radioactive, hazardous?
Quantity/month?
Is there an emergency plan?
Waste minimization

Product
    Extend product life
    Change product materials
    Reduce manufacturing waste

Supplies
    Solvents
    Coolants and lubricants
    Metal finishing
    Painting and coating
Packaging
    Reduction
    Reuse
    Recycling/disposal
Collection/separation, treatment, and storage
    Collection/separation
        Is it pure?

        Central location?
    Treatment
        Volume reduced?
        Detoxified/sterilized?
        Neutralized?
        Biological treatment?
    Storage (where?, how?)
Disposal
    Recycling/reuse
    Incineration
    Landfill
    Sewers

REVIEW QUESTIONS

1. Is radioactive waste considered hazardous waste?
2. Give the four characteristics which can make a waste hazardous.
3. What is the difference between the flash point and the fire point?
4. What is the EPA's cradle-to-grave policy?
5. Is a firm legally responsible for hazardous waste once it has been sold to another firm?
6. Give four different examples of minimizing waste by increasing product life.
7. Give three examples of replacing use of chemical materials with use of a biological or physical process.
8. Give four ways of saving trees in the office.
9. Explain how determining mass balances can reduce waste.
10. Why did an oil firm sweep up the dust in the parking lot?
11. Briefly describe vapor degreasing.
12. What does a biocide do?
13. Why would oil wastes be pasteurized?
14. What is drag-in?
15. Give four ways of minimizing drag-out.
16. What is the transfer efficiency of conventional spray painting?
17. How is the paint that misses the part collected and disposed of?
18. What are VOCs?
19. Give five alternatives to conventional spray painting.
20. For packaging materials, give the waste minimization hierarchy.
21. Give three examples of minimizing the volume (weight) of the package in relation to the volume (weight) of the product.
22. If you had 100 lbs of nonhazardous waste, how many ounces of hazardous waste can you add to it and still call the total nonhazardous waste?
23. How can you design waste baskets to encourage people not to mix other trash with aluminum cans?
24. Why did a company have covers welded over the drains?
25. What is "cuff"?
26. What is secondary containment of hazardous materials?
27. Disposal of liquid wastes at hazardous waste landfills is a) banned, b) encouraged to accelerate biodegrading of material, c) permitted but not encouraged.
28. What is the pH range of nontoxic waste?
29. What do material exchange services do?
30. How were the dead chickens composted?
31. Give the four advantages to incineration.
32. Discuss NIMBY.
33. What does BOD stand for?

PROBLEMS AND PROJECTS

25.1 Visit a factory or warehouse. Is there any hazardous waste? If so, what is the weight/month for each hazardous waste item? What is the firm doing to reduce hazardous waste problems? Do you have any suggestions? What does the supervisor think of your suggestions? (This project may be too big for one person. One person can focus on a department; a team can do the factory.)

25.2 Visit a factory or warehouse. Analyze the nonhazardous waste. What is the weight/month of each major item? Do you have any suggestions? What does the supervisor think of your suggestions?

25.3 How is scrap/waste of the IE department of your university managed? Do you have any suggestions? What does the department head think of your suggestions?

25.4 How are aluminum pop cans disposed of in your living unit? What can be done to encourage recycling to get 100% recycling? How can this be implemented?

REFERENCES

Air pollution control. *Plant Engineering*, Vol. 47, 251–256, July 18, 1991.

Facilities Planning Guide: Service Dept. Shop Building Features, SEBF1604, Peoria, Ill.: Caterpillar Tractor Co., June 1977.

Foam packaging eliminates 1,200 tons of waste annually. *Modern Material Handling*, Vol. 47, 71, Jan. 1992.

Glass, D. *Proceedings of a Conference on the Health Hazards of Cutting Oils and Their Controls*, D. Glass, Ed. Birmingham, UK: Institute of Occupational Health, University of Birmingham, 1989.

Griffin, R. *Principles of Hazardous Materials Management*. Chelsea, Mich.: Lewis Publishers, 1988.

Higgins, D., May, M., Kostrzewa, M., and Edwards, H. Solvent use reduction in microelectronics and metal fabrication industries. Internal report. Fort Collins: Waste Minimization Assessment Center, Colorado State University, 1991.

Higgins, T. *Hazardous Waste Minimization Handbook*. Chelsea, Mich.: Lewis Publishers, 1989.

Holtz, R. EPA regulations on solvent cleaning. *Production*, Vol. 85, 2, 71–75, Feb. 1980.

Kansas Generator of Hazardous Wastes, Manhattan, Kan.: Engineering Extension Programs, Fall 1990.

McCoy, C. Somebody will be raking it in if this new process is successful. *Wall Street Journal*, May 17, 1991.

Raymond, S. Waste minimization for small quantity generators. Chapter 9 in *Hazardous Waste Minimization*, H. Freeman, Ed. New York: McGraw-Hill, 1990.

Roux, M. Setting up a cutting oil reclamation system. *Plant Engineering*, Vol. 37, 3, 68–69, Feb. 3, 1983.

Ruane, P. Control of potential health hazards of formulation practice. *Proceedings of a Conference on the Health Hazards of Cutting Oils and Their Controls*, D. Glass, Ed. Birmingham, UK: Institute of Occupational Health, University of Birmingham, 1989.

Stipp, D. At Cafe Brinton, today's special is chicken a la sawdust. *Wall Street Journal*, July 31, 1991.

Van Valkenburgh, G. Specifying hazardous material storage. *Plant Engineering*, 85–88, Sept. 5, 1991.

Young, A. How to manage underground storage tanks. *Safety and Health*, 30–33, June 1988.

26 TOXICOLOGY, VENTILATION, AND CLIMATE

CHAPTER PREVIEW

Threshold limit values give the permissible exposure level to various compounds. Engineering controls and administrative controls are the preferred approaches; personal protective equipment is a last resort. Toxin control primarily concerns ventilation; local ventilation is preferred over area ventilation. Thermal comfort is determined by air temperature, humidity, and velocity, by radiant temperature, and by the clothing and activity of the person. Tradeoffs are given. Heat and cold stress can be reduced by attention to design.

CHAPTER CONTENTS

1 Toxicology
2 Ventilation for Contaminant Control
3 Interior Climate

KEY CONCEPTS

administrative controls
area (versus local) ventilation
clo values
dermatoses
ducts (round versus rectangular)
engineering controls

exhausting versus blowing
inlet (makeup) air
outside air
percentage of people dissatisfied
psychrometric chart
STEL

threshold limit value (*TLV*)
wet bulb globe temperature
 (WBGT)
wind-chill index
winter (versus summer) zone

1 TOXICOLOGY

1.1 Problem

Toxic compounds are poisons. Almost all problem compounds used in industry are "slow" poisons—their effect takes place over 20 to 30 years. The effect generally is reflected as a small increase in the overall death rate of the affected group. For example, the overall death rate of a group exposed for 25 years may increase by 1% in the 25th year over what it would have been without the exposure. (This increase may be from a doubling of the rate of a specific type of cancer, but the specific type of cancer caused by the toxin causes only a small percentage of the total causes of death.) Thus, concerns about toxicology tend to be relatively low priority to many workers and many managements; the problems seem to be very remote and the solutions proposed are bothersome and expensive.

The potential toxin entrance points to the blood are the skin, the mouth, and the lungs.

Common experience shows that most compounds run off the skin rather than penetrate it. Thus the few compounds that do penetrate the skin are quite dangerous since they conflict with people's expectations. Clothing or shoes wetted with a compound that penetrates the skin increase the danger because they increase contact time. Cuts and abrasions also can permit compounds to penetrate the skin—even if the compound could not penetrate intact skin. Box 26.1 comments on dermatoses.

Toxins entering through the mouth generally come from dust or fumes contaminating food or drink. Thus, ban eating and drinking in the work areas. Require washing of hands to prevent transfer from dirty hands to food, drink, and cigarettes.

However, neither the skin nor the mouth are major problems; most of the toxic input is through the lungs. Thus, much toxicology work concerns ventilation.

1.2 Threshold Limit Values

Threshold limit values (**TLVs**) give the concentration of compound that is not safe. They generally are determined by (1) determining the amount of a compound that will cause a nonreversible functional change in an animal's organ and (2) multiplying that concentration by a safety factor and extrapolating the value to people. Since different animal species respond differently, the human is considered to act as the most sensitive animal. The TLV calculations assume an 8 h day, a 40 h week, and a "long" exposure time (e.g., over 20 years) to healthy adults. There are many judgments involved in setting the TLVs. Thus, although a TLV may be set at 20, do not consider an exposure of 19 as perfectly safe and 21 as immediately fatal. A fetus in the first months of pregnancy is especially at risk. The mother may not even know she is pregnant.

TLVs issued by the federal government are legal requirements. TLVs issued by the American Conference of Governmental Industrial Hygienists (6500 Glenway Ave., Cincinnati, OH 45211) are recommendations only. They often become legal standards when government paperwork and procedures catch up with scientific knowledge.

TLVs are given in (1) parts of vapor or gas per million parts of contaminated air by volume at 25°C and 260 Torr or (2) milligrams of particulate per meter^3 of air. To convert (at 25°C):

$$\text{ppm} = \frac{24.45 \ (\text{mg/m}^3)}{\text{molecular weight}} \qquad (1)$$

Table 26.1 gives some example TLVs and short-term exposure limits (STELs) for gases, dusts, and nonmineral dusts starting with the letter *a*.

The TLV is based on an 8 h exposure. Adjust the value if (1) the concentration varies during the day, (2) the day is not 8 h, or (3) the exposure is to more than one compound. Equivalent exposure is

---

BOX 26.1   *Occupational dermatoses*

Dermatoses fall more into the "common and annoying" categories than the "rare and fatal" categories. Nonetheless, they add up. In any year approximately 1% of the working population suffers occupational skin disease; in 1972 41% of occupational illnesses were skin ailments, causing 25% of lost work days (Birmingham, 1973).

Causes of occupational dermatoses are

Mechanical and physical—abrasions or wounds, sunlight for outdoor workers, fiberglass, and asbestos
Chemical—subdivided into strong irritants (e.g., chromic acid, sodium hydroxide) and marginal irritants (soluble cutting fluids, acetone); marginal irritants take a longer time to have an effect
Plant poisons—sandpapering or polishing of certain woods (West Indian mahogany, silver fir, spruce)
Biological agents—anthrax (handlers of infected skins or hides) and grain/straw itch (food and grain handlers contacting product with mites)
Nickel sensitivity (contact with stainless steel)

Skin cleanliness is the most important measure. Achieve this by preventing initial contact (e.g., wearing gloves, aprons). Cut down the contact time by having adequate washing facilities near the workplace (see Chapter 14). Also cut contact time by requiring thorough laundering of clothes. Many firms provide the workers' clothing and do the laundering to ensure that the clothes are clean. A number of cases have been reported in which family members have died prematurely from the toxins from clothing washed in the family washing machine; beryllium, asbestos, lead, PCBs, and chlorinated hydrocarbons are special problems. Thus, the organization probably should do the laundering of any clothing worn in areas with toxins.

TABLE 26.1 Five example compounds from the 1993–1994 Threshold Limit Values and Biological Exposure Indices (1993). For compounds with the word *skin*, do not exceed this value during any part of the exposure. The word *skin* indicates potential significant contribution to the overall exposure by the cutaneous route (including mucous membranes and the eyes) by direct skin contact or by vapors.

Substance	TWA		STEL	
	ppm	mg/m³	ppm	mg/m³
Acetaldehyde	C25[a], A3[b]	C45[a], A3[b]		
Acetic acid	10	25	15	37
Acetic anhydride	5	21		
Acetone	750	1,780	1,000	2,380
Acrylic acid—skin	2	5.9	—	—

[a]C denotes ceiling limit.
[b]A3—Animal Carcinogen. See Appendix A in TLV/BEI Booklet.

$$TWA = \frac{C_a t_a + C_b t_b + C_c t_c}{8} \qquad (2)$$

TWA = time weighted average (equivalent 8 h exposure)

C = concentration of a, b, c, ppm or mg/m³

t = time or exposure to concentration $a, b, c, \ldots, h$

The TLV for n-amyl acetate is 100 ppm. First, assume a worker is exposed for 4 h at 50 ppm, 2 h at 75 ppm, and 2 h at 150 ppm.

For a first example, $TWA - [50(4) + 75(2) + 150(2)]/8 = 650/8 = 81$. Since 81 is less than the TLV of 100, the exposure is legal.

For a second example, assume a worker was exposed to 90 ppm for a 10 h workday. The $TWA = (90)(10)/8 = 900/8 = 112$. Since 112 is greater than the TLV of 100, the exposure is not legal.

Next, assume exposure to a mixture of contaminants.

$$TWA_{mixture} = \frac{C_1}{TLV_1} + \frac{C_2}{TLV_2} + \frac{C_3}{TLV_3} \qquad (3)$$

where

$TWA_{mixture}$ = equivalent TWA mixture exposure (max of 1 permitted)

$C_{1,2,3}$ = concentration (8 h) of contaminant 1, 2, 3

$TLV_{1,2,3}$ = TLV for contaminant 1, 2, 3

Assume 8 h exposure to 50 ppm of n-amyl acetate and 800 ppm of acetone. Then $TWA_{mixture} = (50/100 + 800/750) = .5 + 1.07 = 1.57$. Since 1.57 is greater than 1, the exposure is not legal.

The short-term exposure limit (**STEL**) is a 15-minute TWA exposure which should not be exceeded at any time during the workday—even if the 8 h TWA is within the TLV-TWA. The $STEL$ exposure should be no more than 4 exposures/day, with at least 60 minutes between exposures.

1.3 Control of Respiratory Hazards

Table 26.2 gives two basic strategies: (1) **engineering controls** and (2) **administrative controls**.

TABLE 26.2 Control for respiratory hazards can be subdivided into engineering controls and administrative controls. Engineering controls are more desirable than administrative controls.

Engineering Controls	*Administrative Controls*
1. Substitute a less harmful material.	1. Screen potential employees.
2. Change the machine or process.	2. Periodically examine existing employees (biological monitoring).
3. Enclose (isolate) the process.	3. Train engineers, managers, and workers.
4. Use wet methods.	4. Reduce exposure time.
5. Provide local ventilation.	
6. Provide general ventilation.	
7. Use good housekeeping.	
8. Control waste disposal.	

1.31 Engineering Controls. A substitution example is methyl chloroform ($TLV = 300$ ppm) as a solvent instead of carbon tetrachloride ($TLV = 10$ ppm). A change process example is reduction of carbon monoxide by using electric lift trucks instead of gasoline lift trucks. An enclosure example is laboratory storage cabinets. A wet methods example is wetting floors before sweeping. A local ventilation example is solder fume removal at the workstation. A general (area) ventilation example is exhausting air from the entire room after the pollutant has dispersed, which is an inefficient strategy (see the following section of this chapter). A good housekeeping example is quick cleanup of spills. A controlled waste disposal example is sealing and monitoring of drains so unauthorized compounds are not dumped down them.

1.32 Administrative Controls. An example of screening potential employees is not permitting people who smoke to work in areas with cotton dust. An example of periodic exams is tests for hearing or for lead in the blood. An example of training would be sending the plant nurse to a course on how to give hearing examinations. An example of reducing exposure time would be scheduling maintenance operations at times when there are fewer fumes.

1.33 Personal Protective Equipment. Personal protective equipment is a last line of defense. Among other disadvantages, it often fits poorly, and workers abuse the equipment. It should be furnished by the organization, since the employees tend to buy the cheapest product with little attention to quality or amount of protection.

2 VENTILATION FOR CONTAMINANT CONTROL

2.1 General Comments

See the last part of this chapter for ventilation for comfort. Ventilation for contaminant control is a complex subject. For detailed information, see *Industrial Ventilation* (1982) and *ASHRAE Systems Handbook* (1988). As was pointed out in the previous section, ventilation for contaminant control complements the strategy of (1) decreasing the concentration of the airborne contaminant (containment, isolation, substitution, and change of operating procedure) and of (2) reducing exposure duration (administrative control).

Figure 26.1 shows a key physical principle concerning ventilation: At a specific distance from the same fan, air velocity for **exhausting** (sucking) is much lower than for **blowing**. The velocity drops exponentially with distance. For example, at a distance of only 1 dia from the exhaust opening, velocity is 10% of face velocity. Fletcher (1982) gives the following formula for people who like to put their calculator to work:

$$\frac{V}{V_0} = \frac{1}{.93 + 8.58 \, C^2} \qquad (4)$$

where

V = centerline velocity at distance x from duct face

V_0 = mean velocity over the duct face

C = coefficient

   $= \dfrac{x \, (R)^B}{A^{.5}}$

x = distance from the duct face

R = aspect ratio of duct = L/W

L = length of duct

W = width of duct

A = area of duct = LW

B = coefficient

   $= .2 \left(\dfrac{A^{.5}}{x}\right)^{.33}$

It is important not to have the **inlet air** entrain (be fed from) the exhaust air. Thus, locate inlets away from exhausts; locate inlets upwind of exhausts; shield inlets from exhausts.

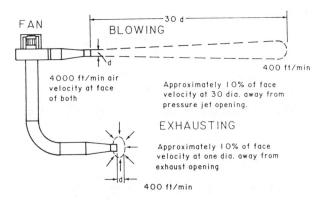

FIGURE 26.1 Exhausting is much weaker than blowing (*Industrial Ventilation*, 1982). Figure courtesy of American Conf. of Governmental Industrial Hygienists.

The goal is to remove as much of the pollutant from the air as possible at minimum cost. The capital cost of most ventilating systems over their life is relatively small compared with the operation cost. The operating cost is twofold: (1) the direct cost of electrical power for the fans, and (2) the hidden cost of replacing the conditioned air (heated or cooled and humidified and purified to desired values) with new, conditioned air. Thus, when large quantities of exhaust (and thus inlet) air are required, use energy-recovery equipment such as rotary wheels, fixed plates, heat pipes, and runaround coils.

Figure 26.2 presents four ventilation strategies. The first two are **area ventilation**; the last two are **local** (hood) **ventilation**. If there is a discrete contaminant source, local ventilation usually is the preferred strategy.

2.2 Area Ventilation

Figure 26.2, view *a*, shows the general concept: a large room with an inlet, person, contaminant source, and exhaust. Area ventilation is used mainly when contaminant sources are diffuse rather than concentrated. Figure 26.3 shows some good and bad examples of the location of the inlet, person, contaminant

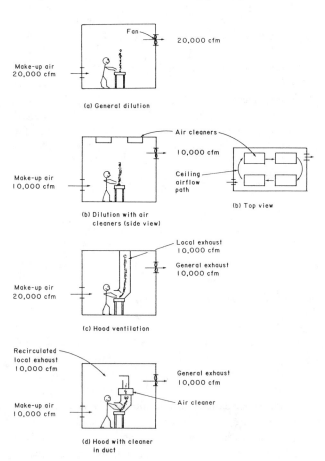

FIGURE 26.2 Four ventilation strategies are illustrated. Specific savings, of course, vary with the contaminant, room dimensions, etc. Dilution (area) ventilation is shown in (*a*). Dilution ventilation with ceiling air cleaners is shown in (*b*). Local (hood) ventilation is shown in (*c*). Local ventilation with a cleaner in the duct is shown in (*d*).

source, and exhaust. The design rule is to keep the contaminant source between the person and the exhaust—that is, keep the source "downwind" from the person. In a multiple-source room, the inlet and exhaust locations usually are fixed. The things that can be varied easily are (1) the orientation of the equipment relative to the exhaust and (2) the location of equipment in the room. Thus, try to orient the workstation so the source is downwind from the operator. Try to locate the source as close to the exhaust as possible.

"Clean rooms" generally are designed in classes according to a dust particle count. Laminar flow air is introduced from the ceiling to the floor on one side of the room. High-efficiency filters remove most of the particles over .3 microns. The air moves with minimum turbulence to the opposite wall, which acts as a large return air grille. For larger rooms, use a ceiling supply of air and a floor (grating) return plenum. The room normally has an independent air-conditioning unit. For even more cleanliness, equip each workstation in the clean room with laminar flow supply air through high-efficiency filters. See Chapter 14.

General ventilation has low effectiveness. It is important to realize that (1) ventilation air is not completely and uniformly dispersed and (2) contaminants are not instantly and completely mixed with the ventilation air and uniformly dispersed throughout the area. In particular, because of the low velocities of general ventilation, the air

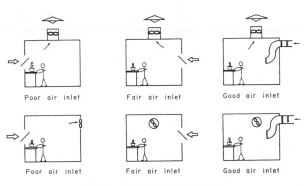

Poor air inlet   Fair air inlet   Good air inlet

Poor air inlet   Fair air inlet   Good air inlet

POOR EXHAUST FAN LOCATIONS

Poor air inlet   Fair air inlet   Good air inlet

Best air inlet   Best air inlet   Best air inlet

GOOD EXHAUST FAN LOCATIONS

Note: inlet air requires tempering during winter months

FIGURE 26.3 Keep the source downwind from the person. Figure courtesy of American Conf. of Governmental Industrial Hygienists.

movement has very little effect in the vicinity of a contaminant source. In addition, the cost of heating the ventilation air during the winter can be very large. Thus, there has been a search for more effective strategies.

The strategy shown in Figure 26.2, view b, is to reduce the contaminant by filtration rather than just diluting it with new makeup air. The amount of new makeup air saved naturally varies with the type of contaminant, amount of contaminant, room characteristics, filter characteristics, etc. Use the outlet air jet from each filtering device to set up a circular flow toward the next filter. The rising contaminant is more likely to become entrained in this circular flow and thus filtered. Flanges are recommended for the inlets. Because the flow is circular and stays away from the walls, the inlets should be located about 10% of room width from the walls (maximum of 5 ft); avoid corners since they are out of the circulating air flow.

2.3 Local (Duct) Ventilation

Figure 26.2, view c, shows a third alternative—direct capture of the contaminant with a local exhaust.

First, remember the drastic drop in velocity as the distance from the exhaust point increases. See equation 1. Place the duct as close as possible to the source (e.g., 12 in.). Since many contaminants are heavier than air, have gravity help the airstream; that is, duct openings below or at the side of the contaminant are more effective than those above. In addition, shield the source from cross-currents by baffles, curtains, or even an enclosure such as painting booth or welding booth. See Figure 26.4. If enclosure on all four sides is not possible, enclose one or two sides. Use flexible plastic strips if vehicles or conveyors need to pass.

Second, the duct should have a flange (flat plate perpendicular to the duct axis). This reduces the flow of air from the rear of the flange and "focuses" the duct flow from the contaminant. Fletcher (1982) reported that, when $x = A^{.5}$, then centerline velocity when a flange is used is increased 25% for a square or round duct ($L/W = 1$) and 55% when $L/W = 16$. He also reported slot-shape ducts are poor—efficiency drops off rapidly as the aspect ratio (L/W) increases. If a slot must be used, such as on tanks when an overhead canopy hood is not possible, then consider a push-pull design: An air jet on operator's side pushes air across the tank, entraining the contaminant and allowing it to be pulled into the duct. See also Chapter 9.

For the duct transporting the air from the hood to the exhaust, **round ducts** have four major advantages over **rectangular ducts**. First, round ducts have less surface area in contact with the air (and thus have less friction and thus require less energy for fans). For example, a 13.5 in. dia round, 12 in. square, and 24 in. × 6 in. rectangular duct all have 144 in.² of surface area; however, the perimeters (friction area) are 43.4 in. (100%), 48 in. (113%), and 60 in. (141%). For equal flow, the square or rectangular duct requires 5% to 25% more power (depending on the system) than the circular. Second, because of lower perimeter

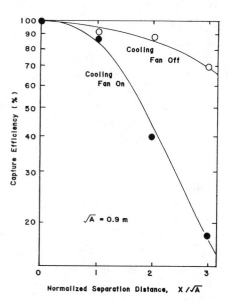

FIGURE 26.4   Avoid cross-currents such as from cooling fans (Ellenbecker et al., 1983). When the cooling velocity was turned on, capture efficiency at 1 m dropped from 90% to 40%. Reprinted with permission from the *American Industrial Hygiene Association Journal.*

area, round ducts transfer less heat from the air in the duct to the air outside the duct. Third, round ducts leak less air than rectangular (typically 50% less) since the fittings between sections are much easier to seal. Rectangular ducts have less depth and thus can be fitted in limited overhead spaces, but this usually is not a problem in industrial applications. Fourth, capital (installed cost) of round duct is less. The penalty increases with aspect ratio. For example, a rectangular duct with aspect ratio of 4 would cost about 45% more than a round duct, whereas a duct with aspect ratio of 7 would cost about 100% more.

For both round and rectangular ducts, minimize the "total equivalent length." This requires the shortest lengths of straight ducts and, most important, eliminating unnecessary elbows, tees, and entries. Specify long-radius elbows and tapered entries and tees.

Naturally you should remember to turn the exhaust off if there is no longer a source of contamination (i.e., machine stops or open-top tank has cover put on it). Another option is to vary ventilation in proportion to the rate of contaminant generation. Melgaard (1981) describes how solvent emission from a batch process baking oven with a 35 min cycle peaks at 5 min. By monitoring the solvent level and varying the ventilation accordingly, there were large savings.

The fourth alternative is shown in Figure 26.2, view *d*—recycle the exhaust air. Here you put the filter directly into the duct (versus the on-the-ceiling approach of view *b*). Naturally the filter must remove the dangerous contaminant—especially the small particles less than 1 micron which enter the lungs. A filter can capture over 98% of a contaminant by weight and still not get any of the particles less than 1 micron. In addition, recycling should consider

system failure, such as a failed filter. Incorporate such safety devices as continuous monitoring of the air downstream of the filter. Also, some nonfilterable toxic gases (such as ozone, oxides of nitrogen) and carbon monoxide will be recirculated and thus accumulate.

2.4 Ventilation in Confined Spaces

Of all occupational fatalities, 2.5% occur from work in confined spaces. Some examples of confined spaces are tanks, bins, sewers, and tunnels. There are three basic hazards: (1) oxygen deficiency, (3) combustibility, and (3) toxicity.

Oxygen content is the first concern. See Figure 26.5.

Combustibles present an explosion problem. Some are heavier than air (e.g., butane, propane) and settle in low spots; others are lighter than air (e.g., methane) and accumulate in high spots. The lower explosive limit (*LEL*) gives the low concentration point at which a given gas will explode in air. The upper explosive limit (*UEL*) defines the high point. See Table 26.3. Ventilate so that spaces have less than 10% of the *LEL*.

Toxicity has short-term and long-term aspects. Certain substances cause problems relatively quickly and have ceiling values and short-term exposure limits. People should not be exposed to ceiling values for even a few seconds. *STEL* values are weighted average limits over 15-minute time periods. For long-term problems (e.g., exposure for 8 h/day for 20 years will increase the risk of cancer), use the threshold limit values. See the first part of this chapter for more on *TLVs* and *STELs*.

Morse and Swift (April 1982) summarize the ANSI Standard Z 117.1–1977 (Safety Requirements for Working in Tanks and Other Confined Spaces):

- Require an entry permit for each admission to a confined space.
- Isolate and lock out the space.
- Clean and purge the space.
- Ventilate the space.
- Consider special hazards. Welding is a problem. Consult *Safety in Welding and Cutting* (ANSI Z 49.1) and

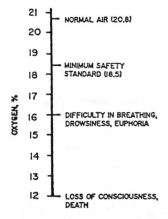

FIGURE 26.5   Oxygen concentration has various effects.

TABLE 26.3 Lower and upper explosive limit of common combustibles (Morse and Swift, Jan. 1982). Oxygen deficiency (less than 10%) may produce an inaccurate combustible gas reading. From "Working Safely in Confined Spaces", *Plant Engineering*, 4/29/82.

Combustible	Lower Explosive Limit (%)	Upper Explosive Limit (%)
Butane	1.6	8.0
Hydrogen[a]	4	75
Methane[a]	5.3	14
Petroleum vapors	1	7.5
Propane	2.1	9.5

[a]Also acts as a simple asphyxiant.

Fire Prevention in Use of Welding and Cutting Processes (ANSI/NFPA 51B).

- Use personal protective equipment. Supplied air devices deliver clean, fresh air from an outside source through a hood or facepiece. Use only in atmospheres not immediately dangerous to life and health (IDLH). Self-contained breathing apparatus (SCBA) can be used in IDLH environments.

- Prepare written procedures.

3 INTERIOR CLIMATE

For sedentary tasks, **outside air** should be furnished at 2.5 L/s for each person if there is no smoking but at 10 L/s if smoking is permitted. For light metabolic rates, outside air should be furnished at 3 to 10 L/s if there is no smoking but at 17 L/s if there is smoking. Thus, nonsmoking areas have much lower ventilation rates. In addition, if the space is not occupied, it does not need ventilation (unless it generates contaminants). Most spaces need to be ventilated only 40 of the week's 168 hours.

3.1 Psychrometric Chart and Comfort

The **psychrometric chart**, Figure 26.6, gives the relations among dry bulb temperature, water vapor pressure (absolute humidity), relative humidity, psychrometric wet bulb temperature, dew point temperature, and effective temperature.

3.2 Comfort for Standard Conditions

The cross-hatched parallelogram gives the American Society of Heating, Refrigeration and Air Conditioning Engineers (ASHRAE) comfort zone. It is for sedentary activity and indoor clothing. Actually there are two zones—a **winter zone** from 20 CET* to 23.9 CET* (68 FET* to 75 FET*) and a **summer zone** from 22.8 CET* to 26.1 CET* (73 FET* to 79 FET*). The different zones compensate for the tendency of people to wear heavier indoor clothing during the winter and lighter indoor clothing during the summer.

It is possible to predict the **percentage of people dissatisfied** (PPD), assuming they are sitting quietly in summer clothing.

If ET^* < 25.3 C

$$CSIG = 10.26 - .477\,(ET^*) \qquad (5)$$

If ET^* ≥ 25.3 C

$$HSIG = -10.53 + .344\,(ET^*) \qquad (6)$$

where

ET^*	=	new effective temperature, C
$CSIG$	=	number of standard deviations from 50% for cold
$HSIG$	=	number of standard deviations from 50% for hot
PPD	=	percentage of people dissatisfied (voting cold, cool, warm, or hot); corresponding cumulative area from negative infinity for $CSIG$ or $HSIG$

At 25.3 CET*, the minimum (6%) are dissatisfied. At 18 CET*, $CSIG$ = +1.67 and, from a normal table, 95% are dissatisfied. At 30 CET*, $HSIG$ = −.21 and, from a normal table, 42% are dissatisfied.

To minimize foot discomfort for people wearing typical indoor footwear, floor temperature should be kept between 18° and 29°C.

3.3 Comfort for Nonstandard Conditions

If the other variables are controlled, comfort conditions do not vary with time of day, season of the year, or individual's gender, age, or nationality. Variables that do affect comfort (in addition to dry bulb temperature and water vapor pressure) are clothing, activity, air velocity, and time of exposure.

Time of exposure is a minor effect and usually is neglected.

Air velocity can be traded off at .1 m/s = .3°C up to a velocity of 1 m/s (Rosen and Konz, 1982). For example, if a ceiling fan increases air velocity at a workstation by .5 m/s, then comfort temperature would be (.5/.1)(.3) = 1.5°C higher. That is, the air conditioning set point could be set 1.5°C higher with no loss in comfort. For intermittent exposure and for medium to high metabolic work rates, velocity on the person up to 2 m/s is effective. Beyond 2 m/s there is little additional heat transfer and the wind pressure becomes annoying. (The threshold of detectability is about .25 m/s, paper blows at .8–1.0 m/s, and 1.5 m/s is a noticeable breeze.) Velocity from an oscillating fan is preferred to steady velocity.

Activity (metabolic rate) can be traded off for light work (up to 225 W) at 30 W = −1.7°C. A seated, sedentary adult has a metabolic rate of about 120 W. Thus if metabolic rate increased to 180 W, comfort temperature would decrease by (60/30) (−1.7) = 3.4°C. See Konz (1990) for metabolic rates of various tasks. (Some examples of 130 W are typing with an electric typewriter and standing relaxed; 160 W is driving a car.)

Clothing is the variable that is easiest to control. The insulation value of clothing (I_d) is given in clo units. (1 clo = .155 m² − C/W). For light activity (less than 225 W) and

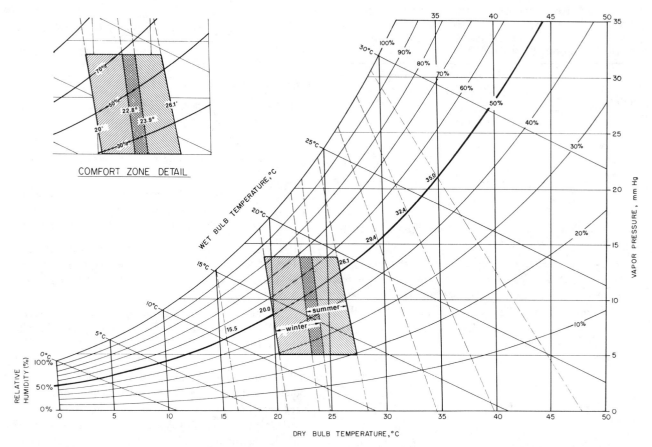

FIGURE 26.6 Psychrometric charts show the relationships of temperature and water vapor pressure. The horizontal axis gives the dry bulb (ordinary thermometer) temperature. The right vertical axis gives water vapor pressure in absolute units, while the left vertical axis gives water vapor pressure in relative units. For example, at 15°C dry bulb and 10-mm Hg vapor pressure, the relative humidity is 80%. The 100% humidity curve also gives three additional indices: dew point temperature, psychrometric wet bulb temperature, and effective temperature. The value of the index is determined by a perpendicular to the dry bulb axis but the index depends upon which direction you came from. For example, for 15°C dry bulb and 10-mm vapor pressure: (a) a horizontal line intersects 100% above 12°C—thus the dew point temperature is 12°C; (b) a slanting solid line intersects 100% above 13°C—thus the psychrometric wet bulb temperature is 13°C; and (c) a dashed slanting line intersects 100% above 14.5°C—thus the effective temperature (ET) is 14.5°C. Any point along the effective temperature line has approximately the same skin wettedness and gives approximately the same comfort. Since 100% humidity is rarely experienced, the effective temperature line has been renumbered by the point where it crosses 50% rh instead of 100% rh. The new index is called new effective temperature (ET*). The dashed line that formerly was labeled ET of 15 is now labeled ET* of 15.5.

"normal" clothing, the tradeoff is .1 clo = − .6°C. That is, if clothing increased from .6 clo to .8 clo, then comfort temperature would decrease $(.2/.1)(− .6) = 1.2°C$. Table 26.4 gives **clo values** of individual work clothing items. See Figure 26.7 for some ensembles. Table 26.5 gives clo values of individual conventional clothing items. For combinations of items

$$ICLO = .82 (\Sigma\ ICLI) \qquad (7)$$

where

$ICLO$ = insulation value of ensemble, clo

$ICLI$ = insulation value of individual items, clo

Thus, from Table 26.5, a man wearing briefs (.06), a T-shirt (.09), a light long-sleeve shirt (.22), light trousers (.26), ankle socks (.04), and oxfords (.04) would have a total clo of $.82(.71) = .58$ clo. Decrease clo by .013 for every increase of air velocity of .1 m/s above .25 m/s (up to a maximum of 2 m/s). If the above ensemble were exposed to a velocity of

.75 m/s, then the clo would be decreased by $(.5/.1)(.013) = .055$, resulting in .51 clo.

A key characteristic of a good industrial clothing ensemble is the ability to change its insulation and moisture-transfer characteristics as conditions change. A simple solution is to add or subtract items (add or remove a sweater, jacket, etc.). Another possibility is to modify the garment itself (open the front of the jacket, open the collar, etc.). Protective clothing tends to be resistant to heat and moisture transfer; a good design will permit opening various vents when there is no need for protection, such as before or after exposure.

3.4 Heat Stress   As there is departure from the comfort zone, the criterion becomes not comfort but the effects on performance and on the individual's health.

3.41 Criterion.   Performance in hot environments that do not immediately threaten health is difficult to

TABLE 26.4 Clo values (ICLI) for work clothing items (McCullough et al., 1982). Multiple values are for different items in the same category.

Item	clo	Item	clo
Coat, aluminized, long with long sleeves	1.20	Trousers, loose	.37
			.32
Coat, aluminized, short with long sleeve	.81	Trousers	.30
			.28
			.28
Coverall, long sleeve	.69		.27
	.69		.24
	.67	Apron, cotton	.19
	.67	Arm protectors, aluminized	.20
Shirt jacket, long sleeve	.54		
	.49	Arm protectors, cotton	.15
		Hip leggings, cotton	.17
Shirt, long sleeve	.44		
	.44	Hood, wool felt	.14
	.41		
	.40		
	.37	Face and head mask	.07
Shirt, short sleeve	.35		
	.33	Gloves, mid-forearm cuff	.07

evaluate, because motivation effects are so important. The use of air conditioning in offices, factories, and schools is an accepted fact, but there are no field studies (as motivation cannot be controlled) showing a change in performance in the hot versus air-conditioned environment. Yet millions of air conditioning systems have been installed, because decision makers know the difference between what it is possible for people to do and what people actually will do. Reduce the effects of heat stress to obtain the best long-term performance of workers.

3.42 Heat Precaution Levels.

As was indicated earlier in this chapter, heat stress results from a complex interaction of temperature, vapor pressure, radiant temperature, clothing, activity, and time. See Konz (1990) for a detailed analysis.

Table 26.6 gives the **wet bulb globe temperature** (**WBGT**) at which precautions (provision of adequate drinking water, annual physical exams, training in emergency aid for heat stroke) should be instituted; they are not limits.

If radiant temperature is close to air temperature:

$$WBGT = .7 \, NWB + .3 \, GT \qquad (8)$$

If radiant temperature is not close to air temperature:

$$WBGT = .7 \, NWB + .2 \, GT + .1 \, DBT \qquad (9)$$

where

NWB = natural wet bulb temperature, C (This is not psychrometric wet bulb (WB) but the temperature of a wet wick exposed to *natural* air currents.)

= WB (for air velocity > 2.5 m/s)

= $.1 \, DBT + .9 \, WB$ (for .3 < V < 2.5 m/s)

GT = globe temperature, C

DBT = dry bulb temperature, C

3.43 Reduction of Heat Stress.

Radiation, convection, and evaporation are the primary heat transfer mechanisms in and out of the body.

Reduce radiant heat by working in the shade since heat radiation resembles light radiation. Clothing (a mobile shield) is the first line of defense. Use hats and long-sleeved shirts. Fixed shields are a second line of defense. Figure 26.8 shows an opaque aluminum shield—if the operator must see through the shield, use coated glass. If objects must move through the shield (e.g., conveyorized objects from an oven), use a screen of hanging chains.

Convective heat transfer is negative (body loses heat) if the air temperature is less than skin temperature (about 35°C). Thus, if there is no radiant heat and insects and social customs are not a problem, minimize the amount of clothing if air temperature is below 35°C. Use roof ventilators to remove the layer of hot stratified air near the ceiling.

Evaporative heat transfer is the most important route. If sweat does not evaporate, it does no cooling. Increase evaporation by (1) decreasing water vapor pressure of the environment (expensive) or (2) increasing air velocity (inexpensive). Keep a fan close, since velocity drops off rapidly with distance from the fan; beyond 2 m/s additional velocity has little benefit.

Large amounts of sweating can cause the body to become low in water and salt. Dehydration is the real problem because the thirst drive is not sufficient to replace the water lost during heavy sweating. Supervisors must insist that workers drink water. Drink from a container, since volume drunk from a water fountain tends to be small, because of ingestion of air. Salt loss is overpublicized. If more salt is needed (it rarely is), eat salty food or add salt to the food at meals. Avoid salt pills—they are too concentrated.

3.5 Cold Stress

As with heat stress, people want one single number that combines all the factors. The **wind-chill index** is the best index. However, it tends to overestimate the effect of air velocity since it is based on cooling a liter bottle of water, not cooling a clothed, living body.

The best defense against cold stress is clothing. The problem areas are the head, hands, and feet. The head does not vasoconstrict and so may lose as much as 50% of total body heat loss. Use stocking caps and mufflers. The hands do vasoconstrict, so protect them with gloves and mittens. Foot cooling causes much discomfort—especially if the feet become wet. Two layers of socks or boots plus overshoes are some alternatives. The torso clothing needs to be adjustable so that the person can vent the hot sweaty air under the clothes rather than let it freeze in the fabric.

Heating the building has become more of a challenge as energy costs have risen. Radiant heaters are popular for exposed areas, such as shipping docks. Generally it is more efficient to heat a facility with several small furnaces (use 1 furnace when it is cool, 2 when it is cold, 3 when it is very cold) than with one large furnace operating at a small

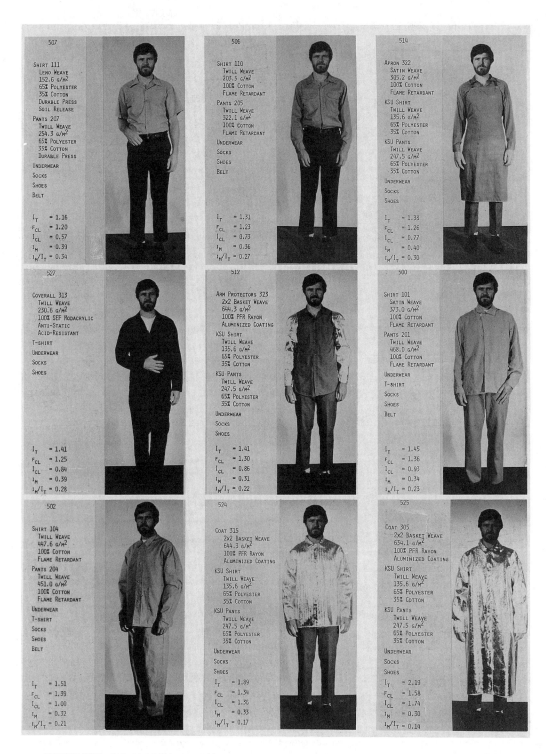

FIGURE 26.7   Industrial clothing ensembles.

fraction of its capacity. There are many devices available to remove heat from the exhaust air and use it to preheat incoming (supply) air or to heat water. The accumulation of hot air near the ceiling (thermal stratification) causes a number of problems: (1) It increases the heat loss through the roof, and (2) it makes ceiling-mounted heaters almost

useless (adding hot air above the cold-air layer is like trying to throw a Ping Pong ball down into water). The air can be destratified by fans connected to ducts which have their outlets at floor level. In general, warm air should enter the room near the floor and cold air should enter near the ceiling.

TABLE 26.5   Clo values for clothing items (Rohles et al., 1980). From *Proceedings of the Human Factors Society*, 1980, 225. Copyright © 1980 by the Human Factors Society, Inc., and reproduced by permission.

Men Item	clo	Women Item	clo
Underwear			
Long underwear upper	.10	Full slip	.19
Long underwear lower	.10	Half slip	.13
T-shirt	.09	Long underwear upper	.10
Sleeveless	.06	Long underwear lower	.10
Briefs	.06	Bra and panties	.05
Shirt			
Heavy long sleeve	.29	Heavy blouse	.29
short sleeve	.25	Light blouse	.20
Long long sleeve	.22		
short sleeve	.14	Heavy dress	.70
(Plus 5% for tie or turtleneck)		Light dress	.22
Heavy vest	.29	Heavy skirt	.22
Light vest	.15	Light skirt	.10
Heavy sweater	.37	Heavy sweater	.37
Light sweater	.20	Light sweater	.17
Heavy jacket	.49	Heavy jacket	.37
Light jacket	.22	Light jacket	.17
Heavy trousers	.32	Heavy slacks	.44
Light trousers	.26	Light slacks	.26
Socks		Stockings	
Knee high	.10	Any length	.01
Ankle length	.04	Pantyhose	.01
Shoes		Shoes	
Boots	.08	Boots	.08
Oxfords	.04	Pumps	.04
Sandals	.02	Sandals	.02

TABLE 26.6   Threshold *WBGT* temperatures (°C) as a function of air velocity and metabolic (basal + activity) rate.

Metabolic Rate (W)	Low Air Velocity (up to 1.5 m/s)	High Air Velocity (1.5 m/s or above)
Light (up to 230)	30	32
Moderate (230–350)	27.8	30.5
Heavy (over 350)	26	28.9

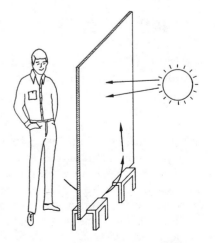

FIGURE 26.8   Heat shields should have legs to permit air flow.

DESIGN CHECKLIST: TOXICOLOGY, VENTILATION, AND CLIMATE

Toxicology
  Dermatosis
  TLV
  Engineering control
  Administrative control
  Personal protective equipment
Ventilation
  Area or local ventilation
  Location of operator/source/exhaust
  Duct ventilation
  Confined spaces

Interior climate
  Smoking or nonsmoking area
  Sedentary work?
  Clothing?
  Comfort zone; percentage of people dissatisfied
  Heat Stress
    Temperatures
    Reduction of stress
  Cold Stress
    Wind-chill index
    Reduction of cold stress

REVIEW QUESTIONS

1. Which of the three entrance points to the body is most important for toxins?
2. Dermatoses cause approximately what percent of lost work days?
3. Who should wash a worker's clothes if there are toxic substances?
4. How are the *TLV*s generally determined?
5. What is the *TLV* for acetone? Is an exposure of 100 ppm for 5 h and 0 ppm for 3 h legal?
6. Is exposure for 8 h to a mixture of 20 ppm of n-amyl acetate and 900 ppm of acetone legal?
7. What *STEL* does the ACGIH recommend for acetone?
8. List three engineering controls and three administrative controls; give an example of each.
9. Why should organizations buy personal protective equipment instead of the employee?
10. Is air velocity 1 m from a fan higher on the inlet or exhaust side?
11. Sketch the four ventilation strategies.
12. Sketch some good and bad examples of location of the inlet, person, contaminant source, and exhaust.
13. Are round or rectangular ducts better in an industrial area? Why?
14. What is the minimum permissible oxygen level?
15. Define the *LEL*.
16. For 25°C and 10 mm Hg, what is (a) relative humidity, (b) psychrometric wet bulb temperature, (c) dew point temperature, and (d) new effective temperature?
17. Calculate the *PPD* at 28 CET*.
18. What is the tradeoff between air velocity and dry bulb temperature?
19. What is the tradeoff between metabolic rate and dry bulb temperature?
20. What is the tradeoff between clothing insulation and dry bulb temperature?
21. Calculate the clothing insulation of the clothing you are wearing.
22. At what *WBGT* temperature should heat stress precautions begin to be taken for light work in low air velocity?
23. Is dehydration or salt loss more of a problem in heat stress?

PROBLEMS AND PROJECTS

26.1 Visit a factory. What are the compounds that they consider toxicology problems? What are they doing to reduce the problems? Any ideas? What does the supervisor think of your ideas?

26.2 How much ventilation volume presently is used for the IE department office at your university? How many hours/week is it on? How much volume of air would be needed if a nonsmoking policy were used? Justify your recommendation to the department head on whether or not to have a nonsmoking policy in the IE department office.

26.3 Using a sling psychrometer, record wet bulb and dry bulb temperature. Calculate relative humidity and new effective temperature.

26.4 Have a debate between two teams on whether "our firm" should give ordinary toxicology protection or special protection to females who might become pregnant. The instructor will give information concerning the firm. There should be a written and oral presentation. Then the class will vote on whether to give normal or special protection to females who might become pregnant.

REFERENCES

ACGIH. Threshold Limit Values and Biological Exposure Indices, 1993–1994, Cincinnati: 1993.

ASHRAE, *HVAC Systems and Applications*. Atlanta: American Society of Heating, Refrigeration, and Air Conditioning Engineers, 1988.

Birmingham, D. Occupational dermatoses: Their recognition, control, and prevention. Chapter 34 in *The Industrial Environment*. Washington, D.C.: Supt. of Documents, 1973.

Ellenbecker, M., Gempel, R., and Burgess, W. Capture efficiency of local ventilation systems. *Am. Ind. Hygiene Assoc. J.*, Vol. 44, 10, 752–755, 1983.

Fletcher, B. Centerline velocity characteristics of local exhaust hoods. *Am. Ind. Hygiene Assoc. J.*, Vol. 43, 8, 626–627, 1982.

Industrial Ventilation: A Manual of Recommended Practice, 17th ed. Ann Arbor, Mich.: American Conference on Governmental Industrial Hygiene, 1982.

Konz, S. *Work Design: Industrial Ergonomics*. Scottsdale, Ariz.: Publishing Horizons, 1990.

McCullough, E., Arpin, E., Jones, B., Konz, S., and Rohles, F. Heat transfer characteristics of clothing worn in hot industrial environments. *ASHRAE Transactions*, Vol. 88, 1, 1982.

Melgaard, H. Monitoring process oven exhausts. *Plant Engineering*, Vol. 35, 26, 91–95, Dec. 23, 1981.

Morse, G., and Swift, R. Working safely in confined spaces: Evaluating the hazards. *Plant Engineering*, Vol. 36, 2, 139–142, Jan. 21, 1982.

Morse, G., and Swift, R. Working safely in confined spaces: Establishing and following correct procedures. *Plant Engineering*, Vol. 36, 8, 94–97, April 29, 1982.

Rohles, F., Konz, S., and Munson, D. Estimating occupant satisfaction from effective temperature. *24th Proc. of Human Factors Society*. Santa Monica, Calif.: Human Factors and Ergonomics Society, 223–227, 1980.

Rosen, E., and Konz, S. Cooling with box fans. *Proceedings of Human Factors Society*. Santa Monica, Calif.: Human Factors and Ergonomics Society, 123–127, 1982.

COPYRIGHTS AND ACKNOWLEDGMENTS

Figure 1.1 From R. Muther and H. Hales, *Systematic Planning of Industrial Facilities, Vol. 1.* © Copyright 1979 by Management and Industrial Research Publications, Kansas City, MO. Reprinted with permission.

Box 2.3 From L. Gould, *Modern Materials Handling*, Vol. 45, 55, Dec. 1990. © Copyright 1990 by Cahners Publishing, Div. of Reed Holdings. Reprinted with permission.

Figures 3.1, 3.11 From R. Muther, *Systematic Layout Planning.* © Copyright 1973, Richard Muther, Management and Industrial Research Publications, Kansas City, MO.

Table 3.2 From D. Cochran and K. Swinehart, The total source of error adjustment model, *International Journal of Production Research.* © Copyright 1991 by Taylor & Francis. Reprinted with permission.

Box 3.2 From D. Lesnet, *Industrial Engineering, Vol. 15*, November 1983. © Copyright 1983 by Institute of Industrial Engineers, 25 Technology Park, Norcross, GA 30092. Reprinted with permission.

Figure 4.10 From D. Heglund, Flexible manufacturing, *Production Engineering.* © Copyright 1981 Production Engineering. Reprinted with permission.

Figure 5.23 From H. Gargano and F. Stewart, Material handling system is key to efficient assembly operation, *Material Handling Engineering.* © Copyright 1975 by Penton Publishing, Cleveland, OH. Reprinted with permission.

Figure 5.25 Figure courtesy of AT&T Istel, Visual Interactive Systems.

Figure 6.7 From R. Francis, L. McGinnis, and J. White, *Facility Layout and Location: An Analytical Approach*, 2e, © Copyright 1992, p. 156. Adapted by permission of Prentice Hall, Englewood Cliffs, NJ.

Figure 7.1 From A. Malde and K. Bafna, Proceedings of the 1986 International Industrial Engineering Conference, Atlanta, Georgia. © Copyright 1986 by Institute of Industrial Engineers. Reprinted with permission.

Table 8.1 From H. Tong, *Plant Location Decisions of Foreign Manufacturing Investors.* © Copyright 1979 by University Microfilms International, Ann Arbor, MI. Reprinted with permission.

Figure 8.1 Reprinted by permission of *The Wall Street Journal*, © 1982 Dow Jones and Company, Inc. All rights reserved worldwide.

Box 8.2 From L. Hales, "Site Selection," in *Management Handbook for Plant Engineers*, E. Lewis, Ed. © Copyright 1977 by McGraw-Hill. Reprinted with permission.

Table 8.2 From J. Huey, *Fortune.* © Copyright 1991 by Time, Inc. All rights reserved.

Table 8.3 From R. Muther and L. Hales, *Systematic Planning of Industrial Facilities (SPIF), Vols. I and II.* © Copyright 1980 by Management and Industrial Research Publications, Kansas City, MO. Reprinted with permission.

Table 9.1 From D. Ernst, How to increase light levels without increasing lighting, *Material Handling Engineering.* © Copyright 1976 by Penson Publishing, Cleveland, OH. Reprinted with permission.

Figure 9.3 Reprinted with permission of *Plant Engineering*, December 13, 1990. © 1990 Cahners Publishing Company.

Table 9.3 Reprinted with permission of *Plant Engineering*, July 18, 1991. © 1991 Cahners Publishing Company.

Table 10.1 From J. Thompkins and J. White, *Facilities Planning.* New York: Wiley, 1984. © Copyright 1984 by John Wiley & Sons. Reprinted by permission of John Wiley & Sons, Inc.

Table 10.6 From J. Templer, *The Staircase: Studies of Hazards, Falls, and Safer Design.* Cambridge, Mass.: MIT Press, 1992. © Copyright 1992 by MIT Press.

Box 10.2; Tables 10.4, 10.5 Reprinted with permission from NFPA 101-1991, Life Safety Code®, Copyright © 1991, National Fire Protection Association, Quincy, MA 02269. This reprinted material is not the complete and official position of the National Fire Protection Association, on the referenced subject which is represented only by the standard in its entirety.

Table 11.2 Reprinted with permission of *Plant Engineering*, May 2, 1991. © 1991 Cahners Publishing Company.

Table 11.3 Reprinted with permission of *Plant Engineering*, July 18, 1991, © 1991 Cahners Publishing Company.

Table 11.8 Reprinted with permission of *Plant Engineering*, July 18, 1991. © 1991 Cahners Publishing Company.

Tables 12.1, 12.2 From J. Thompkins, "Plant Layout," in *Handbook of Industrial Engineering*, G. Salvendy, Ed., Wiley. © Copyright 1982 Institute of Industrial Engineers. Reprinted with permission.

Table 12.3 Reprinted with permission of *Plant Engineering*, July 18, 1991. © 1991 Cahners Publishing Company.

Figure 12.2 Adapted from R. Muther, "Relationship Between Material Handling and Plant Layout," in *Material Handling Handbook*, R. Kulwiec, Ed. © Copyright 1985 by John Wiley & Sons. Reprinted by permission of John Wiley & Sons, Inc.

Table 13.2 Courtesy Steelcase, Inc.

Table 13.3 From W. Wrenell, "Office Layout," in *Handbook of Industrial Engineering*, 2e, G. Salvendy, Ed. © Copyright 1992 by John Wiley & Sons. Reprinted by permission by John Wiley & Sons, Inc.

Tables 13.4, 13.5 From G.F. McVey, Twelve ergonomic guidelines for integrating AV systems into conference room design, *Facilities Planning News*, © Copyright 1991 by Tradeline, Inc., Orinda, CA. Reprinted with permission.

347

INDEX